1501506379

AF606588

Finite-Size Scaling

Current Physics – Sources and Comments

Coordinating editor

H. RUBINSTEIN

1. String theory in four dimensions, *edited by* M. Dine
2. Finite-size scaling, *edited by* J.L. Cardy

NORTH-HOLLAND
AMSTERDAM · OXFORD · NEW YORK · TOKYO

Finite-Size Scaling

Editor

John L. CARDY

Department of Physics, University of California
Santa Barbara, CA 93106, USA

1988

NORTH-HOLLAND
AMSTERDAM · OXFORD · NEW YORK · TOKYO

ISBN: 0 444 87109 8 (hardbound)
0 444 87110 1 (paperback)

Published by:

Physical Sciences & Engineering Division

Elsevier Science Publishers B.V.
P.O. Box 103
1000 AC Amsterdam
The Netherlands

Sole distributors for the USA and Canada:

Elsevier Science Publishing Company, Inc.

52 Vanderbilt Avenue
New York, NY 10017
USA

Library of Congress Cataloging in Publication Data

Finite-size scaling / John L. Cardy, editor.
p. cm. -- (Current physics ; 2)
Includes bibliographies.
ISBN 0-444-87109-8. ISBN 0-444-87110-1 (pbk.)
1. Finite size scaling (Statistical physics) 2. Renormalization.
3. Conformal invariants. I. Cardy, John L. II. Series.
QC174.85.S34F56 1988
530.1'3--dc19 88-25216
CIP

Printed in The Netherlands

Preface

In the past few years, finite-size scaling has become an increasingly important tool in studies of critical systems. This is partly due to an increased understanding of finite-size effects by analytical means, and partly due to our ability to treat larger systems with large computers.

The aim of this volume is to collect those papers which have been important for this progress, and which illustrate novel applications of the method. An emphasis has been placed on relatively recent developments, including the use of the ε-expansion and of conformal methods.

In learning about finite-size scaling, and in compiling this review, I have benefitted from the wisdom of many of my colleagues, in particular Michael Barber, Edouard Brézin, Bernard Derrida, Michael Fisher, Peter Nightingale and Vladimir Privman, and I take this opportunity to thank them. Most of the work for this book was done while visiting the Service de Physique Théorique, CEN-Saclay, and was supported in part by a Guggenheim Fellowship, and also by NSF Grant PHY86-14185.

Acknowledgements

The following articles have been reprinted by kind permission of the publisher, The American Institute of Physics:
From Physical Review:
E. Eisenriegler and R. Tomaschitz, Helmholtz Free Energy of Finite Spin Systems near Criticality, Phys. Rev. B 35 (1987) 4876–4887; K. Binder, M. Nauenberg, V. Privman and A.P. Young, Finite-Size Tests of Hyperscaling, Phys. Rev. B 31 (1985) 1498–1502.
From Physical Review Letters:
M.E. Fisher and M.N. Barber, Scaling Theory for Finite-Size Effects in the Critical Region. Phys. Rev. Lett. 28 (1972) 1516–1519; H.W.J. Blöte, J.L. Cardy and M.P. Nightingale, Conformal Invariance, the Central Charge, and Universal Finite-Size Amplitudes at Criticality, Phys. Rev. Lett. 56 (1986) 742–745; I. Affleck, Universal Term in the Free Energy at a Critical Point and the Conformal Anomaly, Phys. Rev. Lett. 56 (1986) 746–748; F.C. Alcaraz, M.N. Barber and M.T. Batchelor, Conformal Invariance and the Spectrum of the *XXZ* Chain, Phys. Rev. Lett. 58 (1987) 771–774.

The following articles have been reprinted from Journal of Statistical Physics by kind permission of the publisher, Plenum Publishing Corporation:
V. Privman and M.E. Fisher, Finite-Size Effects at First-Order Transitions, J. Stat. Phys. 33 (1983) 385–417; J. Rudnick, H. Guo and D. Jasnow, Finite-Size Scaling and the Renormalization Group, J. Stat. Phys. 41 (1985) 353–373.

The following articles have been reprinted from Zeitschrift für Physik by kind permission of the publisher, Springer-Verlag:
E. Eisenriegler, Finite Size Critical Behavior for Dirichlet Boundary Conditions, Z. Phys. B 61 (1985) 299–309; K. Binder, Finite Size Scaling Analysis of Ising Model Block Distribution Functions, Z. Phys. B 43 (1981) 119–140.

The following articles have been reprinted from Journal de Physique by kind permission of the publisher, Editions de Physique:
E. Brézin, An Investigation of Finite Size Scaling, J. Phys. (France) 43 (1982) 15–22; B. Derrida and L. de Seze, Application of Phenomenological Renormalization to Percolation and Lattice Animals in Dimension 2, J. Phys. (France) 43 (1982) 475–483.

The following articles have been reprinted from Journal of Physics A by kind permission of the publisher, The Institute of Physics:
C.J. Hamer and M.N. Barber, Finite-Size Scaling in Hamiltonian Field Theory, J. Phys. A 13 (1980) L169–174; W. Kinzel and J.M. Yeomans, Directed Percolation: a Finite-Size Renormalization Group Approach, J. Phys. A 14 (1981) L163–168; B. Derrida, Phenomenological Renormalization of the Self-Avoiding Walk in Two Dimensions, J. Phys. A 14 (1981) L5–9; B. Derrida and J. Vannimenus, A. Transfer-Matrix Approach to Random Resistor Networks, J. Phys. A 15 (1982) L557–564; V. Privman and M.E. Fisher, Convergence of Finite-Size

Renormalization Techniques, J. Phys. A 16 (1983) L295–301; M.P. Nightingale and H. Blöte, The Relation Between Amplitudes and Exponents in Finite-Size Scaling, J. Phys. A 16 (1983) L657–664; J.L. Cardy, Conformal Invariance and Universality in Finite-Size Scaling, J. Phys. A 17 (1984) L385–387; M. Kolb and K.A. Penson, Conformal Invariance and the Phase Transition of a Spin Chain with Three Spin Interaction, J. Phys. A 19 (1986) L779–784; J.L. Cardy, Finite-Size Scaling in Strips: Antiperiodic Boundary Conditions, J. Phys. A 17 (1984) L961–964; J.L. Cardy, Logarithmic Corrections to Finite-Size Scaling in Strips, J. Phys. A 19 (1986) L1093–1098; J.L. Cardy, Universal Amplitudes in Finite-Size Scaling: Generalization to Arbitrary Dimensionality, J. Phys. A 18 (1985) L757–760.

Contents

1. Theory of Finite-Size Scaling

Introduction

The singularities in thermodynamic functions associated with a critical point occur only in the thermodynamic limit. This involves allowing all the dimensions of the system under consideration to tend to infinity. If some or all of these dimensions remain finite, the thermodynamic behavior is modified. In the thermodynamic limit, the critical fluctuations are correlated over a distance of the order of the correlation length ξ. This may be defined as the length scale governing the exponential decay $\epsilon^{-r/\xi}$ with distance r of the order parameter correlation function. Other definitions of ξ are possible: for systems with short-ranged interactions they are believed to be equivalent up to constants of order unity. The correlation length ξ diverges at a second-order critical point. If we consider some of the dimensions to be finite, of order L, then as long as ξ/L is small we expect effects only $O(e^{-L/\xi})$ in thermodynamic quantities.

Thus only in systems which are so close to the critical point that $\xi \approx L$ do we expect to see significant finite-size effects. These effects may be described in terms of an effective reduction of dimensionality. If the sample is finite in all directions, for $\xi \gg L$ it is effectively zero-dimensional. Such a system cannot exhibit a thermodynamic singularity. The effect of the finite size is to round the critical point singularities over a region in parameter space for which $\xi \approx L$. If all but one of the dimensions are finite, there will be a crossover to quasi-one-dimensional behaviour: no true singularities, but nevertheless some anomalous behaviour. If d' of the d dimensions are infinite, there will be a crossover to critical behaviour with exponents characteristic of d' dimensions.

Finite-size scaling, as formulated by Fisher [1] and by Fisher and Barber (paper 1.1) concerns itself with the manner in which this rounding or crossover occurs. In this introduction we shall limit ourselves to a brief review of this theory. Readers who wish for more details of the earlier development of the subject should consult the review by Barber [2]. In a finite-size system, there are in principle three length scales involved: ξ, L and the microscopic length a which governs the range of the interactions. Thermodynamic quantities thus may in principle depend on the dimensionless ratios ξ/a and L/a. The finite-size scaling hypothesis assumes that, close to the critical point, the microscopic length drops out. Thus, if we consider a quantity such as the ferromagnetic susceptibility χ, which behaves like $\xi^{\gamma/\nu}$ near the critical point in the infinite system, then in the finite geometry characterized by a size L,

$$\chi = \xi^{\gamma/\nu}\phi(\xi/L), \tag{1.1}$$

where $\phi(u)$ is a scaling function, with the property that $\phi(u) \to$ const. as $u \to 0$. If we use the fact that $\xi \sim |T - T_c|^{-\nu}$, this can be written in the equivalent form

$$\chi = L^{\gamma/\nu}\tilde{\phi}(L^{1/\nu}(T - T_c)), \tag{1.2}$$

where $\tilde{\phi}$ is another scaling function. If the system is quasi-zero- or one-dimensional then χ is supposed to be an analytic function of T. Thus $\tilde{\phi}(v)$ must be a smooth function with a peak at some value $v = v_0$, of width v_1, say. A plot of χ versus T will then have a peak * at $T = T_c + v_0 L^{-1/\nu}$, with a width $v_1 L^{-1/\nu}$, and a height $L^{\gamma/\nu}$. Thus, assuming the validity of the finite-size scaling hypothesis, studies of finite systems of different sizes can give information about the critical exponents of the infinite system.

If the reduced dimensionality d' of the system exceeds the lower critical dimension, then (1.2) still gives useful information. In that case we would expect χ to diverge like $|T - T_c(L)|^{-\gamma'}$ where γ' is the d'-dimensional exponent. Comparing with (1.2) we see that ϕ must have a singularity of the form $|v - v_0|^{-\gamma'}$ for some constant v_0. Thus, close to the singularity,

$$\chi \sim L^{(\gamma-\gamma')/\nu} |T - T_c - v_0 L^{-1/\nu}|^{-\gamma'} \tag{1.3}$$

so that $T_c(L) - T_c(\infty) \propto L^{-1/\nu}$, and the amplitude of the singularity behaves like $L^{(\gamma-\gamma')/\nu}$.

Experimentally, it turns out that finite-size scaling is rarely applicable. The regime where the correlation length is of the same order as one of the linear dimensions is usually complicated by other effects such as impurities, gravity and so on. There are some successes, nevertheless [3]. Its great utility has been as a method of analysing numerical data on finite systems. This includes data derived from Monte Carlo and other kinds of simulation methods, and also from exact methods such as the diagonalization of transfer matrices. One further potential application should be mentioned: to low-temperature quantum systems. So far we have discussed finite-temperature systems, where quantum effects are unimportant near the critical point. However, quantum systems in d dimensions may have critical points at zero temperature. At finite temperature T the Feynman path integral representation for such a system then resembles the partition function for a classical system in $d+1$ dimensions, the extra dimension being periodic with a finite size $\beta = 1/k_B T$. The theory of finite-size scaling may then be taken over to such systems.

In applying finite-size scaling to a particular problem, there are two kinds of question to be answered: (1) is the finite-size scaling hypothesis correct? (2) What is the most efficient way of implementing it? This section is devoted to the first type of question. The second will be addressed when specific applications are discussed in chapter 2.

1.1. Renormalization group and finite-size scaling

We begin by giving an heuristic derivation of the finite-size scaling hypothesis, based on the ideas of the real-space renormalization group (RG) which underpin the modern theory of critical behaviour. Let us first review some of these ideas, as applied to infinite systems. We consider a system defined on a lattice with spacing a, described by a Hamiltonian $\mathscr{H}$ which depends on interaction constants $(K_1, K_2, \ldots)$. As usual a factor of $(k_B T)^{-1}$ is absorbed into these. The idea of the RG is to perform a coarse-graining of the system, so that $a \to ba$ with $b > 1$, in such a way that the probability distribution of fluctuations with wavelengths much larger than a is unaffected. This distribution is assumed to be given by the Boltzmann weights of a Hamiltonian of the same form as $\mathscr{H}$, but with new interactions $(K_1', K_2', \ldots)$. An important assumption is that this new Hamiltonian remains short-ranged. The partition

* In the paper of Fisher and Barber (paper 1.1), the shift exponent was found to be different from $1/\nu$ in two soluble cases: the spherical model with a boundary, and the ideal Bose gas. These were later understood to be pathological: the constraints give an effective long-range interaction. See, e.g., E. Brézin, Ann. NY Acad. Sci. 410 (1983) 339. Note that in some cases the amplitude v_0 may vanish, leading to an effective shift exponent larger than ν^{-1}.

function for the theory with interactions $\{K'_\alpha\}$ will be the same as that for the original theory, up to a multiplicative constant. Thus if we define $f(\{K_\alpha\})$ to be the free energy per unit volume of the theory with interactions $\{K_\alpha\}$, we have

$$f(\{K_\alpha\}) = g(\{K_\alpha\}) + b^{-d} f(\{K'_\alpha\}). \tag{1.4}$$

The function g results from summing over fluctuations with wavelengths between a and ba. It should be an analytic function of the $\{K_\alpha\}$.

If the system is initially close to a second-order critical point, then after a finite number of iterations of the RG we end up in the vicinity of a fixed point $\{K^*_\alpha\}$. Near the fixed point we may choose linear combinations u_i of the deviations $K_\alpha - K^*_\alpha$ which transform simply: $u'_i = b^{y_i} u_i$. These eigenvalues y_i are related to critical exponents. The scaling variables u_i are supposed to be analytic functions of the external parameters. For our prototype ferromagnetic critical point there are two relevant variables (with $y_i > 0$) proportional to deviations t, h in temperature and magnetic field from the critical point. Also there are irrelevant variables u (with $y_i < 0$) which depend analytically on t and h.

In terms of these variables, (1.4) reads

$$f(t, h, u_j) = g(t, h, u_j) + b^{-d} f(b^{y_1} t, b^{y_2} h, b^{y_j} u_j). \tag{1.5}$$

Since g is supposed to be analytic, this shows, for example, that f has a leading singularity of the form t^{d/y_1}, when $h = 0$, if we can safely ignore the u_j dependence on the right-hand side. Similarly, the zero-field susceptibility $\chi \propto \partial^2 f/\partial h^2 |_{h=0}$ has a leading singularity $t^{-\gamma}$ where $\gamma = (2y_2 - d)/y_1$.

These results extend to correlation functions. For example, correlation functions of the local magnetization are given by functional derivatives of f with respect to local magnetic fields $h(\boldsymbol{r})$. These local fields renormalize in the same way as a uniform one, since the RG transformation is local. The two-point correlation function $G^{(2)}$ then satisfies a RG equation

$$G^{(2)}(r, t, h, u_j) = b^{-2d+2y_2} G^{(2)}(r/b, b^{y_1} t, b^{y_2} h, b^{y_j} u_j). \tag{1.6}$$

Let us see how these arguments are modified in a finite geometry. For simplicity, we assume that it has no boundaries (for example, we choose periodic boundary conditions). After rescaling, the coarse-grained system has an effective size L/b. Being local, the RG transformation in the finite system will be identical with that in the infinite one, if L/a is large. Thus we may generalize (1.5):

$$f(t, h, u_j, L^{-1}) = g(t, h, u_j) + b^{-d} f(b^{y_1} t, b^{y_2} h, b^{y_j} u_j, bL^{-1}). \tag{1.7}$$

Therefore L^{-1} may be treated as a scaling variable, with eigenvalue $y = 1$. This observation underlies the finite-size scaling hypothesis. If we define the deficit free energy $\Delta f(L^{-1}) \equiv f(L^{-1}) - f(0)$, it satisfies a homogeneous equation in which g does not appear. From this, scaling laws follow. For example, if we iterate $\sim \ln(L/L_0)/\ln b$ times we find that

$$\Delta f(t, h, u_j, L^{-1}) = (L/L_0)^{-d} \Delta f((L/L_0)^{y_1} t, \ldots, L_0). \tag{1.8}$$

The quantities $(L/L_0)^{y_j}u_j$ are small, and it is usually possible to ignore them. We then find, for example, that, at the critical point $t=h=0$,

$$\Delta f \sim C/L^d, \tag{1.9}$$

where C is a constant. Similarly the quantity $\Delta\chi = \partial^2(\Delta f)/\partial h^2 |_{h=0}$ scales according to

$$\Delta\chi = L^{\gamma/\nu}\,\Delta\tilde{\phi}(L^{1/\nu}t) \tag{1.10}$$

where $\nu = 1/y_1$. From this, and the known behaviour of χ for $L=\infty$, the scaling laws (1.1) and (1.2) follow. Similarly we may consider correlation functions: when $t=h=0$, $G^{(2)}$ satisfies

$$G^{(2)}(r, L) = r^{-2x}\,\psi(r/L), \tag{1.11}$$

where $x = d - y_2$ and ψ is a scaling function. Privman and Fisher [4] have gone further and argued that the quantity C and scaling functions like those of (1.11, 1.12) are universal: they depend only on the shape of the system. Similar results are supposed to hold true for systems with boundaries, although the field-theoretic RG is more suitable than real-space methods for discussing these. In these cases, the universal quantities depend on the nature of the boundary condition, e.g. whether the surface spins are free or fixed. Privman and Fisher argue that the only non-universal factors that enter are the "metrical" ones relating t and h to the physical parameters. These occur already in the infinite system. No new non-universal parameters enter into finite-size scaling formulae.

In deriving the finite-size scaling laws, we have assumed that terms involving irrelevant variables like $(L/L_0)^{y_j}u_j$ in (1.8) may be neglected. This is valid as long as f is analytic in such variables. A classic case when this is not true is when $d>4$ in Ising-like systems. Then a "dangerous irrelevant variable" is responsible for the fact that, when L is infinite, f does not behave like t^{d/y_1}, that is, hyperscaling is violated. We shall see that the same mechanism is responsible for the violation of finite-size scaling for $d>4$.

1.2. Field-theoretic renormalization group

The above ideas are at best heuristic, and need a systematic framework. This is afforded by the field-theoretic RG, which, while not rigorous, predicts scaling laws and relations which are valid beyond the region of accuracy of ε-expansion estimates. Let us see how this method works for the Ising case, in the infinite system. The discrete Ising variables as defined on the lattice are coarse-grained into a continuum field $\phi(\boldsymbol{r})$. The Hamiltonian is replaced by a continuum Landau–Ginzburg–Wilson Hamiltonian of the form

$$\mathcal{H} = \int \left[\tfrac{1}{2}(\nabla\phi)^2 + \tfrac{1}{2}r_0 a^{-2}\phi^2 + \tfrac{1}{4}g_0 a^{d-4}\phi^4\right] \mathrm{d}^d x. \tag{1.12}$$

The dimensionless parameters r_0, g_0 are the analogs of the $\{K_\alpha\}$. The microscopic distance a appears in the explicit dimensional factors in the couplings, and in the measure of functional integration $\int\mathcal{D}\phi$: the Fourier transform $\tilde{\phi}(\boldsymbol{k})$ has support only in $\boldsymbol{k} < a^{-1}$. For $d>4$ this latter a-dependence, which enters only into the loop corrections, is unimportant. We see that there is a fixed point at $r_0 = g_0 = 0$, where r_0 is relevant (and is proportional to t), and g_0 is irrelevant. However, g_0 is a dangerous irrelevant variable, as can be seen from (1.12). If we put $g_0 = 0$, the theory is not defined for $r_0 < 0$. For $d<4$, the loop corrections are important, but the

a-dependence can be controlled. This is because the theory is renormalizable. This means that we can rescale ϕ, r_0 and g_0 by a-dependent constants:

$$\phi_R = Z_1, \quad r_R = Z_2 r_0, \quad g_R = Z_3 g_0$$

in such a way that correlation functions

$$G_R^{(N)} = \langle \phi_R(r_1)\phi_R(r_2)\cdots\phi_R(r_N)\rangle \tag{1.13}$$

are finite as $a \to 0$ when expressed in terms of r_R, g_R. For our purposes this renormalized theory has no physical meaning. However, the fact that

$$a\frac{\partial}{\partial a} G_R^{(N)}|_{r_R, g_R \text{ fixed}} = 0 \tag{1.14}$$

implies the Callan–Symanzik equation for the physical correlation functions:

$$\left[a\frac{\partial}{\partial a} + N\gamma_1(g_0) + \beta(g_0)\frac{\partial}{\partial g_0} + \gamma_3(g_0)r_0\frac{\partial}{\partial r_0}\right]G^{(N)} = 0, \tag{1.15}$$

where

$$\gamma_1(g_0) = a\frac{\partial}{\partial a}\ln Z_1, \quad \beta(g_0) = -g_0\frac{\partial}{\partial a}\ln Z_2, \quad \gamma_3(g_0) = -a\frac{\partial}{\partial a}\ln Z_3.$$

This equation can be solved for $G^{(N)}$. One finds that its behavior as $a \to 0$ is governed by the non-trivial fixed-point value of g_0 where $\beta(g_0) = 0$, $\beta'(g_0) > 0$ and $r_0 = 0$. Thus $G^{(N)} \sim a^{-N\gamma_1^*}$ as $a \to 0$, where γ_1^* is the fixed point value of $\gamma_1(g_0)$.

We are actually interested in the behaviour of $G^{(N)}(\boldsymbol{r}_1, \boldsymbol{r}_2, \ldots)$ as $|\boldsymbol{r}_i - \boldsymbol{r}_j| \to \infty$ at fixed a, which is related to the above by dimensional analysis:

$$\left[a\frac{\partial}{\partial a} + \boldsymbol{r}_i\frac{\partial}{\partial \boldsymbol{r}_i} + \tfrac{1}{2}N(d-2)\right]G^{(N)} = 0. \tag{1.16}$$

Eliminating the term $a\partial/\partial a$ between (1.15) and (1.16), one finds, for example, that $G^{(2)} \sim r^{-d+2+2\gamma_1^*}$ as $r \to \infty$ at the critical point.

How does all this apply to a finite system? This question was first addressed by Brézin (1982; paper 1.2). The existence of a renormalized theory is a statement about short-distance behaviour, and is independent of whether the system is finite or not. Thus (1.15) is still valid. In fact the only difference is the appearance of a term $L\partial/\partial L$ in the differential operator in (1.16). When $a\partial/\partial a$ is a eliminated between (1.15), (1.16) this is completely equivalent to the finite-size scaling result for the correlation functions. Thus finite-size scaling is valid for $d < 4$. However, it was pointed out that the actual calculation of the finite-size scaling functions using the ε-expansion would have some subtleties. This is signalled by the appearance of fractional powers of ε in exact results in the $n \to \infty$ limit.

It was simultaneously realized by two groups of authors [Brézin and Zinn-Justin (paper 1.3); Rudnick, Guo and Jasnow (paper 1.4)] that the key to this problem was the correct treatment of the $\boldsymbol{k} = 0$ mode. In the finite system, the allowed values of $\boldsymbol{k}$ are quantized in units of $2\pi/L$. Those with $\boldsymbol{k} \neq 0$ have an effective mass $\propto L^{-1}$ even at the critical point, and can be treated perturbatively. However the $\boldsymbol{k}=0$ mode must be treated separately. The correct treatment of the problem then involves integrating out the $\boldsymbol{k}\neq 0$ modes perturbatively, leaving an effective Hamiltonian for the $\boldsymbol{k}=0$ mode.

This work has opened the door to many interesting and useful calculations of finite-size scaling functions within the ε-expansion. Eisenriegler and Tomaschitz (paper 1.7) have calculated the probability distribution of the magnetization in a finite region. Their results compare favorably with the Monte Carlo work of Binder (paper 1.6) for $d=3$. Goldschmidt (paper 1.8) and also Niel and Zinn-Justin [5] and Diehl [6] have extended the approach to cover critical dynamics. Once again the agreement with numerical work is good. It covers both the nonlinear and linear relaxation regimes.

The separation of the zero mode also enables one to understand the violation of finite-size scaling for $d>4$. In this case, the perturbative effects of $\boldsymbol{k}\neq 0$ modes can be ignored altogether, leading to an effective partition function

$$Z=\int d\tilde{\phi}\,\exp\left[-L^{d}\left(\tfrac{1}{2}r_0 a^{-2}\tilde{\phi}^{2}+\tfrac{1}{4}g_0 a^{d-4}\tilde{\phi}^{4}\right)\right], \tag{1.17}$$

where now $\tilde{\phi}=\tilde{\phi}(\boldsymbol{k}=0)$. If we calculate the susceptibility $L^{d}\phi^{2}$ directly from this, we find

$$\chi=\xi^{\gamma/\nu}\Phi\left((L/a)^{d}(a/\xi)^{4}g_0^{-1}\right). \tag{1.18}$$

Note the explicit appearance of g_0^{-1} in the argument of the scaling function, showing that g_0 is dangerous, and the explicit appearance of a, showing the breakdown of finite-size scaling.

Eisenriegler has also treated the case of Dirichlet boundary conditions ($\phi=0$ on the boundary). Physically this corresponds to the boundary spins being free. (The reader who finds this statement paradoxical is referred to the excellent review of surface critical behaviour by Diehl [6].) In this case there is of course no zero mode, and the ε-expansion proceeds more straightforwardly. However now one has to deal with surface renormalization effects, and of course the mode expansion is more complicated. We include this article (paper 1.5), because if these calculations are ever to be applied to experimental systems, such complications must be addressed.

We have seen that the RG predicts the validity of both hyperscaling and finite-size scaling for $d<4$. Of course the RG is based on assumptions which may be wrong. Therefore it is important to understand to what extent numerical results on finite-size systems do indeed test hyperscaling. This analysis, based on a particular assumed mechanisms for the violation of hyperscaling, is given in paper 1.10, of Binder, Nauenberg, Privman and Young. It is a salutory warning to all those who attempt to interpret finite-size data!

1.3. First-order transitions

At a first-order transition, the correlation length ξ remains finite, and the finite-size scaling properties in a truly finite geometry are quite simple if $\xi\ll L$. At the critical point there are two or more coexisting phases. If we label the mean values of the order parameter ϕ in these phases by ϕ_1, $\phi_2,\ldots$ then we can write an effective probability distribution proportional to

$$\sum_i \exp\left(-\frac{(\phi-\phi_i)^2}{2\chi_i L^{-d}}\right), \tag{1.19}$$

where χ_i is the susceptibility in the ith pure phase. From this we see that the susceptibility in the finite volume is

$$\chi = L^d(\langle\phi^2\rangle - \langle\phi\rangle^2) \propto L^d \tag{1.20}$$

at the critical point. It is interesting to note that this is consistent with the RG result $\chi \propto L^{2d-y_2}$ if we realize that the first-order transition should be controlled by a discontinuity fixed point at which $y_2 = d$.

Thus finite-size scaling at strong first-order transitions in finite volumes is not very interesting. In quasi-one-dimensional geometries, however, the above formulas do not apply. In this case, we expect the dominant configurations at coexistence to consist of a "gas" of domain walls aligned roughly perpendicular to the infinite direction. The free energy of each wall is σL^{d-1}, where σ is the interfacial free energy and L^{d-1} is the cross-sectional area. Thus we expect a density $\sim \exp(-\sigma L^{d-1}/kT)$ of domain walls per unit length, and a correlation length which is the inverse of this. The question of the power of L in the prefactor multiplying this is more subtle, and is discussed in the papers by Brézin and Zinn-Justin (paper 1.3), and Privman and Fisher (1.11).

References

[1] M.E. Fisher, in: Critical Phenomena, Proc. 51st Enrico Fermi Summer School, Varenna, ed. M.S. Green (Academic Press, New York, 1972).
[2] M.N. Barber, in: Phase Transitions and Critical Phenomena, Vol. 8, eds. C. Domb and J.L. Lebowitz (Academic Press, London, 1983).
[3] B.A. Schneibner, M.R. Meadows, R.C. Mockler and W.J. O'Sullivan, Phys. Rev. Lett. 43 (1979) 590.
M.R. Meadows, B.A. Schneibner, R.C. Mockler and W.J. O'Sullivan, Phys. Rev. Lett. 43 (1979) 592.
[4] V. Privman and M.E. Fisher, Phys. Rev. B 30 (1984) 322.
[5] J.C. Niel and J. Zinn-Justin, Nucl. Phys. B 280[FS18] (1987) 355.
[6] H.W. Diehl, Z. Phys. B 66 (1987) 211.
[7] B. Nienhuis and M. Nauenberg, Phys. Rev. Lett. 35 (1975) 477.

Scaling Theory for Finite-Size Effects in the Critical Region*

Michael E. Fisher and Michael N. Barber
Baker Laboratory, Cornell University, Ithaca, New York 14850
(Received 17 April 1972)

Critical phenomena in films of finite thickness are considered. A detailed scaling theory, with allowance for distinct exponents λ and $\theta = 1/\nu$ for the critical-point shift and rounding, respectively, is confirmed by exact calculations on d-dimensional ferromagnetic spherical models and ideal Bose fluids with various boundary conditions. Ising-model results and existing data on real helium films are consistent with the theory.

Improvements in experimental technique should soon allow the detailed and accurate study of critical phenomena in systems with one or more finite, even though microscopically large, dimensions. Both experimentally and conceptually, the simplest systems to consider are films of finite thickness $L = na$ (where a is an atomic length or lattice spacing), and essentially infinite extent in the remaining $d' = d - 1$ dimensions (with $d = 3$ for real films). In this note we discuss such experiments theoretically.[1]

At the outset one should[2] generally distinguish a fractional *shift* $\epsilon(n)$ in critical temperature (or quasicritical temperature[3]) from a fractional *rounding* $\delta(n)$. If $T_c^{\tau}(n)$ is the critical temperature of the finite-thickness film under boundary conditions denoted by τ (see below), and $T_c = T_c^{\tau}(\infty)$ is the corresponding bulk critical temperature, the fractional shift is defined by

$$\epsilon^{\tau}(n) = [T_c - T_c^{\tau}(n)]/T_c \approx b^{\tau}/n^{\lambda} \text{ as } n \to \infty, \tag{1}$$

where the expected asymptotic behavior for thick films ($n \to \infty$) is characterized by a shift exponent λ. To define the rounding, consider an intensive property Y (such as the specific heat C, or the reduced susceptibility, $\bar{\chi} = k_B T\chi$) which in the bulk system has a critical-point divergence

$$Y_{\infty}(T) \approx A t^{-\psi} \text{ as } T \to T_c +, \tag{2}$$

where $t = (T - T_c)/T_c$ is the reduced temperature deviation (and $\psi = \alpha_d$ or γ_d, if $Y = C$ or $\bar{\chi}$, respectively[4]). In terms of the corresponding finite-thickness property $Y^{\tau}(n, T)$, the rounding may be defined, albeit a little loosely, by

$$\delta^{\tau}(n) = \Delta T^{*\tau}(n)/T_c \approx c^{\tau}/n^{\theta} \text{ as } n \to \infty, \tag{3}$$

where $T^{*\tau}(n) = T_c^{\tau}(n) + \Delta T^{*\tau}(n)$ is the temperature at which $Y^{\tau}(n, T)$ first shows significant (relative order unity) deviations from the bulk limit $Y_{\infty}(T)$. Note that this "rounding" is measured relative to the *shifted* critical temperature $T_c^{\tau}(n)$ and, in effect, measures the region of "crossover" from the bulk behavior (2) to the characteristic $d' = (d-1)$-dimensional film behavior

$$Y^{\tau}(n, T) \approx \dot{A}^{\tau}(n)\dot{t}^{-\dot{\psi}} \text{ as } T \to T_c^{\tau}(n), \tag{4}$$

where the *shifted* reduced temperature deviation is

$$\dot{t} = [T - T_c^{\tau}(n)]/T_c = t + \epsilon^{\tau}(n). \tag{5}$$

The d'-dimensional exponent satisfies $\dot{\psi} = \alpha_{d-1}$ or γ_{d-1} if $Y = C$ or $\bar{\chi}$, respectively.

Fisher and Ferdinand[2] suggested that, in general, the rounding exponent θ in (3) might be larger than λ so that *the shift is asymptotically larger than the rounding.* They argued for the relation

$$\theta = 1/\nu_d, \tag{6}$$

which follows from the hypothesis that rounding should set in when the correlation length $\xi(T) \approx \xi_0/\dot{t}^{\nu_d}$ matches the thickness $L = na$.[5] On the other hand, for free boundaries, they conjectured that λ equals 1, which is generally less than $1/\nu_d$. An exact analysis of the plane-square Ising lattice[2,6] confirmed (6), but could not distinguish the conjecture for λ since $\nu_d = 1$ in that case.

We now report calculations[7] for d-dimensional ferromagnetic spherical models[8] and ideal Bose fluids[5] which confirm the possibility $\theta > \lambda$. Three types of boundary conditions have been considered:

($\tau = 0$) *periodic* (or cyclic), in which the first and nth lattice layers are coupled ferromagnetically as nearest neighbors,[8] or the Bose wave functions satisfy $\Psi_N(x_j + L) = \Psi_N(x_j)$ (all $j = 1, \cdots, N$);

($\tau = \frac{1}{2}$) *antiperiodic*,[9] in which the first and nth lattice layers are coupled *anti*ferromagnetically with the interaction $-J$,[8] or $\Psi_N(x_j + L) = -\Psi_N(x_j)$; and

($\tau = 1$) *free surfaces* (or hard walls), where the first and nth lattice layer each couple to only one adjacent layer, or $\Psi_N(x_j = 0) = \Psi_N(x_j = L) = 0$.

The resulting exact asymptotic behavior of $\epsilon^{\tau}(n)$ and $\delta^{\tau}(n)$ is exhibited in Table I.[10] By comparison with the second column, which lists the

TABLE I. Asymptotic behavior of critical-point shift and rounding for spherical models and ideal Bose fluids. (Note there is no sharp transition when $d=2$.)

Dimension d	Correlation exponent ν_d	Rounding $\delta^\tau(n)$	Shifts $\epsilon^\tau(n)$ $\tau=0$	$\tau=\frac{1}{2}$	$\tau=1$	Surface exponent $\gamma_d^\times$
3	1	$\sim n^{-1}$	b_3^0/n	$b_3^{1/2}/n$	$-\lvert b_3^1\rvert(\ln n)/n$	3 (×log)
4	$\frac{1}{2}(\times\log^{1/2})$	$\sim(\ln n)/n^2$	b_4^0/n^2	$b_4^{1/2}(\ln n)/n^2$	$-\lvert b_4^1\rvert/n$	2 (×log)
5	$\frac{1}{2}$	$\sim n^{-2}$	b_5^0/n^3	$b_5^{1/2}/n^2$	$-\lvert b_5^1\rvert/n$	2
≥6	$\frac{1}{2}$	$\sim n^{-2}$	b_d^0/n^{d-2}	$b_d^{1/2}/n^2$	$-\lvert b_d^1\rvert/n$	2

correlation length exponents ν_d,[8] we see that in all cases the rounding exponent satsifies the matching relation (6). The behavior of the shift is more subtle: For antiperiodic conditions we have $\epsilon^{1/2}(n)\sim\delta^{1/2}(n)$ and $\lambda=\theta=1/\nu_d$, in accord with the analogous matching hypothesis for the shift. Under periodic boundary conditions, however, we find $\lambda=d-2$, so that, for $d\geqslant 4$, the shift is asymptotically much smaller than the rounding. We know of no simple heuristic argument for this effect, although it has also been noticed in series calculations for Ising[11] and Heisenberg[12] ferromagnetic films for $d=3$, where $\lambda\approx 2.0$. Finally, for the more realistic free-surface conditions we find

$$\lambda=1\ (\tau=1,\ \text{all } d), \tag{7}$$

although there is an additional logarithmic factor for $d=3$; the shift $\epsilon^1(n)$ is thus always asymptotically *larger* than the rounding $\delta^1(n)$, as anticipated. Indeed, the value (7) is in accord with the conjecture[2] mentioned, although for $d\geqslant 4$ the *sign* of the shift, which implies $T_c^{\,1}(n)>T_c$ for $n\gg 1$, is in direct contradiction with the "naive mean-field arguments"[2] originally adduced. The rather surprising enhancement of T_c for large n seems to be related to the constraint (imposed for all n) of constant particle density in the ideal Bose fluid, and of constant mean-square spin magnitude in the spherical model. The value $\lambda=1$ may thus not apply to fluid films observed under constant pressure[13] or to more realistic "fixed-spin" models; indeed, numerical evidence for the three-dimensional Ising model indicates $\lambda=1/\nu_3\simeq 1.56$ ($\tau=1$).[1,11]

More detailed predictions and more precise definitions of $\delta(n)$ and $\epsilon(n)$ can be made on the basis of the explicit scaling postulate[1,14]

$$Y^\tau(n,T)\approx n^\omega X^\tau(n^\theta\dot t)\ \text{as}\ n\to\infty,\ \dot t\to 0, \tag{8}$$

where we will accept the matching argument leading to (6). Note that the shifted temperature variable $\dot t$ appears; if this were replaced by t, we would be forced to conclude $\lambda=\theta$. The asymptotic behavior of the scaling functions $X^\tau(x)$ follows from the requirements that (8) reproduce both (2) and (4). This yields

$$X^\tau(x)\approx X_\infty x^{-\psi}\ \text{as}\ x\to\infty, \tag{9}$$

$$\approx X_0^{\,\tau}x^{-\dot\psi}\ \text{as}\ x\to 0, \tag{10}$$

where $X_\infty=A$, and one must have

$$\omega=\theta\psi=\psi/\nu \tag{11}$$

(so that $\omega=2-\eta$ when $Y=\overline{\chi}$). Evidently $X^\tau(x)$ describes the "shape" of the crossover from d-dimensional to d'-dimensional critical behavior as $T\to T_c^{\,\tau}(n)$. We can now make the amplitude prediction

$$\dot A^\tau(n)\approx X_0^{\,\tau}n^{(\psi-\dot\psi)/\nu}\ \text{as}\ n\to\infty. \tag{12}$$

Note that when $\dot\psi=0$ the amplitude $\dot A^\tau(n)$ is just the critical-point value $Y_c^{\,\tau}(n)$.

The case $\psi=0$ usually corresponds to a bulk logarithmic singularity

$$Y_\infty(T)=A\ln(t^{-1})+\cdots\ \text{as}\ t\to 0. \tag{13}$$

To accommodate such a weak singularity, the "background" terms

$$-n^\omega X^\tau(n^\theta\dot t_0)+Y^\tau(n,T_0) \tag{14}$$

should be added to the postulate (8). Here T_0 is a fixed noncritical reference temperature. Then (13) requires $\omega=0$ and the replacement of (9) by $X^\tau(x)\approx A\ln x^{-1}$ ($x\to\infty$). An interesting application can then be made to the specific heat ($Y=C$, $\psi=\alpha=0$) of a planar Ising model[5] and a helium film.[13] In these cases $\dot\psi$ vanishes and we conclude from (8) and (14) that

$$Y_c(n)=Y_{\max}(n)\approx(A/\nu)\ln n+\cdots. \tag{15}$$

This prediction for the specific-heat maximum is confirmed by the Ising-model results[5] ($\nu_2=1$). It is also consistent with the helium data[13] and the theoretically expected value $\nu_3\simeq\frac{2}{3}$, as shown by Moore.[15] However, more extensive experiments are desirable: Indeed, accepting (15) for helium would allow one to determine directly the otherwise inaccessible correlation exponent ν_3.

As $n\to\infty$ at fixed T in a film with free boundaries, a surface contribution can be defined by

$$Y^\times(T) \approx \tfrac{1}{2}n[Y^1(n, T) - Y_\infty(T)], \tag{16}$$

with corresponding exponent $\psi^\times$ and amplitude $A^\times$. This generally entails a higher-order term $X_\infty^\times x^{-\varphi}$ in (9), which then leads to[1] the alternative predictions[16]

$$\psi^\times = \psi + 1, \quad A^\times = -\tfrac{1}{2}\psi b^1 A, \quad \text{if } \lambda = 1 \text{ and } \nu_d < 1, \tag{17}$$

$$\psi^\times = \psi + \nu_d, \quad A^\times = \tfrac{1}{2}X_\infty^\times, \quad \text{if } \lambda > 1. \tag{18}$$

We have checked the predictions (17) in detail for the spherical and ideal Bose gases by direct calculation[6] of the surface susceptibilities $\bar{\chi}^\times(T)$: See the last column of Table I where $\gamma^\times$ is listed. An appropriately adapted scaling theory even accounts correctly for the observed logarithmic factors for $d \leq 4$ and for the leading corrections to $\bar{\chi}^\times(T)$ as $T\to T_c$ [which are generally in accord with (18)]. The surface specific heat of the plane-square Ising model[2,5] also verifies the theory.

Finally, we have been able to check the full scaling postulate (8) by calculating asymptotically the detailed critical behavior in d dimensions as $n\to\infty$. For $d=3$ the postulate applies precisely and the explicit susceptibility scaling functions for the spherical model ($\gamma_3=2$, $\gamma_2=0$)[8] can be given as

$$2JX_3^0(x) = \{2\sinh^{-1}[\tfrac{1}{2}\exp(4\pi K_c x)]\}^{-2}, \tag{19}$$

with $b_3^0 = 0$, and, parametrically for free surfaces,

$$2JX_3^1(x) = y^{-2}[1 - 2y^{-1}\tanh(\tfrac{1}{2}y)], \quad 8\pi K_c x = \ln[(\sinh y)/y], \tag{20}$$

with $b_3^1 = -1/4\pi K_c$, where $K_c = J/k_B T_c$. When $x\to\infty$ these expressions reproduce (9). Furthermore, in the case of free surfaces ($\tau = 1$) the leading correction term is in exact accord with the scaling argument sketched above and the calculated surface susceptibility. Conversely, for periodic boundary conditions the corrections to (9) are exponentially small (for all d). This means, as in the Ising model,[5] that for fixed $T > T_c$, the function $Y^0(n, T)$ approaches its limit $Y_\infty(T)$ as $\exp(-cnt^\nu)$, i.e., *exponentially fast.* This is probably a rather general result for systems with periodic boundary conditions.

A finite Bose or spherical model film in three dimensions has no phase transition (for $T>0$) so $\dot{\psi} = 0$. However, as $x\to-\infty$ the scaling functions (19) and (20) give the correct behavior of $\bar{\chi}(n, T)$ for $n \gg 1$ as $T\to 0$. In this sense the "critical region" of the film extends from $T = T_c$ down to $T=0$. Explicit expressions similar to (19) and (20) have also been found[6] for the scaling functions for the specific heat of ideal Bose films.

In five or more dimensions the detailed scaling predictions are again confirmed. In four dimensions a scaling form based on the variable $x = L/\xi(\dot{t})$ describes the rounding and crossover region correctly. However, the same form fails, by logarithmic corrections, to reproduce precisely the limits $n\to\infty$ at fixed $T > T_c$, or $T\to T_c^\tau(n)$ at fixed n. In view of similar difficulties with thermodynamic scaling[5] for $d = 4$ this is not so surprising. In summary, the overall agreement of the exact calculations with the scaling hypotheses is most gratifying.

We are indebted to Dr. G. A. T. Allan, Dr. B. I. Halperin, Dr. D. S. Ritchie, and Dr. P. G. Watson for helpful discussions and correspondence on phases of this work. The awards of a John Simon Guggenheim Memorial Fellowship (to M.E.F.) and a studentship (to M.N.B.), by the Commonwealth Scientific and Industrial Organization of Australia, are gratefully acknowledged.

---

*Work supported by the Advanced Research Projects Agency through the Materials Science Center at Cornell University, and by the National Science Foundation.

[1] A survey of the previous theoretical situation and more detailed discussion of some of the results reported here will appear in M. E. Fisher, in Proceedings of the Enrico Fermi International School of Physics, Course No. 51 (to be published).

[2] M. E. Fisher and A. E. Ferdinand, Phys. Rev. Lett. 19, 169 (1967). Note the sign of the exponent is in error in Eq. (14).

[3] In the case where, for finite L or n, the system has no sharp transition, one can normally define a pseudocritical point: See also Eq. (20) below.

[4] We use the standard exponent definitions: See, e.g., M. E. Fisher, Rep. Progr. Phys. 30, 615 (1967).

[5] See also A. E. Ferdinand and M. E. Fisher, Phys. Rev. 185, 832 (1969).

[6] M. N. Barber and M. E. Fisher, to be published.

[7] We consider d-dimensional hypercubical lattices with nearest-neighbor ferromagnetic interactions of strength $J>0$: See G. S. Joyce, Phys. Rev. 146, 349 (1966).

[8] For an ideal Bose fluid we may take $a = (h^2/2\pi m k_B T)^{1/2}$; see also J. D. Gunton and M. J. Buckingham, Phys. Rev. 166, 152 (1968).

[9]As will be described elsewhere, antiperiodic conditions are particularly useful for calculating the "helicity modulus" or, for a Bose fluid, the superfluid density $\rho_s(T)$.

[10]These results were announced by the authors in Proceedings of the IUPAP Conference on Statistical Mechanics, Chicago, April 1971 (to be published).

[11]G. A. T. Allan, Phys. Rev. B 1, 352 (1970); for further developments, see Ref. 1. For ideal Bose gas films the asymptotic behavior of $\epsilon^T(n)$ has also been investigated more recently by R. K. Pathria, Phys. Lett. 35A, 351 (1971).

[12]D. S. Ritchie and M. E. Fisher, in *Magnetism and Magnetic Materials—1971*, AIP Conference Proceedings No. 5 (American Institute of Physics, New York, 1972).

[13]D. F. Brewer, J. Low Temp. Phys. 3, 205 (1970).

[14]M. A. Moore, Phys. Lett. 37A, 345 (1971).

[15]We assume a zero ordering field ζ. Nonzero values can be included by allowing X^T to depend also on the standard thermodynamic scaling variable $y = \zeta / t^{\Delta}$, where $\Delta = \beta + \gamma = \beta\delta$.

[16]The proposal (18), but without the condition $\lambda > 1$, was first advanced by P. G. Watson [J. Phys. C: Proc. Phys. Soc., London 1, 268 (1968)], on less general grounds. Watson (private communication) has since withdrawn his claim that it is correct for a spherical model.

Classification
Physics Abstracts
05.20 — 05.50 — 64.70

An investigation of finite size scaling

E. Brézin

Service de Physique Théorique, Orme des Merisiers, 91191 Gif sur Yvette Cedex, France

(Reçu le 23 juillet 1981, accepté le 16 septembre 1981)

Résumé. — De nombreux calculs sur les systèmes finis (ou partiellement infinis), les simulations numériques par la méthode de Monte-Carlo par exemple ou bien la diagonalisation de matrices de transfert finies, sont extrapolés à la limite thermodynamique à l'aide d'hypothèses où les dimensions finies L du système sont incorporées dans des lois d'échelle. Nous montrons ici que ces relations de similitude où apparaît L sont la conséquence des deux faits suivants : (i) dans les équations de groupe de renormalisation la longueur L n'est pas renormalisée ; (ii) il n'y a pas de singularité au point fixe du groupe de renormalisation. Or un examen plus attentif montre que cette deuxième propriété est en défaut à quatre (et au-dessus de quatre) dimensions, et qu'en conséquence les relations de similitude n'y sont pas vérifiées. Pour illustrer le phénomène nous présentons des calculs détaillés relatifs à un modèle de vecteurs classiques à N composantes résolu dans la limite où N tend vers l'infini. Nous montrons également qu'en dessous de quatre dimensions le développement en $\varepsilon = 4 - d$, des fonctions qui apparaissent dans les lois d'échelle est singulier.

Abstract. — Calculations on finite (or partially finite) systems, such as Monte-Carlo simulations or the diagonalization of a finite transfer matrix, are often extrapolated to the thermodynamic limit by the use of scaling assumptions involving the finite sizes L of the system. It is shown here that these finite size scaling laws follow from the fact that : i) in the renormalization group flow equations the length L is not renormalized ; ii) there is no singularity at the fixed point. This second property fails in four (and above four) dimensions and finite size scaling does not hold there. An illustration of these points is presented, with detailed calculations of the large N limit of the N-vector model. As a consequence, the ε-expansion of finite size scaling properties is singular.

1. **Introduction.** — Finite size scaling, formulated several years ago by Fisher [1], has been used increasingly in many ways in order to extrapolate the information available from a finite (or partially infinite) system to the thermodynamic limit. Let us recall here the statements of finite scaling limit. First of all we shall consider two typical geometries : A) a finite system characterized by some length scale L in all dimensions ; B) a system infinite in one of the dimensions but finite with some length scale L in transverse directions. Manifestly we could consider many other situations (such as p infinite dimensions $(d - p)$ finite ([1]) etc...) but we hope our points will be sufficiently clear with these two cases. The geometry B is very often considered : the transfer matrix formalism or the Hamiltonian formulation of lattice field theory deals indeed with one infinite dimension. In any of these geometries, we consider some physical property P of the system (which may be singular at the critical point), such as the specific heat, the magnetic susceptibility, the correlations length ξ [2], etc... ; P is a function of the reduced temperature $t \equiv \dfrac{T - T_c}{T_c}$ which we can calculate for finite L

$$P = P_L(t) .$$

Finite size scaling states that, near T_c, for L large enough compared to the lattice spacing

$$\frac{P_L(t)}{P_\infty(t)} = f\left(\frac{L}{\xi_\infty(t)}\right) \tag{1}$$

in which ξ_∞ is the bulk correlation length. The function f is universal in the sense that it does not depend upon the type of lattice upon irrelevant operators, etc... but it does depend upon the geometry, the observable P, and is not the same for A and B ; it may also depend upon boundary conditions (periodic, free, etc...).

Two asymptotic limits of the function $f(x)$ are known *a priori* :

([1]) However additional interesting features appear when there are two (or three) infinite dimensions [10] since the partially finite system already displays criticality. Thus in addition one sees a crossover from one sort of critical behaviour to another. Such effects are not discussed in this article.

(i) At fixed t, L goes to infinity and P_L approaches P. Thus

$$\lim_{x\to\infty} f(x) = 1 . \tag{2}$$

(ii) At fixed L, t goes to zero. Then either the system is finite (geometry A) or in fact one-dimensional (B) and it has no phase transition. Consequently $P_L(t)$ is not singular at T_c. This implies in particular that, if $P_\infty(t) \underset{t\to 0}{\sim} t^{-\rho}$, the function f should have the behaviour

$$f(x) \underset{x\to 0}{\sim} Cx^{\rho/\nu} , \tag{3}$$

in which ν is the correlation length exponent. In fact there are more constraints since all derivatives of $P_L(t)$ should remain finite. This requirement would be better implemented if we wrote

$$P_L(t) = L^{\rho/\nu} g(tL^{1/\nu}) \tag{4}$$

(which is equivalent to (1)) by the demand that $g(x)$ should be C^∞ at $x = 0$.

Before we question the assumptions let us sketch here one powerful use of property (1), initiated by Nightingale, from which remarkable results have been obtained [3]. Consider the equation (1) for the correlation length itself

$$\frac{\xi_L(t)}{\xi_\infty(t)} = f\left(\frac{L}{\xi_\infty(t)}\right). \tag{5}$$

In geometry B, ξ_L may be obtained from the two largest eigenvalues of the transfer matrix. One thus defines a scale transformation $L \to L'$, $t \to t'$ by the equation (6)

$$\frac{L}{\xi_\infty(t)} = \frac{L'}{\xi_\infty(t')} . \tag{6}$$

For fixed rescaling L'/L this defines a transformation on t which has manifestly T_c as a fixed point (the l.h.s. vanishes at T_c). However since $\xi_\infty(t)$ is unknown it is convenient to use (5) which together with (6) implies

$$\frac{\xi_L(t)}{\xi_\infty(t)} = \frac{\xi_{L'}(t')}{\xi_\infty(t')} . \tag{7}$$

We then obtain from (6) and (7)

$$\frac{\xi_{L'}(t')}{\xi_L(t)} = \frac{L'}{L} ; \tag{8}$$

this is a real space renormalization group transformation which is exact (in the sense that no block coupling has been neglected) and it is entirely calculable from a finite system; T_c is then a fixed point of (8) and the critical exponent ν may be recovered from a linearization around the fixed point

$$\left(1 + \frac{1}{\nu}\right) \mathrm{Log} \frac{L'}{L} = \mathrm{Log} \frac{\dot{\xi}_{L'}(0)}{\xi_L(0)} . \tag{9}$$

If finite size scaling has been widely used it seems that it has been generally regarded as obvious [4]. However we shall see that : *a*) it is true below four dimensions and follows from the usual renormalization group analysis; *b*) it is not true in four and above four dimensions. Consequently it is not possible to compute the universal functions (1) from an ε-expansion : there is a singularity at $\varepsilon = 0$.

This article will be divided as follows. In section 2 we follow the renormalization group treatment, show that it yields finite size scaling laws below four dimensions and explain the difficulties at dimension four. In section 3 we discuss the large N limit (spherical model) in arbitrary dimension above the Curie point and in section 4 we examine the low temperature phase.

2. **Renormalization group and finite size scaling.** — We start from a lattice φ^4-model on a hypercubic d-dimensional lattice (with unit lattice spacing) :

$$\beta H = \sum_{r,\mu} \frac{1}{2} \nabla_\mu \varphi_r \nabla_\mu \varphi_r + \sum_r \left[\frac{m_0^2}{2} \varphi_r^2 + \frac{g_0}{4!} \varphi_r^4\right] \tag{10}$$

in which ∇_μ stands for the lattice gradient

$$\nabla_\mu \varphi_r \equiv \varphi_{r+\hat{\mu}} - \varphi_r ;$$

for simplicity we forget about the indices of φ related to the eventual internal symmetry of the Hamiltonian. Fourier transforms will depend upon the geometry; we shall limit ourselves to periodic boundary conditions on the finite dimensions of the system. For geometry A

$$\tilde{F}(q) = \sum_r F(r) \, e^{i\mathbf{qr}}$$

in which r belongs to a d-dimensional periodic lattice of size L, (11*a*)

$$F(r) = \frac{1}{L^d} \sum_{\mathbf{q}} \tilde{F}(q) \, e^{-i\mathbf{qr}} \qquad q_\alpha = \frac{2\pi}{L} n_\alpha$$

and for B

$$\tilde{F}(q) = \sum_{r_\perp, z} F(r) \, e^{i\mathbf{qr}}$$

in which r varies in a $(d-1)$ periodic lattice of size L and z on an infinite one-dimensional lattice

$$F(r) = \frac{1}{L^{d-1}} \sum_{q_\perp} \int_0^{2\pi} \frac{dq_z}{2\pi} e^{-i\mathbf{qr}} \tilde{F}(q) . \tag{11b}$$

The study of the large L behaviour of the theory relies on the following Poisson identity for periodic functions of a wavenumber $q = \frac{2\pi}{L} p$:

$$G(q) = G(q + 2\pi)$$

$$\frac{1}{L} \sum_{p=0}^{L-1} G(q) = \sum_{n=-\infty}^{+\infty} \int \frac{dq}{2\pi} G(q) \, e^{iq.nL} . \tag{12}$$

(Since it is linear in G, it is sufficient to verify it for e^{imq}, m integer.) The term $n = 0$ is the large L limit of the l.h.s. of (12). There are exponentially small corrections to this limit corresponding to higher values of $|n|$ in the r.h.s.

The key point of the renormalization group (R.G.) analysis is that for dimensions up to four the ultraviolet divergences which appear in the limit of zero-lattice spacing may be absorbed by « counter-terms » which amount to the use of renormalized parameters m and g instead of m_0 and g_0 and to the change of the field strength [5] :

$$\beta H = \sum_{r,\mu} \frac{Z}{2} \nabla_\mu \varphi_r^{R} \nabla_\mu \varphi_r^{R} + \sum_r \frac{\hat{Z}}{2} t(\varphi_r^{R})^2 + + \sum_r \frac{gZ_1}{4!} (\varphi_r^{R})^4 + \frac{\delta m^2}{2} \sum_r (\varphi_r^{R})^2 \quad (13)$$

$$t \equiv \frac{T - T_c}{T_c}.$$

We use in (13) the renormalization constants Z, $\hat{Z}$, Z_1, δm^2 of the infinite volume theory. However these same constants give to the finite (or partially finite) volume theory a well-defined limit when the lattice spacing vanishes. This can be seen from the identity (12) : the ultraviolet divergences which may appear in the l.h.s. of (12) when the lattice spacing vanishes are present only in the $n = 0$ term of the r.h.s. which is precisely the infinite volume limit of the eventual diagram, for which counter-terms are indeed present which perform the required subtractions. Consequently any physical observable such as an N-point correlation function will satisfy the equation for any L (large compared to a), i.e.

$$\lim_{a \to 0} \Gamma_L^{(N)}(m_0, g_0, a)\, Z^{-N/2} = \Gamma_L^{(N)}(m, g)\,. \quad (14)$$

Consequently, following the standard line of argument [5], we obtain for the L-dependent quantity the same renormalization group equations as for the bulk quantity [5] :

$$\Gamma_L^{(N)}(\mu, t, g) = = \exp\left[-\frac{N}{2}\int_1^\lambda \mathrm{d}x\, \eta(g(x))\right] \Gamma_L^{(N)}(\lambda\mu, t(\lambda), g(\lambda)) \quad (15)$$

is which μ is an arbitrary renormalization scale; we then consider the ratio

$$\varphi_L(\mu, t, g) = \Gamma_L^{(N)}/\Gamma_\infty^{(N)}\,, \quad (16)$$

and, manifestly from (15)

$$\varphi_L(\mu, t, g) = \varphi_L(\lambda\mu, t(\lambda), g(\lambda))\,. \quad (17)$$

Using now dimensional analysis, we obtain

$$\varphi_L(\mu, t, g) = \varphi_{\lambda L\mu}\left(1, \frac{t(\lambda)}{\lambda^2\mu^2}, g(\lambda)\right) \quad (18)$$

and we choose the (momentum) dilation λ by the condition

$$\lambda L\mu = 1\,, \quad (19)$$

so that

$$\varphi_L(\mu, t, g) = \varphi_1\left[1, t\left(\frac{1}{L\mu}\right) L^2, g\left(\frac{1}{L\mu}\right)\right]. \quad (20)$$

In the limit of large L (compared to the lattice spacing, or to the finite distance scale μ^{-1}) $g(1/L\mu)$ is driven to the infrared stable fixed point g^* of the model. The behaviour of $L^2\, t(1/L\mu)$ is obtained from the differential equation [5]

$$\lambda \frac{\mathrm{d}t}{\mathrm{d}\lambda} = t(\lambda)\left[\frac{1}{\nu(g(\lambda))} - 2\right] \quad (21)$$

which leads to

$$t(\lambda) \underset{\lambda \to 0}{\sim} t\lambda^{1/\nu - 2}$$

or

$$t\left(\frac{1}{L\mu}\right) L^2 \underset{L\mu \gg 1}{\sim} tL^{1/\nu}\,. \quad (22)$$

Consequently we have shown that, in the scaling limit $L\mu \gg 1$, $t/\mu^2 \ll 1$

$$\varphi_L(\mu, t, g) = \varphi_1(1, tL^{1/\nu}, g^*) = f(t^\nu L) \quad (23)$$

which proves finite size scaling for any multiplicatively renormalizable quantity ([2]) as well as the universality of f.

However, as usual, in the R. G. there is the implicit assumption that the functions are not singular at the fixed point. Below four dimensions, in which case g^* is non-zero, there is no special reason to question this assumption. But in four dimensions we can see immediately that we are going to face a difficulty : indeed since $g(1/L\mu)$ vanishes in the large L limit, the limiting theory is simply mean field theory. However mean field theory has the bad feature of predicting a phase transition even for a one-dimensional system, or for a system of finite volume. Therefore we should expect $\xi_L(T_c)$ for instance to be infinite. In order to understand in more detail the situation we have made exact calculations in arbitrary dimension for the (large N) N-vector model.

3. Correlation length in the large N-limit ($T > T_c$). —

We consider the classical N-vector model on a lattice, with short-range translation invariant, ferromagnetic interactions [6]

$$\beta H = -N \sum_{i,j} J_{ij}\, \mathbf{S}_i . \mathbf{S}_j \quad (24)$$

$$\mathbf{S}_i^2 = (S_i^1)^2 + \cdots + (S_i^N)^2 = 1\,. \quad (25)$$

([2]) Therefore these arguments apply to the order parameter correlation functions, to the correlation length or to the magnetic susceptibility for instance. However if one takes the specific heat the R.G. equation is inhomogeneous [5] and as a consequence the finite size scaling form for the specific heat per unit volume has to be written $C_L(t) - C_\infty(t) = L^{\alpha/\nu} f(L/\xi_\infty(t))$. This form cannot be reduced to (1), because of the constant part of $C_\infty(t)$.

We obtain equations of motion for the correlation functions by performing an infinitesimal rotation at site i

$$S_i^a \to S_i^a + \omega^{ab}(i)\, S_i^b \qquad \omega^{ab} + \omega^{ba} = 0 \tag{26}$$

which leaves the integration measure over the sphere invariant. Expanding the identity obtained by performing (26) on $\langle S_i^a S_j^b \rangle$ we obtain at first order in ω

$$0 = \delta_{jk}\,\delta_{bc} \langle S_i^a S_j^d \rangle - \delta_{jk}\,\delta_{bd} \langle S_i^a S_j^c \rangle + \delta_{ik}\,\delta_{ac} \langle S_i^d S_j^b \rangle - \delta_{ik}\,\delta_{ad} \langle S_i^c S_j^b \rangle + 2N \sum_l J_{kl} \langle S_i^a S_j^b (S_k^d S_l^c - S_k^c S_l^d) \rangle . \tag{27}$$

Defining

$$G(i,j) = \langle \mathbf{S}_i . \mathbf{S}_j \rangle \tag{28a}$$

$$G(i,i) = 1 \tag{28b}$$

we obtain from (27)

$$(N-1)\, G(i,j)\,(\delta_{jk} - \delta_{ik}) + 2N \sum_l J_{kl} \langle (\mathbf{S}_i \mathbf{S}_k)(\mathbf{S}_j \mathbf{S}_l) - (\mathbf{S}_i \mathbf{S}_l)(\mathbf{S}_k \mathbf{S}_j) \rangle = 0 . \tag{29}$$

In the large N limit invariant correlation functions factorize [7] : if A and B are rotationally invariant

$$\langle AB \rangle = \langle A \rangle \langle B \rangle + 0\left(\frac{1}{N}\right). \tag{30}$$

We then obtain from (29) a closed equation for G in the large N limit

$$0 = (\delta_{jk} - \delta_{ik})\, G(i,j) + 2 \sum_l J_{kl} [G(i,k)\, G(j,l) - G(i,l)\, G(j,k)] . \tag{31}$$

After Fourier transformation

$$G(q) = \sum_j e^{i\mathbf{q}(\mathbf{r}_j - \mathbf{r}_i)}\, G(i,j) \tag{32a}$$

$$J(q) = \sum_j e^{i\mathbf{q}(\mathbf{r}_j - \mathbf{r}_i)}\, J_{ij} \tag{32b}$$

we obtain from (31)

$$G(q_1) - G(q_2) + 2(J(q_2) - J(q_1))\, G(q_1)\, G(q_2) = 0 , \tag{33}$$

i.e.

$$\frac{1}{G(q)} + 2\, J(q) = C ,$$

in which C is q-independent.

Therefore we end with the results

$$G(i,j) = \frac{1}{L^d} \sum_q \frac{e^{i\mathbf{q}.\mathbf{r}_{ij}}}{C - 2\,J(q)} \quad \text{for (A)} \tag{34a}$$

$$G(i,j) = \frac{1}{L^{d-1}} \sum_{q_\perp} \int \frac{dq_\parallel}{2\pi} \frac{e^{i\mathbf{q}.\mathbf{r}_{ij}}}{C - 2\,J(q)} \quad \text{for (B)}. \tag{34b}$$

The constant C is determined by the requirement (28b). Specializing for definiteness to nearest neighbour interaction (but as usual it is only the short-range, translationally invariant character which matters)

$$J_{ij} = \begin{cases} \beta/2 & i, j \text{ nearest neighbours} \\ 0 & \text{otherwise} , \end{cases}$$

we have

$$J(q) = \beta \sum_{\mu=1}^{d} \cos q_\mu . \tag{35}$$

It will be convenient to define m_L^2 by

$$C = 2\, J(0) + \beta m_L^2 . \tag{36}$$

Then (28b) yields

$$\beta = \frac{1}{L^d} \sum_q \frac{1}{2 \sum_1^d (1 - \cos q_\mu) + m_L^2} \tag{37}$$

for (A) and a similar equation for (B).

3.1 Correlation length for the geometry B. — Let us consider in more detail the geometry (B); we can there define a true correlation length by studying correlations along the infinite z-direction

$$G(z) = \frac{1}{\beta} \frac{1}{L^{d-1}} \sum_{q_\perp} \int_0^{2\pi} \frac{dq_\parallel}{2\pi} \frac{e^{iq_\parallel z}}{2 \sum_1^d (1 - \cos q_\mu) + m_L^2} \tag{38a}$$

$$G(0) = 1 . \tag{38b}$$

When z goes to infinity the leading term in the r.h.s. of (38a) corresponds to $q_\perp = 0$, q_z small

$$G(z) \simeq \frac{1}{\beta} \frac{1}{L^{d-1}} \int \frac{dq_\parallel}{2\pi} \frac{e^{iq_\parallel z}}{q_\parallel^2 + m_L^2}$$

which falls-off exponentially as $e^{-m_L|z|}$. Then the correlation length is given by

$$\xi_L = m_L^{-1} . \tag{39}$$

Therefore the correlation length ξ_L is related to the temperature and to L by the equation

$$\beta = \frac{1}{L^{d-1}} \sum_{q_\perp} \int_0^{2\pi} \frac{dq_\parallel}{2\pi} \frac{1}{2 \sum_1^d (1 - \cos q_\mu) + \xi_L^{-2}} . \tag{40}$$

The bulk correlation length ($T > T_c$) is of course defined by

$$\beta = \int_0^{2\pi} \cdots \int_0^{2\pi} \frac{d^d q}{(2\pi)^d} \left[2 \sum_1^d (1 - \cos q_\mu) + \xi_\infty^{-2} \right]^{-1} \tag{41}$$

and the critical temperature corresponds to $\xi_\infty^{-2} = 0$ (we consider dimensions d larger than two).

In order to compare ξ_L to ξ_∞ we transform (40) by the identity (12) for the $(d-1)$ transverse directions

$$\beta = \sum_{\mathbf{n}_\perp} \int \frac{d^{d-1} q_\perp}{(2\pi)^{d-1}} \int \frac{dq_\parallel}{2\pi} \left[2 \sum_1^d (1-\cos q_\mu) + \xi_L^{-2} \right]^{-1} e^{i\mathbf{q}_\perp \cdot \mathbf{n}_\perp L} \tag{42}$$

from which we isolate the $\mathbf{n}_\perp = 0$ term and subtract the infinite volume limit (41). This yields

$$(\xi_\infty^{-2} - \xi_L^{-2}) \int \frac{d^d q}{(2\pi)^d} [\xi_\infty^{-2} + 2\sum (1-\cos q_\mu)]^{-1} [\xi_L^{-2} + 2\sum(1-\cos q_\mu)]^{-1} + \\ + \sum_{\mathbf{n}_\perp \neq 0}' \int \frac{d^d q}{(2\pi)^d} \left[2\sum_1^d (1-\cos q_\mu) + \xi_L^{-2} \right]^{-1} e^{i\mathbf{q}_\perp \cdot \mathbf{n}_\perp L} = 0 . \tag{43}$$

We want to study the limit in which L goes to infinity, ξ_L^{-1} and ξ^{-1} go to zero (L large, $T - T_c$ small). In this limit we can write (for $\mathbf{n}_\perp \neq 0$)

$$\int \frac{d^d q}{(2\pi)^d} \left[2\sum_1^d (1-\cos q_\mu) + \xi_L^{-2} \right]^{-1} e^{i\mathbf{q}\cdot\mathbf{n}_\perp L} \simeq \int \frac{d^d q}{(2\pi)^d} \frac{e^{i\mathbf{q}\cdot\mathbf{n}_\perp L}}{\xi_L^{-2} + \mathbf{q}^2} = L^{2-d} \left(\frac{1}{4\pi}\right)^{d/2} \int_0^\infty e^{-t(L/\xi_L)^2 - \mathbf{n}_\perp^2/4t}\, t^{-d/2}\, dt . \tag{44}$$

Defining

$$g(t) = \sum_{n_\perp}' e^{-n_\perp^2/4t} \tag{45}$$

$$g(t) \underset{t\to 0}{\sim} 2(d-1)\, e^{-1/4t}, \qquad g(t) \underset{t\to\infty}{\sim} (4\pi t)^{(d-1)/2} \tag{46}$$

and putting

$$\begin{aligned} L/\xi_\infty &= x \\ L/\xi_L &= y \end{aligned} \tag{47}$$

we obtain from (43)

$$(x^2 - y^2) \int \frac{d^d q}{(2\pi)^d} [\xi_\infty^{-2} + 2\sum(1-\cos q_\mu)]^{-1} [\xi_L^{-2} + 2\sum(1-\cos q_\mu)]^{-1} + \\ + L^{4-d} \frac{1}{(4\pi)^{d/2}} \int_0^\infty e^{-ty^2} g(t)\, t^{-d/2}\, dt = 0 . \tag{48}$$

It is necessary now to discuss (48) according to the dimension :

a) $d < 4$.

The first integral of the l.h.s. of (48) becomes

$$\int_{\mathbb{R}^d} \frac{d^d q}{(2\pi)^d} (\xi_\infty^{-2} + q^2)^{-1} (\xi_L^{-2} + q^2)^{-1} = \xi_\infty^{4-d} \left(\frac{1}{4\pi}\right)^{d/2} \Gamma(2 - d/2) \int_0^1 du \left[u + \left(\frac{y}{x}\right)^2 (1-u) \right]^{\frac{d}{2}-2}$$

and therefore we obtain a relation between x and y

$$(x^2 - y^2)\, \Gamma(2 - d/2) \int_0^1 du \left[u + \left(\frac{y}{x}\right)^2 (1-u) \right]^{\frac{d}{2}-2} + x^{4-d} \int_0^\infty e^{-ty^2} g(t)\, t^{-d/2}\, dt = 0 . \tag{49}$$

Finite size scaling is thus indeed valid : if we write

$$\frac{\xi_L}{\xi_\infty} = f(L/\xi_\infty) \quad \text{then} \quad f(L/\xi_\infty) = \frac{x}{y} .$$

It is easy to verify that f has all the desired properties ; let us in particular consider $\xi_L(T_c)$: i.e.

$$x = 0 \qquad \frac{\xi_L(T_c)}{L} = \alpha . \tag{50}$$

We obtain from (49)

$$\alpha^{2-d} \frac{\Gamma(2 - d/2)}{\frac{d}{2} - 1} = \int_0^\infty e^{-t/\alpha^2} g(t)\, t^{-d/2}\, dt \tag{51}$$

which defines α.

In three dimensions this gives

$$\alpha^{-1} = \sum_{n_1 n_2}' (n_1^2 + n_2^2)^{-1/2} \exp - [\sqrt{n_1^2 + n_2^2}/\alpha]$$

$$\alpha \simeq 0.6614,$$

but near four dimensions the l.h.s. of (51) has a pole, α increases and $g(t)$ in the r.h.s. of (51) may be replaced by its large t approximation (46). We thus obtain that in dimension $d = 4 - \varepsilon$

$$\xi_L(T_c)_{\varepsilon \to 0} = (4 \pi^2 \varepsilon)^{-1/3} L . \tag{52}$$

We do see that a singular behaviour is to be expected in dimension four.

b) $d = 4$.

The first integral of (48) has to be calculated with a fixed lattice spacing cut-off and therefore we obtain

$$- x^2 \ln\left(\frac{\xi_\infty}{a}\right)^2 + y^2 \ln\left(\frac{\xi_L}{a}\right)^2 = \int_0^\infty \frac{dt}{t^2} e^{-ty^2} g(t) . \tag{53}$$

Manifestly because of these logarithms we cannot put (53) into a scaling form. Another way of visualizing this breakdown of scaling is to consider $\xi_L(T_c)$ which is no longer linear in L; it is given by the $x = 0$ limit of (53) :

$$\left(\frac{L}{\xi_L}\right)^2 \ln\left(\frac{\xi_L}{a}\right)^2 = 8 \pi^2\left(\frac{\xi_L}{L}\right),$$

or

$$\xi_L(T_c) = L\left(\frac{1}{4 \pi^2} \ln \frac{L}{a}\right)^{1/3} . \tag{54}$$

c) $d > 4$.

In the regime in which L, ξ_L and ξ_∞ are simultaneously large compared to the lattice spacing, we can simply set in the first integral of equation (48) $\xi_\infty^{-1} = \xi_L^{-1} = 0$, without any restriction on the size of the ratios x and y since the integral

$$\int_0^{2\pi} \frac{d^d q}{(2 \pi)^d} \left(2 \sum_1^d (1 - \cos q_\mu)\right)^{-2} = \sigma \tag{55}$$

is convergent. Therefore (48) gives

$$\sigma(x^2 - y^2) + \frac{L^{4-d}}{(4 \pi)^{d/2}} \int_0^\infty e^{-ty^2} g(t)\, t^{-d/2}\, dt = 0 . \tag{56}$$

Again, because of the explicit L which remains present in (56) we see that scaling is broken, unless it is modified in the new form $\xi_L/\xi_\infty = f\left(\frac{L}{\xi_\infty} L^{d-4}\right)$; at T_c we now obtain

$$\xi_L(T_c) = L(2 \sigma L^{d-4})^{1/3} . \tag{57}$$

In particular when d increases $\xi_L(T_c)$ diverges : this is to be expected since for large d mean field theory becomes exact and $\xi_L(T_c)$ is predicted to become infinite for large d.

3.2 GEOMETRY (A). — We simply briefly mention the modifications. Firstly there is no « true » correlation length. One can use as a substitute the second moment of the spin-spin correlation function

$$\xi_L^2 = \frac{\sum_q \sum_r r^2 e^{i\mathbf{qr}} \left(m_L^2 + 2 \sum_1^d (1 - \cos q_\mu)\right)^{-1}}{\sum_q \sum_r e^{i\mathbf{qr}} \left(m_L^2 + 2 \sum_1^d (1 - \cos q_\mu)\right)^{-1}} . \tag{58}$$

From this definition ξ_L is bounded by L. If we use instead as before the quantity m_L^{-1}, which is now the magnetic susceptibility, then it is still determined by the condition (37) and the whole procedure may be repeated. There is no preferred direction now and the only change is the asymptotic behaviour of $g(t)$ (46), which is now proportional to $t^{d/2}$ for large t.

Consequently we end with the same conclusions : breakdown of finite size scaling for $d \geqslant 4$, and

$$m_L^{-1}(T_c) \propto \begin{cases} \varepsilon^{-1/4} L & \varepsilon \to +0 \quad (d = 4 - \varepsilon) \\ L(\ln L)^{1/4} & d = 4 \\ L(L)^{\frac{d-4}{4}} & d > 4 \end{cases} \tag{59}$$

One may wonder about what is specific in the previous calculations to the large N-limit. For instance when we found that $\xi_L(T_c)/L$ diverges as $\varepsilon^{-1/3}$ (Eq. (52)), could it be that for N finite this ratio would diverge with a different power of ε ? In fact a direct analysis near four dimensions [8] reveals that the power $-1/3$ is indeed independent of N. The whole picture that we found for the spherical model is qualitatively unchanged for finite N.

4. **Large N-limit ($T < T_c$).** — Below T_c the long distance properties are very different if Goldstone modes are present, i.e. if we are dealing with a broken continuous symmetry. For an Ising-like system it is expected [9] that $\xi_L(T)$ diverges for large L exponentially below the Curie point. However for N-vector models we shall see that $\xi_L(T)$ behaves as L^{d-1} for fixed T below T_c.

Again we perform the calculation for the large N-limit : from the normalization condition (42) we substract now

$$\beta_c = \int_0^{2\pi} \frac{d^d q}{(2 \pi)^d} \left[2 \sum_1^d (1 - \cos q_\mu)\right]^{-1}$$

and obtain (for geometry B)

$$\beta - \beta_c = -\frac{1}{\xi_L^2}\int_0^{2\pi}\frac{d^d q}{(2\pi)^d}[2\sum(1-\cos q_\mu)]^{-1}[2\sum(1-\cos q_\mu)+\xi_L^{-2}]^{-1} +$$

$$+\sum_{n_\perp}{}'\int\frac{d^d q}{(2\pi)^d}e^{i\mathbf{q}_\perp \mathbf{n}_\perp . L}[2\sum(1-\cos q_\mu)+\xi_L^{-2}]^{-1}. \quad (60)$$

The first term of the r.h.s. of (60) is negligible in the large L limit as we shall see below. Assuming it for the time being, we can deal with the sum over $\mathbf{n}_\perp$ as before and for $\xi_L/L \gg 1$, we obtain

$$\beta - \beta_c = \tfrac{1}{2}\xi_L L^{1-d} \quad\text{or}\quad \xi_L(T) \underset{T<T_c}{=} 2(\beta-\beta_c)L^{d-1}. \quad (61)$$

We now consider the first integral of the r.h.s. of (60); for $d > 4$ it behaves as ξ_L^{-2} and for $d < 4$ as $\xi_L^{-1(d-2)}$ which are negligible for fixed $\beta - \beta_c$.

Again one may try to see what in (61) is particular to the large N limit and what is general. A renormalization group analysis of the non-linear σ-model [8] shows that for T smaller than T_c we also have a finite size scaling law (in the vicinity of two dimensions)

$$\xi_L(T) = Lf(L(T_c - T)^\nu) \quad (62)$$

which is indeed compatible with (61) provided

$$f(x) \underset{x\to\infty}{\sim} x^{d-2} \quad (63)$$

(one should remember that for the large N limit $\nu(d-2) = 1$).

Since the behaviour $\xi_L(T) \propto L^{d-1}$ (at fixed $T < T_c$) follows from a straightforward spin wave analysis, the behaviour (63) should be valid for any N and hence the general law is

$$\xi_L(T) \underset{T<T_c}{\propto} (T_c - T)^{\nu(d-2)} L^{d-1}. \quad (64)$$

5. **Conclusion.** — We have seen that finite size scaling follows directly from the invariance in a renormalization group flow of the length L (if L is much larger than a). However in addition to the flow equations a non singular behaviour at the fixed point is necessary and is not fulfilled in four and above four dimensions. Consequently : *a*) finite size scaling does not hold for $d \geqslant 4$; *b*) the limit $\varepsilon \to +0$ ($d = 4 - \varepsilon$) is singular.

Acknowledgments. — I have benefited greatly from several discussions with Dr. B. Derrida who introduced me to the subject, and explained to me his ideas. I have also enjoyed several very useful discussions with Dr. J. Zinn-Justin on various aspects of this problem. Enlightening comments from Pr. M. E. Fisher are gratefully acknowledged.

References

[1] FISHER, M. E., in *Critical Phenomena*, Proc. 51st Enrico Fermi Summer School, Varena, edited by M. S. Green (Academic Press, N.Y.) 1972.
FISHER, M. E. and BARBER, M. N., *Phys. Rev. Lett.* **28** (1972) 1516.
Some aspects of this problem, such as the size dependence of thermo-dynamic quantities at the critical point, were discussed in the article of IMRY, Y. and BERGMAN, D., *Phys. Rev.* **A 3** (1971) 1416.

[2] Some words should be added to the definition of the correlation length for finite L. In the geometry B we can consider a correlation function between two spins, for which the vector joining the two points is parallel to the infinite dimension. Then ξ_L characterizes the exponential fall-off of the correlation in that direction. For a finite system we can define ξ_L from the second moment of the two point correlation function $G_L(r)$:

$$\xi_L^2 = \sum_r r^2 G_L(r) / \sum_r G_L(r).$$

[3] NIGHTINGALE, M. P., *Physica* **83** (1976) A 561.
DERRIDA, B., *J. Phys. A* **14** (1981) L-5.
BLÖTE, H., NIGHTINGALE, M. and DERRIDA, D., *J. Phys. A* **14** (1981) L-45.
HAMER, C. J. and BARBER, M. N., *J. Phys. A* **14** (1981) 241 and 259.
PICHARD, J. L. and SARMA, G., *J. Phys.* **C 14** (1981) L-127 and L-617.

[4] There are some earlier attempts to understand finite size scaling from the renormalization group, in particular SUZUKI, M., *Progr. Theor. Phys.* **58** (1977) 1142.

[5] The formalism and the notations follow closely the article by BRÉZIN, E., LE GUILLOU, J. C. and ZINN-JUSTIN, J., in *Phase Transitions and Critical Phenomena*, vol. VI, C. Domb and M. Green eds. (Academic Press, N.Y.) 1976, p. 127.

[6] The finite size scaling properties of the spherical model have been studied in great detail in the very extensive work of BARBER, M. N. and FISHER, M. E., *Ann. Phys.* **77** (1973) 1, who have discovered similar logarithmic difficulties in four dimensions (but not above four). We have presented here the factorization property of the large N limit [7] which is powerful and simple.

[7] WITTEN, E., in *Recent Developments in Gauge theories*, Proceedings of the Cargèse Summer School, 1979 (Plenum Press, N.Y.) 1980, G't Hooft *et al.* eds. This factorization property has been used also in a different context by MIGDAL, A. A.

[8] This was established in collaboration with Dr. J. ZINN-JUSTIN.

[9] In the proceedings of the 1968 IUPAP Stat. Mech. Conf. (Kyoto) published in *J. Phys. Soc. Japan Suppl.* **26** (1969) 87, Pr. M. E. FISHER gives the relation :

$$\xi_L \sim \exp - \beta L^{d-1} \Sigma(T)$$

in which $\Sigma(T)$ is the interfacial free energy generating the surface tension.

I thank Dr. B. DERRIDA for explaining to me why this was expected.

[10] A physically interesting example of this crossover is treated in the article by :

BARBER, M. N. and FISHER, M. E., *Phys. Rev.* **A 8** (1973) 1124.

See also the early article by :

IMRY, Y., DEUTSCHER, G., BERGMAN, D. and ALEXANDER, S., *Phys. Rev.* **A 7** (1972) 744.

Nuclear Physics B257 [FS14] (1985) 867–893

FINITE SIZE EFFECTS IN PHASE TRANSITIONS

E. BRÉZIN, J. ZINN-JUSTIN

Service de Physique Théorique, CEN-Saclay, 91191 Gif-sur-Yvette Cedex, France

Received 4 March 1985
(Revised 6 May 1985)

We develop a systematic approach to the calculations of finite size effects in phase transitions. The method consists of constructing an effective hamiltonian for the homogeneous modes, obtained by tracing out all other degrees of freedom. These modes are obtained by averaging the order parameter over the finite dimensions of the system. These techniques, together with the renormalization group, lead to explicit calculations of universal finite size scaling functions, under the form of $(2+\varepsilon)$ or singular $(4-\varepsilon)$ expansions. Some simple universal results above the upper critical dimension are presented. Simple and universal properties of the rounding of first order transitions are derived.

1. Introduction

Formulated for the first time in 1971 by Fisher [1], the finite size scaling (FSS) properties in the vicinity of a critical point, or at first order transitions, are more and more studied and used for practical purposes. It is almost invariably used as an efficient extrapolation procedure for numerical simulations of Monte Carlo type limited in general to rather small samples. It is also the basis of the powerful phenomenological renormalization group method [2]. A recent review article on FSS by Barber [3] describes many other aspects of the subject.

In this work we consider some new aspects of FSS; in particular we discuss analytical means of calculating the size-dependent universal scaling functions through ε-expansions, a problem left unanswered by previous renormalization group attempts to understand FSS [4–6]. Indeed in ref. [4] it was shown that, above the upper critical dimension, four for our purposes here, since the lattice spacing cannot be removed from the problem – there is a trivial critical point, but no scaling limit – this additional length scale must be included in the finite size formulation. This fact, equivalent to Fisher's dangerous irrelevant variables [7] description, had the more serious consequence that an attempt to expand below four dimensions in ε yielded a singularity, a power $\varepsilon^{1/3}$ in the case considered in ref. [4], and it was not at all clear that any expansion could be envisaged. In this work we describe a systematic method applicable below the upper critical dimension. We have focused our attention on two different geometrical shapes though the method could easily be extended to different boundary conditions, different shapes etc.... The first case corresponds to a periodic cube of volume L^d (it would be trivial to modify it for a large rectangular box with sizes going to infinity simultaneously); this case is encountered in numerical simulations. The second case corresponds to a cylindrical geometry: a tube of "square" basis (or L^{d-1} in general dimension) infinite along one direction, called "time" in this work since it corresponds to a hamiltonian (or transfer matrix) formulation of the problem; periodic boundary conditions are imposed on the "square". In the first case we use discrete Fourier analysis along the d dimensions and the leading long-distance singularities correspond to the zero-momentum component $\varphi \equiv \Phi(\boldsymbol{q}=0)^{\star}$. All non-zero modes may be treated perturbatively (within the ε-expansion) and we compute an effective action for this mode φ by tracing out all the $\boldsymbol{q}$ non-zero modes

$$\exp(-S_{\text{eff}}[\varphi]) = \sum_{\{\Phi_q\}}' \exp(-S[\Phi]) .$$

The averages are finally performed with the Boltzmann weight $\exp(-S_{\text{eff}}[\varphi])$. This turns out to lead to a systematic expansion in powers of $\varepsilon^{1/2}$.

In the cylindrical case, after Fourier analysis over the $(d-1)$ finite dimensions, we can isolate the leading long-distance modes $\varphi(\tau) = \Phi(\tau, \boldsymbol{q}_{\perp}=0)$, which depend upon the position τ, along the axis of the cylinder. The same procedure, namely tracing out all modes except $\varphi(\tau)$, yield an effective action $S_{\text{eff}}[\varphi(\tau)]$ which may be written, for large L, as the integral over a local "lagrangian" density:

$$S_{\text{eff}}[\varphi(\tau)] = \int_0^{\infty} d\tau \, L_{\text{eff}}(\varphi, \dot{\varphi}) .$$

The remaining sum over the $\varphi(\tau)$ modes corresponds to an (imaginary time) Feynman path integral and may be conveniently replaced by a Schrödinger equation, in the usual way. The spectrum of this Schrödinger operator provides the FSS functions that we are looking for. The results take the form of an expansion in powers of $\varepsilon^{1/3}$. The calculations can be performed for an arbitrary number n of components of the order parameter. In the large-n limit one recovers the $\varepsilon^{1/3}$ singularity discussed in ref. [4], but with now a systematic handle on the corrections.

As a byproduct we have also considered the situation above the upper critical dimension, in which some very simple results, which do not seem to have been noticed up to now, are derived. They are based upon the same technique, which simplifies considerably since corrections to the tree-level approximation for the effective action of the φ or $\varphi(\tau)$ modes respectively, are irrelevant. As a result we have obtained some universal answers above dimension four. For instance we have considered in the cubic geometry the ratio M_4/M_2^2 (in which $M = (1/N)\sum_1^N S_i$, and $M_n = \langle M^n \rangle$), which is related at T_c to the renormalized coupling constant. Above four dimensions, for an Ising-like system for instance, one obtains at T_c

$$\frac{M_4}{M_2^2} = \frac{1}{8\pi^2}[\Gamma(\tfrac{1}{4})]^4 \simeq 2.188440 .$$

* Other boundary conditions would lead to different calculations and different results, even if translation invariance is not broken (twisted periodic conditions for instance).

We also considered in this work some aspects of FSS for broken continuous symmetries below T_c, which has recently been extensively studied by Fisher and Privman [8]. In these problems the quasi-Goldstone modes dominate the long distance properties in the infinite volume limit, at fixed temperature below T_c. The FSS analysis of the spin wave fluctuations (i.e. the non-linear σ model) leads to a zero-temperature fixed point. The properties may thus be related to a low temperature expansion and studied in a $(2+\varepsilon)$ expansion.

We have considered FSS properties at Ising-like first-order transitions [9] within the same technique of isolating the non-perturbative long-distance modes. Although the correlation length of the infinite system remains finite at and below the first order transition, simple and universal results for the rounding of these transitions may be obtained. This universality results from the divergence of the finite size correlation length with the size of the system.

Finally some additional results about rounding in n-vector models below T_c are discussed.

The paper is organized as follows. In sect. 2 we review briefly the renormalization group set-up and its consequence in the presence of boundaries. In sect. 3 we discuss FSS in a cubic geometry; results for dimensions larger than four are derived, and afterwards we discuss the principles - and compute the first terms - of the $\varepsilon^{1/2}$ expansion. Sect. 4 deals with the cylinder $L^{d-1}\times\infty$ and the $\varepsilon^{1/3}$ expansion. Sect. 5 is devoted to spin waves and the corresponding $(2+\varepsilon)$ expansion. Finally in sect. 6 we discuss some properties of FSS for first-order transitions.

2. Renormalization group and FSS

Let us briefly recall here the renormalization group formalism in the presence of (periodic) boundaries and its application to FSS [4]. It is well-known that there are many ways of writing renormalization group transformations (though, naturally, the physics is independent of this arbitrariness) and it is convenient in order to describe FSS to use a continuum description. In this scaling limit (or renormalized theory), all the renormalizations, namely of the magnitude of the order parameter, of the coupling constant, the shift of the critical temperature, are defined in the infinite volume bulk theory. This scaling limit exists only below the upper critical dimension. In this continuum theory or scaling limit, the lattice spacing has completely disappeared; the integrations over the wave vectors of the fluctuations are performed without cut-off and they are convergent. If some of the dimensions of the sample are finite the integrations over the corresponding momenta are replaced by discrete sums. These sums still extend to infinity, since we have taken a zero lattice spacing, but they converge exactly as the infinite volume integrals. The sides of the sample are finite if expressed in units of the dimensional scale of the renormalized theory, which is infinite in lattice spacing units. We therefore limit ourselves to a regime in which L/a and ξ/a go simultaneously to infinity, however their ratio L/ξ is finite and remains at our disposal. This is precisely the regime in which FSS is expected to hold. Since there is no need for any new renormalization in this scheme (if we use periodic boundary conditions) we obtain the usual renormalization group equations, without any renormalization of L. (The situation is very similar to that of a field theory, such as QED, at finite temperature. The integrals over frequencies are replaced by discrete sums but the renormalizations are temperature independent [10]).

In a ϕ^4, Landau-Ginzburg-Wilson formalism [11], the observables depend upon the temperature $t=(T-T_c)/T_c$, the dimensionless coupling constant g, a length l in order to fix up the renormalization procedure, and the finite size parameter L. For an observable X of (length) dimension p_x, we conclude from the previous analysis [12] that, for a dilatation ρ

$$X[t, g; l; L]=\zeta(\rho)X[t(\rho), g(\rho), l\rho; L], \tag{2.1}$$

which summarizes the renormalization group transformation and the non-renormalization of L. From the bulk problem one knows that for large dilatations $g(\rho)$ approaches the infrared stable fixed point g^*,

$$t(\rho)\approx t\rho^{(1/\nu)-2} \qquad \text{and} \qquad \zeta(\rho)\approx \rho^{(\gamma_x/\nu)-p_x},$$

in which γ_x is the bulk exponent of $X: X \underset{t\to 0}{\approx} t^{-\gamma_x}$. Using now dimensional analysis, together with eq. (1), one obtains

$$X[t, g; l; L]=(\rho l)^{p_x}\zeta(\rho)X[t(\rho)(l\rho)^2, g(\rho), 1; L/l\rho]. \tag{2.2}$$

Choosing $\rho=L/l$, which is a large parameter, we obtain the well-known FSS result

$$X[t, g; l; L]=t^{-\gamma_x}f[tL^{1/\nu}], \tag{2.3}$$

in which the function f is universal (up to scales fixing).

Let us next consider the FSS renormalization group for an n-vector model below T_c. The long-distance properties of the bulk system are governed by massless spin wave excitations, whose fluctuation properties are reproduced by the non-linear σ-model [13, 14]. Observables are now functions of a coupling constant t, equal to kT divided by the (renormalized) spin stiffness constant, of a renormalization scale l, and of the sizes L. Instead of eq. (2), one obtains in a similar way,

$$X[t; l; L]=(\rho l)^{p_x}\tilde{\zeta}(\rho)X[t(\rho); 1; L/l\rho]. \tag{2.4}$$

With the same choice for ρ, we have to consider the behaviour of $t[L/l]$. From the β-function we know that there is an ultraviolet stable fixed point at the critical temperature, and an infra-red stable fixed point at $t=0$ (zero temperature). In the large-ρ limit $t(\rho)$ is thus governed by the zero-temperature fixed point. The large-ρ behavior of $t(\rho)$ is given by the lowest term of the β-function [14] and

$$t(\rho) \underset{\rho\to\infty}{\sim} \rho^{-(d-2)}. \tag{2.5}$$

For large L, $\tilde{\zeta}(L/l)$ has a limit which is a function of t, and in the spin wave regime (fixed temperature below T_c) one recovers from (2.4) the FSS property

$$X[t; l; L] = X_\infty[t]\left(\frac{L}{\xi(t)}\right)^{p_x} f\left(\frac{\xi(t)}{L}\right), \tag{2.6}$$

in which $f(u)$ is calculable in a low-temperature perturbation theory.

The correlation length $\xi(t)$ is equal to [14]

$$\xi(t) = l t^{1/(d-2)} \exp\left\{\int_0^t \mathrm{d}t'\left[\frac{1}{\beta(t')} - \frac{1}{(d-2)t'}\right]\right\} \tag{2.7}$$

with, at one-loop order,

$$\beta(t) = (d-2)t - \frac{n-2}{2\pi} t^2 + \mathrm{O}(t^3) . \tag{2.8}$$

Near the critical point $\xi(t)$ diverges as $(t_c - t)^{-\nu}$ (with $\nu = -\beta'(t_c)$); at low temperature (t small),

$$\left(\frac{\xi(t)}{l}\right)^{d-2} = t\left[1 + \frac{n-2}{d-2}\frac{t}{2\pi} + \mathrm{O}(t^2)\right]. \tag{2.9}$$

This implies that the low-temperature scaling variable is proportional to

$$tL^{-(d-2)}\left[1 + \frac{n-2}{d-2}\frac{t}{2\pi} + \mathrm{O}(t^2)\right].$$

In a magnetic field similar arguments lead to

$$X[t, H; l; L] = L^{p_x} f\left[\frac{\xi(t)}{L}, \frac{H\sigma(t)\xi^d(t)}{t}\right], \tag{2.10}$$

in which $\sigma(t)$ is the spontaneous magnetization [14]

$$\sigma(t) = \exp\left\{-\frac{1}{2}\int_0^t \mathrm{d}t' \frac{\zeta(t')}{\beta(t')}\right\} \tag{2.11}$$

with, at one-loop order,

$$\zeta(t) = (n-1)\frac{t}{2\pi} + \mathrm{O}(t^2) . \tag{2.12}$$

The magnetization vanishes with a power β at T_c and may be expanded at low temperature as

$$\sigma(t) = \left[1 - \frac{1}{2}\frac{n-1}{d-2}\frac{t}{2\pi} + \mathrm{O}(t^2)\right]. \tag{2.13}$$

In sect. 5 we compute in the $(d-2)$ expansion the correlation length ξ_L in zero field, for a cylinder geometry.

3. FSS for a periodic cube

In the critical domain we can replace the spin hamiltonian, by a Landau-Ginzburg-Wilson [11] model, for an n-component order parameter

$$S[\Phi]=\int_v d^d x\left\{\tfrac{1}{2}(\nabla\boldsymbol{\phi})^2+\tfrac{1}{2}r_0\boldsymbol{\phi}^2+\frac{1}{4!}u_0(\boldsymbol{\phi}^2)^2\right\} \tag{3.1}$$

with some ultra-violet cut-off $\Lambda\approx a^{-1}$ (a is the lattice spacing). In a finite periodic box we expand Φ in Fourier modes

$$\boldsymbol{\Phi}(\boldsymbol{x})=\sum_q e^{i\boldsymbol{q}\cdot\boldsymbol{x}}\boldsymbol{\varphi}_q \tag{3.2}$$

in which each component of $\boldsymbol{q}$ runs over discrete values, multiples of $2\pi/L$, and is still cut-off at a^{-1}. As long as L is finite, the oscillations in the q non-zero modes will be damped, even at bulk T_c, by the q^2 restoring force of the kinetic energy. However there is nothing to prevent the mode

$$\boldsymbol{\varphi}\equiv\boldsymbol{\varphi}(\boldsymbol{q}=0) \tag{3.3}$$

from growing at bulk T_c, and therefore it cannot be treated perturbatively. In other words the propagator of Φ has an isolated pole at $q=0$, and diagrams involving a φ-loop would be divergent. The method consists of computing the effective action for φ defined by tracing out all modes except φ in perturbation theory (or rather in the loop expansion). Physical observables are then calculated, without any further approximation, since φ is certainly not small, with the averaging weight $\exp[-S_{\text{eff}}(\varphi)]$. Finally the renormalization group equations of sect. 2 are used. Let us first illustrate this program in the simple case of dimensions larger than four.

3.1. $d\geqslant 4$

Splitting as usual r_0 into t, which vanishes at T_c, and a counterterm (r_0-t), we obtain the effective action at the tree level, by ignoring the $\Phi_{q\neq 0}$ modes and the mass counterterm since they would both contribute only to loops. This procedure will be justified below. At the tree level we thus obtain

$$S_{\text{eff}}(\varphi)=L^d\left\{\frac{1}{2}t\boldsymbol{\varphi}^2+\frac{1}{4!}u_0(\boldsymbol{\varphi}^2)^2\right\}. \tag{3.4}$$

Let us note that $\boldsymbol{\varphi}$ is nothing but the total spin per unit volume

$$\boldsymbol{\varphi}=\frac{1}{v}\int_v \mathrm{d}^d x\, \boldsymbol{\Phi}(x)\,. \tag{3.5}$$

We have kept only one collective mode, a stochastic uniform magnetization of the system, and this explains why one finds a factor of the volume in front of the action (3.4). This collective mode is treated exactly whereas the others may be eliminated by a perturbative scheme.

We now compute averages of $\boldsymbol{\varphi}$, such as

$$M_{2p}=\langle(\boldsymbol{\varphi}^2)^p\rangle=\frac{\int \mathrm{d}^n\boldsymbol{\varphi}(\boldsymbol{\varphi}^2)^p \exp(-S_{\mathrm{eff}})}{\int \mathrm{d}^n\boldsymbol{\varphi} \exp(-S_{\mathrm{eff}})}\,. \tag{3.6}$$

A simple rescaling of $\boldsymbol{\varphi}$

$$\boldsymbol{\varphi}\to(u_0L^d)^{-1/4}\boldsymbol{\varphi} \tag{3.7}$$

leads to the scaling relation

$$M_{2p}=(u_0L^d)^{-p/2}f_{2p}[tL^{d/2}u_0^{-1/2}]\,, \tag{3.8}$$

in which f_{2p} is given by the ratio of two single integrals:

$$g_{2p}(x)=\int_0^\infty \mathrm{d}\varphi\, \varphi^{2p+n-1}\exp(-(\tfrac{1}{2}x\varphi^2+\tfrac{1}{24}\varphi^4)) \tag{3.9}$$

with

$$f_{2p}(x)=\frac{g_{2p}(x)}{g_0(x)}\,. \tag{3.10}$$

Eq. (3.8) shows that, above dimension four, the usual FSS does not hold. Consider for instance the magnetic susceptibility

$$\chi=\frac{1}{r}\int \mathrm{d}^d x\langle\boldsymbol{\phi}(x)\cdot\boldsymbol{\phi}(0)\rangle=\frac{L^d}{n}M_2\,. \tag{3.11}$$

From (3.8) one obtains

$$\chi(r,L)=L^{d/2}f(tL^{d/2}) \tag{3.12}$$

and since t vanishes linearly with the temperature we can write (3.12) as

$$\chi(t,L)=t^{-1}\tilde{f}(L^{(d-4)/4}\times L/\xi) \tag{3.12$'$}$$

which exhibits the classical exponent $\gamma=1$, together with the breakdown of FSS, identical to that found in ref. [4] in the large-n limit and suggested for finite n in ref. [9].

The universal character of the result (3.8) above dimension four is even more explicit if we work at bulk T_c ($t=0$) and consider dimensionless ratios which eliminate any dependence upon u_0. For instance the ratios

$$r_p(T)=\frac{M_{2p}}{(M_2)^p} \tag{3.13}$$

are given in terms of the scaling variable $x=rL^{d/2}u_0^{-1/2}$ as

$$r_p(T)=\frac{g_{2p}(x)g_0^{p-1}(x)}{(g_2(x))^p} \tag{3.13$'$}$$

in which $g(x)$ is defined in eq. (3.9).

At T_c ($x=0$) an elementary calculation gives for an n-vector model,

$$r_p(T_c)=\frac{\Gamma(\frac{1}{4}(2p+n))[\Gamma(\frac{1}{4}n)]^{p-1}}{[\Gamma(\frac{1}{4}(n+2))]^p}. \tag{3.14}$$

The renormalized coupling constant is related to

$$r_2(T_c)=\frac{M_4}{M_2^2}=\frac{\langle(\varphi^2)^2\rangle}{\langle\varphi^2\rangle^2}=\frac{n+2}{3n}\frac{\langle\varphi_1^4\rangle}{\langle\varphi_1^2\rangle^2}$$

and from (3.14) we obtain

$$r_2(T_c)=\frac{n}{4}\frac{\Gamma^2(\frac{1}{4}n)}{\Gamma^2(\frac{1}{4}n+\frac{1}{2})}. \tag{3.15}$$

For Ising-like models in particular, we predict from (3.15) that for any $d\geq 4$

$$r_2(T_c)=\frac{\Gamma^4(\frac{1}{4})}{8\pi^2}\simeq 2.1884396\,. \tag{3.16}$$

Recent Monte Carlo measurements by Binder et al. [15, 16] on a five-dimensional Ising system, for lattices of linear sizes up to $L=7$, give an estimate $r_2(T_c)=2.0$.

What is the effect of loop corrections that we have neglected in (3.4)? Apart from a shift of the critical temperature they yield a series of diagrams, which are not singular for large L, according to the well-known power counting arguments. Two kinds of terms are generated by these loop corrections. The first type corresponds to operators which would contribute to the infinite volume limit: they lead to a finite renormalization of u_0 and to higher powers of φ^2:

$$\delta S^{(1)}=L^d(\delta u_0(\varphi^2)^2+v(\varphi^2)^3+\cdots). \tag{3.17a}$$

Performing again the rescaling (3.7) of φ one finds immediately that higher operators such as φ^6 are down at least by powers of $(a/L)^{d/2}$.

In addition there are small corrections, vanishing for $d=4$ in the infinite volume limit, which come from the sum over the lowest discrete modes. As an example let us compute the one-loop corrections to $S_{\text{eff}}(\varphi)$ coming from the gaussian integration over the $\phi_{q\neq 0}$ modes:

$$\sum_{q\neq 0} \ln\left(1+\frac{u_0}{2(q^2+t)}\varphi^2\right);$$

(for simplicity we have restricted ourselves to $n=1$). This leads to operators of the form considered in (3.17a) and to corrections coming from the lowest non-zero momenta:

$$\delta S^{(2)} = c_1 L^2 \varphi^2 + c_2 (L^2 \varphi^2) + \cdots . \tag{3.17b}$$

The same rescaling shows that these terms contribute to corrections proportional to $(a/L)^{(d-4)/2}$. Consequently, as was said earlier, the tree level approximation which led to the scaling relations (3.8) is exact for large L. In addition there are correction terms multiplied by the small factor $(a/L)^{(d-4)/2}$ which are of the same scaling form (3.8). In particular the coefficient of φ^2 in (3.17b) can be absorbed into a shift of the temperature at which the total coefficient of φ^2 vanishes. This defines, as in ref. [15], an effective critical temperature $T_c(L)$ which differs from the bulk T_c by an amount proportional to $L^{-(d-2)}$ (this conflicts with ref. [16], in which this shift was assumed to be proportional to $L^{-d/2}$).

3.2. $d = 4-\varepsilon$

At the tree level (or lowest order in ε) the same effective action (3.4) will govern the long distance limit and at this order we recover the same form (3.8) for the result. Furthermore, in less than four dimensions, loop corrections are responsible of strong long-distance singularities, that are dealt with through the renormalization group formalism and the ε-expansion. Using eq. (2.2), we know that the dimensionless ratios r_p of eq. (3.13) satisfy

$$r_p(t, u_0, L) \underset{L\ \text{large}}{\sim} r_p(t(L)L^2, u_0^*) ; \tag{3.18}$$

the fixed point u_0^* being of order ε, one notes already from (3.18) and (3.8) the presence of an $\varepsilon^{-1/2}$ singularity (analogous to the $\varepsilon^{-1/3}$ singularity discussed in ref. [4]). However this singularity is explicit in terms of the integrals (3.9) and we can still perform loop corrections by integrating perturbatively on the non-zero q modes. At one-loop this integration generates a shift of T_c, a change of u_0, and new operators involving powers of φ larger than four. We first consider the shift of t, and the renormalization of u_0 and argue afterwards that the other terms lead to (calculable) higher order corrections.

From now on we will work within the renormalized theory. The renormalized coupling constant u_R is expressed in terms of the dimensionless coupling constant

$$g = l^{\varepsilon} u_R , \tag{3.19}$$

in which l is an arbitrary length scale. The minimal subtraction scheme [13] is used throughout all these calculations. In this scheme, the counterterms of the massless theory including the φ^2 insertion counterterm are introduced; the one-loop counterterm for the coupling constant and the φ^2 insertion will be the only one relevant in the lowest corrections.

3.3. SHIFT OF T_c

The simple diagram

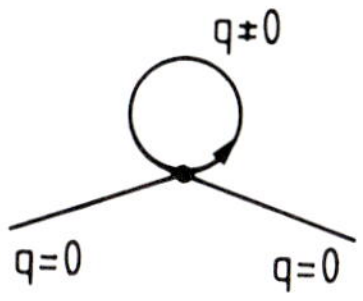

q≠0

q=0 q=0

$$= gl^{-\varepsilon}\frac{n+2}{6}\sum_q{}' \frac{1}{(q^2+t)}, \tag{3.20}$$

combined with the one-loop φ^2 counterterm yields a finite shift of t (we now use the system of units in which $l=1$):

$$t \to \tilde{t} = t + \tfrac{1}{12}(n+2)\hat{g}t \ln t + \tfrac{1}{3}(n+2)\hat{g}L^{-2}\int_0^\infty du\, e^{-utL^2/4\pi^2}[A^4(u) - 1 - \pi^2/u^2] + O(\hat{g}^2), \tag{3.21}$$

in which

$$\hat{g} = g\frac{2\pi^{d/2}}{\Gamma(\frac{1}{2}d)(2\pi)^d} = \frac{g}{8\pi^2}[1 - \tfrac{1}{2}\varepsilon(C - 1 - \ln 4\pi) + O(\varepsilon^2)], \tag{3.22}$$

$$A(u) \equiv \sum_{-\infty}^{+\infty} e^{-un^2}. \tag{3.23}$$

For large u, $A(u)-1$ decreases exponentially and the integral (3.21) converges at infinity, even at T_c ($t=0$). For small u, the Poisson transformation of (3.23)

$$A(u) = \left(\frac{\pi}{u}\right)^{1/2} \sum_{-\infty}^{+\infty} e^{-n^2\pi^2/u} = \left(\frac{\pi}{u}\right)^{1/2} A(\pi^2/u), \tag{3.23'}$$

shows that (3.21) converges.

Let us show in more detail how one goes from (3.20) to (3.21). We first write the integral representation

$$\frac{1}{L^d}\sum_q{}' \frac{1}{q^2+t} = \frac{1}{L^d}\sum_q{}' \int_0^\infty du\, e^{-u(t+q^2)}$$

$$= L^{-d}\int_0^\infty du\, e^{-ut}\left[\left(\sum_{-\infty}^{+\infty} e^{-u(4\pi^2 n^2/L^2)}\right)^d - 1\right]. \tag{3.24}$$

An analytic continuation in d is required to give a meaning to the integral (3.24) which diverges at small u (i.e. in the ultraviolet) for $\mathrm{Re}\, d > 2$. Subtracting and adding the asymptotic behaviour at small u, given by (3.23′), one obtains

$$L^{-d}\sum_q{}' \frac{1}{q^2+t}=\frac{L^{2-d}}{4\pi^2}\int_0^\infty \mathrm{d}u\, \mathrm{e}^{-u(tL^2/4\pi^2)}\left[A^d(u)-1-(\pi/u)^{d/2}\right]$$
$$+\frac{L^{2-d}}{4\pi^2}\pi^{d/2}\Gamma(1-\tfrac{1}{2}d)\left(\frac{tL^2}{4\pi^2}\right)^{d/2-1} \qquad (3.25)$$

The second term of (3.25) is continued analytically above Re $d=2$; it has a pole at $\varepsilon=0$. Therefore keeping the pole and the finite part one obtains

$$\frac{1}{L^d}\sum{}'\frac{1}{q^2+t}=-\frac{t}{8\pi^2\varepsilon}+\frac{t}{16\pi^2}[\ln t-(\ln 4\pi-C+1)]$$
$$+\frac{1}{4\pi^2L^2}\int_0^\infty \mathrm{d}u\,\mathrm{e}^{-utL^2/4\pi^2}\left[A^4(u)-1-\pi^2/u^2\right]+\mathrm{O}(\varepsilon)\,. \qquad (3.26)$$

The φ^2 insertion counterterm replaces t by tZ_{φ^2} [12]; at one-loop order, in the minimal scheme, it leads to the replacement $t\to t\,[1+\hat{g}(n+2)/6\varepsilon]$. Therefore the one-loop corrections lead to

$$t\to\tilde{t}=t[1+\hat{g}\tfrac{1}{6}(n+2)]+\frac{g}{L^d}\frac{n+2}{6\varepsilon}\sum{}'\frac{1}{(q^2+t)}\,. \qquad (3.27)$$

Using (3.26) one first verifies that, as renormalization theory implies, the $1/\varepsilon$ pole of the counterterm cancels the pole of the diagram. The finite terms add up to give the eq. (3.21).

3.4. RENORMALIZATION OF THE COUPLING CONSTANT

In a similar way, we add the one-loop counterterm for the coupling constant to the one-loop diagram

$$=\frac{1}{L^d}\sum_q{}'\frac{1}{(q^2+t)^2}$$

and replace, as in (3.27), g by

$$\tilde{g}=g\left[1+\hat{g}\frac{n+8}{6\varepsilon}\right]-g^2\frac{n+8}{6}\frac{1}{L^d}\sum_q{}'\frac{1}{(q^2+t)^2}\,. \qquad (3.28)$$

Similar algebraic transformations lead, after cancellation of the $1/\varepsilon$ poles, to

$$\tilde{g}=g\Big[1+\hat{g}\tfrac{1}{12}(n+8)(\tfrac{1}{2}+\ln t)-\tfrac{1}{12}(n+8)\hat{g}\int_0^\infty \mathrm{d}u\,u^{-utL^2/4\pi^2}$$
$$\times[A^4(u)-1-\pi^2/u^2]+\mathrm{O}(\hat{g}^2)\Big]\,. \qquad (3.29)$$

Finally, since no higher-order operator contributes at this order, we can use the eq. (3.8), with the same function $f_{2p}(x)$ provided $tL^{d/2}g^{-1/2}$ is replaced by $\tilde{t}L^{d/2}\tilde{g}^{-1/2}$ with $\tilde{t}$ and $\tilde{g}$ given by eqs. (3.21) and (3.29). At the fixed point

$$\hat{g}^* = \frac{6\varepsilon}{n+8} + \mathrm{O}(\varepsilon^2)\,, \tag{3.30}$$

one verifies the FSS, namely all the logarithms do exponentiate provided one introduces the scaling variable

$$y = tL^{1/\nu}\,. \tag{3.31}$$

A few lines of algebra lead indeed to,

$$\begin{aligned}\tilde{t}L^{d/2}\tilde{g}^{-1/2}\big|_{\text{fixed-point}} = (g^*)^{-1/2}\bigg\{ & y - \tfrac{1}{8}\varepsilon y + \frac{n-4}{2(n+8)}\varepsilon y \ln y \\ & + \frac{1}{4}\varepsilon y \int_0^\infty \mathrm{d}u\, u\, \mathrm{e}^{-uy/4\pi^2}\,[A^4(u) - 1 - \pi^2/u^2] \\ & + \varepsilon\frac{2(n+2)}{(n+8)} \int_0^\infty \mathrm{d}u\, \mathrm{e}^{-uy/4\pi^2}\,[A^4(u) - 1 - \pi^2/u^2] + \mathrm{O}(\varepsilon^2)\bigg\}\,.\end{aligned} \tag{3.32}$$

Finally the moments M_{2p}, and the ratios r_p, are given by the same formulae (3.8), (3.13′) in which one substitutes to $tL^{d/2}u_0^{-1/2}$ the expression (3.32). In particular at T_c ($t=0$, hence $y=0$) we obtain for the ratios $r_p(T_c)$ the result (3.13′) in which we substitute for x the value of (3.32) at T_c ($y=0$)

$$\begin{aligned} x_0 &= (g^*)^{-1/2}\left[\frac{2\varepsilon(n+2)}{n+8}\int_0^\infty \mathrm{d}u(A^4(u) - 1 - \pi^2/u^2) + \mathrm{O}(\varepsilon^2)\right] \\ &= \varepsilon^{1/2}\left\{\frac{1}{2}\frac{n+2}{\sqrt{3(n+8)}}\frac{I}{\pi} + \mathrm{O}(\varepsilon)\right\}, \end{aligned} \tag{3.33}$$

in which

$$\frac{I}{\pi} = \frac{1}{\pi}\int_0^\infty \mathrm{d}u(A^4(u) - 1 - \pi^2/u^2) = -1.7650848012\ldots. \tag{3.34}$$

Therefore at this order we obtain for $r_2(T_c)$ for instance instead of (3.14)

$$r_2(T_c) - \frac{g_4(x_0)g_0(x_0)}{(g_2(x_0))^2}\,. \tag{3.35}$$

Expanding in ε (x_0 small) one finds

$$r_2(T_c)=\frac{n}{4}\frac{\Gamma^2(\frac{1}{4}n)}{\Gamma^2(\frac{1}{4}(n+2))}\left\{1-x_0\sqrt{6}\left[\frac{\Gamma(\frac{1}{4}(n+6))}{\Gamma(\frac{1}{4}(n+4))}+\frac{\Gamma(\frac{1}{4}(n+2))}{\Gamma(\frac{1}{4}n)}-2\frac{\Gamma(\frac{1}{4}(n+4))}{\Gamma(\frac{1}{4}(n+2))}\right]\right.$$
$$\left.+6x_0^2\left[\frac{\Gamma(\frac{1}{4}(n+6))\Gamma(\frac{1}{4}(n+2))}{\Gamma(\frac{1}{4}(n+4))\Gamma(\frac{1}{4}n)}+\frac{3\Gamma^2(\frac{1}{4}(n+4))}{\Gamma^2(\frac{1}{4}(n+2))}-n-1\right]+O(\varepsilon^{3/2})\right\}, \tag{3.36}$$

in which x_0 is given by (3.33).

Setting $\varepsilon=1$ in (3.35) or (3.36) we obtain three-dimensional estimates for an ($n=1$) Ising-like system which are

$$r_2(T_c)=1.800 \quad \text{and} \quad 1.786, \quad \text{respectively}.$$

(instead of (2.19) at order zero). This ε-expansion estimate is not very far from the Monte Carlo measurement of Pearson et al. [18] which gives $g_{\text{Ren}}=-1.4$, i.e. $r_2(T_c)=1.6$.

3.5. HIGHER-ORDER CORRECTIONS

In this calculation we have neglected higher loops, which would lead to corrections of order $\varepsilon^{k+1/2}$ with $k\geq 1$, and higher operators such as φ^6, coming from the determinant of gaussian fluctuations, $\sum'_q \log(1+g\varphi^2/(q^2+t))$ (for simplicity written only for $n=1$), expanded in powers of φ. Such higher operators do contribute to the FSS functions, but they are multiplied by higher powers of g (g^3 for φ^6, etc...) and are negligible at this order of the ε-expansion.

4. FSS for a cylinder

In the transfer matrix (or hamiltonian) formulation of the problem, one deals with the geometry of a cylinder (or ribbon, or tube, according to the dimension) infinite along one dimension called time, and finite and periodic in the $(d-1)$ other dimensions. We assume for simplicity that the transverse sizes are all equal to the same length L. The fields are expanded in Fourier modes of fluctuations in the $(d-1)$ directions perpendicular to the axis of the cylinder

$$\Phi(\boldsymbol{x}_\perp,\tau)=\sum_{\boldsymbol{q}_\perp} e^{i\boldsymbol{q}_\perp\cdot\boldsymbol{x}_\perp}\varphi_{q_\perp}(\tau) \tag{4.1}$$

in which again the components of $q_\perp$ are quantized, in units of $2\pi/L$. The discussion of the sect. 3, may be repeated here: the dangerous modes, which cannot be treated perturbatively, correspond to $\boldsymbol{q}_\perp=0$; however now instead of one single mode, arbitrary functions of τ have to be considered, since nothing prevents

$$\varphi(\tau)\equiv\varphi_{q_\perp=0}(\tau) \tag{4.2}$$

from taking arbitrary large values (at T_c).

The effective action $S_{\text{eff}}[\boldsymbol{\varphi}(\tau)]$, defined by tracing out all the $q_\perp \neq 0$ modes, can be written as the time integral of a local density

$$S_{\text{eff}}[\varphi(\tau)] = \int \mathrm{d}\tau \, L_{\text{eff}}[\varphi(\tau), \dot{\varphi}(\tau)] \, ; \tag{4.3}$$

L_{eff}, the euclidean lagrangian of field theory is nothing but the matrix element of the transfer matrix between two infinitesimally close "times". Again let us begin the discussion with the simpler case of a spin system above its upper critical dimension.

4.1. $d \geqslant 4$

Neglecting again loop contributions, one reads immediately the effective action from the LGW formalism (3.1)

$$S_{\text{eff}}[\varphi] = L^{d-1} \int \mathrm{d}\tau \left\{ \tfrac{1}{2}(\dot{\boldsymbol{\varphi}})^2 + \tfrac{1}{2} t \boldsymbol{\varphi}^2 + \frac{u_0}{4!}(\varphi^2)^2 \right\}. \tag{4.4}$$

Averages are now performed with the weight $\exp[-S_{\text{eff}}(\varphi)]$ by summing over all possible trajectories $\boldsymbol{\varphi}(t)$. This is nothing but a Feynman path integral corresponding to an n-dimensional, imaginary time, quantum mechanical problem. Indeed let us recall that Feynman's formulation for the hamiltonian

$$H = \frac{1}{2m} \boldsymbol{p}^2 + V(\boldsymbol{q}) \tag{4.5}$$

gives the evolution operator

$$\langle \boldsymbol{q}_2 | \mathrm{e}^{-TH} | \boldsymbol{q}_1 \rangle = \int_{q(0)=q_1}^{q(T)=q_2} \mathrm{D}q(\tau) \exp\left(- \int_0^T \mathrm{d}\tau [\tfrac{1}{2} m (\dot{\boldsymbol{q}})^2 + V(q)] \right). \tag{4.5'}$$

Consequently identifying $\boldsymbol{\varphi}(\tau)$ of (4.4) with $\boldsymbol{q}(\tau)$ of (4.5′) we can associate to the action (4.4) an hamiltonian in n dimensions (note that the mass parameter of (4.5′) is L^{d-1}):

$$H(p, q) = \frac{\boldsymbol{p}^2}{2L^{d-1}} + L^{d-1}\left[\frac{t}{2} \boldsymbol{q}^2 + \frac{u_0}{24} (\boldsymbol{q}^2)^2 \right]. \tag{4.6}$$

The association of a quantum mechanical hamiltonian in $(d-1)$ dimensions to a classical statistical mechanics problem in d dimensions is well-known. Here in addition we have kept only the zero-momentum mode of this $(d-1)$-dimensional hamiltonian. The spectrum of the low-lying states of the hamiltonian (4.6) will govern the exponential decay of correlations along the axis of the cylinder. In particular the gap $(E_1 - E_0)$ between the lowest two levels gives the correlation length

$$\xi_L = (E_1 - E_0)^{-1}, \tag{4.7}$$

according to standard transfer matrix arguments (correlations along the axis of the cylinder correspond to large time separations). The spectrum of this n-dimensional anharmonic hamiltonian (4.6) has to be found by solving a Schrödinger equation. The lowest states being $0(n)$ rotationally invariant, one has to solve a radial one-dimensional problem. However the scaling properties of the eigenlevels may be obtained very simply without any numerical work. Indeed the dilatation

$$q \to L^{-1/3(d-1)} u_0^{-1/6} q, \qquad p \to L^{1/3(d-1)} u_0^{1/6} p \tag{4.8}$$

shows that the hamiltonian (4.6) may be written as

$$H(p,q) = u_0^{1/3} L^{-(d-1)/3} [\tfrac{1}{2}\boldsymbol{p}^2 + \tfrac{1}{24}(\boldsymbol{q}^2)^2 + \tfrac{1}{2} t u_0^{-2/3} L^{2/3(d-1)} \boldsymbol{q}^2]\,. \tag{4.9}$$

Therefore any eigenvalue of (4.6) satisfies the scaling property

$$E_\alpha(t, L, u_0) = u_0^{1/3} L^{-(d-1)/3} f_\alpha[t L^{2/3(d-1)} u_0^{-2/3}]\,. \tag{4.10}$$

Consequently the correlation length ξ_L scales as

$$\xi_L(t, L, u_0) = u_0^{-1/3} L^{(d-1)/3} f[t L^{2/3(d-1)} u_0^{-2/3}] \tag{4.11}$$

in which $f(x)$ is the inverse of the gap of the hamiltonian

$$h = \tfrac{1}{2}\boldsymbol{p}^2 + \tfrac{1}{2}x\boldsymbol{q}^2 + \tfrac{1}{24}(\boldsymbol{q}^2)^2\,. \tag{4.12}$$

Let us stress again the breakdown of FSS exhibited by (4.11) which may be written as

$$\xi_L = \xi_\infty(t) g\left[\frac{L}{\xi_\infty(t)} L^{(d-4)/3}\right] \tag{4.13}$$

(with $\xi_\infty(t)$ given by the mean-field result $\xi_\infty \propto t^{-1/2}$), in agreement with the large-n result of ref. [4]. Note also the "dangerous irrelevant character" of u_0, manifest in (4.11).

Let us finally show that loop corrections are truly irrelevant. The discussion is completely parallel to the one of sect. 3. Higher-order operators, such as φ^6, will add terms proportional to $L^{d-1}(\boldsymbol{q}^2)^3$ to the hamiltonian (4.6). The rescaling (4.7) shows that such terms provide corrections of relative order $(a/L)^{-2/3(d-1)}$ to the eigenvalues (4.10).

Finally, as in (3.17b), we have to consider the corrections due to the discrete nature of the sums involved in the loop expansion. In order to avoid redundancy in the presentation, this question will be answered after the discussion of the expansion in $\varepsilon = 4 - d$.

4.2. $d=4-\varepsilon$

The discussion follows closely the lines of sect. 3. The shift of T_c given by the diagram

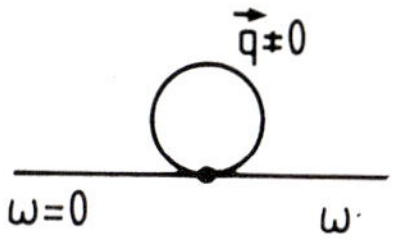

$\vec{q}\neq 0$

$\omega=0$ ω

$$= \int \frac{\mathrm{d}\omega}{2\pi} \sum_{q_\perp}{}' \frac{1}{q^2+\omega^2+t},$$

together with the φ^2 insertion counterterm leads, through similar steps to

$$t \to \tilde{t} = t + \hat{g}\tfrac{1}{12}(n+2)t \ln t + \hat{g}\tfrac{1}{3}(n+2)\frac{\sqrt{\pi}}{L^2} \times \int_0^\infty \frac{\mathrm{d}u}{\sqrt{u}} \mathrm{e}^{-utL^2/4\pi^2}[A^3(u)-1-(\pi/u)^{3/2}] + \mathrm{O}(\hat{g}^2)\,. \tag{4.14}$$

The modification of the coupling constant, together with the one-loop counterterm leads to

$$g \to \tilde{g} = g\Big[1+\tfrac{1}{12}(n+8)\hat{g}(1+\ln t) - \tfrac{1}{12}(n+8)\hat{g}\frac{1}{\pi^{3/2}} \times \int_0^\infty \mathrm{d}u\sqrt{u}\, \mathrm{e}^{-utL^2/4\pi^2}(A^3(u)-1-(\pi/u)^{3/2}) + \mathrm{O}(\hat{g}^2)\Big]. \tag{4.15}$$

Neglecting, for the same reasons, irrelevant operators (which are required at higher orders in ε), we substitute in eq. (4.11) $\tilde{t}$ and $\tilde{g}$ to t and u_0 respectively. This gives a scaling variable

$$x = \tilde{t}(\tilde{g})^{-2/3} L^{2/3(d-1)}, \tag{4.16}$$

which, at the fixed point $\hat{g}^*$, taking into account (4.14) and (4.15), takes the expected scaling form in terms of

$$y = tL^{1/\nu},$$

$$x = (g^*)^{-2/3}\Big\{ y(1-\tfrac{1}{3}\varepsilon) + \frac{\varepsilon}{6}\frac{n-10}{n+8} y \ln y + y\frac{1}{3\pi^{3/2}} \int_0^\infty \mathrm{d}u\, \sqrt{u}\, \mathrm{e}^{-uy/4\pi^2}[A^3(u)-1-(\pi/u)^{3/2}] + \varepsilon \frac{2(n+2)}{n+8}\sqrt{\pi} \int_0^\infty \frac{\mathrm{d}u}{\sqrt{u}} \mathrm{e}^{-uy/4\pi^2}[A^3(u)-1-(\pi/u)^{3/2}] + \mathrm{O}(\varepsilon^2)\Big\}. \tag{4.17}$$

Consequently we obtain for ξ_L the scaling result

$$\frac{\xi_L}{L} = (g^*)^{-1/3} f[x](1+\mathrm{O}(\varepsilon)), \tag{4.18}$$

in which $f(x)$ is the (inverse) gap of the hamiltonian (4.12) for the variables x given in (4.17).

In particular at T_c $(y=0)$, one obtains

$$\frac{\xi_L}{L}=\left(\frac{48\pi^2}{n+8}\varepsilon\right)^{-1/3} f\left[\left(\frac{6\varepsilon}{n+8}\right)^{1/3}\frac{n+2}{12}\pi^{-5/6}I\right](1+O(\varepsilon)) \qquad (4.19)$$

with

$$I=\int_0^\infty \frac{du}{\sqrt{u}}[A^3(u)-1-(\pi/u)^{3/2}]=-5.0289788 \qquad (4.20)$$

(in which $A(u)$ was defined in (3.23)).

In the large-n limit the result (4.19) agrees with the ε-expansion of the equation for ξ_L/L given in ref. [4].

Finally let us consider corrections to this result. The effective action (4.4) is modified by the inclusion of loop corrections given by integrating out the $q_\perp$ non-zero modes; at one-loop order (for simplicity we deal with the $n=1$ case)

$$S_{\text{eff}}^{(1)}[\varphi]=L^{d-1}\int d\tau\left\{\tfrac{1}{2}\dot{\varphi}^2+\tfrac{1}{2}t\varphi^2+\frac{u_0}{4!}\varphi^4\right\}$$
$$+\tfrac{1}{2}\sum_{q_\perp}\operatorname{tr}\ln\left[-\frac{d^2}{d\tau^2}+t+\tfrac{1}{2}u_0\varphi^2+\boldsymbol{q}_\perp^2\right]. \qquad (4.21)$$

A rescaling of φ and τ:

$$\varphi\to L^{-1/3(d-1)}u_0^{-1/6}\varphi,$$
$$\tau\to L^{1/3(d-1)}u_0^{-1/3}\tau, \qquad (4.22)$$

yields (at T_c, for instance)

$$S_{\text{eff}}^{(1)}(\varphi)=\int d\tau\{\tfrac{1}{2}\dot{\varphi}^2+\tfrac{1}{24}\varphi^4\}$$
$$+\tfrac{1}{2}\sum_{n_\perp}\operatorname{tr}\ln\left[-\frac{d^2}{d\tau^2}+\tfrac{1}{2}\varphi^2+L^{2/3(d-4)}\,4\pi^2u_0^{-2/3}\boldsymbol{n}_\perp^2\right]. \qquad (4.23)$$

Expanding the second term of (4.23) in powers of φ^2 we have to consider the propagator

$$\langle\tau_1|\left(-\frac{d^2}{d\tau^2}+4\pi^2L^{2/3(d-4)}\,u_0^{-2/3}n_\perp^2\right)^{-1}|\tau_2\rangle. \qquad (4.24)$$

Above four dimensions, for large L, the propagator is proportional to $L^{-2/3(d-4)}$ $\delta(\tau_1-\tau_2)$; this gives a negligible modification of u_0 proportional to $L^{-4/3(d-4)}$, a small φ^6 correction, etc... Below four dimensions we have to use the renormalization group equations, which relate the theory at L with coupling u_0, to the theory at

$L=1$ with coupling $u_0(L) \simeq u_0^*$, which is proportional to ε. Therefore again the "mass" of the propagator (4.24) is large, for small ε, and this propagator is proportional to $\varepsilon^{2/3}\delta(\tau_1-\tau_2)$: this gives (calculable) $\varepsilon^{4/3}$ correction to u_0, a φ^6 operator proportional to ε^2 etc.... Therefore the results (4.18)-(4.19) are correct up to order $\varepsilon^{2/3}$.

5. Low-temperature expansion of FSS

When a continuous transition is spontaneously broken, below the critical temperature (or the transition temperature of a first-order transition) massless Goldstone modes give singularities at large distances for any value of the temperature. The thermodynamics of these "spin waves" modes is described by the non-linear σ-model [13, 14], which is nothing but a model for an order parameter with a fixed magnitude varying over a sphere (for the single $O(n)$ case). In a finite volume these modes have a small mass and we consider the corresponding non-linear σ-model in a low-temperature expansion, which is valid up to T_c in a $(d-2)$ expansion. Note that the renormalization group analysis of sect. 2 revealed that for $T<T_c$ the large-L properties are related to the zero-temperature fixed point.

Several authors, in particular Lüscher [20] and Floratos and Petcher [21] have already considered the two-dimensional case ($n>2$), in order to deduce the mass gap from a low-temperature expansion. The method that we have followed here for $T<T_c$, $d>2$ has many features in common with the work of Floratos and Petcher [21].

In the cubic geometry the final integration over the $q=0$ mode projects out $O(n)$-invariant observables. (All the other ones vanish, as expected since for L finite the symmetry is unbroken.) The invariant observables can be calculated from perturbation theory with a finite zize propagator $[(2\pi \boldsymbol{m}/L)^2]^{-1}$ in which the zero mode $m=0$ is omitted. As long as L is finite there is an infrared cut-off, but even in the infinite-L limit, the low-temperature expansion of the invariant observables remains finite [19] (whereas non-invariant observables are infrared singular in the large-L limit).

In the cylindrical geometry, let us consider this problem in more details. The general method of sect. 4, which leads to an effective Schrödinger equation for a quantum rotor, may be applied. However, without going to this hamiltonian formulation, it is also possible to calculate directly with the path integral. The model in the geometry $L^{d-1}xT$ is described by

$$S[\phi]=\frac{1}{2t}\int_{L^{d-1}} \mathrm{d}^{d-1}x \int_0^T \mathrm{d}\tau[(\nabla_\perp \boldsymbol{\phi})^2+(\dot{\boldsymbol{\phi}})^2] \tag{5.1a}$$

with the constraints

$$\phi_1^2(x)+\cdots+\phi_n^2(x)=1\,. \tag{5.1b}$$

An ultraviolet regularization is meant; the dimensional regularization, which does not break the O(n) symmetry, is particularly convenient. The parameter t is proportional to the temperature divided by the spin-wave stiffness constant.

Let us consider a situation with periodic boundary conditions over the L^{d-1} sides of the cylinder and

$$\begin{aligned} \boldsymbol{\phi}(\tau=0, \boldsymbol{x}_\perp) &= \boldsymbol{u}_1 , \\ \boldsymbol{\phi}(\tau=T, \mathbf{x}_\perp) &= \boldsymbol{u}_2 , \end{aligned} \tag{5.2}$$

in which $\boldsymbol{u}_1$ and $\boldsymbol{u}_2$ are two given unit vectors with

$$\boldsymbol{u}_1 \cdot \boldsymbol{u}_2 = \cos\theta . \tag{5.3}$$

The corresponding partition function takes the form

$$\begin{aligned} Z(T,\theta) &= \langle \boldsymbol{u}_2 | \, \mathrm{e}^{-TH} | \boldsymbol{u}_1 \rangle \\ &= \sum_l \delta_l p_l(\cos\theta)\, \mathrm{e}^{-T\varepsilon_l} , \end{aligned} \tag{5.4}$$

in which the ε_l are the eigenvalues of the hamiltonian, and $p_l(\cos\theta)$ are orthogonal polynomials with the measure $\sin^{n-2}\theta$ and δ_l is the degeneracy of the lth level. One can project any eigenvalue ε_l, by multiplying Z by the appropriate polynomial p_l and integrating over θ.

Let us parametrize the fields $\boldsymbol{\Phi}$ in a frame in which

$$\begin{aligned} \boldsymbol{u}_1 &= [1, 0; \mathbf{0}] , \\ \boldsymbol{u}_2 &= [\cos\theta, \sin\theta; \mathbf{0}] , \end{aligned} \tag{5.5}$$

$$\begin{aligned} \Phi_1 &= \varphi_1 \cos(\theta\tau/T) + \varphi_2 \sin(\theta\tau/T), \\ \Phi_2 &= -\varphi_1 \sin(\theta\tau/T) + \varphi_2 \cos(\theta\tau/T), \end{aligned} \tag{5.6}$$

and $(n-2)$ components $\boldsymbol{\Phi}_\perp$, in which φ_1, φ_2 and $\boldsymbol{\Phi}_\perp$ are functions of $\boldsymbol{x}$, and τ, satisfying the boundary conditions

$$\Phi_\perp = \varphi_2 = 0 , \qquad \varphi_1 = 1 \tag{5.7}$$

at $\tau = 0$ and $\tau = T$ together with the constraint

$$\varphi_1^2 + \varphi_2^2 + \boldsymbol{\Phi}_\perp^2 = 1 . \tag{5.8}$$

Note that $(\varphi_1 - 1)$, φ_2 and $\boldsymbol{\Phi}_\perp$ are the deviations from the minimum energy configuration with interpolates between $\boldsymbol{u}_1$ and $\boldsymbol{u}_2$. With this parametrization

$$\dot{\boldsymbol{\phi}}^2 = \frac{\theta^2}{T^2}(\varphi_1^2 + \varphi_2^2) + \dot{\varphi}_1^2 + \dot{\varphi}_2^2 + (\dot{\boldsymbol{\phi}}_\perp)^2 \tag{5.9}$$

and with the constraint (5.8)

$$\dot{\boldsymbol{\phi}}^2 = \frac{\theta^2}{T^2}(1 - \boldsymbol{\phi}_\perp^2) + \dot{\varphi}_1^2 + \dot{\varphi}_2^2 + \dot{\boldsymbol{\phi}}_\perp^2 . \tag{5.10}$$

The low-temperature expansion is obtained as usual, by solving for φ_1

$$\varphi_1 = (1-\varphi_2^2 - \boldsymbol{\Phi}_\perp^2)^{1/2} \tag{5.11}$$

and expanding in powers of φ_2 and $\boldsymbol{\Phi}_\perp$. At leading order we keep the quadratic terms in φ_2 and $\boldsymbol{\Phi}_\perp$ and

$$S(\varphi_2, \boldsymbol{\Phi}_\perp) = \frac{\theta^2}{2tT^2} L^{d-1} + \frac{1}{2t}\int \mathrm{d}^{d-1}x\,\mathrm{d}\tau \times\left[-\frac{\theta^2}{T^2}\boldsymbol{\Phi}_\perp^2 + \dot{\boldsymbol{\Phi}}_\perp^2 + (\nabla\boldsymbol{\Phi}_\perp)^2 + \dot{\varphi}_2^2 + (\nabla\varphi_2)^2\right] + S_{\text{int}}(\varphi_2, \boldsymbol{\Phi}_\perp)\,, \tag{5.12}$$

in which S_{int} contains terms of degree higher than two in the $(n-1)$ fields φ_2, $\boldsymbol{\Phi}_\perp$. The integral over φ_2 is independent of θ, and gives a normalization constant in front of Z. The integral over $\boldsymbol{\Phi}_\perp$ gives the $-\frac{1}{2}(n-2)$-th power of the determinant of the quadratic form. Hence, at one-loop order of the low-t expansion

$$\ln\frac{Z(T,\theta)}{Z(T,0)} = -\frac{\theta^2}{2tT}L^{d-1} - \tfrac{1}{2}(n-2)\left[\operatorname{tr}\ln\left(-\frac{\theta^2}{T^2} - \nabla_\perp^2 - \frac{\mathrm{d}^2}{\mathrm{d}\tau}\right) - \operatorname{tr}\ln\left(-\nabla_\perp^2 - \frac{\mathrm{d}^2}{\mathrm{d}t^2}\right)\right]. \tag{5.13}$$

The eigenmodes of the operator $-\nabla_\perp^2 - \mathrm{d}^2/\mathrm{d}t^2$, given the boundary conditions (5, 7) at $\tau=0$ and T and periodic boundary conditions on $\boldsymbol{x}$, are

$$\left(\frac{2\pi}{L}\boldsymbol{p}\right)^2 + \frac{m^2\pi^2}{T^2} \tag{5.14}$$

in which $p_1, \ldots, p_{n-2}$ are algebraic integers and m is a positive integer:

$$\ln\frac{Z(T,\theta)}{Z(T,0)} = \theta^2\left\{-\frac{L^{d-1}}{2tT} + \frac{n-2}{2T^2}\sum_{\boldsymbol{p}}\sum_{m=1}^{\infty}\frac{1}{(4\pi^2/L^2)\boldsymbol{p}^2 + \pi^2 m^2/T^2}\right\} + \mathrm{O}(\theta^4)\,, \tag{5.15}$$

in which we have expanded (5.13) around $\theta=0$, since the relevant values of θ needed to project out the eigenvalues ε_l are peaked around $\theta=0$ in the small-t domain. In order to evaluate the rhs of (5.15), in the large-T limit it is necessary to separate the $\boldsymbol{p}=0$ contribution

$$\sum_{\boldsymbol{p}}\sum_{m}\frac{1}{(4\pi^2/L^2)\boldsymbol{p}^2 + \pi^2 m^2/T^2} \underset{T\to\infty}{\simeq} \frac{T^2}{6} + \sum_{\boldsymbol{p}}{}'\frac{TL}{4\pi|\boldsymbol{p}|}\,. \tag{5.16}$$

The sum over $\boldsymbol{p}$ in the r.h.s. of (5.16) is dimensionally regularized

$$\sum_{\boldsymbol{p}}\frac{1}{|\boldsymbol{p}|} = \frac{1}{\sqrt{\pi}}\int_0^\infty \mathrm{d}u\, u^{-1/2}\sum{}'\mathrm{e}^{-u\boldsymbol{p}^2} = \frac{1}{\sqrt{\pi}}\int_0^\infty \mathrm{d}u\, u^{-1/2}\{[A(u)]^{d-1}-1\}\,, \tag{5.17}$$

in which $A(u)$ is the function defined in (3.23). The integral (5.17) converges for $d<2$; an expansion in powers of $\varepsilon=(d-2)$ gives, after some labour,

$$\sum_p \frac{1}{|\boldsymbol{p}|} = -\frac{2}{\varepsilon} + C - \ln 4\pi + \mathrm{O}(\varepsilon)\,, \tag{5.18}$$

in which C is Euler's constant ($C=-\Gamma'(1)$). Consequently

$$\ln\frac{Z(T,\theta)}{Z(T,0)} = \theta^2\left\{\frac{n-2}{12}\frac{L^{d-1}}{2tT} + \frac{n-2}{4T}\frac{L}{2\pi}\left(-\frac{2}{\varepsilon}+C-\ln 4\pi\right)+\mathrm{O}(\varepsilon)\right\} + \mathrm{O}(\theta^4;\, t,\, 1/T^2)\,. \tag{5.19}$$

In order to project out the lowest eigenvalues ε_0 and ε_1 defined in (5.4), we multiply successively Z by one and $\cos\theta$ and integrate (with the $\sin^{n-2}\theta$ measure). This gives the correlation length

$$\frac{1}{\xi_L} = \varepsilon_0 - \varepsilon_1 = \frac{n-1}{2}\frac{t}{L^{d-1}}\left\{1+\frac{t(n-2)}{4\pi}L^{-\varepsilon}\left(-\frac{2}{\varepsilon}+C-\ln 4\pi+\mathrm{O}(\varepsilon)\right)+\mathrm{O}(\varepsilon^2)\right\}. \tag{5.20}$$

The $1/\varepsilon$ poles disappear, as they should, when one introduces the renormalized temperature t_R, with

$$t = t_\mathrm{R} + \frac{n-2}{2\pi\varepsilon}t_\mathrm{R}^2 + \mathrm{O}(t_\mathrm{R}^3) \tag{5.21}$$

taken from ref. [14].

The result satisfies the expected scaling relation

$$\xi_L[t_\mathrm{R}, L] = L\xi_L[t_\mathrm{R}(L), 1] \tag{5.22}$$

with, at this order,

$$\frac{1}{t_\mathrm{R}} = \frac{1}{t_\mathrm{R}^*} + L^{-\varepsilon}\left(\frac{1}{t_\mathrm{R}(L)} - \frac{1}{t_\mathrm{R}^*}\right), \tag{5.23}$$

$$t_\mathrm{R}^* = \frac{2\pi\varepsilon}{n-2} + \mathrm{O}(\varepsilon^2)\,. \tag{5.24}$$

This allows one finally to write (5.20) under the scaling form

$$\frac{\xi_L}{L} = \frac{2}{(n-1)t_\mathrm{R}(L)}\left[1-(C-\ln 4\pi)\frac{n-2}{4\pi}t_\mathrm{R}(L)+\mathrm{O}(t_\mathrm{R}^2)\right]. \tag{5.25}$$

For large L and $T<T_\mathrm{c}$, $t_\mathrm{R}(L)$ vanishes (eq. (2.5)) and from (2.7) one obtains

$$t_\mathrm{R}(L) = \left(\frac{\xi(t_\mathrm{R})}{L}\right)^{d-2}\left[1-\frac{1}{2\pi}\frac{n-2}{d-2}\left(\frac{\xi(t_\mathrm{R})}{L}\right)^{d-2}+\mathrm{O}\left(\frac{1}{L^{2(d-2)}}\right)\right]. \tag{5.26}$$

Thus from (5.25) one has

$$\frac{\xi_L}{L}=\frac{2}{n-1}\left(\frac{L}{\xi}\right)^{d-2}\left[1+\frac{n-2}{2\pi}\left(\frac{\xi}{L}\right)^{d-2}\left(\frac{1}{d-2}+\tfrac{1}{2}\ln(4\pi)-\tfrac{1}{2}C\right)+O\left(\frac{1}{L^{2(d-2)}}\right)\right]. \tag{5.27}$$

Irrelevant operators of the non-linear σ-model generate additional $1/L^2$ corrections to (5.27).

For large L and at low temperature (since from (2.9) $\xi(t)\propto t^{1/(d-2)}$) one recovers a result of Fisher and Privman [8].

In dimension two, this result has been already derived by a different method by M. Lüscher [20], and by Floratos and Petcher [21] up to two-loop.

5.1. MAGNETIZATION IN A BOX

For an n-vector model below T_c the spontaneous magnetization in a finite box vanishes. In a field a magnetization $M[H, t, L]$ is induced, which vanishes if H goes to zero first but not if L goes first to infinity. This cross-over is contained in the same renormalization group equations, which for large L reads

$$M[H, t, L]=\sigma(t)M\left[\frac{H\sigma(t)}{t}\xi^{d-2}(t)L^2, \left(\frac{\xi(t)}{L}\right)^{d-2}, 1\right], \tag{5.28}$$

in which $\sigma(t)$ and $\xi(t)$ are the infinite system spontaneous magnetization (per unit volume) and correlation length given by eqs. (2.11) and (2.7). For large L, fixed t, since the effective temperature $t(L)\simeq(\xi(t)/L)^{d-2}$ is small, the magnetization of the r.h.s. of (6.7) can again be calculated in a low-t expansion. At lowest order in t, one keeps only the uniform mode and the partition function is

$$Z(H, t, L=1)=\int d^n\hat{\varphi}\exp\left(\frac{H\varphi}{t}\right)=C\int_0^{\pi}d\theta\sin^{n-2}\theta\exp\left(\frac{H\cos\theta}{t}\right) \tag{5.29}$$

(a modified Bessel function). Therefore the magnetization of the r.h.s. is given by a logarithmic derivative of Z and one substitutes for H/t the variable

$$x=\frac{H(L)}{t(L)}=\frac{H\sigma(t)}{t}\xi^{d-2}(t)L^2\Big/\left(\frac{\xi(t)}{L}\right)^{d-2}=\frac{H\sigma(t)L^d}{t}. \tag{5.30}$$

The final formula is

$$M[H, t, L]=\sigma(t)\frac{\partial}{\partial x}\ln\int_0^{\pi}d\theta\sin^{n-2}\theta\, e^{x\cos\theta} \tag{5.31}$$

with x given by (5.30), in agreement with the formula given by Fisher and Privman [8]. The technique of this section is also useful in the context of lattice gauge theories [22].

6. First-order transitions and FSS

In this section we shall study *uniform* first-order transitions in which the order parameter jumps from a constant value to another constant value throughout the sample, excluding thereby transitions such as crystallization. The FSS properties at first-order transitions have been considered from a transfer matrix formalism by Privman and Fisher [9]. We want now to show that the method that we have applied to continuous transitions explains also the rounding of first-order transitions and leads to a simple derivation of some universal properties. Thus many of the formulae that we shall derive are already given in ref. [9]. Again we can introduce one single $\boldsymbol{q}=0$ mode to describe the system. In the case of a cubic L^d geometry, one defines

$$\varphi=\frac{1}{L^d}\int_v \Phi(x)\,\mathrm{d}^d x\,, \tag{6.1a}$$

and for a cylinder $L^{d-1}\times\infty$

$$\varphi(\tau)=\frac{1}{L^{d-1}}\int \Phi(\boldsymbol{x}_\perp,\tau)\,\mathrm{d}^{d-1}x_\perp\,; \tag{6.1b}$$

periodic boundary conditions are used in all this section. Tracing out the $q\neq 0$ modes, one obtains

$$S_{\mathrm{eff}}(\varphi,L)=L^d s(\varphi) \tag{6.2a}$$

or

$$S_{\mathrm{eff}}(\varphi,L)=L^{d-1}\int \mathrm{d}\tau\, s(\varphi(\tau))\,. \tag{6.2b}$$

for the two geometries respectively.

At a first-order transition the correlation length remains finite, the integrations over the $q\neq 0$ modes are non-singular, therefore the coefficients of $s(\varphi)$ are regular functions of the physical parameters, such as the temperature, the chemical potentials etc. . . .

The physical observables will be defined through an integration over φ (or $\varphi(\tau)$). The Boltzmann weight $(\exp(-S_{\mathrm{eff}}))$ is governed by the L^d (or L^{d-1}) factor in front of s, and these last integrations can be performed, exactly in the large-L limit, by a saddle-point approximation. This yields some universal formulae for the rounding of first-order jumps independent of all the details of the actual interaction. Let us distinguish now several cases.

6.1. CUBIC GEOMETRY, ISING-LIKE SYSTEMS

In the first case the zero-field action (6.2) is symmetric under $\varphi\to-\varphi$. We consider, specifically an Ising-like system below T_c in a small external field H. When H changes from plus zero to minus zero the infinite volume magnetization jumps from $+M_0$ to $-M_0$. This jump is rounded at finite L, since M now vanishes with L. In the presence of a field

$$S_{\text{eff}}(\varphi, H, L) = L^d[s(\varphi) - \beta H\varphi] \tag{6.3}$$

and $s(\varphi)$ is an even function of φ, with two minima at $+M_0$ and $-M_0$. For small H (finite HL^d) the minima of (6.3) are close to M_0 and $-M_0$ and we can expand the Boltzmann weight near these two points [9, 23]:

$$\begin{aligned} e^{-S_{\text{eff}}(\varphi)} \propto & \exp\left(-L^d[\tfrac{1}{2}s''(M_0)(\varphi - M_0)^2 - \beta H\varphi]\right) \\ & + \exp\left(-L^d[\tfrac{1}{2}s''(M_0)(\varphi + M_0)^2 - \beta H\varphi]\right). \end{aligned} \tag{6.4}$$

With the weight (6.4) one can calculate any average; for instance the magnetization is given by

$$M = \langle\varphi\rangle = M_0 \tanh(\beta H L^d M_0) + \frac{\beta H}{s''(M_0)}, \tag{6.5}$$

$$\chi = \frac{\beta}{s''(M_0)} \tag{6.6}$$

is the bulk zero-field magnetic susceptibility.

For H small, HL^d finite, the second term is negligible and the rounded magnetization is given by the well-known [9], but indeed universal formula

$$M = M_0 \tanh(\beta H L^d M_0). \tag{6.7}$$

6.2. CUBIC GEOMETRY, NO SYMMETRY

A first-order transition often takes place for an effective action, which has no symmetry (such as $\varphi \to -\varphi$) but simply a minimum at $\varphi = M$ which becomes lower (for $T < T_c$) than the minimum at $\varphi = M_0$. Let us define the "driving parameter"

$$t = s(M_1) - s(M_0), \tag{6.8}$$

which is proportional to $T - T_c$. The statistical weight or the order parameter, becomes proportional to

$$\exp(-S_{\text{eff}}) \propto \{\exp(-\tfrac{1}{2}L^d s''(M_1)(\varphi - M_1)^2) + \exp(+tL^d - \tfrac{1}{2}L^d s''(M_0)(\varphi - M_0)^2)\} \tag{6.9}$$

and averages such as the magnetization $M = \langle\varphi\rangle$ are easily calculated. One finds

$$M = M_0 + \frac{M_1 - M_0}{1 + e^{+tL^d}\sqrt{s''(M_1)/s''(M_0)}} \tag{6.10}$$

which describes the rounding of the jump of M for t small, L large, finite tL^d; $s''(M_1)/s''(M_0) = \chi(M_0)/\chi(M_1)$ is the ratio of bulk susceptibilities in the two phases.*

* We thank Dr. M.E. Fisher for pointing out to us that this normalization depends on the definition (6.8) of the parameter t and is therefore non-universal.

6.3. CYLINDER, ISING-LIKE SYSTEMS

The effective action (6.2b) is the (euclidean) weight of a Feynman path integral for a quantum mechanical problem. Below T_c, the zero-field action has two (degenerate) minima. The quantum mechanical problem has an exponentially small splitting between the symmetric and antisymmetric lowest states. This splitting is exponentially small in $\hbar$, which is replaced here by $1/L^{d-1}$. A semi-classical computation of this splitting, valid for small $\hbar$ i.e. large L, may be done easily by a saddle-point expansion (instanton calculus [24]). The saddle-point itself corresponds to trajectories which interpolate between the two minima, a simple picture of a dilute gas of domain walls. The result takes the form for the correlation length (the inverse of the energy splitting) at finite T below T_c

$$\xi_L = A(T)L^{-(d-3)/2} \exp(\beta L^{d-1}\sigma(T)) . \tag{6.11}$$

This result is valid below the transition temperature for Ising-like systems, namely symmetric under $\varphi \to -\varphi$, whether the transition is first order or continuous. The exponential factor in (6.11) is typical of a WKB behaviour; the coefficient $\beta\sigma(T)$ of L^{d-1} is the value of the effective action for the kink solution and coincides with the surface tension [25]. The power of L in the prefactor of (6.11) is in fact a delicate problem. We believe that it is correctly given by (6.11) for arguments exposed below, but we have not made a systematic investigation of higher order corrections. The first contribution to this prefactor is nothing but the WKB normalization $L^{-(d-1)/2}$. In the semi-classical language it comes from an integration over the location of the center of the kink. An extra-factor L^{-1} comes from one loop fluctuations [26], a contribution from the modes $q_\perp \neq 0$, or in physical terms a result of the fuziness of the interface, i.e. of the capillary modes of the interface. It is worth noting that the result (6.11) is in agreement with exact results on the two-dimensional Ising model in a strip [27], which lead after some calculations to

$$\xi_L = \sqrt{\frac{\pi L}{2 \sinh m}}\, e^{Lm} \tag{6.12a}$$

with

$$m = -2(\beta J) - \ln \operatorname{th}(\beta J) , \tag{6.12b}$$

in units in which the lattice spacing is equal to one. In the vicinity of a second-order transition temperature T_c, $A(T)$ and $\sigma(T)$ may be expanded in powers of ε along the lines of sect. 4 [28].

7. Conclusion

We have presented a general approach to FSS for a phase transition. To illustrate the method we have presented some calculations, but it is clear that it would not be difficult to compute many other physical quantities of interest. It would also be feasible to calculate higher orders in the ε-expansions for the quantities considered in this article, such as the correlation length or the renormalized coupling constant.

Similarly one may consider different geometrical shapes, or new situations in which for instance two dimensions are infinite and $(d-2)$ remain finite. This would be relevant to the study of the dimensional cross-over between a 3D and a 2D critical behaviour. Indeed it would involve the solution of a two-dimensional field theory, but in a non-critical region accessible to numerical methods.

Another extension of this approach would be the study of the dynamics of a finite system. Technically this is similar to the cylindrical geometries considered in this work, but the real time and the space propagations are now anisotropic. This would be relevant to the study of the critical slowing down in Monte Carlo measurements.

Finally, another possible use of this approach would be the study of three-dimensional systems at their critical point. In the infinite volume limit, perturbation theory at T_c breaks down at finite order because of infrared singularities but L provides an infrared cut-off. This could lead to calculations, similar in their spirit, although quite different in practice, to those proposed by Parisi [29] and extended in refs. [30].

It seems clear that, in spite of the extensive literature on the subject, there is still a lot to say about finite size effects.

It is a pleasure to thank for stimulating discussions Drs. B. Derrida, K. Binder, and I. Affleck, and Drs. M.E. Fisher and M. Lüscher for an helpful correspondence.

References

[1] M.E. Fisher, in Critical phenomena, Proc. 51st Enrico Fermi Summer School, Varena, ed. M.S. Green (Academic Press, NY, 1972); M.E. Fisher and M.N. Barber, Phys. Rev. Lett. 28 (1972), 1516;
Y. Imry and D. Bergman, Phys. Rev. A3 (1971) 1416

[2] M.P. Nightingale, Physica 83 (1976) A561;
B. Derrida, J. Phys. A14 (1981) L5;
H. Blöte, M. Nightingale and B. Derrida, J. Phys. A14 (1981) L45;
C.J. Hamer and M.N. Barber, J. Phys. A14 (1981) 241 and 259

[3] M.N. Barber, in Phase transitions and critical phenomena, vol. VIII, p. 146 ed. C. Domb and J. Lebowitz (Academic Press, NY, 1984)

[4] E. Brézin, J. de Phys. 43 (1982) 15

[5] M. Suzuki, Progr. Theor. Phys. 58 (1977) 1142

[6] J.-M. Luck, Phys. Rev. B31 (1985) 3069

[7] M.E. Fisher, in Renormalization group in critical phenomena and quantum field theory, ed. by J. Gunton and M.S. Green (Temple Univ., Philadelphia, 1974)

[8] M.E. Fisher and V. Privman, First-order transitions breaking O(n) symmetry: finite-size scaling, Caltech preprint (1985)

[9] V. Privman and M.E. Fisher, J. Stat. Phys. 33 (1983) 385

[10] L. Dolan and R. Jackiw, Phys. Rev. 12 (1974) 75

[11] K.G. Wilson and J. Kogut, Phys. Reports 12 (1974) 75

[12] E. Brézin, J.-C. Le Guillou and J. Zinn-Justin, in Phase transitions and critical phenomena, vol. VI, p. 127, ed. C. Domb and M.S. Green (Academic Press, NY, 1976)

[13] A.M. Polyakov, Phys. Lett. 59B (1975) 79
[14] E. Brézin and J. Zinn-Justin, Phys. Rev. Lett. 36 (1976) 691 and Phys. Rev. B14 (1976) 3110
[15] K. Binder, M. Nauenberg, V. Privman and A.P. Young, Finite size tests of hyperscaling, Phys. Rev. B31 (1985) 1498
[16] K. Binder, Critical properties and finite-size effects of the five-dimensional Ising model, Univ. of Mainz preprint (1985)
[17] G. 't Hooft, Nucl. Phys. B61 (1973) 455
[18] R.B. Pearson, Phys. Reports 103 (1984) 185
M.N. Barber, R.B. Pearson, D. Toussaint and J. Richardson, Finite size scaling in the 3-D Ising model, UCSB preprint, ITP 83-144
[19] S. Elitzur, IAS preprint (1979); A. Jevicki, Phys. Lett. 71B (1977) 327;
F. David, Comm. Math. Phys. 81 (1981) 149
[20] M. Lüscher, Phys. Lett. 118B (1982) 391; Nucl. Phys. B219 (1983) 233
[21] E.G. Floratos and D. Petcher, Phys. Lett. 133B (1983) 206 and Nucl. Phys. B252 (1985) 689
[22] T. Jolicoeur and A. Morel, Finite size effects due to Goldstone particles in lattice gauge theory, Saclay preprint SPhT/85/45 (1985)
[23] K. Binder and D.P. Landau, Univ. of Mainz, preprint (1984)
[24] J. Zinn-Justin, Lectures at Les Houches Summer School 1982, ed. by J-B. Zuber and R. Stora, p. 39 (North-Holland, Amsterdam, 1984)
[25] M.E. Fisher, J. Phys. Soc. Japan, Suppl. 26 (1969) 87
[26] N.J. Günther, D.A. Nicole and D.J. Wallace, J. Phys. A13 (1980) 1755
[27] Ref. [9], appendix
[28] E. Brézin and S. Feng, Phys. Rev. B29 (1984) 472
[29] G. Parisi, method presented at the Temple University Conference in 1973; J. Stat. Phys. 23 (1981) 49
[30] G.A. Baker, B.G. Nickel and D.I. Meiron, Phys. Rev. B17 (1978) 1365;
J.C. Le Guillou and J. Zinn-Justin, Phys. Rev. B21 (1980) 3976

Journal of Statistical Physics, Vol. 41, Nos. 3/4, 1985

Finite-Size Scaling and the Renormalization Group

Joseph Rudnick,[1] Hong Guo,[2] and David Jasnow[2]

Received May 7, 1985

Renormalization group calculations in $d=4$ and $d=4-\varepsilon$ are performed for a system of finite size. A form of mean-field theory is used which yields a rounded transition for a finite system, and this allows a sensible expansion in fluctuations. A combination of Ewald and Poisson sum techniques is used to produce explicit numerical results for the specific heat in $d=4$ which, with the setting of two nonuniversal metrical factors and the fourth-order coupling constant may be compared with simulations. The numerical visibility of logarithmic corrections is investigated. The universal scaling function for the specific heat to relative $O(\varepsilon)$ is also evaluated numerically.

KEY WORDS: Finite size; scaling; critical phenomena; logarithmic corrections.

1. INTRODUCTION

Monte Carlo methods seem destined to play an increasingly prominent role in the study of equilibrium and nonequilibrium problems in statistical physics. This is due in large measure to the greater availability of powerful, high-speed computers and to the ongoing development of special purpose processors. Even with the best computing systems, however, simulations are still restricted essentially to microscopic systems. Studies of phase transitions are thus made problematical; the thermodynamic singularities that are the hallmark of a phase transition occur only in the infinite system limit. The central task in the interpretation of Monte Carlo data, when one is studying phase transitions, is to extrapolate to the thermodynamic limit from results for relatively small systems.

[1] Department of Physics, U.C.L.A., Los Angeles, California 90049.
[2] Department of Physics and Astronomy, University of Pittsburgh, Pittsburgh, Pennsylvania 15260.

0022-4715/85/1100-0353$04.50/0

The most widely used method of extrapolation to large size is based on phenomenological finite-size scaling.[1] In its traditional and simplest form a thermodynamic function, denoted generically by A, which is a function of reduced temperature $t=(T-T_c)/T_c$ and (linear) system size L, is assumed to take the asymptotic form.

$$A(t, L) \sim t^{\tau} X(t^{\nu} L) \tag{1.1}$$

where τ is the bulk critical exponent associated with the particular function A, and ν is the standard correlation length exponent. Such a form is expected to hold in the domain $L \gg a$, $\xi \sim t^{-\nu} \gg a$, where a is the microscopic length or lattice spacing. In general the function $X(x)$ is expected to have a singularity as $x \to 0$ which causes the function A to behave smoothly as $t \to 0$ for fixed L. On the other hand $X(x \to \infty)$ approaches a constant yielding the proper singular behavior in t for an infinite system.

To extract, say, the critical exponent τ (or more precisely τ/ν) from a Monte Carlo simulation one typically might evaluate $A(t=0, L)$, assuming the bulk critical temperature is known or can be estimated. Then one extrapolates it according to $A(t=0, L) \sim L^{-\tau/\nu}$, which comes from the limiting form $X(x) \sim x^{-\tau/\nu}$ as $x \to 0$. Many successful evaluations of the usual critical exponents[2] along with such less well-studied exponents as that for the surface tension[3] have been accomplished using this approach.

Recently, Brézin[4] has shown that the form (1.1) appears within the field theoretic renormalization group structure. Brézin also performed calculations on the isotropic N-vector model in the spherical model ($N \to \infty$) limit. Other spherical model calculations have also been carried out by Fisher and Barber,[5] Singh and Pathria,[6] and Rudnick.[7]

Special subtleties appear in $d \geq 4$ because of the breakdown of hyperscaling, and some aspects of this behavior may be seen within the context of spherical models. In the upper critical dimension $d_c = 4$ one expects logarithmic corrections to mean-field-like critical singularities. Such corrections have remained experimentally elusive (for example, in tricritical behavior in $d_c = 3$), and it is reasonable to ask whether simulations, with proper guidance as to the detailed size dependences to be expected, can be used to extract logarithms. At present no explicit calculations of scaling functions for Ising-like systems (or more generally for finite N) have been caried out.

We thus feel it instructive to carry out an explicit finite-system calculation in $d=4$ to see what subtleties simulators may have to deal with to achieve a proper extrapolation. A proper renormalization group calculation in $d=4$ is presumably asymptotically exact in the case of the Ising model and, in principle, may be compared to a standard Ising simulation with the adjustment of two nonuniversal amplitudes and the fourth-order coupling constant. As will be seen graphically below, without knowledge of the true infinite system (second-order transition) behavior, one might easily conclude from simulations that the ordering transition is first-order or, perhaps, that there is no transition at all. Using the results of this calculation one can also attempt to determine whether logarithmic behavior is visible in the finite-size region.

A fundamental objection to the application of field theoretical techniques—and the epsilon expansion in particular—to the study of finite systems has been raised by Brézin.[4] He has noted that strict mean field theory, the asymptotically correct description of the $d=4$ critical point and the theory about which one expands in $4-\varepsilon$ dimensions, predicts a sharp phase transition in *all dimensions* for *infinite* as well as *finite* systems. Such a prediction for a finite system is clearly pathological since all singularities are rounded. We have managed to overcome this objection by using a variant of mean field theory[7] that rounds properly when the system is finite. We find that hyperscaling breaks down when $d \geqslant 4$ as it does in the thermodynamic limit, but there are otherwise no difficulties in principle or practice in our approach.

The outline of this paper is as follows. In Section 2 we present a version of man-field theory for a finite system, while in Section 3 we evaluate a one-loop graph for such a system. A combination of Ewald and Poisson sum methods is required for good convergence. In Section 4 the proper renormalization group calculation is performed, while in Section 5 numerical evaluations and plots are presented. Calculations for $d=4-\varepsilon$ are presented in Section 6 and concluding remarks and summary appear in Section 7. Some further details are included in an Appendix.

2. MEAN-FIELD THEORY FOR A FINITE SYSTEM

We begin, as usual, with a Landau–Ginzburg–Wilson description of the system in which case the Hamiltonian divided by $k_B T$ is taken to be

$$H=\int_{\Omega} d^d x \left[\frac{1}{2}(\nabla s)^2+\frac{1}{2} r s^2+\frac{1}{4!} u s^4-h s\right] \tag{2.1}$$

where the spatial integration is over a system of linear extent L in each of its d dimensions. Other shapes may also be considered and generalization to an N-component order parameter is straightforward. As usual, the parameter $r \propto T-T_0$, where T_0 is a reference temperature. The statistical mechanics of the system follow from the partition function

$$Z=\int Ds\, e^{-H} \tag{2.2}$$

The system is confined to a finite box with periodic boundary conditions. One writes $s = m + \sigma$, where it is understood that $m = \text{const}$, and σ contains no $k = 0$ part. Then

$$H = L^d\left(\frac{1}{2}rm^2 + \frac{1}{4!}um^4 - hm\right) + \int_\Omega d^dx\left[\frac{1}{2}(\nabla\sigma)^2 + \frac{1}{2}\left(r + \frac{um^2}{2}\right)\sigma^2 + \frac{um\sigma^3}{6} + \frac{1}{4!}u\sigma^4\right] \tag{2.3}$$

The additional term involving $\int d^dx\sigma$ vanishes since σ has no $k = 0$ part. Hence the partition function (2.2) may be written

$$Z = \int dm\, e^{-H_{mf}(m)}\, e^{-\Gamma_1(m)} \tag{2.4}$$

where

$$\exp \Gamma_1(m) \equiv \int d\sigma\, e^{-H(\sigma;m)} \tag{2.5}$$

with $H(\sigma; m)$ the σ-dependent part of (2.3), and $H_{mf}(m)$ is the (mean-field) first term of that equation. $\Gamma_1(m)$ will be evaluated as usual via a diagram expansion; since there is no $k = 0$ term, the expansion will take the form

$$\Gamma_1(m) = \frac{1}{2}\sum_{\mathbf{k}\neq 0} \ln(k^2 + r + um^2/2) + \cdots \tag{2.6}$$

where the dots represent terms higher order in u. We shall return to this term below. The terms in $\Gamma_1(m)$ are at least $O(u^0)$, whereas the mean field part gives terms of order u^{-1}.

Neglecting fluctuctions completely one has the mean-field contribution to the free energy (in zero field)

$$\ln Z_{mf} = \ln\left\{\int dm \exp[-L^d(rm^2/2 + um^4/4!)]\right\} \tag{2.7}$$

from which the entropy and spcific heat follow by differentiation with respect to r. Notice that even at $r = 0$ moments of m, i.e., $\langle m^{2j}\rangle$, are perfectly finite for finite L so long as $u > 0$ (which we, of course, assume). Within mean field theory the moments have the scaling form

$$\langle m^{2j}\rangle \equiv Z_{mf}^{-1}\int dm\, m^{2j} \exp[-H_{mf}(m)] = (uL^d)^{-j/2} D_j(rL^{d/2}/u^{1/2}) \tag{2.8}$$

The heat capacity then has the form

$$C = L^{-4} \frac{\partial^2 \ln Z_{mf}}{\partial r^2} = \frac{1}{4}[\langle m^4 \rangle - \langle m^2 \rangle^2] \tag{2.9}$$

The functions D_j are perfectly well behaved through zero argument; from (2.7) it is seen (from steepest descent arguments) that a singularity can develop in the limit of infinite L. In Fig. 1 we display the specific heat C/L^d as a function of "temperature" r for various values of L. Notice the sharpening up and the appearence of a discontinuous jump as $L \to \infty$. It is true, however, that such a discontinuity will appear for all $d > 0$, which is a well-known defect of mean-field theories. In our calculation in $d = 4$ moments of the form (2.8) will play a central role.

We turn now to evaluation of a basic fluctuation integral in $d = 4$.

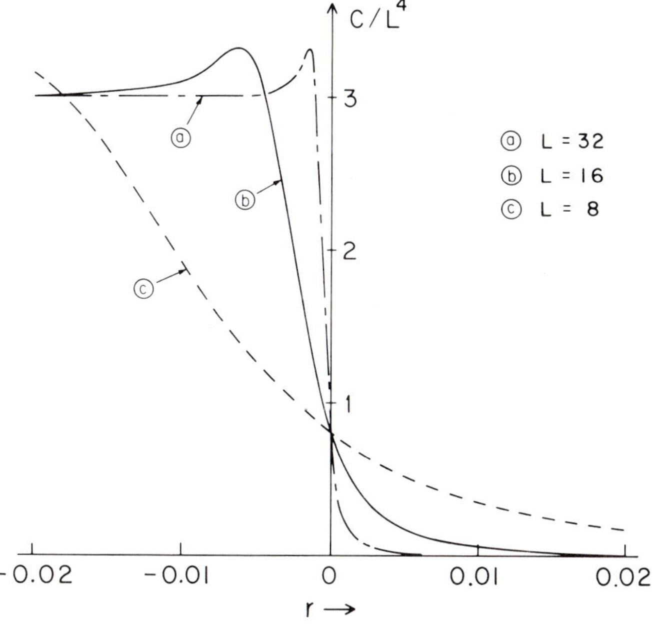

Fig. 1. Specific heat as a function of reduced temperature from mean-field calculation for various system sizes, L. The coupling constant u has been set to unity.

3. ONE-LOOP TERM

An essential ingredient of the finite-size calculation is the evaluation of the term

$$I(r) = \sum_{\mathbf{k} \neq 0} \frac{1}{r + k^2} \tag{3.1}$$

from which one may obtain the lowest-order fluctuation term in the free energy by integration, namely,

$$\sum_{\mathbf{k} \neq 0} \ln \frac{k^2 + T}{k^2} = \int_0^T dr\, I(r) \tag{3.2}$$

In (3.1) and (3.2) the wave vectors are appropriate to periodic boundary conditions,

$$\mathbf{k}(\mathbf{n}) = \frac{2\pi}{L} \mathbf{n}, \qquad n_i = 0, \pm 1, \pm 2, \ldots \tag{3.3}$$

for $i = 1, 2, \ldots, d = 4$. Then $I(r)$ may be written

$$I(r) = \sum_{\mathbf{k} \neq 0} \left[\int_0^B d\alpha\, e^{-\alpha(r + k^2)} + \int_B^\infty d\alpha\, e^{-\alpha(r + k^2)} \right] \tag{3.4}$$

where we shall let $B = c(L/2\pi)^2$ and choose the arbitrary constant c for good convergence. (The choice $c = \pi$ will finally be made.) In the first term of (3.4) Poisson summation techniques are used below. But for the second term we have

$$\begin{aligned} I_2 &= \frac{cL^2}{(2\pi)^2} \sum_{\mathbf{n} \neq 0} \int_1^\infty dx\, e^{-rc(L/2\pi)^2 x}\, e^{-cn^2 x} \\ &= \frac{cL^2}{(2\pi)^2} \int_1^\infty dx\, e^{-rc(L/2\pi)^2 x} [X^4(cx) - 1] \end{aligned} \tag{3.5}$$

where

$$X(y) \equiv \sum_{n=-\infty}^{\infty} e^{-yn^2} \tag{3.6}$$

Now consider the first term I_1 of (3.4) and use the Poisson formula[8]

$$\sum_{\mathbf{n}} f[\mathbf{k}(\mathbf{n})] = (L/2\pi)^4 \sum_{\mathbf{m}=-\infty}^{\infty} \int d^4k\, f(\mathbf{k})\, e^{i\mathbf{m} \cdot \mathbf{k}L} \tag{3.7}$$

to find

$$I_1 = -\frac{1}{r}[1 - e^{-rc(L/2\pi)^2}] + \frac{\pi^2}{c}\left(\frac{L}{2\pi}\right)^2 \int_1^\infty dx\, e^{-rc(L/2\pi)^2 x^{-1}} \left[X^4\left(\frac{\pi^2 x}{c}\right) - 1\right]$$
$$+ \left(\frac{L}{2\pi}\right)^4 \int d^4k \frac{1 - e^{-c(L/2\pi)^2(r+k^2)}}{r+k^2} \tag{3.8}$$

Combining I_1 and I_2 one has

$$\sum_{\mathbf{k}\neq 0} \frac{1}{r+k^2} = \left(\frac{L}{2\pi}\right)^4 \int \frac{d^4k}{r+k^2} + c\left(\frac{L}{2\pi}\right)^2 \int_1^\infty dx\, e^{-rc(L/2\pi)^2 x}[X^4(cx) - 1]$$
$$+ \frac{\pi^2}{c}\left(\frac{L}{2\pi}\right)^2 \int_1^\infty dx\, e^{-rc(L/2\pi)^2 x^{-1}} \left[X^4\left(\frac{\pi^2 x}{c}\right) - 1\right]$$
$$+ \pi^2 \left(\frac{L}{2\pi}\right)^4 r\, e^{-rc(L/2\pi)^2} \int_0^\infty \frac{e^{-x}\, dx}{x + rc(L/2\pi)^2}$$
$$- \frac{\pi^2}{c}\left(\frac{L}{2\pi}\right)^2 e^{-rc(L/2\pi)^2} + \frac{1}{r}(e^{-rc(L/2\pi)^2} - 1) \tag{3.9}$$

The first term is immediately recognizable as the bulk contribution. All the other terms are finite-size corrections in which $r \to 0$ may be taken. Clearly from the structure of (3.9) only usual bulk counterterms are required to remove the high momentum cutoff to infinity. This feature has previously been noted by Brézin.[4] Note also that formally the finite-size terms come in at $O(u^0)$.

Straightforward manipulations are required for effecting the r integration as in (3.2). The results may be expressed in terms of the exponential integral[9]

$$E_1(z) = \int_z^\infty dx \frac{e^{-x}}{x} \tag{3.10}$$

The most direct expression is valid for the case

$$T \equiv r + \frac{1}{2} u m^2 > 0 \tag{3.11}$$

although from (2.4) one would apparently need negative values. For r somewhat negative, but $r + (2\pi/L)^2 > 0$, an alternate representation can be constructed. It turns out to be sufficient to consider the case $T>0$ (see below), in which case (choosing $c=\pi$)

$$\frac{1}{2L^4}\sum_{\mathbf{k}\neq 0}\ln\frac{k^2+T}{k^2}=\frac{1}{2}\int\frac{d^4k}{(2\pi)^4}\ln\frac{k^2+T}{k^2}$$
$$+\frac{1}{2L^4}\int_1^\infty\frac{1}{x}(1-e^{-\omega x})[X^4(\pi x)-1]\,dx+\frac{1}{64\pi^2}T^2E_1(\omega)$$
$$+\frac{1}{2L^4}\int_1^\infty x(1-e^{-\omega x^{-1}})[X^4(\pi x)-1]\,dx-\frac{1}{4L^4}(1-e^{-\omega})$$
$$-\left(\frac{\omega}{4L^4}\right)e^{-\omega}-\frac{1}{2L^4}(E_1(\omega)-\ln(\omega)+\gamma) \tag{3.12}$$

where $\gamma=0.57721$ is Euler's constant and, for convenience, $\omega=L^2T/4\pi$. Notice the function X, defined in (3.6), has an argument at least equal to π in (3.12) and is rapidly convergent. Likewise E_1 has an accurate polynomial approximant; hence the representation (3.12) is free of numerical or convergence difficulties. It forms the core of the full renormalization group calculation to which we now turn.

4. RENORMALIZATION GROUP APPROACH

Standard bulk counterterms are sufficient to renormalize the theory.[10] They are introduced into (2.1) in which case the full (renormalized) exponential, $L^{-4}(H_{\mathrm{mf}}+\Gamma_1)$, becomes

$$\Gamma_R(t,m,u,\kappa,L)=\Gamma_{\mathrm{bare}}(r,m,u,\Lambda,L)-\Gamma_{\mathrm{bare}}\Big|_{\substack{t=0\\m=0}}$$
$$-t\frac{\partial\Gamma_{\mathrm{bare}}}{\partial t}\Big|_{\substack{t=0\\m=0}}-\frac{1}{2}t^2\frac{\partial^2\Gamma_{\mathrm{bare}}}{\partial t^2}\Big|_{\substack{t=\kappa^2\\m=0}} \tag{4.1}$$

where, as usual, the following replacements are made in the unrenormalized or bare quantities r, m, u on the right-hand side:

$$r\to Z_{\phi^2}t+\delta t$$
$$u\to\kappa^\varepsilon Z_u u \tag{4.2}$$
$$m\to Z_\phi^{1/2}m$$

and from now on t, u, m, etc. all refer to renormalized quantities. As usual, $\varepsilon=4-d$. The three additive renormalizations are the same as for the bulk system, and Γ_{bare} is calculated diagrammatically from (2.1), the leading contributions being the mean field part $H_{mf}(m)$ and the loop in (2.6). All momentum integrals may be regularized with a hard cutoff Λ which is eventually removed to infinity, and κ in (4.1) introduces the basic (inverse) length scale into the renormalized theory. Then acting with $\kappa d/d\kappa$ on (4.1) and introducing $S_d = [2^{d-1}\pi^{d/2}\Gamma(d/2)]^{-1}$ as the surface of a unit sphere in d dimensions, one finds the inhomogeneous RG equation,

$$\left\{\kappa\frac{\partial}{\partial\kappa} + W(u)\frac{\partial}{\partial u} - \frac{1}{2}\eta(u)m\frac{\partial}{\partial m} + [-2+\gamma_{\phi^2}(u)]t\frac{\partial}{\partial t}\right\}\Gamma_R$$

$$= -\frac{S_d}{8v}\left(1-\frac{\varepsilon}{2}+\frac{5}{4}S_d u\right)t^2\kappa^{-(4-d)} \tag{4.3}$$

which is solved in the usual fashion.[3] Note that the inhomogeneous term is appropriate to two loops, which is necessary in getting all $O(\varepsilon^0)$ contributions to the free energy. One finds in $d=4$

$$\Gamma_R = \Gamma_R(\rho^2 t(\rho), m(\rho), u(\rho), \kappa\rho, L)$$

$$+\frac{S_d}{8v}\int_\rho^1 \frac{dx}{x}\left[1-\frac{\varepsilon}{2}+\frac{5}{4}S_d u(x)\right]t^2(x)(\kappa x)^{-(4-d)} \tag{4.4}$$

where the last term is the trajectory term and formally is $O(u^{-1})$. The flow equations are, as usual,[10]

$$\rho\frac{dt(\rho)}{d\rho} = [-2+\gamma_{\phi^2}(u(\rho))]\,t(\rho)$$

$$\rho\frac{du(\rho)}{d\rho} = W(u(\rho)) \tag{4.5}$$

$$\rho\frac{dm(\rho)}{d\rho} = -\frac{1}{2}\eta(u(\rho))\,m(\rho)$$

with, in $d=4$,

$$W(u)=\tfrac{3}{2}S_d u^2, \qquad \gamma_{\phi^2}(u)=\tfrac{1}{2}S_d u, \qquad \eta(u)=\tfrac{1}{24}S_d u^2$$

Generalization to an N-component system is straightforward. Solutions of the flow equations yield

$$u(\rho)=\frac{u}{1-(3/2)S_d u\ln\rho}$$

$$t(\rho)=\frac{t}{[1-(3|2)\,S_d u\ln\rho]^{1/3}} \tag{4.6}$$

$$m(\rho)=m\exp\{1/72[u-u(\rho)]\}$$

[3] For future reference, the form of general d has been written.

which may be substituted into (4.4) to give a complete representation of the finite-size free energy. The partition function (2.4) becomes

$$Z=\int_{-\infty}^{\infty} dm\, e^{-L^4\Gamma_R} \tag{4.7}$$

For evaluation of the free energy a choice of $\rho=\rho^*$ will be required. This will be discussed below.

At lowest order, often refered to as renormalized man field theory (Rmf), and which yields the free energy correctly to leading logarithms, one neglects loop contributions to Γ_R but retains the trajectory term in (4.4). Hence at this order, which formally keeps terms $O(u^{-1})$,

$$\begin{aligned}\Gamma_{\rm Rmf}(t, m, u; \kappa, L)=&\frac{1}{2}\rho^{*2}t(\rho^*)\, m^2(\rho^*)+\frac{1}{4!}u(\rho^*)\, m^4(\rho^*)\\ &+\frac{t^2}{16\pi^2 u}\left[1-\left(1-\frac{3}{2}S_d u\ln\rho^*\right)^{1/3}\right]\end{aligned} \tag{4.8}$$

The specific heat follows from (4.8) according to

$$\begin{aligned}\frac{C}{L^4}&\equiv L^{-4}\frac{\partial^2}{\partial t^2}\ln Z_{\rm Rmf}\\ &=L^4\left\langle\left(\frac{\partial\Gamma_{\rm Rmf}}{\partial t}\right)^2\right\rangle-L^4\left\langle\frac{\partial\Gamma_{\rm Rmf}}{\partial t}\right\rangle^2-\left\langle\frac{\partial^2\Gamma_{\rm Rmf}}{\partial t^2}\right\rangle\end{aligned} \tag{4.9}$$

where averages are evaluated according to the probability distribution in (2.7) with

$$r\to\rho^{*2}t(\rho^*),\qquad u\to u(\rho^*) \tag{4.10}$$

(Henceforth all averages are considered to be taken by such procedure.) Note that in (4.9) only *explicit* temperature derivatives are taken; the implicit t dependence of ρ^* is ignored. The results are formally independent of ρ^* and taking derivatives of ρ^* formally yields higher-order contributions.[11] Note also that all averages entering (4.9), namely, $\langle m^4(\rho^*)\rangle$ and $\langle m^2(\rho^*)\rangle$ depend on the combination $wv^{-1/2}$ where $w=L^2\rho^{*2}t(\rho^*)$ and $v=u(\rho^*)$. These same variables enter the loop corrections. A more explicit evaluation of (4.9) is provided in the Appendix.

The choice of ρ^* must be made for detailed evaluations. The choice we have made is to determine ρ^* from

$$\rho^{*2}t(\rho^*)+\frac{1}{2}u(\rho^*)\langle m^2(\rho^*)\rangle+\frac{1}{L^2}=(\kappa\rho^*)^2$$

where, as noted, $\langle m^2(\rho^*)\rangle$ is determined as follows from (2.7) with $r \to \rho^{*2}t(\rho^*)$ and $u \to u(\rho^*)$. The choice has the following interpretation. The quantity $\rho^{*2}t(\rho^*)+1/2u(\rho^*)\,m^2(\rho^*)$ has the interpretation of the "mass" appearing in the renormalized Hamiltonian. If the temperature $t \sim T-T_c$ is such that the bulk correlation length ξ is *large* compared to the system size L, renormalization must end when the effective block spin size $[\sim(\kappa\rho^*)^{-1}]$ is on the order of L. This is the strongly finite-size regime and $\rho^* \approx (\kappa L)^{-1}$. On the other hand if L is *much larger* than the bulk correlation length, renormalization transformations are continued until the block spin size is on the order of ξ. This corresponds to $\rho^{*2}t(\rho^*)+1/2u(\rho^*)\,m^2(\rho^*)=(\rho^*\kappa)^2$, the usual bulk-system determination of ρ^*. Equation (4.13) interpolates between these regimes. To find $\rho^*(t, L)$ we must numerically solve (4.11). This must be done iteratively since ρ^* is required for $\langle m^2(\rho^*)\rangle$. The starting choice $\langle m^2(\rho^*)\rangle = -6\rho^{*2}t(\rho^*)/u(\rho^*)$ allows convergence after a small number of iterations. Numerical results will be displayed in the next section.

Loop corrections may now be handled, in fact, iteratively. Formally one must insert the renormalized Γ_1 which is given in the Appendix into (4.7). However, order-by-order one may make the following expansion:

$$Z=\int_{-\infty}^{\infty} dm\,\exp[-L^4(\Gamma_{\mathrm{Rmf}}+\Gamma_{R,1})]$$

$$=\int_{-\infty}^{\infty} dm\,\exp\{-L^4[\Gamma_{\mathrm{Rmf}}(m^2(\rho))+\Gamma_{R,1}(\langle m^2(\rho)\rangle)+\Delta]\} \quad (4.12)$$

where

$$\Delta \equiv \Gamma_{R,1}(m^2(\rho))-\Gamma_{R,1}(\langle m^2(\rho)\rangle)$$

and averages are computed with respect to the (renormalized) mean field theory. It is to be understood that $\rho=\rho^*$ has been chosen in what follows. Then

$$Z=\exp\{-L^4\Gamma_{R,1}(\langle m^2(\rho)\rangle)\}\cdot Z_{\mathrm{Rmf}}\cdot\langle\exp\{-L^4\Delta\}\rangle$$

which is expanded in terms of cumulants, so that

$$\ln Z=\ln Z_{\mathrm{Rmf}}-L^4\Gamma_{R,1}[\langle m^2\rho)\rangle]+\sum_{n=1}^{\infty}\frac{(-1)^n}{n!}\langle(L^4\Delta)^n\rangle_c \quad (4.13)$$

where $\langle\cdots\rangle_c$ are cumulant averages. Recall that we are searching for terms of order u^0 in the specific heat [leading terms being $O(u^{-1})$]. Temperature differentiation promotes terms so some care must be taken. It is easy to verify using scaling relations like (2.8) that

$$\langle L^4 \Delta \rangle_c = L^4 \sum_{p=1}^{\infty} a_p \langle (m^2(\rho) - \langle m^2(\rho) \rangle)^p \rangle$$
$$\simeq L^4 a_2(\langle m^4(\rho) \rangle - \langle m^2(\rho) \rangle^2) \tag{4.14}$$

where of course the coefficients a_p arise from the expansion of $\Gamma_{R,1}(m^2(\rho))$ about $\Gamma_{R,1}(\langle m^2(\rho) \rangle)$. The same reasoning allows the replacement

$$\langle (L^4 \Delta)^2 \rangle_c \simeq (L^4 a_1)^2 [\langle m^4(\rho) \rangle - \langle m^2(\rho) \rangle^2] \tag{4.15}$$

all other cumulants yield contributions beyond the (one-loop) order of the calculation.

Substitution of (4.14) and (4.15) into (4.13) yields the free energy consistently so that the specific heat is correctly given to $O(u^0)$. The only two expansion coefficients which enter are

$$\begin{aligned} a_1 &\equiv \left. \frac{\partial \Gamma_{R,1}}{\partial m^2(\rho)} \right|_{m^2(\rho) = \langle m^2(\rho) \rangle} \\ a_2 &\equiv \left. \frac{1}{2} \frac{\partial^2 \Gamma_{R,1}}{\partial (m^2(\rho))^2} \right|_{m^2(\rho) = \langle m^2(\rho) \rangle} \end{aligned} \tag{4.16}$$

and $\Gamma_{R,1}(m^2(\rho))$ may be found in the Appendix. The specific heat follows from $d^2 \ln Z/dt^2$ keeping ultimately terms $O(u^0)$.

The form of the scaling can be explicitly discussed at this point. At renormalized mean field level one finds directly that

$$\ln Z(t, u; \kappa, L) = \bar{F}[w, v] \qquad [w = L^2 \rho^2 t(\rho), v = u(\rho)] \tag{4.17}$$

where $\bar{F}$ is a dimensionless function. (A smooth background term of the form $L^4 t^2/u$ and additional background constants which can be removed by normalization have been omitted.) There is further simplification here since $\bar{F}(w, v)$ takes the form

$$\bar{F}_{\mathrm{Rmf}}(w, v) = f(z = wv^{-1/2})$$
$$= z^2 + \ln \int_{-\infty}^{\infty} dy \exp \left\{ -\left(\frac{1}{2} zy^2 + \frac{1}{4!} y^4 \right) \right\} \tag{4.18}$$

This form contains the *leading* logarithms in $d = 4$.

More generally

$$\ln Z(t, u; \kappa, L) = F\left[\frac{t(\rho)}{\kappa^2}, u(\rho), \rho \kappa L \right] \tag{4.19}$$

which form reproduces (4.17). This is a general statement of finite size scaling; the loop correction discussed above takes this form. Such scaling forms have been discussed by Brézin.[4] In usual cases in which $u(\rho) \to u^* \neq 0$ as $\rho \to 0$ one may show that a relationship like (4.19) produces the intuitive sort of finite-size scaling discussed in the Introduction.

Because of the marginality of u in $d = 4$, $u(\rho)$ iterates slowly to zero as $\rho \to 0$ as indicated in (4.6). This will yield additional L dependence and explicit dependence on the coupling constant u.

5. NUMERICAL RESULTS

It is instructive to plot numerical results for the specific heat which is given in the Appendix (A2). First, however, we return to the pure mean-field theory discussed in Section 2. The results have been shown in Fig. 1. Note in particular that for larger sizes the specific heat maximum does not change significantly; the curve merely sharpens up.

The situation with respect to renormalized mean field theory is shown in Fig. 2. Recall that the central ingredient is given in (4.8). A more explicit

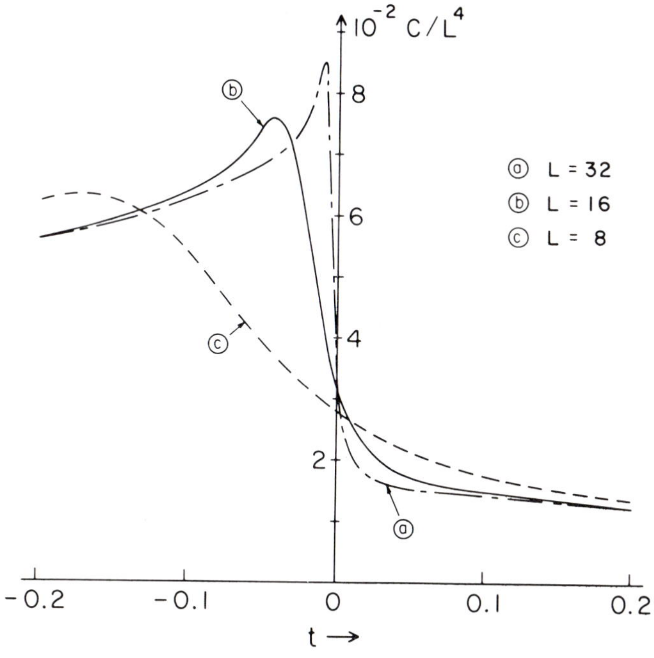

Fig. 2. Specific heat from renormalized mean-field theory at $d = 4$ for various system sizes L. The characteristic inverse length κ and uS_4 have been set to unity.

822/41/3-4-2

form is shown in (A.2) where renormalized mean field theory contains all terms without $\Gamma_{R,1}$. It is easy to show that $C(t=0)$ diverges as $(\ln \kappa L)^{1/3}$ as $L \to \infty$ although such slow dependence is barely visible in the plots. One may also demonstrate that the specific heat peak diverges with the same L dependence. By direct computation for the infinite system, the per unit volume specific heat diverges as $C \sim |\ln |t|/\kappa^2|^{1/3}$, so that the mechanism for the divergence is that the values at the peak and at $t=0$ diverge together.

It is interesting to consider numerically $C(t_{\max})/L^4$, i.e., the peak height, vs. system size, since this is what a simulator might do to look for logarithmic corrections. At present simulations involving 10^6 Ising spins are possible, and a signature of the logarithm may be possible to detect if sufficient precision in a narrow temperature range can be maintained. However even with less precision it might to be possible to detect the increase in peak height itself as a sign of logarithmic corrections. The ability to make definitive statements is restricted because the strength of the logarithmic corrections depends crucially on the strength of the coupling constant u. If u were reduced, the peak structure would remain

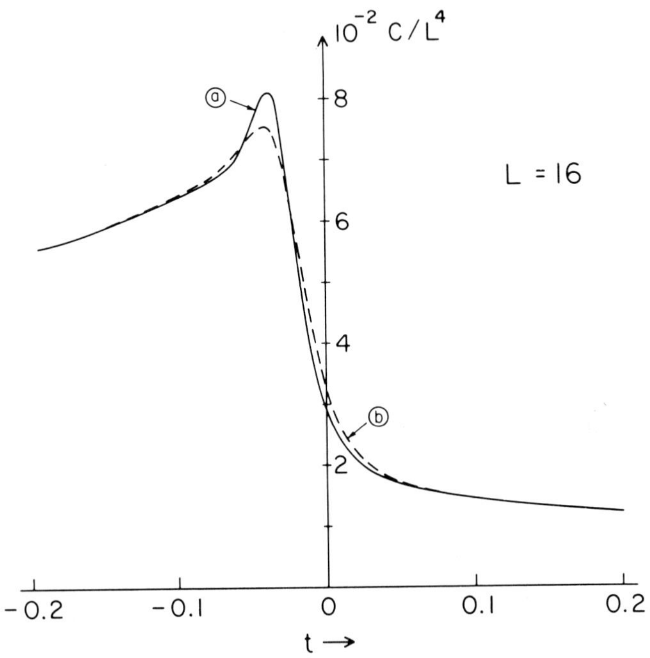

Fig. 3. Specific heat at $d=4$, $L=16$, showing (b) the renormalized mean-field theory and (a) the result including the loop correction. The characteristic inverse length κ and the coupling constant uS_4 have been set to unity.

approximately the same, but the overall magnitude of the specific heat would increase. This would make such logarithmic effects more difficult to observe in an hypothetical simulation of a system characterized by a small fourth-order coupling. Additional numerical and analytic results suggest that the position of the maximum $t_{\max}$ goes to zero with size rather rapidly, i.e., $|t_{\max}| \sim [L^2 \ln^{1/6}(\kappa L)]^{-1}$.

In Fig. 3, the effect of the loop correction to renormalized mean field theory is shown. As might be expected, the specific heat peak is enhanced and sharpened.

6. RESULTS IN $4-\epsilon$ DIMENSIONS

The analysis presented above may be carried over directly to $d=4-\varepsilon$ dimensions. In the preceding sections we have concentrated on $d=4$ because an asymptotically correct finite size calculation is possible. Here one may calculate order-by-order in ε in analogy with ordinary bulk phenomena. Note, however, that since the finite-size corrections are $O(1)$, a consistent calculation must include bulk contributions to the trajectory integral to two-loop order. This gives $O(1)$ bulk contributions correctly.

One must solve the renormalization group flow equations (4.5) in which case now one finds

$$\begin{aligned} u(\rho) &= \frac{u\rho^{-\varepsilon}}{Q(\rho)}, \qquad Q(\rho) = 1 + \frac{3}{2}\frac{u}{\varepsilon}(\rho^{-\varepsilon} - 1) \\ t(\rho) &= tQ(\rho)^{-1/3} \\ m(\rho) &= m \exp\{1/72[u - u(\rho)]\} \end{aligned} \tag{6.1}$$

consistent to one-loop order. The expressions for $u(\rho)$, $t(\rho)$, and $m(\rho)$ reduce to those of (4.6) when $\varepsilon \to 0$. Hence the renormalized mean-field contribution (including the trajectory integral) is

$$\begin{aligned} \Gamma_{\mathrm{Rmf}} &= \frac{1}{2}\rho^{*2} t(\rho^*)\, m^2(\rho^*) + \frac{1}{4!}(\kappa\rho)^{\varepsilon}\, u(\rho^*)\, m^4(\rho^*) \\ &\quad - \frac{t^2\kappa^{-\varepsilon}}{32\pi^2}\int_{\rho^*}^{1} \frac{dx}{x^{1+\varepsilon}}[Q(x)]^{-2/3} \end{aligned} \tag{6.2}$$

which goes over to the four-dimensional result (4.8) as $\varepsilon \to 0$. Since the loop integral may actually be calculated in $d=4$ consistent to one loop order, the one-loop part $\Gamma_{R,1}$ properly goes to the previous result as $\varepsilon \to 0$.

On the other hand for $\varepsilon>0$ one may set $u=u^*$ and observe that $m(\rho)\approx m$ to this order. Then

$$L^d\Gamma_R(t, m, u, \kappa, L)=L^d\Gamma_{R\mathrm{mf}}+L^d[\Gamma_{R,1}]_{d=4-\varepsilon} \tag{6.3}$$

where the product $L^d[\Gamma_{R,1}]_{d=4-\varepsilon}$ is identical to $L^4[\Gamma_{R,1}]_{d=4}$, where $[\Gamma_{R,1}]_{d=4}$ is given in (A1), except that the *bulk* term

$$\frac{1}{64\pi^2}L^4T^2(\rho)\left[\ln\frac{L^2T(\rho)}{(\kappa\rho L)^2}+\frac{1}{2}\right]-\frac{1}{4}S_d\rho^4L^4t^2(\rho)$$

is divided by $(\kappa\rho L)^\varepsilon$. With these modifications it is clear that the free energy takes the form (within additive constant)

$$\ln Z=\ln\int dm\exp\{-L^d\Gamma_R\}=L^d[\mathrm{T.I.}]+\ln\int dx\, e^{-A} \tag{6.4}$$

where A depends only on the combination $L^2\rho^{*2}t(\rho^*)+\frac{1}{2}u^*x^2$, and T.I. is the trajectory term of (6.2). The final free energy depends only on the combination $L^2\rho^{*2}t(\rho^*)$. To see the scaling form one sets $\kappa\rho^*L=1$ so that $L^2\rho^{*2}t(\rho^*)=(\rho^*)^{2-1/\nu}L^2t=(t/\kappa^2)(\kappa L)^{1/\nu}$. Hence the full free energy $F=\ln Z$ is a function of this variable and

$$\begin{aligned}L^{-d}\ln Z&=L^{-d}F[(t/\kappa^2)(\kappa L)^{1/\nu}]\\&=\kappa^d(t/\kappa^2)^{d\nu}\tilde{F}[(t/\kappa^2)(\kappa L)^{1/\nu}]\end{aligned}$$

Differentiation twice with respect to temperature yields the heat capacity. Further details are provided in the Appendix.

For the above discussion it is clear that a scaling form

$$\frac{\kappa^\varepsilon C}{L^d}=(\kappa^{-2}t)^{-\alpha}X[\kappa^{-2}t(\kappa L)^{1/\nu}] \tag{6.5}$$

holds for the specific heat. As such the function $X(y)$ is not itself universal; the limiting form $X(y\to 0)$ essentially yields the nonuniversal specific heat amplitude for the bulk system, $(T>T_c)$. Given an arbitrary overall amplitude the function $X(y)$ is then universal, and one may compare two differential systems in the same universality class. [Alternatively $\tilde{X}(y)=X(y)/X(\infty)$ is expected to be universal.] In Fig. 4 the approximate $(d=3)$ form for $X(y)$ is given by setting $\varepsilon=1$ in the numerical evaluation of Eq. (A4).[4] The scaling function is evaluated by fixing κL and evaluating the

[4] The limiting values of function $X(y)$, as $y\to\pm\infty$, give the specific heat amplitudes of the bulk system. The amplitude ratio of the bulk system is

$$\frac{A^+}{A^-}=\frac{2^\alpha}{4}\frac{1+(7/12)\varepsilon}{1-(5/12)\varepsilon}=\frac{2^\alpha}{4}(1+\varepsilon)+O(\varepsilon^2)$$

and our function $X(y)$ gives the value consistent with the first equality evaluated at $\varepsilon=1$.

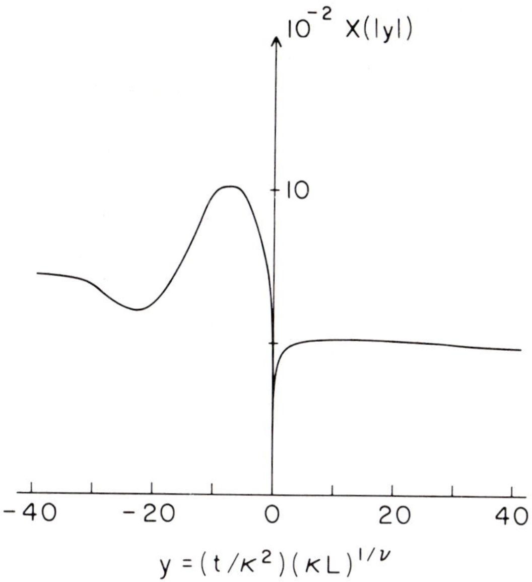

Fig. 4. The scaling function $X(|y|)$ at $d=4-\varepsilon$ evaluated at $\varepsilon=1$, where $y=(t/\kappa^2)(\kappa L)^{1/\nu}$. The two branches for $t \gtrless 0$ have been shown.

specific heat at values of t/κ^2. The argument y is extracted using $\nu = (1/2+1/12\varepsilon) \rightarrow 0.58$ at $\varepsilon=1$. (Alternatively, one may use the appropriate values of ν and α for $d=3$ to obtain a "hybrid" estimate which may offer better agreement with other schemes.) To compare $X(y)$ with a simulation or experiment, in which one would plot

$$\frac{C}{L^d(\Delta T/T_c)^{-\alpha}} \text{ vs. } L^{1/\nu}\left(\frac{\Delta T}{T_c}\right)$$

one must allow for a rescaling of the magnitude X as well as the independent variable y. The latter rescaling allows for a nonuniversal "metrical" factor associated with the combination $tL^{1/\nu}$.

If $u \neq u^*$ ordinary corrections to scalling get mixed in with finite size corrections. Keeping $u \neq u^*$ allows one to consider the crossover from "Gaussian" to "Wilson–Fisher" fixed points and investigate circumstances in which the crossover is or is not completed before the thermodynamic functions are rounded. Experimentally one typically sees "nonclassical" exponents before the rounding regime is reached, but on small systems there may in principle be interference.

7. CONCLUDING REMARKS

The application of finite-size scaling ideas has become widespread since it has become appreciated that, for determining critical properties of systems which undergo second-order phase transitions, some sort of extrapolation with size is essential. Using ideas of finite-size scaling a simulator can infer behavior of the infinite system from "rounded" data. Brézin[(4)] has shown that the phenomenological form of finite-size scaling follows from renormalization group equations. The essential point is that counterterms for the infinite system suffice to renormalize the theory for finite L.

In this paper we have shown how to compute explicitly thermodynamic functions in the scaling regime $L/a \gg 1$, $ta^2 \ll 1$, where a represents the microscopic length scale of the system, say, the lattice spacing. The computation in $d=4$ is exact to leading logarithms and so provides a potentially useful benchmark for simulations. By adjusting two "metrical" factors, corresponding to a $T-T_c$ scale dilation or contraction and an amplitude for, say, the specific heat, along with the nonuniversal parameter u, the theoretical form discussed in the Appendix is expected to agree in the scaling regime with a simulation. (Of course only if the u which one obtains by such a fitting is small can one expect one-loop corrections to give a reasonable representation in a region about criticality.)

The calculation in $d=4$ also has the advantage of addressing the question of visibility of logarithmic corrections in a rounded regime. This information is carried in the increase of the peak height shown in Fig. 2. For the Ising case the increase is proportional to $\ln^{1/3}(L/a)$. Our numerical estimates indicate that if data in the range $L/a=8$ to $L/a=32$ (or 64) were available and if specific heat peaks were determined to $\pm 5\%$, then a case might be made for direct observation of the logarithmic corrections. (These statements assume a reasonable value of $uS_d \approx 1$.)

Below $d=4$ we cannot calculate the asymptotically correct result without working order by order in $\varepsilon = 4-d$. However, to one-loop order we have produced the finite-size scaling function for the specific heat. With the adjustment of two metrical factors (now $u \to u^*$ and need not be adjusted) it would be interesting to see how results at $\varepsilon = 1$ compare with Ising simulations in three dimensions.

After the completion of this work we learned that Brézin and Zinn-Justin[(12)] have recently shown that an ε expansion for finite-size systems can be constructed. Their scheme for expanding about mean-field theory in finite systems is essentially the same as ours.

ACKNOWLEDGMENTS

The authors are grateful to the National Science Foundation for partial support of this work. For D. J. and J. R., respectively, the support has been through the Division of Materials Research under grant Nos. DMR-83-02326 and DMR81-15542. Part of this work was done while D. J. was a visitor at the Institute for Theoretical Physics, University of California, Santa Barbara. He gratefully acknowledges the hospitality and support of I.T.P.

8. APPENDIX

Some technical details are presented in this Appendix.

The renormalized $\Gamma_{R,1}(t(\rho), m(\rho), u(\rho), \kappa\rho, L)$ is given in $d=4$ by

$$
\begin{aligned}
[\Gamma_{R,1}]_{d=4} = {} & \frac{S_d}{8} T^2 \left[\frac{1}{2} - \ln \frac{T}{(\kappa^2 \rho^2)} \right] - \frac{S_d}{4} \rho^4 t^2(\rho) \\
& + \frac{S_d}{8} T^2 E_1(\omega) - \frac{1}{2L^4} [E_1(\omega) + \ln \omega + \gamma] - \frac{1}{4L^4} (1 - e^{-\omega}) \\
& - \frac{\omega}{4L^4} e^{-\omega} + \frac{1}{2L^4} \int_1^\infty \frac{dx}{x} (1 - e^{-\omega x}) [X^4(\pi x) - 1] \\
& + \frac{1}{2L^4} \int_1^\infty dx\, x (1 - e^{-\omega/x}) [X^4(\pi x) - 1]
\end{aligned}
\tag{A1}
$$

where

$$
T \to T(\rho) \equiv \rho^2 t(\rho) + \frac{1}{2} u(\rho)\, m^2(\rho)
$$

$$
\omega \equiv L^2 T(\rho)/4\pi
$$

and where $X(x)$ has been defined in (3.6), $E_1(z)$ has been given in (3.10) and γ is Euler's constant. Derivatives

$$
a_1 = \frac{\partial \Gamma_{R,1}}{\partial m^2(\rho)} \qquad \text{and} \qquad a_2 = \frac{\partial^2 \Gamma_{R,1}}{\partial [m^2(\rho)]^2}
$$

evaluated at $m^2(\rho^*) = \langle m^2(\rho^*) \rangle$ are required for the specific heat; the choice of ρ^* has been discussed in Section 4. These derivatives are easily converted to derivatives with respect to $T(\rho)$. Then the singular part of the specific heat, which is evaluated numerically in Section 5, has the form

$$\frac{C}{L^4}=\frac{1}{4}\left[\frac{u(\rho^*)}{u}\right]^{2/3} L^4[\langle m^4(\rho^*)\rangle-\langle m^2(\rho^*)\rangle^2]$$
$$+\frac{1}{u}\left[\left(1-\frac{3}{2}S_d u\ln\rho^*\right)^{1/3}\right]-\left[\frac{u(\rho^*)}{u}\right]^{2/3}\left[\frac{\partial^2\tilde{\Gamma}_{R,1}}{\partial T(\rho)^2}\bigg|_*\right]$$
$$-\frac{S_d}{2}\left[\frac{u(\rho)}{u}\right]^{2/3}+\frac{1}{16}u^2(\rho^*)\left[\frac{u(\rho^*)}{u}\right]^{2/3}$$
$$\times\left\{\frac{1}{2}\left[L^2\frac{\partial\tilde{\Gamma}_{R,1}}{\partial T(\rho)}\right]^2-\frac{\partial^2\tilde{\Gamma}_{R,1}}{\partial T(\rho)^2}\right\}_* L^8\langle[m^2(\rho^*)]^4\rangle_c \tag{A2}$$

where $\tilde{\Gamma}_{R,1}$ is $\Gamma_{R,1}$ without the term $S_d\rho^4t^2/4$. In the final expression derivatives, as indicated by $|_*$, are evaluated at

$$T(\rho^*)=\rho^{*2}t(\rho^*)+\tfrac{1}{2}u(\rho^*)\langle m^2(\rho^*)\rangle$$

and $\langle\ \rangle_c$ means cumulant average. The terms without $\tilde{\Gamma}_{R,1}$ belong to the renormalized mean field theory. Finally it should be recalled that

$$\langle m^{2j}(\rho^*)\rangle=[u(\rho^*)L^4]^{-j/2}D_j\left[\frac{\rho^{*2}t(\rho^*)L^2}{u(\rho^*)^{1/2}}\right] \tag{A3}$$

as is (2.7).

In Section 6 the modifications for dealing with $d=4-\varepsilon$ were described. A general formulation is possible (as noted) which goes over to the $d=4$ results. For simplicity, as long as one does not intend to study the limiting behavior as $\varepsilon\to 0$, one may set $u=u^*$, the fixed-point value. Then in place of (A2) one finds the dimensionless quantity

$$\frac{\kappa^\varepsilon C}{L^d}=(\rho^*\kappa L)^\varepsilon(\rho^*)^{-\alpha/\nu}\left\{\frac{1}{4}[\langle x^4\rangle-\langle x^2\rangle^2]-L^{d-4}\frac{\partial^2\Gamma_{R,1}}{\partial T(\rho)^2}\bigg|_*\right.$$
$$\left.+\frac{u^{*2}}{16}\left\{\frac{1}{2}\left[L^{d-2}\frac{\partial^2\Gamma_{R,1}}{\partial T(\rho)}\right]^2-L^{d-4}\frac{\partial^2\Gamma_{R,1}}{\partial T(\rho)^2}\right\}_*\cdot\langle(x^2)^4\rangle_c\cdot(\rho^*\kappa L)^{2\varepsilon}\right\}$$
$$+\frac{S_d}{4\alpha}\left(1+\frac{\varepsilon}{3}\right)(\rho^*)^{-\alpha/\nu} \tag{A4}$$

Once again the symbol * implies derivatives are evaluated at

$$T(\rho^*)=\rho^{*2}t(\rho^*)+\tfrac{1}{2}u^*(\kappa\rho^*)^\varepsilon\langle m^2(\rho^*)\rangle$$

It is straightforward to verify

$$\left[L^{d-2}\frac{\partial\Gamma_{R,1}}{\partial T(\rho)}\right] \quad\text{and}\quad \left[L^{d-4}\frac{\partial^2\Gamma_{R,1}}{\partial T(\rho)^2}\right] \tag{A5}$$

became functions of $L^2\rho^{*2}t(\rho^*)$ alone. Explicitly the averages $\langle x^n \rangle$ are computed according to the probability distribution

$$\exp\left\{-\frac{1}{2}L^2\rho^{*2}t(\rho^*)x^2 - \frac{1}{4!}u^*(\rho^*\kappa L)^\varepsilon x^4\right\} \tag{A6}$$

so that the cumulants also depend only on that combination. Substitution of $\rho^* = (\kappa L)^{-1}$ yields the scaling form given in Eq. (6.5). The explicit form of the scaling function is found numerically using, as in (4.11),

$$\rho^{*2}t(\rho^*)L^2 + \frac{u^*}{2}(\rho^*\kappa L)^\varepsilon \langle x^2 \rangle + 1 = (\rho^*\kappa L)^2 \tag{A7}$$

which implies

$$\rho^* = (\kappa L)^{-1} R(tL^{1/\nu})$$

The "universal" scaling function $X(y)$ for the specific heat has been plotted in Fig. 4. See the text for further discussion.

REFERENCES

1. M. E. Fisher and M. N. Barber, *Phys. Rev. Lett.* **28**:1516 (1972).
2. K. Binder, ed., *Monte Carlo Methods in Statistical Physics* (Springer, New York, 1979), and references therein.
3. K. K. Mon and D. Jasnow, *Phys. Rev. A* **30**:670 (1984).
4. E. Brézin, *J. Phys. (Paris)* **43**:15 (1982).
5. M. N. Barber and M. E. Fisher, *Ann. Phys.* **77**:1 (1973).
6. S. Singh and R. K. Pathria, *Phys. Rev. B* **31**:4483 (1985).
7. J. Rudnick, unpublished.
8. L. Schwartz, *Mathematics for the Physical Sciences* (Addison-Wesley, Reading, Massachusetts, 1966), Chap. 5.
9. Milton Abramowitz and Irene A. Stegun, eds. *Handbook of Mathematical Functions with Formulae, Graphs, and Mathematical Tables* (Dover, New York, 1972).
10. E. Brézin, J. C. LeGuillou, and J. Zinn-Justin, in *Phase Transitions and Critical Phenomena*, Vol. 6, C. Domb and M. S. Green, eds. (Academic, New York, 1976); D. J. Amit *Field Theory*, the *Renormalization Group* and *Critical Phenomena* (McGraw-Hill, New York, 1978).
11. J. Rudnick and D. Nelson, *Phys. Rev. B* **13**:2208 (1976).
12. E. Brézin and J. Zinn-Justin, preprint.

Z. Phys. B – Condensed Matter 61, 299–309 (1985)

Condensed Matter
Zeitschrift für Physik B

Finite Size Critical Behavior for Dirichlet Boundary Conditions

E. Eisenriegler
Institut für Festkörperforschung der Kernforschungsanlage Jülich, Federal Republic of Germany

Received June 7, 1985

The behavior, near the upper critical dimension $d=4$, of finite size properties at bulk criticality for n-vector models is shown to depend *qualitatively* on the type of boundary condition (bc). Contrary to the more complicated behavior which holds for periodic bc's, there exists an $\varepsilon=4-d$ expansion for Dirichlet (or free) bc's with only integer powers in ε. Several universal finite size amplitude ratios are calculated for systems with Dirichlet bc's and spherical shape. This concerns ratios at bulk criticality as well as ratios related to the crossover into the region above the bulk critical temperature where the bulk correlation length is small compared to the diameter of the system and where the same simple type of ε-expansion holds. The general finite size scaling form of the free energy for Dirichlet bc's is compared with conjectures by Privman and Fisher.

1. Introduction

Modification of the critical behavior of an infinite system due to finite sample size is, of course, a subject of considerable importance. The corresponding finite size scaling was originally proposed by Fisher and has been reviewed recently by Barber [1]. Apart from its importance for the behavior of real systems (see e.g. [2]) a detailed understanding of finite size scaling is indispensable for extracting meaningful results from numerical methods, such as Monte Carlo simulations, on finite systems.

Consider a completely finite sample of a ferromagnet with only one characteristic length L, e.g. a sphere with radius L or a cube with edge length L. As has been argued by Privman and Fisher [3] the sample's total free energy $(k_B T)^{-1}\mathscr{F}(H,T,L)$ in units of $k_B T$ shows a leading singular behavior with the form

$$(k_B T_{cb})^{-1}\mathscr{F}^{(s)}(H,T,L)=Y(\tilde{v},\tilde{w}) \tag{1.1a}$$

if $H\to 0$, $T\to T_{cb}$ and $L\to\infty$. Here H is the magnetic field, T_{cb} the critical temperature of the *infinite* (bulk) system and

$$\tilde{v}=\sigma_H\cdot H\cdot L^{\Delta/\nu} \tag{1.1b}$$

$$\tilde{w}=\sigma_T\cdot(T-T_{cb})L^{1/\nu} \tag{1.1c}$$

with Δ and ν the usual exponents of the infinite system. σ_H and σ_T are nonuniveral scales depending on lattice structure, number of interacting neighbors, etc. The function Y is believed to be universal. Apart from the usual dependence on space dimensionality d and spin dimensionality n, Y depends also on the shape of the sample and the type of boundary condition on the surface.

The validity of (1.1) and the explicit form of Y has been mainly investigated for periodic boundary conditions. This concerns most of the Monte Carlo work, see [1], but also important recent analytic work by Brézin and Zinn-Justin [4, 5]. For the more realistic free boundary conditions only few investigations have been performed [6], see especially the Monte Carlo work of Landau [6] on $d=2$ and $d=3$ Ising models on finite lattices with the shape of a square or a cube, respectively.

In the present paper the n-component Ginzburg-Landau model in the interior of a d-dimensional sphere with radius L is investigated. Dirichlet boundary conditions are imposed, i.e. the order parameter vanishes at the surface of the sphere. This system is described by a Hamiltonian

$$\mathscr{H}=\int d^d r\left\{-h\phi_1+\frac{1}{2}(\nabla\phi)^2+\frac{t}{2}\phi^2+\frac{g}{24}(\phi^2)^2\right\} \tag{1.2a}$$

with

$$\phi(\mathbf{r})=0 \quad \text{if } r=L. \tag{1.2b}$$

Here and in the following, space integrals $\int d^d r$ are confined to the sphere $r<L$. h and t are model parameters proportional to H and $T-T_{cb}$, respectively. One can argue [7] that Dirichlet and free boundary conditions should lead to the same function Y in (1.1a).

The starting point of the usual fluctuation- (or loop-) expansion, i.e. mean-field-theory, unphysically predicts a phase transition in a completely finite system, see e.g. [4]. In the case of periodic boundary conditions this transition occurs at the (mean field-) transition temperature of the infinite system which leads [4, 5] to a *singular* dependence on $\varepsilon=4-d$, involving $\varepsilon^{1/2}$, of derivatives of Y in (1.1a) at bulk criticality $h=t=0$.

For Dirichlet boundary conditions, however, the mean field transition occurs at a finite negative value of the mean field counterpart $\mathring{w}$ of $\tilde{w}$ in Eq. (1.1c) [see e.g. (2.16) below]. Therefore at bulk criticality (as well as for $t>0$) the usual loop expansion makes sense and Y displays an ε-expansion which involves only integer powers in ε [8]. This expansion will be used in the present paper to get estimates for Y in $d=3$ if $t\geqq 0$. For $t<0$, however, there exists a region where the usual loop expansion is inapplicable and where an approach similar to the one in [5] would be necessary.

The influence of boundary conditions on the behavior at bulk criticality for $d\lesseqgtr 4$ (as discussed above) can be seen explicitly in the large n limit. In this case the model (1.2) with Dirichlet boundary conditions shows a local susceptibility of the central spin with the form [9]

$$\int d^d r\langle\phi_1(\mathbf{r})\phi_1(0)\rangle=L^2\cdot[d(d-2)]^{-1}. \tag{1.3}$$

The amplitude multiplying L^2 in (1.3) can be expanded in integer powers of $\varepsilon=4-d$. This should be compared with Brézin's result [4] for periodic boundary conditions (for a cube) in the large n limit where the amplitude diverges $\sim\varepsilon^{-1/2}$.

Chapter 2 presents a few mean field properties of the model described by (1.2). In Chap. 3 the finite size scaling properties given in (1.1) are confirmed for Dirichlet boundary conditions by renormalization group arguments. The present formalism also confirms the existence of universal relations [3] between the arguments of the function Y in (1.1a) and the ratio L/ξ with ξ the correlation length in the infinite (bulk) system. These relations can be used to normalize the arguments of the function Y which is then uniquely defined.

As already pointed out by Symanzik [10], counterterms related to the curvature of the sample's surface are necessary in the (field-theoretic) renormalization process. These lead to regular finite size contributions of $\mathscr{F}$, proportional to L^{d-2}. Their existence was already suspected in [3].

In Chap. 4, explicit one-loop expressions are presented and allow for ε-expansion estimates of universal amplitude ratios at bulk criticality ($L/\xi=0$) as well as of ratios related to the crossover to $L/\xi\gg 1$, above T_{cb}. In the latter case one expects that

$$Y\to\tilde{w}^{\nu d}\cdot y_b(\tilde{v}/(\tilde{w}^{\Delta}))+\tilde{w}^{\nu(d-1)}\cdot y_s(\tilde{v}/\tilde{w}^{\Delta})) \tag{1.4}$$

shows bulk ($\sim L^d$) and surface excess ($\sim L^{d-1}$) behavior. Apart from scales, the functions y_b and y_s are known from previous work on bulk systems [11] and on semi-infinite systems with a plane surface (and Dirichlet boundary conditions), respectively. For the latter type of systems see [12] and the reference given in [7]. Note that y_s would vanish for periodic boundary conditions.

2. Mean-Field Approximation

In the symmetric case $h=0$, the order parameter configuration which minimizes $\mathscr{H}$ in (1.2a) and respects the boundary condition (1.2b) is obviously

$$\phi(\mathbf{r})=0 \tag{2.1}$$

if t is positive. This holds even for a certain range of negative t. To see this, one may evaluate the smallest eigenvalue κ_0 of the quadratic part of (1.2a). Denoting the corresponding eigenvector by $\varphi_0(\mathbf{r})$, one finds κ_0 from

$$\tfrac{1}{2}(-\Delta_r+t)\varphi_0(\mathbf{r})=\kappa_0\varphi_0(\mathbf{r}) \tag{2.2}$$

and the boundary condition (1.2b) for φ_0. Since the smallest eigenvalue occurs for a spherically symmetric φ_0, one finds

$$\varphi_0(r)\sim r^{-1+\varepsilon/2}J_{1-\varepsilon/2}([2\kappa_0-t]^{1/2}\cdot r) \tag{2.3}$$

with

$$2\kappa_0=t+[\zeta_1(\varepsilon)/L]^2. \tag{2.4}$$

Here $\zeta_1(\varepsilon)$ is the smallest nonvanishing positive zero of the Bessel function $J_{1-\varepsilon/2}(x)$. Note that $\zeta_1(0)\approx 4$. Since in the symmetric case considered

$$\mathscr{H}\geqq\int d^d r\left\{\kappa_0\phi^2+\frac{g}{24}(\phi^2)^2\right\} \tag{2.5}$$

and $g \geqq 0$, the order parameter configuration (2.1) is the absolute minimum of $\mathscr{H}$ in the range

$$-[\zeta_1(\varepsilon)]^2 < tL^2 < \infty \tag{2.6}$$

(where $\kappa_0 > 0$) and the mean field scaling functions are analytic in this range, see e.g. (2.16) below. This behavior has to be compared with that for periodic boundary conditions, for a cube of length L say, where the mean field scaling functions have a singularity at $tL^2 = 0$.

The propagator of Gaussian fluctuations around the minimum (2.1) (or linear response of the order parameter to a space dependent magnetic field, for $h = 0$ and in mean field approximation) is given by

$$\langle \phi_i(\mathbf{r}) \phi_j(\mathbf{r}') \rangle_0 = \delta_{ij} \mathring{G}(\mathbf{r}, \mathbf{r}') \tag{2.7}$$

where the subscript o on the l.h.s. denotes a thermal average as implied by (1.2) with $h = g = 0$. This propagator will be used later for the loop expansion. It satisfies

$$(-\Delta_{\mathbf{r}} + t) \mathring{G}(\mathbf{r}, \mathbf{r}') = \delta(\mathbf{r} - \mathbf{r}') \tag{2.8a}$$

together with the boundary condition

$$\mathring{G}(\mathbf{r}, \mathbf{r}') = 0 \quad \text{if } r = L \text{ or } r' = L. \tag{2.8b}$$

Using standard procedures, compare e.g. [13], one finds

$$\mathring{G}(\mathbf{r}, \mathbf{r}') = \mathring{G}_b(\mathbf{r} - \mathbf{r}') - \mathring{G}_s(\mathbf{r}, \mathbf{r}'). \tag{2.9}$$

Here

$$\mathring{G}_b = (2\pi)^{-2+\varepsilon/2} |\mathbf{r} - \mathbf{r}'|^{-2+\varepsilon} z^{1-\varepsilon/2} K_{1-\varepsilon/2}(z) \tag{2.10a}$$

is the well-known propagator in infinite space [14] with K a modified Bessel function,

$$z = t^{1/2} \cdot |\mathbf{r} - \mathbf{r}'| \tag{2.10b}$$

and $d = 4 - \varepsilon$. The second contribution in (2.9) is due to the surface and is given by

$$\mathring{G}_s = \tfrac{1}{2} \pi^{-2+\varepsilon/2} \Gamma(1 - \varepsilon/2) (rr')^{-1+\varepsilon/2} \cdot \sum_l \lambda \cdot C_l^{(1-\varepsilon/2)}(\eta) I_\lambda(\vartheta r) I_\lambda(\vartheta r') \cdot K_\lambda(\vartheta L)/I_\lambda(\vartheta L) \tag{2.11a}$$

where l runs over all nonnegative integers 0, 1, 2, 3, ..., Γ denotes the gamma function, I a modified Bessel function and C Gegenbauer- (or ultraspherical) polynomials in the notation of [15] with the argument

$$\eta = (\mathbf{r} \cdot \mathbf{r}')/(rr') \tag{2.11b}$$

and the abbreviations

$$\lambda = l + 1 - \varepsilon/2, \qquad \vartheta = t^{1/2}. \tag{2.11c}$$

We note a few properties of the propagator (2.9). *(i)* for $L \to \infty$ with r, r', η fixed, the propagator $\mathring{G} \to \mathring{G}_b$. *(ii)* For $t \to 0$, equations (2.9)–(2.11) lead to

$$\mathring{G} = \tfrac{1}{4} \pi^{-2+\varepsilon/2} \Gamma(1 - \varepsilon/2) \left\{ |\mathbf{r} - \mathbf{r}'|^{-2+\varepsilon} - L^{-2+\varepsilon} \left[1 - 2\frac{\mathbf{r}\mathbf{r}'}{L^2} + \frac{r^2 r'^2}{L^4} \right]^{-1+\varepsilon/2} \right\} \tag{2.12}$$

The second term in curly brackets is related to the first one basically by a reciprocal radius transformation as known from problems in electrostatics, see the reference given in [9] for more details. *(iii)* Contrary to $\mathring{G}_b$, $\mathring{G}$ is analytic in t at $t = 0$. Using Gegenbauer's addition theorem [15], one notices that $\mathring{G}_b$ can also be written in the form (2.11a) with L replaced by $r_> \equiv \max(r, r')$. This allows for the form

$$\mathring{G} = \tfrac{1}{2} \pi^{-2+\varepsilon/2} \Gamma(1 - \varepsilon/2) (rr')^{-1+\varepsilon/2} \cdot \sum_l \lambda \cdot C_l^{(1-\varepsilon/2)}(\eta) \frac{\pi}{2\sin(\lambda\pi)} \cdot \left[I_\lambda(\vartheta r_<) I_{-\lambda}(\vartheta r_>) - I_\lambda(\vartheta r) I_\lambda(\vartheta r') \frac{I_{-\lambda}(\vartheta L)}{I_\lambda(\vartheta L)} \right] \tag{2.13}$$

with $r_< \equiv \min(r, r')$. Equation (2.13) shows explicitly that the small t-expansion involves only powers t^n with $n = 0, 1, 2, \ldots$. *(iv)* For negative t within the range (2.6), $\mathring{G}$ is also given by (2.13) if I is replaced by J and ϑ is replaced by $(-t)^{1/2}$. Note the appearance of $J_{1-\varepsilon/2}((-t)^{1/2} L)$ in the denominator in square brackets in this equation for $l = 0$ which causes a divergence of $\mathring{G}$ if t approaches the lower limit of the interval given by (2.6).

Finally, we present the mean field expression, $\mathring{\Xi}$, of the total susceptibility which is given generally as

$$\Xi(h, t, L) = \partial_h \langle M \rangle = \langle M^2 \rangle - \langle M \rangle^2 \sim (-) \partial_H^2 \mathscr{F} \tag{2.14a}$$

with

$$M = \int d^d r \, \phi_1(r) \tag{2.14b}$$

and the brackets in (2.14a) denote a thermal average with $\mathscr{H}$ in (1.2). Note that $\langle M \rangle$ vanishes for $h = 0$ and t arbitrary in a completely finite sample. In mean field approximation and for $h = 0$ and t within the range (2.6) one finds for the local susceptibility

$$\mathring{\chi}_{\text{loc}}(r, t, L) = \int d^d r' \, \mathring{G}(\mathbf{r}, \mathbf{r}') = t^{-1} [1 - (L/r)^{1-\varepsilon/2} I_{1-\varepsilon/2}(\vartheta r)/I_{1-\varepsilon/2}(\vartheta L)] \tag{2.15}$$

and for the total susceptibility

$$\mathring{\Xi}(0, t, L) = \int d^d r \, \mathring{\chi}_{\text{loc}} = L^2 V_d \mathring{w}^{-1} \cdot [1 - d \cdot \mathring{w}^{-1/2} I_{d/2}(\mathring{w}^{1/2})/I_{d/2-1}(\mathring{w}^{1/2})]. \tag{2.16}$$

Here V_d is the volume of our spherical sample and

$$\mathring{w}=tL^2 \tag{2.17}$$

is the mean field scaling variable. The square bracket in (2.16) can also be written as $I_{d/2+1}/I_{d/2-1}$. Note that the form of (2.16) is consistent with (1.1a), (2.14a) since $2\Delta/\nu=d+2$ in mean field approximation. Here it is assumed that only the singular part $\mathscr{F}^{(s)}$ contributes in (2.14a). This point will become clearer within the renormalization treatment presented in the next section. The form of (2.16) clearly shows the behavior of the mean field approximation for Dirichlet boundary conditions as mentioned in the Introduction. The divergence of $\mathring{\Xi}$ for $\mathring{w}$ approaching the lower limit of the range (2.6) is unphysical. However, the behavior for larger $\mathring{w}$, e.g. for $\mathring{w}\geqq 0$, is qualitatively correct (and may be systematically improved by a loop expansion together with the renormalization group). Firstly, it is analytic in $\mathring{w}$ as it should, even when $L/\mathring{\xi}=0$, i.e. when $\mathring{w}=0$. Here $\mathring{\xi}$ is the mean field correlation length of the infinite system. Secondly, for $L/\mathring{\xi}\gg 1$, i.e. $\mathring{w}\gg 1$, Eq. (2.16) leads to a form

$$\mathring{\Xi}\rightarrow V_d\mathring{\chi}+S_d\mathring{\chi}_s+(S_d/L)\mathring{\chi}_c+\ldots \tag{2.18}$$

and reproduces correctly the bulk susceptibility

$$\mathring{\chi}_b=t^{-1} \tag{2.19}$$

and the surface excess susceptibility [16]

$$\mathring{\chi}_s=-t^{-3/2} \tag{2.20}$$

of a semi-infinite system with a plane surface.

$$S_d=L^{d-1}\Omega_d, \quad \Omega_d=2\cdot\pi^{d/2}/\Gamma(d/2) \tag{2.21}$$

are the surfaces of our spherical sample and that of the d-dimensional unit sphere, respectively. The third term on the r.h.s. of (2.18) is determined by

$$\mathring{\chi}_c=\frac{d-1}{2}t^{-2} \tag{2.22}$$

and is due to the *curvature* of the surface. The occurrence of such a term is easy to understand if one adopts the picture that for $\xi\ll$ sample diameter the local susceptibility for Dirichlet boundary conditions essentially vanishes within a layer of thickness $\xi'\sim\xi$ at the surface and equals the bulk susceptibility otherwise (see Fig. 1). In the case of a convex surface (Fig. 1a), as in our spherical sample, this leads to a total surface excess susceptibility per surface element, dS, of magnitude

$$\begin{aligned}&-\chi_b(dS/L^{d-1})\int_{L-\xi'}^{L}dr\,r^{d-1}\\&=-\chi_b(dS)\xi'[1-\tfrac{1}{2}(d-1)(\xi'/L)+\ldots].\end{aligned} \tag{2.23}$$

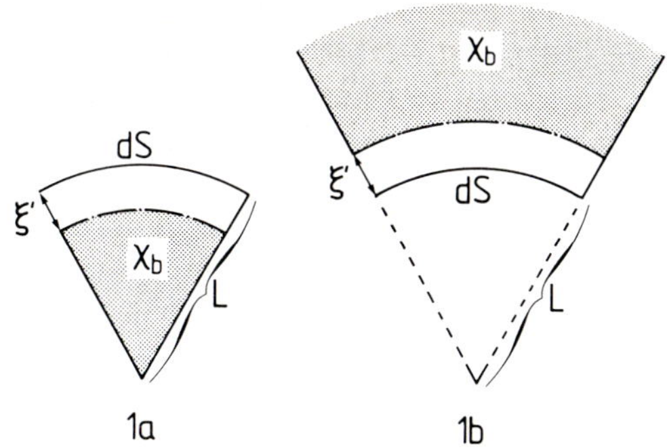

Fig. 1a and b. Convex surface element (**a**) from spherical sample and concave surface element (**b**) from infinite system with spherical hole, with radius of curvature L large compared to the bulk correlation length ξ

The two terms on the r.h.s. of (2.23) indeed show the structure of the second and third terms on the r.h.s. of (2.18). The picture clearly shows a susceptibility enhancement due to surface curvature, for a convex surface. For a concave surface (Fig. 1b) the limits of integration on the l.h.s. of (2.23), $L-\xi'$ and L, have to be changed to L and $L+\xi'$, respectively. This changes the sign of the second term on the r.h.s. of (2.23) and leads to a susceptibility decrease due to surface curvature.

To check this prediction within the mean field approximation, one may consider the situation of a spherical hole in an otherwise infinite system. The propagator $\mathring{G}$ in this case is also given by (2.9)–(2.11), the symbols I and K have to be interchanged in the expression of $\mathring{G}_s$ on the r.h.s. of (2.11a), however. The corresponding local susceptibility, $\mathring{\chi}_{\text{hole}}$, is given by (2.15) with the symbols I replaced by K's. Thus one finds for the total surface excess suceptibility

$$\begin{aligned}\mathring{\Xi}^{(\text{excess})}_{\text{hole}}&=\Omega_d\int_L^{\infty}dr\,r^{d-1}[\mathring{\chi}_{\text{hole}}(r)-\mathring{\chi}_b]\\&=-L^2V_d\,d\,\mathring{w}^{-3/2}K_{d/2}(\mathring{w}^{1/2})/K_{d/2-1}(\mathring{w}^{1/2})\end{aligned} \tag{2.24}$$

which has the form of the second term in (2.16) with I's replaced by K's. Expanding (2.24) for $\mathring{w}\gg 1$, one obtains again the second and third terms in (2.18) with $\mathring{\chi}_s$ and $\mathring{\chi}_c$ from (2.20) and (2.22), apart from a change in sign of the curvature term as expected from the discussion above.

3. Renormalization, Universality and Scaling

Renormalization for field theoretic models with surfaces has been discussed by Diehl and Dietrich [12] (see also Diehl [17]) and by Symanzik [10]. The latter considered also the effect of surface curvature upon renormalization. The reasoning of [10] and [17] can be applied to the Hamiltonian in (1.2) and to the resulting divergencies in Feynman diagrams with the propagator in (2.12). Considering the deviation, δF, of the sample's total free energy from the one at $h=t=0$ (the latter would require a separate investigation), we study a renormalized quantity, with divergencies removed by counterterms, of the form

$$\delta F^R(u, \{Q^R\}, \Lambda) = -\ln\left\langle \exp\int d^d r\left[h\phi_1 - \frac{t}{2}\phi^2\right]\right\rangle_c + \mathcal{H}_{AC}. \tag{3.1}$$

Here $\langle\ \rangle_c$ is a thermal average with $\mathcal{H}$ in (1.2) for $h=t=0$ and Q runs over the set of variables h, t. The relation between g, $\{Q\}$, L and u, $\{Q^R\}$, Λ is

$$g=\mu^{\varepsilon}(4\pi)^{2-\varepsilon/2}Z_u u \tag{3.2a}$$

$$h=\mu^{3-\varepsilon/2}Z_{\phi}^{-1/2}h^R \tag{3.2b}$$

$$t=\mu^2 Z_t t^R \tag{3.2c}$$

$$L=\mu^{-1}\Lambda \tag{3.2d}$$

with the Z-factors from [12]. Note that no Z-factor is associated with L. The quantity $\mathcal{H}_{AC}$ is given by

$$\mathcal{H}_{AC}=\int d^d r\{\mu^d\tfrac{1}{2}(t^R)^2\cdot a+\delta(L-r)\mu^{d-2}L^{-1}t^R\cdot k\}, \tag{3.3}$$

μ is the arbitrary inverse length scale [11] and the quantities a and k have the structure

$$a=\varepsilon^{-1}a_I(u)+\varepsilon^{-2}a_{II}(u)+\ldots,\quad \text{etc.} \tag{3.4}$$

where $a_m(u)$ is independent of ε and starts with u^{m-1}. As in [10, 12] we use dimensional regularization [18]. In connection with (3.1)-(3.4) a few remarks are in order.

(i) The counterterms from the reparametrizations in (3.2) and the counterterm containing a in (3.3) cancel bulk divergencies and are known from previous work [11] on translational invariant systems.

(ii) The present treatment is confined to a discussion of *global* properties (total susceptibility, total specific heat, etc.) which are well defined for a completely finite system. In addition, we are aiming only at the *asymptotic* scaling behavior for free surfaces and use Dirichlet boundary conditions. Therefore, there is no need to introduce space-dependent sources or local surface sources and all surface-counterterms known from semi-infinite systems with a plane surface [10, 12] do not occur. An instructive example for the points *(i)*, *(ii)* is the cancellation of divergencies for the renormalized total susceptibility

$$\begin{aligned}\Xi^R &= -\partial^2\delta F^R/\partial(h^R)^2\\ &=\mu^{6-\varepsilon}Z_{\phi}^{-1}\int d^d r\, d^d r'\\ &\quad\cdot[\langle\phi_1(r)\phi_1(r')\rangle-\langle\phi_1(r)\rangle\langle\phi_1(r')\rangle]\end{aligned} \tag{3.5}$$

which for $h^R=t^R=0$ has been checked explicitly in two loop order [19]. Here $\langle\ \rangle$ denotes a thermal average with $\mathcal{H}$ in (1.2).

(iii) The counterterm containing k in (3.3) has the L-dependence, L^{d-2}, of a surface curvature contribution (as the last term in (2.18)). This term does not occur, indeed, for a plane surface [12]. A calculation of the quantity

$$E^R=\partial\delta F^R/\partial t^R \tag{3.6}$$

as given by (3.1)-(3.3)

$$\begin{aligned}E^R &= \mu^2 Z_t\int d^d r\tfrac{1}{2}\langle\phi^2(\mathbf{r})\rangle\\ &\quad+\mu^d t^R a V_d+\mu^{d-2}kL^{-1}S_d\end{aligned} \tag{3.7}$$

should illustrate this statement. Consider $u=h=t=0$ for simplicity. Since

$$\lim_{\mathbf{r}\to\mathbf{r}'}\mathring{G}_b(\mathbf{r}-\mathbf{r}')=0\qquad \text{if } t=0 \tag{3.8}$$

in dimensional regularization [11], Eq. (2.12) leads to

$$\begin{aligned}\langle\phi^2(\mathbf{r})\rangle &= -(n/4)\pi^{-2+\varepsilon/2}\Gamma(1-\varepsilon/2)\\ &\quad\cdot L^{-2+\varepsilon}[1-(r/L)^2]^{-2+\varepsilon}.\end{aligned} \tag{3.9}$$

Then the first term, $\mu^2 E_0$, on the r.h.s. of (3.7) for $u=h=t=0$ becomes

$$\begin{aligned}\mu^2 E_0 &= -(n/16)(\mu L)^2\pi^{-2+\varepsilon/2}\cdot\Omega_d\\ &\quad\cdot\Gamma(1-\varepsilon/2)\Gamma(2-\varepsilon/2)\Gamma(-1+\varepsilon)/\Gamma(1+\varepsilon/2)\end{aligned} \tag{3.10}$$

with a pole term which can indeed be absorbed by k if one chooses

$$k_I=-n/(4\pi)^{d/2}+O(u). \tag{3.11}$$

$\mu^2 E_0$ contains no pole term with an L-dependence $\sim S_d$, i.e. no pole term contributing to the excess quantity per surface area

$$e_s^R=S_d^{-1}[E^R-V_d e_b^R] \tag{3.12}$$

in the limit of a plane surface as realized for $L\to\infty$ in (3.12). Here e_b^R is the density of E^R in the infinite bulk. The absence of this pole term for a plane surface may also be read off, of course, from the form of the integrand, Eq. (3.9). For a plane surface at $z=0$, e_s^R contains an integrand which follows from (3.9) by letting $L\to\infty$ for $z=L-r$ fixed and has the form

$$\langle \phi^2(\mathbf{r}) \rangle \to -(n/4)\cdot(2\pi)^{-2+\varepsilon/2}\Gamma(1-\varepsilon/2)z^{-2+\varepsilon} \tag{3.13}$$

which is well-known from previous investigations of semi-infinite systems. The corresponding integral in e_s^R extends from $z=0$ to $z=\infty$ and contains no pole term, it actually vanishes in dimensional regularization.

In the asymptotic region where $h^R, t^R \to 0$ and $\Lambda \to \infty$, the quantity δF^R in (3.1) shows a leading singular contribution $\delta F^{(s)}$ of the form

$$\delta F^{(s)} = X(D_h h^R \Lambda^{\Delta/\nu}, D_t t^R \Lambda^{1/\nu}). \tag{3.14}$$

Here $D_{h,t}$ are u-dependent (nonuniveral) scale factors. The function X is independent of u, however. Its explicit form can be obtained most easily from a renormalized perturbation expansion of δF^R at the fixed point value $u=\overset{*}{u}$, see Eq. (B9) below, by means of the relation

$$\begin{aligned}\delta F^R(\overset{*}{u}, h^R, t^R, \Lambda) &= X(h^R\Lambda^{\Delta/\nu}, t^R\Lambda^{1/\nu})\\ &+ (V_d/L^d)[\overset{*}{\bar{a}}_I/(-d+2/\nu)]\tfrac{1}{2}(t^R)^2\Lambda^d\\ &+ \Omega_d[\overset{*}{\bar{k}}_I/(-d+2+1/\nu)]t^R\Lambda^{d-2}.\end{aligned} \tag{3.15}$$

Here $\overset{*}{\bar{a}}_I$, $\overset{*}{\bar{k}}_I$ denote values at $u=\overset{*}{u}$ of

$$(\bar{a}_I, \bar{k}_I) = (1+u\partial_u)(a_I, k_I) \tag{3.16}$$

with a_I, k_I from (3.3) and (3.4). The results (3.14)–(3.16) follow from well-known procedures as outlined in Appendix A.

The last term on the r.h.s. of (3.15) represents a case where an additive renormalization (of the curvature type) causes a regular contribution to the finite size free energy which is not present in the infinite bulk limit. Such a term invalidates the procedure to isolate the singular part of the finite size free energy as proposed in (3.1) of [3]. The existence of a term of this type was already suspected in [3].

We note the form of X in the limit $\Lambda \to \infty$ with $(h^R, t^R) \neq (0,0)$ fixed. Consider e.g. $t^R > 0$. Then the existence of a bulk free energy

$$\delta f_b^R(u, h^R, t^R) = \lim (\mu^d V_d)^{-1}\delta F^R \tag{3.17}$$

and of a surface free energy

$$\begin{aligned}&\delta f_s^R(u, h^R, t^R)\\ &= \lim(\mu^{d-1}S_d)^{-1}[\delta F^R - \mu^d V_d \delta f_b^R]\end{aligned} \tag{3.18}$$

implies a behavior

$$\begin{aligned}X(v,w) &\to (V_d/L^d)w^{\nu d}x_b(v/(w^\Delta))\\ &+ (S_d/L^{d-1})w^{\nu(d-1)}x_s(v/w^\Delta))\end{aligned} \tag{3.19}$$

for $w \to +\infty$ and $v/(w^\Delta)$ fixed. This result is related to (1.4), compare Eqs. (3.30) and (3.32) below. The explicit form of $x_{b,s}$ again follows from calculations at $u=\overset{*}{u}$ by means of

$$\begin{aligned}\delta f_b^R(\overset{*}{u}, h^R, t^R) &= (t^R)^{\nu d}x_b(h^R/(t^R)^\Delta)\\ &+ [\overset{*}{\bar{a}}_I/(-d+2/\nu)]\tfrac{1}{2}(t^R)^2\end{aligned} \tag{3.20}$$

and

$$\delta f_s^R(\overset{*}{u}, h^R, t^R) = (t^R)^{\nu(d-1)}x_s(h^R/(t^R)^\Delta). \tag{3.21}$$

The existence of a curvature contribution

$$\begin{aligned}\delta f_c^R(u, h^R, t^R) &= \lim(\mu^{d-2}S_d/L)^{-1}\\ &\cdot[\delta F^R - \mu^d V_d \delta f_b^R - \mu^{d-1}S_d\delta f_s^R]\end{aligned} \tag{3.22}$$

which is present in the mean field approximation (see (2.18)) and very plausible in general (compare Fig. 1), would imply a further contribution to X in (3.19) of the form

$$\Omega_d w^{\nu(d-2)}x_c(v/(w^\Delta)) \tag{3.23}$$

and

$$\begin{aligned}\delta f_c^R(\overset{*}{u}, h^R, t^R) &= (t^R)^{\nu(d-2)}x_c(h^R/(t^R)^\Delta)\\ &+ [\overset{*}{\bar{k}}_I/(-d+2+1/\nu)]t^R.\end{aligned} \tag{3.24}$$

As argued in [3], the two arguments of X in (3.14) can be expressed in a universal manner by the ratio L/ξ with ξ the correlation length in the infinite bulk. This is particularly easy to see in the present formalism (compare Appendix A) and leads to

$$D_h h^R \Lambda^{\Delta/\nu} = [\overset{*}{\xi}_h]^{\Delta/\nu}\cdot[L/\xi_h]^{\Delta/\nu} \tag{3.25}$$

and

$$D_t t^R \Lambda^{1/\nu} = [\overset{*}{\xi}_+]^{1/\nu}\cdot[L/\xi_+]^{1/\nu}. \tag{3.26}$$

Here ξ_h and ξ_+ denote ξ within the scaling region for $t=0$, h finite and for $h=0$, $t>0$, respectively. The quantities $\overset{*}{\xi}_{h,+}$ are pure numbers. If one defines the correlation length for arbitrary h, t in the usual form

$$[\xi(g,h,t)]^2 = (2d)^{-1}M_2/M_0 \tag{3.27}$$

via moments

$$M_{2j} = \int_\infty d^d r(\mathbf{r}^2)^j G_{c,b}(r) \tag{3.28}$$

of the two point cumulant $G_{c,b}$ in the infinite bulk, one finds e.g.

$$\overset{*}{\xi}_+ = 1 + \frac{\varepsilon}{4}\frac{n+2}{n+8}(1-C_E) \tag{3.29}$$

to first order in ε, with $C_E = 0.577$ the Euler constant.

It is convenient to define the function $\delta Y = Y - Y(0,0)$ with Y of (1.1a) in the particular form

$$\delta Y(\tilde{v}, \tilde{w}) = X([\overset{*}{\xi}_h]^{\Delta/\nu} \cdot \tilde{v}, [\overset{*}{\xi}_+]^{1/\nu} \cdot \tilde{w}). \tag{3.30}$$

Then the leading singular part

$$(k_B T_{cb})^{-1} \delta \mathscr{F}^{(s)}(H, T, L) = \delta F^{(s)} \tag{3.31}$$

of the sample's total free energy $\delta \mathscr{F} = \mathscr{F} - \mathscr{F}(0, T_{cb}, L)$ (in units of $k_B T$) can be expressed in the form

$$(k_B T_{cb})^{-1} \delta \mathscr{F}^{(s)} = \delta Y([L/\xi_h]^{\Delta/\nu}, [L/\xi_+]^{1/\nu}). \tag{3.32}$$

Since δY is a universal function which can be computed in an ε-expansion (see Chap. 4) and since its variables

$$\tilde{v} = [L/\xi_h]^{\Delta/\nu} \sim H \tag{3.33}$$

$$\tilde{w} = [L/\xi_+]^{1/\nu} \sim T - T_{cb} \tag{3.34}$$

are well defined (nonuniversal) multiples of H and $T - T_{cb}$ which can, in principle, be determined quantitatively from bulk properties, Eq. (3.32) makes an explicit prediction for the finite size free energy $\mathscr{F}^{(s)}$ for any system in the universality class.
As a simple application of (3.32) one finds universal ratios at bulk criticality with the form

$$[H/\tilde{v}]^N [(T - T_{cb})/\tilde{w}]^M (k_B T_{cb})^{-1} \mathscr{F}^{(s)}_{N,M}(0, T_{cb}, L) = Y_{N,M}(0,0) \tag{3.35}$$

for any $N + M > 0$. Here $\tilde{v}, \tilde{w}$ are given by (3.33), (3.34). The subscripts N, M denote the number of derivatives with respect to H, T and with respect to $\tilde{v}, \tilde{w}$ in the case of $\mathscr{F}^{(s)}$ and Y, respectively. Note that these derivatives exist for a finite system (compare e.g. Eq. (2.13)). For illustration we consider the case $N = 0$, $M = 2$ more closely. With the usual definition

$$C = -T \partial_T^2 \mathscr{F} \tag{3.36}$$

for the total specific heat, one obtains a finite size ratio

$$\begin{aligned} \mathscr{X}_{fs} &= [(T/T_{cb}) - 1]^2 [\xi_+/L]^{2/\nu} \cdot \alpha \cdot C^{(s)}(0, T_{cb}, L)/k_B \\ &= (\nu d - 2) \cdot Y_{0,2}(0,0) \end{aligned} \tag{3.37}$$

at bulk criticality. Here $\alpha = 2 - \nu d$ is the bulk exponent of the specific heat. $\mathscr{X}_{fs}$ may be compared with a bulk ratio $\mathscr{X}_{SFW}$ introduced in (3a) of [20] by Stauffer, Ferer and Wortis. In our notation

$$\begin{aligned} \mathscr{X}_{SFW} &= (\nu d - 2)(L^d/V_d) \lim_{\tilde{w} \to \infty} \tilde{w}^{2 - \nu d} Y_{0,2}(0, \tilde{w}) \\ &= (\nu d - 2)[\overset{*}{\xi}_+]^d (t^R)^{2 - \nu d} \partial_{tR}^2 (t^R)^{\nu d} x_b(0) \end{aligned} \tag{3.38}$$

with x_b from (3.20).

4. Universal Ratios

We turn now to an explicit evaluation of several universal amplitude ratios of finite size scaling. Apart from specific heat ratios such as $\mathscr{X}_{fs}$ in (3.37), susceptibility ratios are also of interest. We consider in particular

$$\begin{aligned} \mathscr{D} &= (T - T_{cb})(\xi_+/L)^{1/\nu}(-)\mathscr{F}^{(s)}_{2,1}(0, T_{cb}, L)/\mathscr{F}^{(s)}_{2,0}(0, T_{cb}, L) \\ &= -Y_{2,1}(0,0)/Y_{2,0}(0,0), \end{aligned} \tag{4.1}$$

compare Eq. (3.35) and, for $T > T_{cb}$

$$\begin{aligned} \mathscr{A} &= \lim_{L \to \infty} [L/\xi_+]^{\gamma/\nu} \cdot \mathscr{F}^{(s)}_{2,0}(0, T, L)/\mathscr{F}^{(s)}_{2,0}(0, T_{cb}, L) \\ &= \lim_{\tilde{w} \to +\infty} \tilde{w}^{\gamma} Y_{2,0}(0, \tilde{w})/Y_{2,0}(0,0), \end{aligned} \tag{4.2}$$

compare Eqs. (3.19), (3.30). These susceptibility ratios contain information about the temperature derivative at bulk criticality and about the crossover to asymptotic bulk behavior, respectively. Ratios were chosen, where the correlation length ξ_h drops out. Note that $\mathscr{A} \cdot \mathscr{D}^{\gamma}$ is a susceptibility ratio which involves no correlation length at all. The values of both $\mathscr{D}$ and $\mathscr{A}$ can be obtained from the renormalized susceptibility, Eq. (3.5), via

$$\begin{aligned} &Y_{2,0}(0, [\overset{*}{\xi}_+]^{-1/\nu} t^R \Lambda^{1/\nu})/Y_{2,0}(0,0) \\ &= \Xi^R(\overset{*}{u}, 0, t^R, \Lambda)/\Xi^R(\overset{*}{u}, 0, 0, \Lambda) \end{aligned} \tag{4.3}$$

where (3.30) and (3.15) have been used. Similarly the specific heat ratio $\mathscr{X}_{fs}$ in (3.37) follows from $\partial_{tR}^2 F^R$ by means of

$$\begin{aligned} &(\nu d - 2) \cdot Y_{0,2}(0, [\overset{*}{\xi}_+]^{-1/\nu} t^R \Lambda^{1/\nu}) \\ &= [\overset{*}{\xi}_+]^{2/\nu} \Lambda^{-2/\nu} [\Lambda^d \overset{*}{a}_I (V_d/L^d) \\ &\quad + (d - 2/\nu) \partial_{tR}^2 F^R(\overset{*}{u}, 0, t^R, \Lambda)] \end{aligned} \tag{4.4}$$

Here the form of

$$a_I = \frac{n}{(4\pi)^{d/2}} + O(u^2) \tag{4.5}$$

and thus via (3.16) the form of

$$\overset{*}{a}_I = \frac{n}{(4\pi)^{d/2}} + O(\varepsilon^2) \tag{4.6}$$

follows from a calculation of the bulk free energy δf_b^R in (3.20) to two loop order [21].
The quantity Ξ^R in (4.3) is given by

$$\begin{aligned} &\Xi^R(u, 0, t^R, \Lambda) \\ &= \mu^{6-\varepsilon} \left\{ \Xi(0, t, L) - \frac{g}{2} \frac{n+2}{3} \Omega_d \int_0^L dr r^{d-1} [\overset{\circ}{\chi}_{loc}]^2 B \right\} \end{aligned} \tag{4.7}$$

apart from terms of $O(u^2)$. Here $\mathring{\Xi}$ and $\mathring{\chi}_{loc}$ are the quantities introduced in (2.16) and (2.15), respectively, and

$$B = \langle \phi_1^2(r) \rangle_0 = \lim_{\mathbf{r}' \to \mathbf{r}} \{ \mathring{G}(\mathbf{r}, \mathbf{r}') - [\mathring{G}_b(\mathbf{r} - \mathbf{r}')]_{t=0} \} \tag{4.8}$$

with sub- and superscripts o denoting $h = g = 0$ as in Chap. 2. As shown by (3.6) and (3.7), $\partial_{tR}^2 F^R$ in (4.4) to one loop order can be expressed by B in (4.8) as well. With the explicit form of

$$B = B_b - B_s \tag{4.9}$$

according to (2.9), (2.10) and (2.11) where

$$B_b = (4\pi)^{-2+\varepsilon/2} t^{1-\varepsilon/2} \Gamma(-1+\varepsilon/2) \tag{4.10}$$

and

$$B_s = \tfrac{1}{2}\pi^{-2+\varepsilon/2} r^{-2+\varepsilon} [\Gamma(1-\varepsilon/2)/\Gamma(2-\varepsilon)] \cdot \sum_l \lambda \cdot [\Gamma(l+2-\varepsilon)/\Gamma(l+1)] \cdot [I_\lambda(\vartheta r)]^2 K_\lambda(\vartheta L)/I_\lambda(\vartheta L) \tag{4.11}$$

in the notation of (2.11), one may calculate the ratios in (4.1), (4.2) and (3.37). Details are given in Appendix B. The results are

$$\mathscr{D} = \frac{1}{16}\left\{1 + \varepsilon\left[\frac{3}{8} + \frac{n+2}{n+8}\left(\ln 2 - C_E - \pi^2 + \frac{281}{24}\right)\right]\right\} \tag{4.12}$$

and

$$\mathscr{A} \cdot \mathscr{D}^\gamma = \frac{3}{2}\left\{1 - \varepsilon\left[\frac{1}{24} + \frac{n+2}{n+8}\left(\ln 2 + C_E + \pi^2 - \frac{263}{24}\right)\right]\right\} \tag{4.13}$$

as well as

$$\mathscr{X}_{fs} = \frac{n}{64}\left\{1 + \varepsilon\left[\frac{\pi^2}{6} - \frac{5}{2} + \frac{n+2}{n+8}\left(2\ln 2 - 2C_E - \frac{\pi^2}{3} + 8\right)\right]\right\} \tag{4.14}$$

apart from terms of $O(\varepsilon^2)$.

5. Conclusion

In the present paper, the field theoretic methods of [10, 12] have been used to obtain information about finite size critical behavior in systems with Dirichlet (or free) boundary conditions. For systems with spherical shape, some representative universal amplitude ratios defined in (3.37), (4.1) and (4.2) have been calculated in an $\varepsilon = 4 - d$ expansion with results given in (4.14), (4.12) and (4.13), respectively. Other interesting ratios at bulk criticality such as $\mathscr{F}_{0,3}^{(s)} \cdot \mathscr{F}_{2,0}^{(s)} / [\mathscr{F}_{0,2}^{(s)} \cdot \mathscr{F}_{2,1}^{(s)}]$ or [3, 5] $k_B T_{cb} \mathscr{F}_{4,0}^{(s)} / [\mathscr{F}_{2,0}^{(s)}]^2$ could be calculated along similar lines.

The reason for considering systems with spherical shape more closely was twofold. (i) Results at bulk criticality are connected even for $d > 2$ by conformal mapping to corresponding ones in the much investigated semi-infinite geometry. (ii) The spherical geometry is the simplest where effects due to the curvature of the surface can be studied. One finds firstly that curvature leads to an additive counterterm [10] which implies a regular finite size contribution $\sim L^{d-2}$ for the total free energy. Secondly, a curvature contribution to $\mathscr{F}^{(s)}$ in the limit $L/\xi \gg 1$ with the form given by (3.19), (3.23), (3.30) and (3.32) is extremely plausible. Explicit results for this contribution could also be obtained by the present technique and might possibly contribute to a quantitative understanding of the Tolman-effect [22] in nucleation phenomena (in an infinite system) in close proximity of the critical point.

The present technique can be applied also to other geometries, e.g. a film of width L with all other $(d-1)$ dimensions infinite. This geometry is widely used in the context of the phenomenological renormalization group method [23], in $d = 2$. Consider again Dirichlet boundary conditions and denote by $\xi_\parallel(L)$ the correlation length within a plane parallel to and a distance $L/2$ away from the film's surfaces, for values $H = 0$, $T = T_{cb}$ where the bulk $(L = \infty)$ is critical. Then the universality (and L-independence) of the ratio $L/\xi_\parallel(L)$ follows for arbitrary [24] d from arguments (using renormalization group equations) which are completely similar to those in Chap. 3. A calculation of the ε-expansion for $L/\xi_\parallel$ is in progress.

It is a pleasure to thank K. Binder, A. Bringer, T. Burkhardt, and H.W. Diehl for useful discussions.

Appendix A

The first term on the r.h.s. of (3.1) only depends on the four bare parameters g, h, t, L and remains unchanged by varying the five quantities μ, u, h^R, t^R, Λ at fixed bare parameters, compare (3.2). This leads to a renormalization group equation (RGE) for δF^R with the form

$$D \delta F^R = D \mathscr{H}_{AC} \tag{A.1}$$

where

$$D = \beta \frac{\partial}{\partial u} - \sum_Q d_Q \frac{\partial}{\partial \ln Q^R} + \frac{\partial}{\partial \ln \Lambda}. \tag{A.2}$$

The quantities

$$\beta(u) = -\varepsilon/(\partial \ln Z_u/\partial u) \quad \text{(A.3)}$$
$$= -\varepsilon u + f(u) \quad \text{(A.4)}$$

with f independent of ε, and

$$d_Q(u) = d_Q^0 + \eta_Q(u) \quad \text{(A.5)}$$

are well known from previous work [11] on translational invariant systems. Here

$$d_h^0 = 3 - \varepsilon/2, \quad d_t^0 = 2 \quad \text{(A.6)}$$

and

$$\eta_h = \beta \partial \ln Z_\phi^{-1/2}/\partial u \quad \text{(A.7)}$$
$$\eta_t = \beta \partial \ln Z_t/\partial u \quad \text{(A.8)}$$

Equations (3.3), (3.4), (A.2) and (A.4) show that the r.h.s. of (A.1) can be written as

$$D\mathscr{H}_{AC} = -\bar{\mathscr{H}}_{I,AC} \quad \text{(A.9)}$$

since it must not contain pole terms in ε. The quantity $\bar{\mathscr{H}}_{I,AC}$ has the form of $\mathscr{H}_{AC}$ in (3.3) with the pair (a, k) replaced by the pair $(\bar{a}_I, \bar{k}_I)$ which is defined in (3.16).
The RGE given by (A.1), (A.9) and (3.3) is solved by the usual method of characteristics, see e.g. [11]. One introduces running variables $\tilde{u}(l)$, $\tilde{Q}^R(l)$ and $\Lambda \cdot l$ defined by

$$\int_{\tilde{u}(l)}^{u} du'/\beta(u') = -\ln l \quad \text{(A.10)}$$

and

$$\tilde{Q}^R(l) = C_Q(l) Q^R l^{-d_Q^*} \quad \text{(A.11)}$$

with

$$C_Q(l) = \exp \int_{\tilde{u}(l)}^{u} du' [d_Q(u') - d_Q^*]/\beta(u') \quad \text{(A.12)}$$

and $d_Q^* = d_Q(\overset{*}{u})$. Obviously $\tilde{u} = u$, $\tilde{Q}^R = Q^R$ if $l = 1$. In terms of these variables the RGE has the form

$$l \frac{d}{dl} \delta F^R[l] = -\bar{\mathscr{H}}_{I,AC}[l] \quad \text{(A.13)}$$

with the solution

$$\delta F^R[1] = \delta F^R[l] - \int_l^1 dl' (l')^{-1} \bar{\mathscr{H}}_{I,AC}[l]. \quad \text{(A.14)}$$

Here the notation

$$[l] = (\tilde{u}(l), \{\tilde{Q}^R(l)\}, \Lambda \cdot l) \quad \text{(A.15)}$$

was used for the arguments of δF^R and $\bar{\mathscr{H}}_{I,AC}$. With the choice

$$\Lambda \cdot l = 1 \quad \text{(A.16)}$$

for l and in the asymptotic region (where $\Lambda \to \infty$, $l \to 0$ and $\tilde{u} \to \overset{*}{u}$) Eq. (A.14) leads to an asymptotic behavior

$$\delta F^R \to \delta F^{as} \quad \text{(A.17)}$$

with

$$\delta F^{as} = \delta F^R(\overset{*}{u}, v, w, 1) - A_V^{as} - A_c^{as}. \quad \text{(A.18)}$$

Here v, w are the two arguments of the function X on the r.h.s. of (3.14). The two quantities D_Q in v and w are given by the r.h.s. of (A.12) with $\tilde{u}(l)$ replaced by $\overset{*}{u}$. The two exponents in v and w follow from the quantity d_Q in (A.5) via

$$\Delta/v = d_h^*, \quad 1/v = d_t^*. \quad \text{(A.19)}$$

The A's in (A.18) are given by

$$A_V^{as} = (V_d/L^d)[\overset{*}{\bar{a}}_I/(-d + 2d_t^*)]$$
$$\cdot \tfrac{1}{2}\{w^2 - (t^R)^2 \Lambda^d P_V(u)\} \quad \text{(A.20)}$$

and

$$A_c^{as} = \Omega_d[\overset{*}{\bar{k}}/(-d + 2 + d_t^*)] \cdot \{w - t^R \Lambda^{d-2} P_c(u)\}. \quad \text{(A.21)}$$

$P_{V,c}$ are nonuniversal amplitudes. For $u = \overset{*}{u}$

$$D_Q(\overset{*}{u}) = P_{V,c}(\overset{*}{u}) = 1. \quad \text{(A.22)}$$

The r.h.s. of (A.20) and (A.21) are both composed of two sorts of terms. *(i)* A singular term containing $w \sim L^{d_t^*}$. *(ii)* A regular term where the exponents d_Q^*, which contain anomalous dimensions, do not enter. The relative magnitude of *(i)* vs *(ii)* for $L \to \infty$ is of some interest. It depends upon the sign of $-d + 2d_t^*$ and $-d + 2 + d_t^*$ in the cases of A_V and A_c, respectively. In the case of A_V, *(i)* wins against *(ii)* for spin dimensionality $n = 1$ but the situation changes for large n. In the case of A_c, *(i)* dominates *(ii)* for all n. Note that we have neglected terms in (A.20) and (A.21) (as well as in the remaining contributions to the r.h.s. of (A.18)) which are smaller by a factor of $\Lambda^{-\omega}$ than the leading singular terms and whose amplitudes vanish for $u = \overset{*}{u}$. Here ω is the correction to scaling exponent.
The singular contribution $\delta F^{(s)}$ of δF^{as} in (A.18) has the form given in (3.14) with

$$X(v, w) = \delta F^R(\overset{*}{u}, v, w, 1) - A_V^{(s)} - A_c^{(s)} \quad \text{(A.23)}$$

and $A_{V,c}^{(s)}$ denotes the singular part of the quantities in (A.20) and (A.21). The solution of the RGE (A.1),

(A.2), (A.9) for $u=\mathring{u}$ is as given in (3.15) and is consistent with (A.17), (A.18), (A.20) and (A.21) if (A.22) is used.

Finally we make a few remarks concerning the relations (3.25), (3.26) between the variables of X on the r.h.s. of (3.14) and the bulk correlation lengths ξ_h, ξ_+ in the scaling region. One first notices that it does not matter whether bare or renormalized functions $G_{c,b}$ are used in (3.27) and thus

$$\xi^R(u,\{Q^R\})=\mu\cdot\xi(g,\{Q\}) \tag{A.24}$$

is finite for $\varepsilon\to 0$ for u, Q^R fixed. A RGE for the renormalized bulk function $G^R_{c,b}$ leads to the well-known relation

$$\xi^R(u,\{Q^R\})=l^{-1}\,\xi^R(\tilde{u}(l),\{\tilde{Q}^R(l)\}) \tag{A.25}$$

with $\tilde{u}, \tilde{Q}^R$ from (A.10), (A.11). In order to see e.g. the relation (3.26) one sets $h^R=0$ in (A.25) and chooses l such that

$$\tilde{t}^R(l)=1 \tag{A.26}$$

In the asymptotic limit this leads to

$$D_t\cdot t^R=[\xi^R(\mathring{u},0,1)/\xi^R(u,0,t^R)]^{d_t^*} \tag{A.27}$$

and with (3.2d), (A.19), (A.24) leads to the desired relation (3.26). Here we have identified

$$\xi_h=\xi^R(\mathring{u},1,0),\qquad \xi_+=\xi^R(\mathring{u},0,1) \tag{A.28}$$

and $\xi_{h,+}$ in (3.25), (3.26) are appropriate limits of $\xi(g,\{Q\})$ within the scaling region.

Appendix B

The form (2.13) of $\mathring{G}$ shows that the quantity B in (4.8) can be expanded in a Taylor series in t

$$B(r,t,L)=B_0+t\,B_1+t^2B_2+\dots \tag{B.1}$$

with B_i independent of t. One finds

$$B_0=-b\cdot L^{\varepsilon-2}[1-q]^{-2+\varepsilon} \tag{B.2}$$

with

$$q=(r/L)^2,\qquad b=\tfrac{1}{4}\pi^{-2+\varepsilon/2}\Gamma(1-\varepsilon/2) \tag{B.3}$$

and

$$B_1=\tfrac{1}{4}b\cdot r^{\varepsilon}\sum_l\frac{(l+2-\varepsilon)}{\Gamma(2-\varepsilon)\Gamma(l+1)}\cdot\left\{\frac{1}{\lambda+1}-\frac{1}{\lambda-1}+q^{\lambda}\left[q^{-1}\left(\frac{1}{\lambda+1}+\frac{1}{\lambda-1}\right)-\frac{2}{\lambda+1}\right]\right\} \tag{B.4}$$

and a similar expression for B_2. Here $l=0,1,2,\dots$ and λ is given in (2.11c). One finds the ε-expansion

$$B_1=\tfrac{1}{4}bL^{\varepsilon}\{-2\varepsilon^{-1}+q^{-1}[1-q-(2-q^{-1}+q)\ln(1-q)]\}+O(\varepsilon) \tag{B.5}$$

and

$$B_2=-2^{-6}\pi^{-2}L^2q^{-1}(1-q)^2\cdot\left\{q^{-1}+3/2+q^{-2}\ln(1-q)+q^{-1}\int_0^q dp\,p^{-1}\ln(1-p)\right\}+O(\varepsilon) \tag{B.6}$$

From (3.6), (3.7) and (B.4) one may compute $\partial^2_{t^R}F^R$ at $t^R=0$ in $O(u^0,\varepsilon^0)$. It turns out, that the r-integration and the ε expansion of the first term in (3.7) commute to this order, i.e. one may use the expansion (B.5) of the integrand. The result is

$$[\partial^2_{t^R}F^R]_{t^R=0}=n\Lambda^4\cdot\left[-\ln\Lambda+\ln 2-\tfrac{1}{2}C_E-\frac{\pi^2}{6}+\frac{13}{4}\right]/32 \tag{B.7}$$

Together with (4.4), (4.6) this leads to the result (4.14) for the ratio defined in (3.37).

Similarly, the small t^R-expansion of Ξ^R in (4.7) follows from (B.2), (B.5) and (2.13), (2.15) with the result

$$\Xi^R=\Omega_d\Lambda^{6-\varepsilon}\frac{1}{96}\cdot\left\{1+\frac{2}{3}\varepsilon-t^R\Lambda^2\left(1+\frac{25}{24}\varepsilon\right)/16+u\frac{n+2}{3}\cdot\left[\frac{3}{4}+t^R\Lambda^2\left(\ln\Lambda-\ln 2+\frac{1}{2}C_E+\pi^2-\frac{287}{24}\right)/16\right]\right\} \tag{B.8}$$

apart from terms of $O(u^2, u\varepsilon, \varepsilon^2)$. Inserting the fixed point value

$$\mathring{u}=\frac{3}{n+8}\varepsilon+O(\varepsilon^2) \tag{B.9}$$

of u and using (3.5), (3.15) and (3.30), one finds the result (4.12) for the ratio $\mathcal{D}$ defined in (4.1).

For the ratio $\mathcal{A}$ in (4.2) one needs the behavior of $Y_{2,0}$ or Ξ^R in the bulk limit, i.e. for $h^R=0$, $t^R>0$ and $L\to\infty$. Within one loop order one may convince oneself that the explicit form of the expression (4.7) indeed shows a leading and next to leading behavior of the form

$$\Xi^R\to V_d\mu^d\chi_b^R+S_d\mu^{d-1}\chi_s^R \tag{B.10}$$

in this limit, in conformity with (1.4) or (3.19). Here χ_b^R and χ_s^R are the renormalized bulk susceptibility density and excess susceptibility per unit area for a plane surface, respectively, well known from previous work, see [11] and [7, 12, 17]. The explicit form

$$\mu^{-2}\chi_b^R = \overset{\circ}{\chi}_b + \overset{1}{\chi}_b + O(u^2) \qquad \text{(B.11)}$$

with $\overset{\circ}{\chi}_b$ from (2.19) and

$$\overset{1}{\chi}_b = -\frac{g}{2}\frac{n+2}{3} t^{-1-\varepsilon/2}\Gamma(-1+\varepsilon/2)\cdot(4\pi)^{-2+\varepsilon/2} \qquad \text{(B.12)}$$

together with (B.10), (B.8), (4.2) leads to the result (4.13) if use is made of (3.15), (3.30) and (4.12).

References

1. Barber, M.N.: In: Phase transitions and critical phenomena. Domb, C., Lebowitz, J.L. (eds.), Vol. 8. New York: Academic Press 1983
2. Scheibner, B.A., Meadows, M.R., Mockler, R.C., O'Sullivan, W.J.: Phys. Rev. Lett. **43**, 590 (1979); Meadows, M.R., Scheibner, B.A., Mockler, R.C., O'Sullivan, W.J.: Phys. Rev. Lett. **43**, 592 (1979)
3. Privman, V., Fisher, M.E.: Phys. Rev. B **30**, 322 (1984)
4. Brézin, E.: J. Phys. (Paris) **43**, 15 (1982)
5. Brézin, E., Zinn-Justin, J.: Nucl. Phys. B [FS] (to be published)
6. Binder, K., Rauch, H., Wildpaner, V.: J. Phys. Chem. Solids **31**, 391 (1970) Binder, K., Hohenberg, P.C.: Phys. Rev. **B9**, 2194 (1974) Landau, D.P.: Phys. Rev. B **13**, 2997 (1976); Phys. Rev. B **14**, 255 (1976)
7. The argument holds only for free boundary conditions without a surface field and is completely similar to that for surface excess quantities in the case of semi-infinite systems as given in Diehl, H.W., Gompper, G., Speth, W.: Phys. Rev. B **31**, 5841 (1985)
8. An expansion of this type even holds for periodic boundary conditions, provided $tL^{1/\nu} \gtrsim 1$, see Nemirovsky, A.M., Freed, K.F.: J. Phys. A **18**, L319 (1985). However, this expansion breaks down at bulk criticality and is unable to describe crossover properties of the type mentioned in front of (1.4)
9. Note that the result (1.3) is in accord with (1.1) since $(2\Delta/\nu)-d=2$ for $n=\infty$. Equation (1.3) follows from Bray and Moore's $n=\infty$ correlation function for a half space with Dirichlet boundary conditions when conformally mapped unto the spherical geometry of (1.2), see Burkhardt, T., Eisenriegler, E.: J. Phys. A **18**, L83 (1985)
10. Symanzik, K.: Nucl. Phys. B **190** (FS3), 1 (1981)
11. Brézin, E., Le Guillou, J., Zinn-Justin, J.: In: Phase transitions and critical phenomena. Domb, C., Green, M.S. (eds.), Vol. 6. New York: Academic Press 1976 Amit, D.: Field theory, the renormalization group, and critical phenomena. New York: McGraw Hill 1978
12. Diehl, H.W., Dietrich, S.: Z. Phys. B – Condensed Matter **42**, 65 (1981)
13. Sommerfeld, A.: Lectures in theoretical physics. Vol. 6. Leipzig: Akademische Verlagsgesellschaft 1966
14. Ma, S.K.: The modern theory of critical phenomena. New York: Benjamin 1976
15. Abramowitz, M., Stegun, I.: Handbook of mathematical functions. New York: Dover 1965
16. Binder, K.: In: Phase transitions and critical phenomena. Domb, C., Lebowitz, J. (eds.), Vol. 8. New York: Academic Press 1983
17. Diehl, H.W.: J. Appl. Phys. **53**, 7914 (1982)
18. t'Hoofft, G., Veltman, M.: Nucl. Phys. B **44**, 189 (1972)
19. Eisenriegler, E.: Unpublished work
20. Stauffer, D., Ferer, M., Wortis, M.: Phys. Rev. Lett. **29**, 345 (1972)
21. The corresponding ε-expansion of $\mathscr{X}_{\mathrm{SFW}}$, Eq. (3.38), is given in Eq. (24) of Hohenberg, P.C., Aharony, A., Halperin, B.I., Siggia, E.D.: Phys. Rev. B **13**, 2986 (1976)
22. See e.g. Heermann, D.W.: J. Stat. Phys. **29**, 631 (1982). I am grateful to K. Binder for bringing the Tolman-effect to my attention
23. Nightingale, M.P.: J. Appl. Phys. **53**, 7927 (1982)
24. For $d=2$ this ratio has been evaluated in Cardy, J.L.: J. Phys. A **17**, L385 (1984). Compare also Ref. 3 for a conjecture concerning a similar ratio in a cylindrical geometry for arbitrary d

E. Eisenriegler
Institut für Festkörperforschung
Kernforschungsanlage Jülich GmbH
Postfach 1913
D-5170 Jülich 1
Federal Republic of Germany

Z. Phys. B - Condensed Matter 43, 119-140 (1981)

Finite Size Scaling Analysis of Ising Model Block Distribution Functions

K. Binder

Institut für Festkörperforschung, Kernforschungsanlage Jülich, Jülich, Federal Republic of Germany

Received April 25, 1981

The distribution function $P_L(s)$ of the local order parameters in finite blocks of linear dimension L is studied for Ising lattices of dimensionality $d=2,3$ and 4. Apart from the case where the block is a subsystem of an infinite lattice, also the distribution in finite systems with free $[P_L^{(f)}(s)]$ and periodic $[P_L^{(p)}(s)]$ boundary conditions is treated. Above the critical point T_c, these distributions tend for large L towards the same gaussian distribution centered around zero block magnetization, while below T_c these distributions tend towards two gaussians centered at $\pm M$, where M is the spontaneous magnetization appearing in the infinite systems. However, below T_c the wings of the distribution at small $|s|$ are distinctly nongaussian, reflecting two-phase coexistence. Hence the distribution functions can be used to obtain the interface tension between ordered phases.

At criticality, the distribution functions tend for large L towards scaled universal forms, though dependent on the boundary conditions. These scaling functions are estimated from Monte Carlo simulations. For subsystem-blocks, good agreement with previous renormalization group work of Bruce is obtained.

As an application, it is shown that Monte Carlo studies of critical phenomena can be improved in several ways using these distribution functions: *(i)* standard estimates of order parameter, susceptibility, interface tension are improved *(ii)* T_c can be estimated independent of critical exponent estimates *(iii)* A Monte Carlo "renormalization group" similar to Nightingale's phenomenological renormalization is proposed, which yields fairly accurate exponent estimates with rather moderate effort *(iv)* Information on coarse-grained hamiltonians can be gained, which is particularly interesting if the method is extended to more general Hamiltonians.

I. Introduction

In the statistical mechanics of many-body systems, it is a familiar concept to divide the system in "cells" or "blocks" of finite linear dimension L. This concept has been applied to understand phase coexistence in the van der Waals fluid [1], to derive scaling laws between critical exponents [2], and to justify the use of a coarse-grained free energy functional useful for both the description of nucleation and spinodal decomposition [3-5], and as a starting point for the renormalization group theory of critical phenomena [6-9]. For a d-dimensional Ising system, the magnetization s_i of the i'th cell (cf. Fig. 1) can be defined as

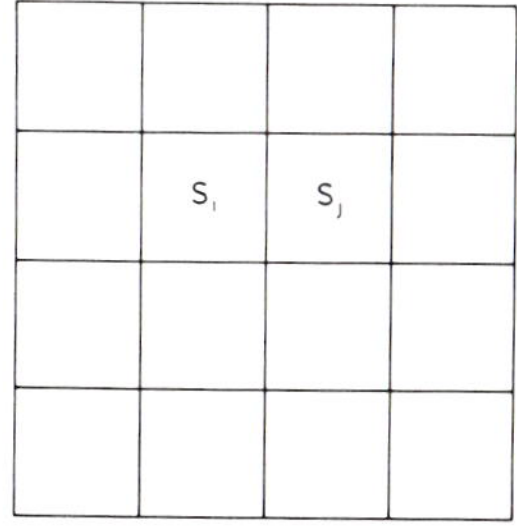

Fig. 1. Division of the system into blocks of linear dimension L, s_i representing the degrees of freedom of the i-th block

0340-224X/81/0043/0119/$04.40

$$s_i = (1/L^d) \sum_{l \in i^{\text{th}} \text{cell}} S_l, \quad S_l = \pm 1. \tag{1}$$

It is then assumed that the Boltzmann factor $(1/Z)\exp[-\mathscr{H}_{\text{Ising}}/k_B T]$, with

$$\mathscr{H}_{\text{Ising}} = -\sum_{l \neq l'} J_{ll'} S_l S_{l'} - H \sum_l S_l, \tag{2}$$

where $J_{ll'}$ are the exchange constants between spins at lattice sites l, l', and H is the magnetic field, is replaced by the probability for the field s_i,

$$P_L(\{s_i\}) = (1/Z)\exp[-\mathscr{H}_{\text{GLW}}(\{s_i\})]. \tag{3}$$

Here factors of $1/k_B T$ are for convenience absorbed in the coefficients of the Ginzburg-Landau-Wilson Hamiltonian $\mathscr{H}_{\text{GLW}}(\{s_i\})$, which usually is expanded as

$$\mathscr{H}_{\text{GLW}}(\{s_i\}) = \sum_i (h_L s_i + r_L s_i^2 + u_L s_i^4 + v_L s_i^6 + \dots) + \sum_{\langle i,j \rangle} C_L (s_i - s_j)^2 + \dots, \tag{4}$$

with $h_L, r_L, u_L, v_L, \dots, C_L, \dots$ appropriate coefficients, and $\langle i,j \rangle$ denotes a summation over nearest-neighbor cells (Fig. 1). In order that an expansion such as (4) is valid, where only low-order terms are kept and the coefficients $r_L, u_L, v_L, \dots, C_L, \dots$ depend on temperature (and other external parameters) in a non-singular way, it is crucial that $L \ll \xi$, the correlation length of order parameter fluctuations [5–8]. Close enough to T_c, where ξ is arbitrarily large, one can choose $L \gg 1$ (measuring lengths in units of the lattice spacing) at the same time, and one can proceed further, replacing (4) by a continuum approximation.

Although the step from (2) to (4) is of crucial conceptual importance, it is hardly ever carried out explicitly. Thus, the parameters $r_L, u_L, v_L, \dots, C_L, \dots$ can not be explicitly related to the parameters of the microscopic hamiltonian (exchange constants $J_{ll'}$ in the present example). Hence the resulting theories of critical behavior (and also of first-order transition kinetics [4]) can predict the universal properties of the system only, information on non-universal properties is lost. In more complicated cases it may happen, that the same type of hamiltonian can exhibit critical behavior belonging to different universality classes, depending on the values of the interaction parameters. Then the renormalization group analysis of the resulting coarse-grained hamiltonian $\mathscr{H}_{\text{GLW}}$ will exhibit several (nontrivial) fixed points [7] - but it may be unclear to which of them a particular given hamiltonian belongs.

While it is hardly possible to perform the above coarse-graining exactly, it is possible to obtain at least explicit numerical results by Monte Carlo methods [9], sampling the distribution function $P_L(\{s_i\})$. As a first step of such a program, the present paper studies the reduced distribution function of one block,

$$P_L(s_i) = \int \prod_{j \neq i} ds_j P_L(\{s_j\}). \tag{5}$$

In addition, we consider the related distribution functions of finited systems of linear dimension L, such as $P_L^{(f)}(s)$ for isolated blocks with free boundary conditions and $P_L^{(p)}(s)$ for isolated blocks with periodic boundary conditions. Section II describes some general results on these functions, including properties for $L \gg \xi$, which is a situation not normally considered in the renormalization group approach. A scaling assumption, which can be justified by the latter [10], is exploited. Section III presents numerical results, and the resulting universal scaling functions are estimated. In addition, it is shown how these distribution functions can be used to obtain improved estimates of order parameter, susceptibility, and interface tension, i.e. quantities often obtained by standard Monte Carlo analysis in a different way. Section IV then proposes to use the distribution functions to construct a phenomenological "Monte Carlo renormalization group (MCRG)", similar to Nightingale's finite size renormalization group [11–13]. As an example, this method is applied to Ising lattices for $d = 2, 3$ and 4 dimensions, and we compare our approach to other versions of MCRG [14–23]. Section V contains our conclusions, and briefly discusses generalizations to other systems.

II. General Properties of the Block Distribution Functions

We consider the situation where the magnetic field $H = 0$. Then $\mathscr{H}_{\text{Ising}}$ (2) is symmetric with respect to a change of sign of all the spins, and hence we obtain a symmetric block distribution function

$$P_L(s) = P_L(-s). \tag{6}$$

Above T_c we then have, from (1)

$$\langle s^2 \rangle_L = L^{-2d} \sum_{l, l' \in i^{\text{th}} \text{cell}} \langle S_l S_{l'} \rangle_T \equiv L^{-d} k_B T \chi_L, \tag{7}$$

where χ_L is an estimate for the susceptibility χ of the total system

$$k_B T \chi = \frac{1}{N} \sum_{l, l'} \langle S_l S_{l'} \rangle_T, \quad N \to \infty, \; T > T_c. \tag{8}$$

It is clear that for values of L much larger than the correlation length ξ of the order-parameter correlation function $\langle S_l S_{l'}\rangle_T$ the difference between χ_L and χ will be small, involving only boundary terms of relative magnitude $\sim L^{-1}$,

$$\chi_L = \chi - \chi_b L^{-1}, \quad L \gg \xi. \tag{9}$$

Of course, similar relations are already familiar for finite systems with periodic or free boundary conditions [24]; for subblocks, however, the significance of χ_b is different.

Next we consider the higher moments and cumulants of the distribution, which we denote as U_L, V_L, etc.,

$$\langle s^k\rangle_L = \int ds\, s^k P_L(s), \quad k=2,4,6,\ldots, \tag{10}$$

$$U_L = 1 - \langle s^4\rangle_L/(3\langle s^2\rangle_L^2), \tag{11}$$

$$V_L = 1 - \langle s^4\rangle_L/(2\langle s^2\rangle_L^2) + \langle s^6\rangle_L/(30\langle s^2\rangle_L^3), \ldots \tag{12}$$

Since

$$\langle s^4\rangle_L = L^{-4d} \sum_{l,l',l'',l'''\in i'\text{th cell}} \langle S_l S_{l'} S_{l''} S_{l'''}\rangle$$
$$= 3\langle s^2\rangle_L^2 + L^{-4d} \sum_{l l' l'' l'''} (\langle S_l S_{l'} S_{l''} S_{l'''}\rangle$$
$$- \langle S_l S_{l'}\rangle \langle S_{l''} S_{l'''}\rangle - \langle S_l S_{l''}\rangle \langle S_{l'} S_{l'''}\rangle$$
$$- \langle S_l S_{l'''}\rangle \langle S_{l'} S_{l''}\rangle) \equiv 3\langle s^2\rangle_L^2 - L^{-3d}(k_B T)\chi_L^{(2)}. \tag{13}$$

One finds that

$$U_L \equiv L^{-d}\chi_L^{(2)}/3\chi_L^2 \xrightarrow[L\to\infty]{} L^{-d}\chi^{(2)}/3\chi^2, \tag{14}$$

where $\chi^{(2)} = \lim_{L\to\infty} \chi_L^{(2)}$ is defined in terms of (13). For large L, thus $U_L \propto L^{-d}$, and similarly one finds that V_L vanishes proportional to L^{-2d} for large L, etc. Away from the critical point, the correlation length ξ is finite, and hence the "susceptibilities" $\chi_L, \chi_L^{(2)}$, etc. tend to finite values for $L\to\infty$. As expected, all cumulants become negligible for large L, and $P_L(s)$ is a gaussian

$$P_L(s) = L^{d/2}(2\pi k_B T\chi_L)^{-1/2} \exp[-s^2 L^d/(2k_B T\chi_L)], \tag{15a}$$

or

$$P_L(s) \approx L^{d/2}(2\pi k_B T\chi)^{-1/2} \exp[-s^2 L^d/(2k_B T\chi)]. \tag{15b}$$

Of course, this result could have been justified at once referring to the central limit theorem of probability theory [25], but here we are interested also in the deviations from gaussian behavior, as measured by (11)–(14).

Below T_c a spontaneous magnetization $\pm M$ appears in the thermodynamic limit, and any state with a spontaneous magnetization does no longer have the symmetry property (6). Rather we must distinguish the probability $P_L^{(+)}(s)$ for finding the block magnetization s in a block L^d for positive spontaneous magnetization from the probability $P_L^{(-)}(s)$ for negative magnetization, and (6) is replaced by

$$P_L^{(+)}(s) = P_L^{(-)}(-s). \tag{16}$$

In the following we are interested in the symmetrized distribution,

$$P_L^{(s)}(s) = [P_L(s) + P_L(-s)]/2, \quad T<T_c, \tag{17}$$

which no longer distinguishes the sign of the magnetization. The reason for this choice is that it can be meaningfully compared to the distributions of finite systems $\{P_L^{(f)}(s)$ and $P_L^{(p)}(s)$, as introduced above$\}$, for which there is no spontaneous magnetization. Thus also for arbitrarily large but finite N (6) still holds, and $P_L(s)$ smoothly develops towards $P_L^{(s)}(s)$ [rather than either $P_L^{(+)}(s)$ or $P_L^{(-)}(s)$] for $N\to\infty$. The block magnetization $\langle|s|\rangle_L$,

$$\langle|s|\rangle = \int_{-\infty}^{+\infty} ds|s| P_L^{(s)}(s) = L^{-d} \sum_{l\in i'\text{th cell}} \langle|S_l|\rangle_T \tag{18}$$

then tends smoothly towards the spontaneous magnetization M as $N\to\infty$, and similar to (7) one can define a susceptibility χ_L,

$$k_B T\chi_L = L^d(\langle s^2\rangle_L - \langle|s|\rangle_L^2) \xrightarrow[L\to\infty]{} k_B T\chi$$
$$= \sum_{i(\neq j)} [\langle S_i S_j\rangle_T - M^2], \quad T<T_c. \tag{19}$$

Thus, although there is never a spontaneous magnetization in a finite system, there is no difficulty in estimating both M and χ for $T<T_c$, as long as $L\gg\xi$. Of course, this fact has been utilized in Monte Carlo calculations for a long time already [26].

From the fact that $k_B T\chi$ for $T<T_c$ is finite, we conclude that the "variance" $\langle s^2\rangle_L - \langle|s|\rangle_L^2 \approx k_B T\chi/L^d$ is very small for $L\gg\xi$, and hence $P_L(s)$ must be sharply peaked at $s\approx\pm\langle|s|\rangle_L$. To leading order, the cumulant U_L then becomes

$$U_L \approx \tfrac{2}{3} - \tfrac{4}{3}\langle|s|\rangle_L^{-2} L^{-d} k_B T\chi_L + O(L^{-2d}), \tag{20}$$

and similarly V_L tends towards $V_\infty = 8/15$.

As a first attempt of constructing $P_L(s)$ for $T<T_c$ explicitly, it is tempting to use two displaced gaussians,

$$P_L(s) \cong \tfrac{1}{2} L^{d/2}(2\pi k_B T\chi_L)^{-1/2}$$
$$\{\exp[-(s-\langle|s|\rangle_L)^2 L^d/(2k_B T\chi_L)]$$
$$+\exp[-(s+\langle|s|\rangle_L)^2 L^d/(2k_B T\chi_L)]\}. \tag{21}$$

It must be realized, however, that (21) represents the

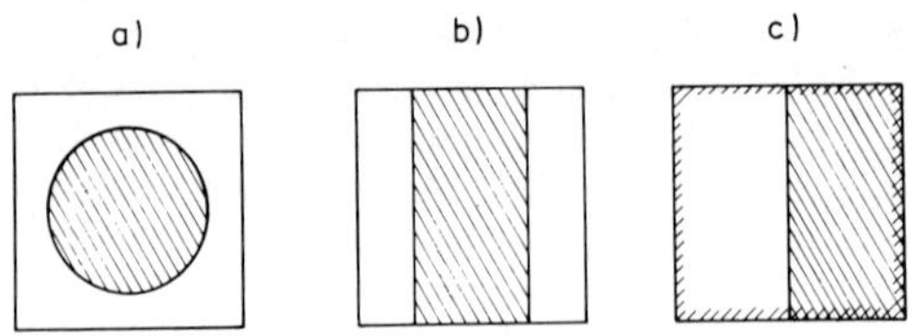

Fig. 2. Typical configurations of twodimensional blocks for $L \gg \xi$, $s \approx 0$, in the cases where the block is a subsystem of a large system **(a)** or where it is an isolated finite system with periodic **(b)** or free **(c)** boundary conditions. Shaded areas indicate domains of negative magnetization, white areas have positive magnetization

distribution reasonably even for $L \gg \xi$ only near the peaks of the distribution, but not in its wings. In the regime $-\langle|s|\rangle_L \ll s \ll +\langle|s|\rangle_L$ (21) underestimates distinctly the actual $P_L(s)$. This fact is appreciated by noting that in this regime $P_L(s)$ will be dominated by configurations corresponding to two-phase coexistence within the cell L^d, Fig. 2. Consider for the moment finite blocks with free or periodic boundary conditions. Their free energy is given by

$$Z_L = \exp[-F_L(H)/k_B T] \equiv Tr \exp[-\mathscr{H}/k_B T] = \int_{-\infty}^{-\infty} ds \exp[-F_L(s)/k_B T + HL^d s/k_B T], \tag{22}$$

where we have allowed for $H \neq 0$, and $F(s)$ is defined by the constrained partition function [27]

$$\exp[-F_L(s)/k_B T] = Tr \exp\left[\frac{J}{k_B T} \sum_{\langle l,l'\rangle} S_l S_{l'}\right] \delta(sL^d - \sum_l S_l). \tag{23}$$

The probability distributions $P_L^{(p,f)}(s)$ hence can be related to the constrained free energy $F(s)$,

$$P_L^{(p,f)}(s) = \exp[-F_L^{(p,f)}(s)/k_B T]/Z_L \tag{24}$$

Also $P_L(s)$ can be associated to a free energy $F_L(s)$ of a finite block which is the subsystem of a large system. The free energies $F_L(s)$, $F_L^{(p)}(s)$ and $F_L^{(f)}(s)$ have their minima at $S_{\max}$ where $P_L(s)$ is maximal, which is in the vicinity of $s = \pm M$, while the free energy cost of having states with $s \approx 0$ is due to interface contributions [27, 28]. For large enough subsystem blocks, the free energy $F_L(s \approx 0)$ will be dominated by the domain configuration which has the minimum interface area, which is a spherical domain. We estimate the radius R of this domain putting

$$L^d/2 = V_d R^d, \tag{25}$$

denoting volume (and surface area) of the d-dimensional unit sphere by V_d (and S_d, respectively). Hence the surface area of the domain in Fig. 2a is estimated as

$$A = S_d \left(\frac{|M-s|}{2MV_d}\right)^{(d-1)/d} L^{d-1}, \tag{26}$$

and thus we estimate the free energy cost of the domain in Fig. 2a (for $H=0$) as

$$F_L(s \approx 0) - F_L(s = s_{\max}) = Af_s = S_d \left(\frac{|M-s|}{2MV_d}\right)^{(d-1)/d} L^{d-1} f_s, \tag{27}$$

where f_s is an appropriate surface tension. Since the interface entropy of a finite spherical droplet is expected to differ from that of an infinite planar interface [28], f_s is itself expected to be weakly dependent on L.

For $F_L^{(p)}(s)$ the minimum free energy excess is again needed for the spherical domain configuration as long as $|s| > s_{\text{crit}}$

$$s_{\text{crit}} = M[1 - 2V_d(2/S_d)^{d/(d-1)}] = \begin{cases} M(1-2/\pi) \approx 0.363\,M & (d=2) \\ M(1 - 4/(3(2\pi)^{1/2})) \approx 0.468\,M & (d=3) \end{cases} \tag{28}$$

while for $|s| < s_{\text{crit}}$ the minimum free energy excess is obtained for a rectangular domain extending throughout the block, making use of the periodic boundary condition. While (27) then again holds for $|s| > s_{\text{crit}}$, for $|s| < s_{\text{crit}}$ it is replaced by [see Fig. 2b]

$$F_L(s \approx 0) - F_L(s - s_{\max}) = 2L^{d-1} f_s^{(p)}, \tag{28}$$

which is independent of s. The surface tension which enters in (28) will again depend weakly on L, since contributions of fluctuations with wavelengths exceeding L are not contributing to it due to the periodic boundary condition. Similarly, the minimum free energy excess for $F^{(f)}(s)$ is again needed for the spherical domain configuration as long as $|s| > s'_{\text{crit}}$,

$$s'_{\text{crit}} = M[1 - 2V_d(1/S_d)^{d/(d-1)}] = \begin{cases} M\left(1 - \dfrac{1}{2\pi}\right) \approx 0.841\,M & (d=2) \\ M\left(1 - \dfrac{2}{3(4\pi)^{1/2}}\right) \approx 0.812\,M & (d=3), \end{cases} \tag{29}$$

while for $|s| < s'_{\text{crit}}$ the minimum free energy excess is obtained by creating just one wall, Fig. 2c, and hence

$$F_L(s \approx 0) - F_L(s = s_{\max}) = L^{d-1} f_s^{(f)}. \tag{30}$$

In this expression, it is assumed that the free surface contributions to $F(s \approx 0)$ and $F(s = s_{\max})$, which would be also of order L^{d-1}, are essentially the same and thus cancel out. We expect that

$$\lim_{L\to\infty} f_s(L) = \lim_{L\to\infty} f_s^{(p)}(L) = \lim_{L\to\infty} f_s^{(f)}(L) = F_s, \tag{31}$$

with F_s being the usual interface tension between two phases of opposite magnetization separated by an infinitely large interface. Equations (24), (31) suggest that the asymptotic decay of $P_L(s)$ for $-M \ll s \ll +M$ is given by $\exp(-\text{const}\,L^{d-1})$ rather than $\exp(-\text{const}\,L^d)$ as suggested by the two-gaussian approximation (21). In addition (31) suggests that $P_L(s)$ can be used to estimate the interface tension F_s.

Next we turn to the behavior of $P_L(s)$ in the immediate vicinity of the critical point. We assume that for $T\to T_c$ and $L\to\infty\, P_L(s)$ satisfies a finite size scaling hypothesis, analogous to usual finite size scaling assumptions [29]

$$P_L(s) = L^x \hat{P}\tilde{P}(a s L^y, \xi/L), \tag{32}$$

where $\hat{P}$ and a are constants setting the scales for the variables, $\tilde{P}(z,z')$ is a universal scaling function, and the exponents x, y will be estimated below [30]. Equation (32) is equivalent to scaling assumptions for multispin-correlation functions, which enter the moments of $P_L(s)$ as discussed above, and which are more familiar from the literature [2, 31]. Although the scaled universal structure of (32) can also be justified from renormalization group arguments [10], doubts in (32) can be raised on the grounds of a possible violation of hyperscaling relations for the three-dimensional Ising model [32, 33]. If one accepts these arguments indicating what might go wrong in applying field theory to the Ising model, one would expect that "weak scaling" theories [34] apply, which "allow for the possibility that one or more additional lengths, such as the width of the interfacial boundary between two coexisting phases, become important in the critical region" [33]. As we have argued above that $P_L(s)$ directly reflects interfacial phenomena, the structure of (32) would be more complicated.

We use here (32) as a working hypothesis and exploit its properties. First we note that the normalization of probability relates the constants $\hat{P}$ and a, as well as the exponents x and y

$$1 \equiv \int_{-\infty}^{+\infty} ds P_L(s) = L^x \hat{P} \int_{-\infty}^{+\infty} ds \tilde{P}(a s L^y, \xi/L)$$
$$= L^{x-y} \frac{\hat{P}}{a} \int_{-\infty}^{+\infty} dz \tilde{P}(z, \xi/L). \tag{33}$$

We conclude $x=y$ and note that the (universal) function $\int_{-\infty}^{+\infty} dz \tilde{P}(z,\xi/L)$ must acutally be a constant, which then also is universal, $C_0 = \int_{-\infty}^{+\infty} dz \tilde{P}(z,\infty)$. Thus

$$\hat{P} = a/C_0. \tag{34}$$

Next we calculate the moments $\langle s^k \rangle_L$ as

$$\langle s^k \rangle_L = L^y \frac{a}{C_0} \int_{-\infty}^{+\infty} ds\, s^k \tilde{P}(a s L^y, \xi/L)$$
$$= L^{-ky}(a^k C_0)^{-1} \int_{-\infty}^{+\infty} dz\, z^k \tilde{P}(z, \xi/L)$$
$$\equiv L^{-ky}(a^k C_0)^{-1} f_k(\xi/L), \tag{35}$$

and thus define (universal) functions $f_k(\xi/L)$. Matching this expression with (7), (9) we conclude that $f_2(\xi/L) = c_2(\xi/L)^{\gamma/\nu}$ for $\xi/L\to 0$, as $\chi \propto \xi^{\gamma/\nu}$ for $T\to T_c$. Thus we find, matching the powers of L,

$$d = \frac{\gamma}{\nu} - 2y, \quad y = \frac{d\nu-\gamma}{2\nu} = \frac{\beta}{\nu}. \tag{36}$$

Here the hyperscaling relation $d\nu = \gamma + 2\beta$ is explicitly used. Next we consider the cumulant

$$U_L = 1 - C_0 \frac{f_4(\xi/L)}{3 f_2^2(\xi/L)} \xrightarrow[T\to T_c]{} U^* \equiv 1 - \frac{C_0 f_4(\infty)}{3 f_2^2(\infty)}, \tag{37}$$

which is a universal function of ξ/L in the critical region and tends to a universal finite constant, which is nontrivial in contrast to the limiting behavior for $L\to\infty$ off T_c. Similar universal properties hold for the higher order cumulants, too.

Comparing (32), (15b) we find that for $\xi/L \ll 1$ $\tilde{P}(z,z')$ takes the form

$$\tilde{P}(z,z') = \frac{C_0}{\sqrt{\pi}} z'^{-\gamma/2\nu} \exp(-z^2 z'^{-\gamma/\nu}), \; z' \to 0, \tag{38}$$

the scale factor a being related to the critical amplitudes $\hat{\chi}^+$ and $\hat{\xi}^+$ of susceptibility and correlation length $[\chi = \hat{\chi}^+(T/T_c - 1)^{-\gamma}, \xi = \hat{\xi}^+(T/T_c - 1)^{-\nu}]$ as

$$a = [(\hat{\xi}^+)^{\gamma/\nu}(2k_B T_c \hat{\chi}^+)^{-1}]^{1/2}. \tag{39}$$

Below T_c a comparison of (32), (21) yields, for $z'\to 0$,

$$\tilde{P}(z,z') = \frac{C_0 A_1}{2\sqrt{\pi}} z'^{-\gamma/2\nu}\{\exp[-A_1^2 z'^{-d}(z z'^{\beta/\nu} - A_2)^2]$$
$$+ \exp[-A_1^2 z'^{-d}(z z'^{\beta/\nu} + A_2)^2]\}, \tag{40}$$

which is valid in the regime where either $|z z'^{\beta/\nu} - A_2| \lesssim z'^{d/2}/A_1$ or $|z z'^{\beta/\nu} + A_2| \lesssim z'^{d/2}/A_1$, respectively. The universal constants A_1, A_2 can again be expressed in terms of critical amplitudes $[\chi = \hat{\chi}^-(1-T/T_c)^{-\gamma}$, $\xi = \hat{\xi}^-(1-T/T_c)^{-\nu}$, $M = \hat{M}(1-T/T_c)^{\beta}]$ as

$$A_1=\sqrt{\frac{\hat{\chi}^+}{\hat{\chi}^-}\left(\frac{\hat{\xi}^+}{\hat{\xi}^-}\right)^{-\gamma/\nu}}, \quad A_2=\hat{M}(\hat{\xi}^-)^{\beta/\nu}a. \tag{41}$$

In the normalization of (21), (40) it is assumed that the wings of the distribution – where these equations are no longer valid due to the phase coexistence phenomena as discussed above – make a negligible contribution to the area under $P_L(s)$, for $z'=\xi/L\to 0$.

Finally we compare (32) to the expression

$$P_L(0)=P_L(s_{\max})$$
$$\exp[-S_d(2V_d)^{-(d-1)/d}L^{d-1}F_s/k_BT],\ \xi/L\to 0, \tag{42}$$

which follows from (24), (27) and where (21) yields $P_L(s_{\max})=L^{d/2}(2\pi k_BT\chi)^{-1/2}/2$. In this limit $\tilde{P}(0,z')$ becomes

$$\tilde{P}(0,z')=\frac{C_0A_1}{2\sqrt{\pi}}z'^{-\gamma/2\nu}\exp[-A_3(z')^{1-d}],\ z'\to 0, \tag{43}$$

the (universal) constant A_3 being related to the critical amplitude of the interface tension $\{F_s/k_BT_c=\hat{F}_s(1-T/T_c)^{(d-1)\nu}\}$ as

$$A_3=\hat{F}_sS_d[2V_d/(\hat{\xi}^-)^d]^{-(d-1)/d}. \tag{44}$$

III. Monte Carlo Results for Block Distribution Functions of Ising Systems at Two, Three- and Four Dimensions

By standard Monte Carlo simulation [9] $P_L(s)$ was obtained for a variety of temperatures in the critical region for Ising square, cubic and hypercubic lattices of linear dimension $N=60\,(d=2)$, $N=24\,(d=3)$ and $N=12\,(d=4)$, respectively. Periodic boundary conditions were used throughout. The possible values of L consistent with these choices of N are $L=2, 3, 4, 5, 6, 10, 12, 15, 20, 30$ ($d=2$); $L=2, 4, 6, 8, 12$ ($d=3$); $L=2, 3, 4, 6$ ($d=4$). For $d=3$, we have also obtained $P_L^{(f)}(s)$ and $P_L^{(p)}(s)$, for all L from $L=2$ to $L=12$. $P_L^{(f)}(s)$ and $P_L^{(p)}(s)$ are less convenient for Monte Carlo study, since each choice of L requires a separate run, while the above set of subsystem-block sizes is generated in one run simultaneously. We feel that a study of subsystem blocks should also be advantageous for standard Monte Carlo finite size scaling analysis [10, 24, 35], where one tries to fit magnetization M_L (and other observables) of finite blocks to scaling forms such as $M_L(T)=\hat{M}(1-T/T_c)^{\beta}\tilde{M}\{L(1-T/T_c)^{\nu}\}$, by simultaneously adjusting the exponents β, ν and T_c such that the set of curves $M_L(T)/(1-T/T_c)^{\beta}$ falls onto a single curve $\tilde{M}$: The analysis of Sect. II shows that a scaling of this type holds for the subsystems of a large system as well, and hence one can obtain all the requested information from one run for a sufficiently large system. Every single spin flip in the system yields an event for all studied block sizes simultaneously. Since the system contains $(N/L)^d$ blocks, the number of Monte Carlo steps (MCS)/spin contributing to $P_L(s)$ is enhanced by the factor $(N/L)^d$ in comparison to that for the total system. Thus a number of about 10^4 MCS/spin for the total system yielded rather satisfactory accuracy, while for $P_L^{(f)}(s)$ and $P_L^{(p)}(s)$ about 10^5-10^6 MCS/spin were needed to reach comparable accuracy. Still the results presented below are based on a total computing effort which is more than an order of magnitude less than what is needed for other simulation purposes (such as polymer studies [36], etc.) – thus it would be feasible to even improve the precision of the results significantly.

Figures 3 and 4 show typical "raw data" of the simulation for $P_L(s)$ [38] at $d=2$ and $d=3$ (Results for $d=4$, as well as results for $P_L^{(p)}(s)$ and $P_L^{(f)}(s)$ are qualitatively similar). Below T_c a distinct peak develops increasing in size with L, as expected. The shape of this peak is rather asymmetric, reflecting the expected structure due to two-phase coexistence for $s<M$. This will be analyzed in more detail below. For $d=2$, a double-peaked structure results even at temperatures slightly above T_c, for not too large L. For temperatures far above T_c, there is a single peak at $s=0$; but even then the distribution does not resemble a gaussian for small L.

Of course, in an Ising system $P_L(s)$ as defined by (1), (3) is defined only at a set of discrete points s_k,

$$s_k=1-2k/L^d, \quad k=0, 1, \ldots, L^d. \tag{45}$$

Thus the curves drawn in Figs. 2, 3 through the values $P_L(s_k)$ are just smooth continuations of the points to guide the eye. Remarkably enough, even for small L very smooth functions result from this continuation, and hence it is clearly reasonable to approximate the actual quasi-continuous $P_L(s_k)$ by a continuous function, as done in Sect. II. In addition, for large L the weight of $P_L(s_k)$ at $s_k=\pm 1$ becomes negligibly small, and then it is legitimate to replace the limits of integration $\int_{-1}^{+1} ds\ldots$ by $\int_{-\infty}^{+\infty} ds\ldots$, as done above.

The insert of Fig. 4b illustrates the use of the block distribution for obtaining a more reliable estimate for the spontaneous magnetization appearing in the thermodynamic limit: One obtains sequences of estimates $\langle|s|\rangle_L, \sqrt{\langle s^2\rangle_L}, \ldots$, as well as $(s_{\max})_L$, which all must extrapolate towards the *same value* M for $L\to\infty$. The fact that such extrapolations yield results

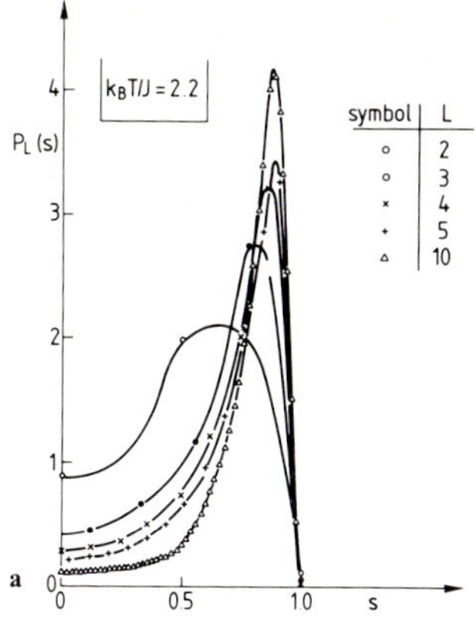

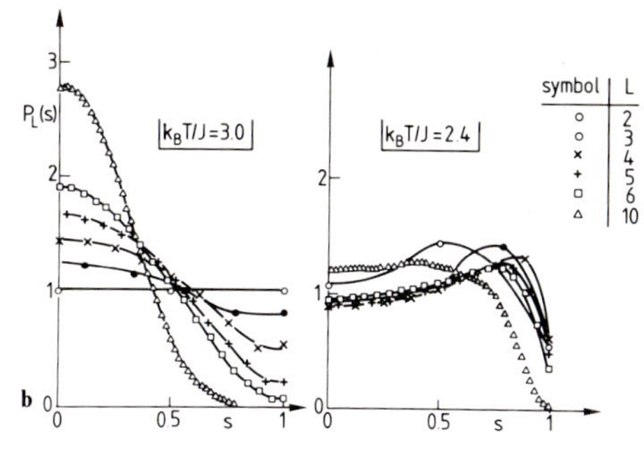

Fig. 3. Block distribution function of the twodimensional Ising square lattice at temperatures below T_c **(a)** and above T_c **(b)** (Note $k_B T_c/J \cong 2.269$ [37])

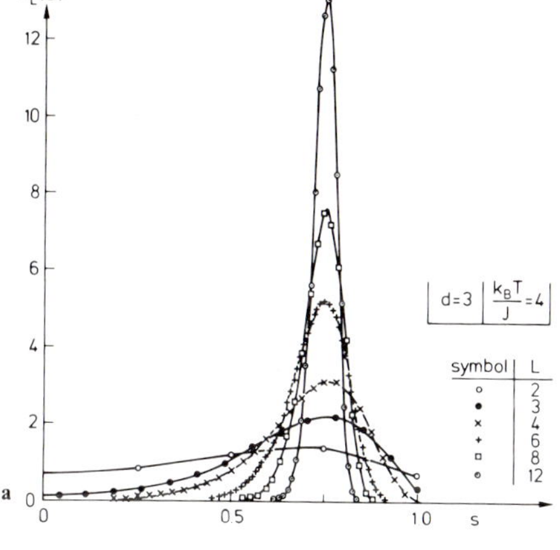

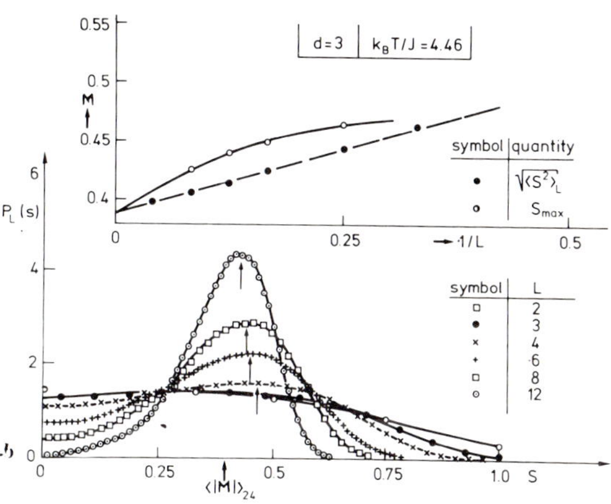

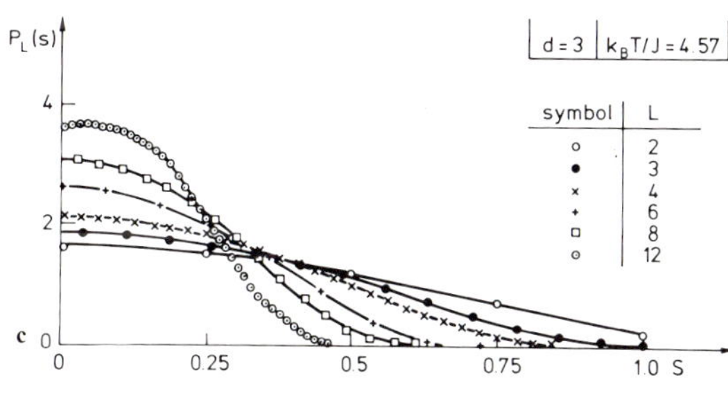

Fig. 4. Block distribution function $P_L(s)$ of the three-dimensional simple cubic Ising lattice at temperatures distinctly below T_c **(a)** slightly below T_c **(b)** (from high temperature series extrapolation, $k_B T_c/J \cong 4.510$ [39] and slightly above T_c **(b)**: arrows in Fig. 4b show estimates for $(s_{max})_L$. Insert of this figure shows an extrapolation yielding the spontaneous magnetization

consistent with each other is an important consistency check on the accuracy of the calculation, and gives an idea of finite size effects also for temperatures farther away from T_c, where a standard finite size scaling analysis is not warranted. We also note that very close to T_c (or above T_c) spurious nonzero estimates for M result from finite lattices, because both $\langle |s| \rangle$, and $\sqrt{\langle s^2 \rangle_L}$ then differ from zero appreciably due to the broad width of the distribution. In this case extrapolation of $(s_{max})_L$ may yield more reliable results.

Figure 5 shows various ways to estimate the susceptibility using the knowledge on the block distribution function. Above T_c, one may either use $\langle s^2 \rangle_L$

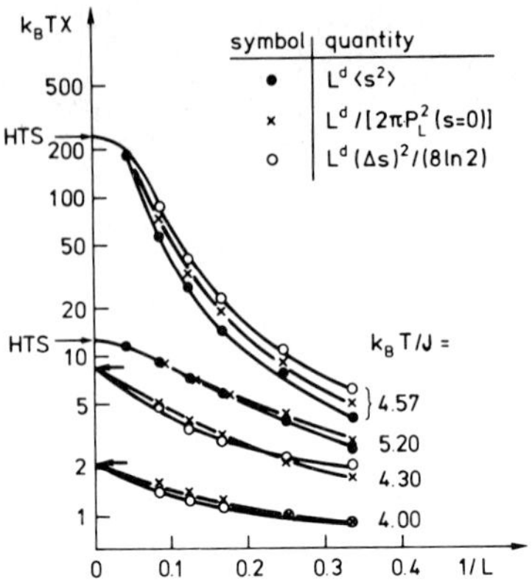

Fig. 5. Various extrapolations of block observables yielding the susceptibility of the three-dimensional Ising model at several temperatures. High temperature series estimates (HTS) [39] are indicated by arrows

(7), (9), $P_L(s=0)$ (15) or the half width Δs of the distribution, which we define from

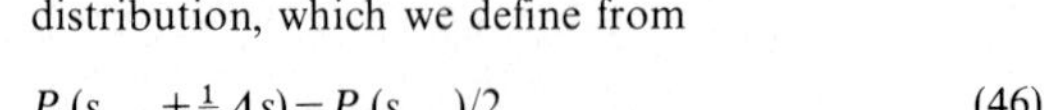

$$P_L(s_{\max} \pm \tfrac{1}{2}\Delta s) = P_L(s_{\max})/2, \tag{46}$$

and use (15) again; below T_c, one may analogously use (19), (20), $P_L(s_{\max})$ (21) or again Δs. One notes from (5) that often various estimates are more or less nicely consistent with each other, though significantly different from the result for χ (known from high temperature series [39]). This fact indicates that the gaussian approximations (15a), (21) are reasonable, but it is important to take the difference between χ_L and χ into account (9), and hence find the true χ from an extrapolation linear in $1/L$, both above and below T_c. Only for the temperature closest to T_c shown here ($k_BT/J = 4.57$), the different methods for estimating χ_L disagree more distinctly, indicating that the distribution is distinctly nongaussian for the values of L considered, and hence here we are in a regime $L \leqq \xi$ rather than $L \gg \xi$ needed for the linear extrapolation (9) to be valid. At this temperature, one hence cannot reliably extrapolate the data to find χ – but this fact is apparent from the behavior of the data themselves.

Figure 6 shows the temperature variation of $P_L(0)$. Since for large enough L $P_L(0)$ must increase with L for $T > T_c$ [Eq. (15)] but decrease with L for $T < T_c$ (42), a (rough) estimate of T_c is obtained from the temperature where these curves intersect. With free boundary conditions, the "effective" T_c of small blocks is shifted to distinctly lower temperatures, as expected [24, 29].

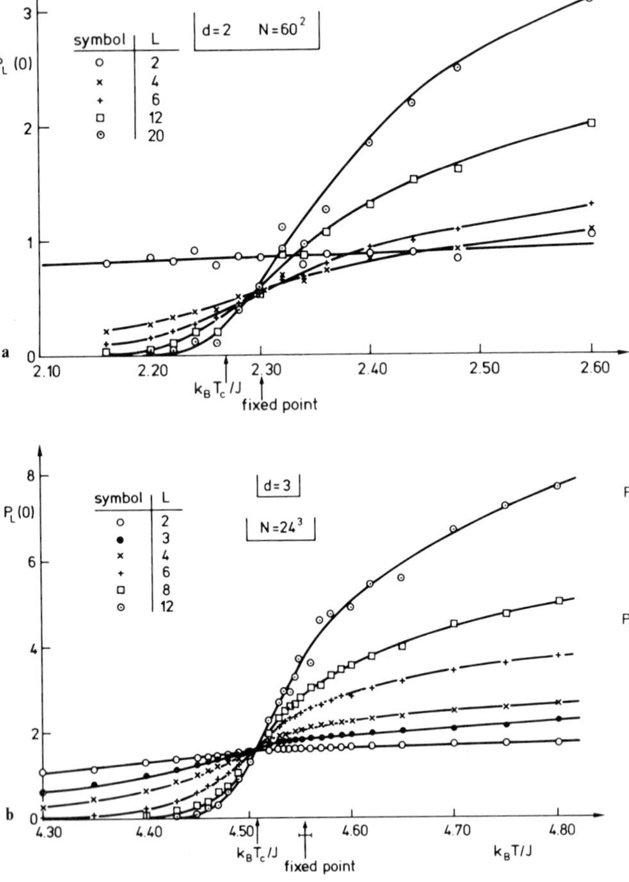

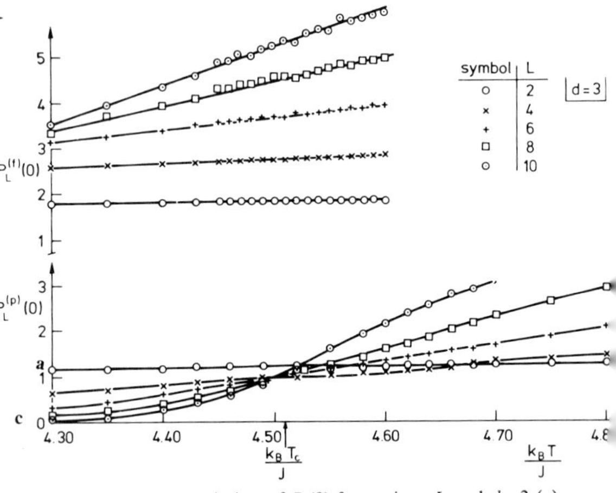

Fig. 6. Temperature variation of $P_L(0)$ for various L and $d=2$ **(a)**, $d=3$ **(b)**, as well as corresponding data for $P_L^{(f)}(0)$ and $P_L^{(p)}(0)$ **(c)**. Estimates for T_c as well as estimates for the temperatures at which $P_L(s)$ exhibits fixed point behavior (Sect. IV) are included

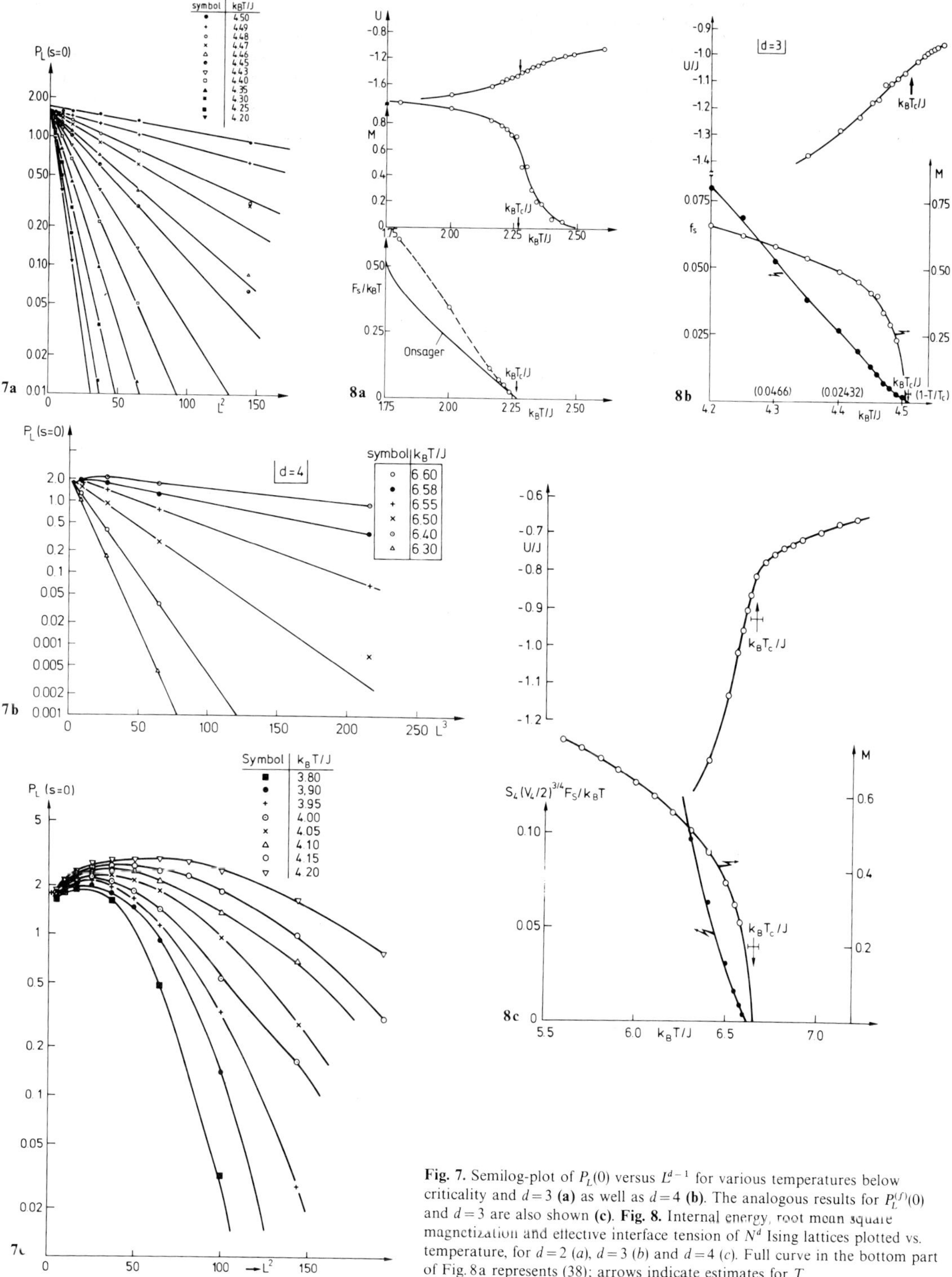

Fig. 7. Semilog-plot of $P_L(0)$ versus L^{d-1} for various temperatures below criticality and $d=3$ **(a)** as well as $d=4$ **(b)**. The analogous results for $P_L^{(f)}(0)$ and $d=3$ are also shown **(c)**. **Fig. 8.** Internal energy, root mean square magnetization and effective interface tension of N^d Ising lattices plotted vs. temperature, for $d=2$ (*a*), $d=3$ (*b*) and $d=4$ (*c*). Full curve in the bottom part of Fig. 8a represents (38); arrows indicate estimates for T_c

For finite systems with free or periodic boundary conditions, $P_L^{(f,p)}(s)$ as well as its moments are analytic at T_c, and this fact is also quite obvious from the data (Fig. 6c). For subsystem-blocks, however, the moment $\langle s^2\rangle_L$ must have an energy-like singularity, since for L fixed it contains short-ranged correlations $\langle s_l s_{l'}\rangle_T$ only, (7), and hence (35) implies

$$\langle s^2\rangle_L = L^{2\beta/\nu}(a^2 C_0)^{-1}$$
$$[f_2(\infty) - g_2(\xi/L)^{-(1-\alpha)/\nu} + - \ldots] \quad (47a)$$
(subblocks)

where g_2 is a constant and α the specific heat exponent, while for free or periodic blocks analyticity in temperature implies

$$\langle s^2\rangle_L^{(f,p)} = L^{2\beta/\nu}(a^2 C_0)^{-1}[f_2^{(f,p)}(\infty)$$
$$-g_2^{(f,p)}(\xi/L)^{-1/\nu} + - \ldots] \quad (47b)$$
(isolated blocks)

This difference, which will turn out to be important for later analysis, shows up in the structure of $P_L(0)$ also, where the rapid variation with temperature in Fig. 6a, b reflects the existence of this energy-like singularity [which is somewhat smoothed out due to the finiteness of N, of course], in contrast to the case of isolated blocks studied in Fig. 6c.

The rapid decay of $P_L(s=0)$ for $T<T_c$ with decreasing temperature and/or increasing L reflects the exponential variation (42). This is seen more clearly in Fig. 7 where $\log P_L(s=0)$ is plotted vs. L^{d-1}. It appears, notably for $d=3$ and $d=4$, that (42) is accurate down to very small L (Fig. 7a, b). The same behavior was found to be true for periodic blocks, while for free blocks the behavior is more complicated due to the interfering free surface contributions and effects due to edges and corners, Fig. 7c.

Fitting straight lines to the points in Fig. 7a, b, one may extract an effective interface tension F_s from their slope. The temperature variation of this effective interface tension is shown in Fig. 8, together with the temperature variation of more common quantities such as (root mean square-) magnetization M or internal energy U of the system. The temperature where F_s vanishes gives again an estimate of T_c, which is of comparable accuracy as the more standard procedures of locating T_c by estimating where M vanishes or where U has an inflection point [10]. For $d=2$ and $d=3$, all these estimates are consistent with the accepted results $k_B T_c/J \cong 2.269$ ($d=2$) [37] and $k_B T_c/J \cong 4.510$ ($d=3$) [39], while for $d=4$ the smallness of N in this case ($N=12$) already leads to an appreciable shift of T_c, our estimate being $k_B T_c/J \cong 6.65$ instead of $k_B T_c/J \cong 6.68$ [40]. In view of this fact, it seems doubtful to us whether previous studies of the $d=4$ Ising which did not take any shift of T_c into account [41] could verify the presence of logarithmic corrections to the mean-field power law behavior [42].

For $d=2$ the "effective interface tension" F_s estimated as described above can be compared with the exact result [37]

$$F_s = 2J - k_B T \ln\{(1+e^{-2J/k_BT})/(1-e^{-2J/k_BT})\}. \quad (48)$$

While the agreement between the data and (48) is reasonable in the critical region, the effective surface tension estimated here is distinctly enhanced in comparison with (48) at lower temperatures. A similar enhancement of the surface tension of small droplets was found in direct simulations of the coexistence between small "critical clusters" and surrounding supersaturated lattice gas [28]. A more detailed analysis of this problem, as well as methods for obtaining reliably the standard interface tension associated with the infinite flat interface between bulk phases will be given elsewhere [43].

We next turn to a check of the finite size scaling assumption (32). We first consider the scaling of $P_L(0)$. Taking $\beta=0.325$, $\nu=0.630$ [8] and $k_B T_c=4.51$ [39] and plotting $P_L(0)L^{-\beta/\nu}$ vs. $|1-T/T_c|^{-\nu}/L$, the set of curves in Fig. 6b should "collapse" to a single curve $\tilde{P}(0, z')$. Figure 9a shows that this holds rather approximately, however, and systematic deviations are apparent. These systematic deviations may be entirely due to the use of too small L. We suspect, however, and this is corroborated by the analysis of Sect. IV, that at least part of the systematic deviation is due to the finiteness of N, which leads to slight shifts of T_c. The data shown in Fig. 9a correspond to the case $\xi/L>1$, and hence corrections ξ/N cannot be completely disregarded. Anticipating the result of Sect. IV that the main result of these corrections is a 1%-shift of the effective T_c from $k_B T_c/J \cong 4.51$ to $k_B T_c^{\text{eff}}/J \cong 4.55$, Fig. 9b shows that now the data points scatter around a single curve, as expected from the scaling hypothesis.

It is also interesting to estimate the function $\tilde{P}(z,\infty)$, since a treatment of Bruce [10] using Wilson' approximate recursion relations [6] predicts pronounced dimensionality effects for $\tilde{P}(z,\infty)$. Figure 10 shows that for $d=3$ our results for $\tilde{P}(z,\infty)$ are not much affected by the uncertainty about T_c^{eff} - temperatures close to T_c do yield all (Fig. 10a, b) roughly the same scaled function, and the agreement with the prediction of Bruce is excellent! For $d=2$, the agreement is not as good (Fig. 10c): although we confirm the main point, that $\tilde{P}(z,\infty)$ has a pronounced double-peak structure, we do not find as

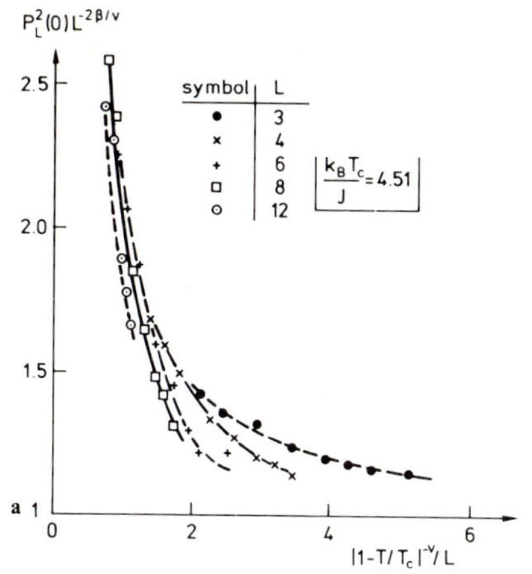

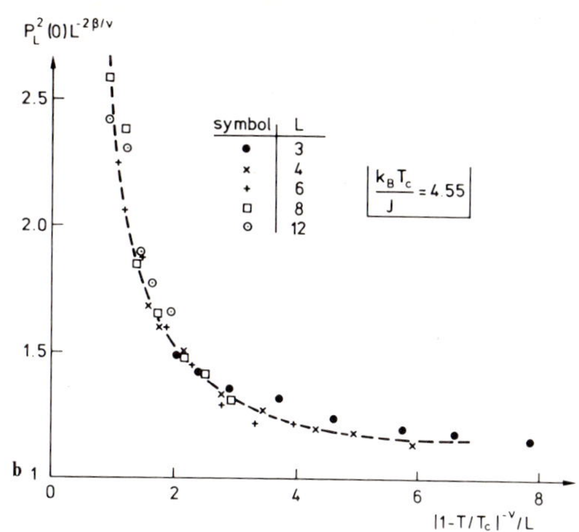

Fig. 9. Finite size scaling plot of $P_L^2(0)L^{-2\beta/\nu}$ versus $|1-T/T_c|^{\nu}/L$, for three-dimensional subblocks using either $k_BT_c/J=4.51$ **(a)** or $k_BT_c^{\mathrm{eff}}/J=4.55$ **(b)**

rich a structure for small z, and also both peak height and position are somewhat different. This discrepancy may be due to one (or more) of the following reasons: *(i)* Due to the finite size of the lattice ($N=60$) and/or finite computing time our results are not yet representative enough for the critical-point behavior of an infinite system. *(ii)* For $d=2$ the methods of Reference 10 are not as reliable as for $d=3$ *(iii)* The block distribution function at the critical point depends on the shape of the block. Reference 10 treats spherical blocks in momentum space, while we treat quadratic blocks in real space.

The last explanation seems the most likely to us, since we find that the *block distribution function at criticality significantly depends on the boundary conditions of the block*, Fig. 11. Due to the infinite correlation length at T_c, the effects due to the boundaries never become negligible, irrespective of the size of the block: thus the limit of the distribution function for $L\to\infty$ for subblocks, free and periodic blocks are all different, and it is plausible that different block shapes would also yield different block distribution functions. Such a behavior is possible since the limits $T\to T_c$ and $L\to\infty$ are not interchangeable. As a result, the "universality" of the finite size-scaling function $\tilde{P}(z,z')$ (32) is true only in a very restricted sense!

IV. Monte Carlo Renormalization Group (MCRG) Based on Block Distribution Functions

There have been a vast number of suggestions to utilize Monte Carlo methods for real-space renormalization group treatments [14–23, 44, 45]. The usual treatment again divides the system into blocks, but rather than allowing the block variable S_i to be quasi-continuous {as it actually is, see Sect. III, if one defines it by (1)}, one introduces a block-variable S_i' which has again Ising-like character, $S_i'=\pm 1$. Thus S_i' is not obtained by a coarse-graining as in (1) but by some more or less arbitrary projection operator $P(\{S_l\},\{S_i'\})$ (majority rule, etc. [4–6]). Thus a new Hamiltonian $\mathscr{H}'(\{S_i'\})$ is constructed,

$$\exp[-\mathscr{H}'(\{S_i'\})/k_BT] = \operatorname*{Tr}_{\{S_l\}} P(\{S_l\},\{S_i'\})\exp[-\mathscr{H}(\{S_l\})/k_BT]. \qquad (49)$$

Taking the trace in (49) one has to consider both the contribution of spins in the same block, as well as spins in different blocks producing an interaction between the blocks. A first step towards a MCRG is to do the first part (trace over spins within one block) by Monte Carlo, and treat the block-block interaction by cumulant expansion [45] or mean-field like approximations [44]. This approach has the merit that it is straightforward to apply to $d=3$ or even higher dimensionality as well to arbitrary spin dimensionality [44], but clearly a meaningful accuracy has not been obtained so far.

More powerful is to consider the block-block interaction exactly for a system containing of two blocks only, where one can proceed to large blocks and perform an extrapolation to infinite block sizes [15]. Related extrapolations have also been performed for rather special renormalization group transformations appropriate for percolation [16] and random walk problems [21, 22].

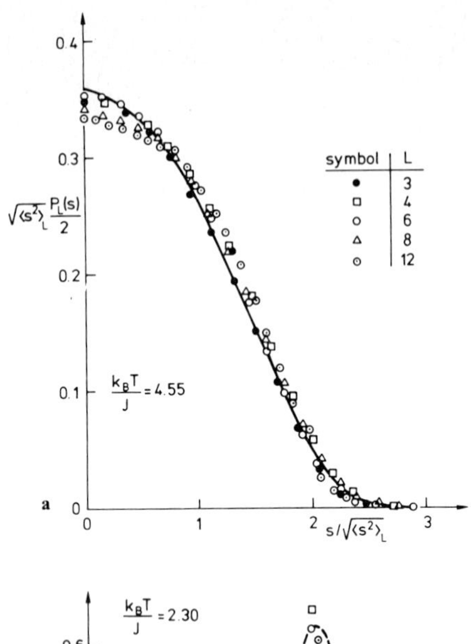

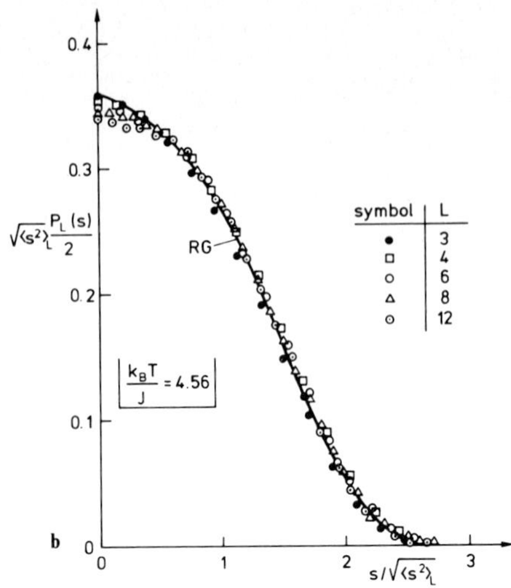

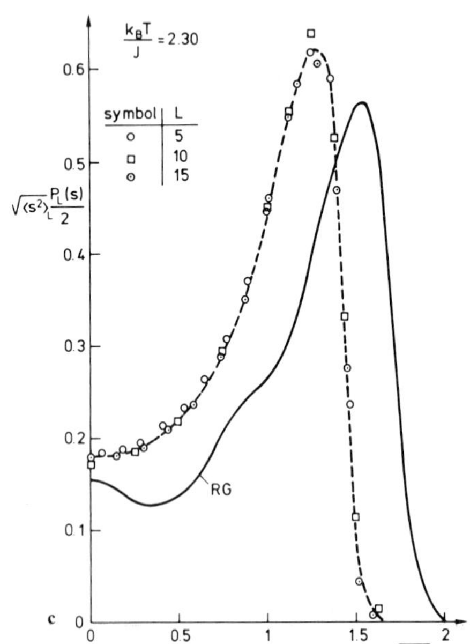

Fig. 10. Scaled block distribution function at the critical point for $d=3$, estimated from data taken for a lattice with 24^3 spins at $k_BT/J=4.55$ **(a)** and $k_BT/J=4.56$ **(b)**, and for $d=2$ **(c)**. Full curves are the renormalization group prediction of [10], broken curve is drawn to guide the eye

The most ambitious approach clearly is due to Ma [14] who tries to fit the fixed-point form of the block spin Hamiltonian, and includes also dynamic critical properties. However, the most successful applications are due to the approach of Swendsen [17–20], where one is concerned directly with the eigenvalues of the linearized renormalization group transformation near the fixed point Hamiltonian rather than with the fixed point Hamiltonian itself. Both approaches [14, 17–20] include all block-spin interactions (compatible with the finite size of the lattice), only in the analysis yielding the eigenvalues there is always some ambiguity concerning which block-spin interactions are included. In addition, the results are somewhat dependent on both the size of the blocks and the projection $P(\{S_i\}, \{S'_i\})$ [17], and hence it is hard to estimate reliably the errors of this procedure – though the errors seem to be encouragingly small [17–20].

Thus there is still some interest in constructing alternative MCRG procedures, particularly since none of the existing real-space renormalization group studies

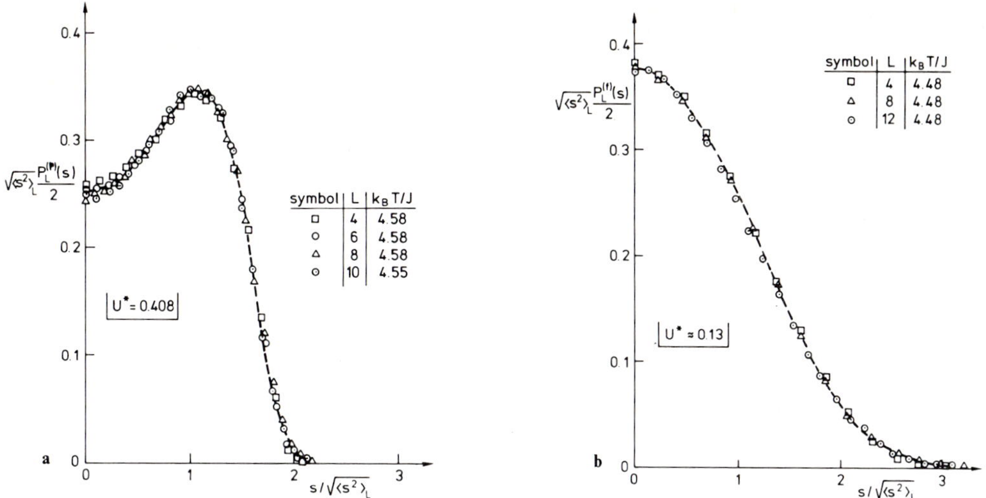

Fig. 11. Scaled block distribution function at criticality for three-dimensional blocks with periodic **(a)** and free **(b)** boundary conditions. The temperatures were chosen such that U_L equals the fixed-point value U^* expected from the treatment of Sect. IV

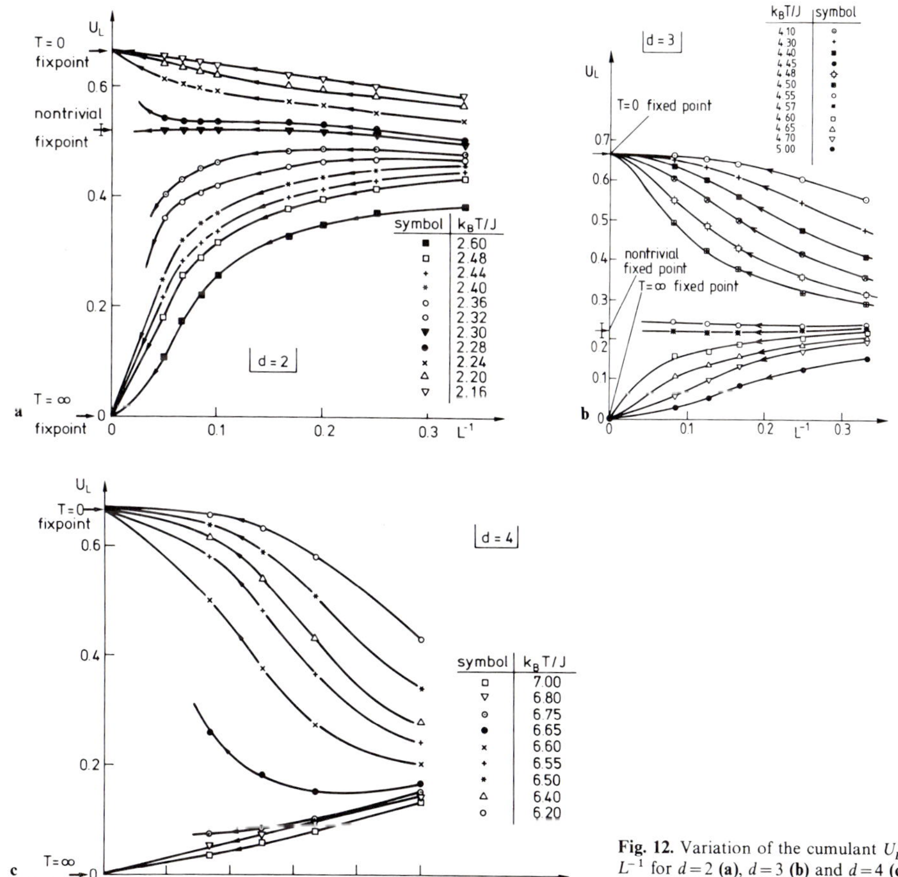

Fig. 12. Variation of the cumulant U_L with inverse block size L^{-1} for $d=2$ **(a)**, $d=3$ **(b)** and $d=4$ **(c)** for various temperatures

allow for the occurrence of truly gaussian fixed points (apart from self-avoiding walk studies [21], where gaussian behavior trivially occurs for the non-self-avoiding walk). In addition, it is not clear that the existing schemes [14–23] are also powerful for systems with continuous degrees of freedom, such as the n-vector model, or even systems belonging to the Ising universality class as liquid-gas systems, or systems with structural phase transitions, which would also be interesting candidates for renormalization group studies [47].

The most complete approach along the lines developed in this paper would be to obtain the set of parameters $\{h_L, u_L, v_L, \ldots, C_L \ldots\}$ introduced in (4) and study their change upon a change of length scale L, and analyze the approach of $\mathscr{H}_{\mathrm{LGW}}$ to a nontrivial fixed point. This ambitious program cannot be based on a study of $P_L(s_i)$ alone, however, it would require studying the joint distribution function of two neighboring blocks i, j

$$P_L(s_i, s_j) = \int \prod_{k \neq i,j} ds_k P_L(\{s_k\}). \tag{50}$$

Of course, also $P_L(s)$ could be represented formally in a structure analogous to (3), (4),

$$P_L(s) = (1/Z') \exp[-\mathscr{H}'(s)],$$
$$\mathscr{H}'(s) = h'_L s + r'_L s^2 + v'_L s^4 + \ldots, \tag{51}$$

where $Z' = \int_{-\infty}^{+\infty} ds \exp[-\mathscr{H}'(s)]$. Thus one can study the change of the parameters $\{h'_L, r'_L, v'_L, \ldots\}$ as a function of L. Although the nontrivial fixed point which is eventually reached when transformations from the length scale L to another length scale L' are performed is not the fixed point of the renormalization group transformation of $\mathscr{H}_{\mathrm{LGW}}$, one nevertheless can obtain the exponents from the behavior near the fixed point in an analogous fashion. Of course, the parameters $\{h'_L, r'_L, u'_L, v'_L, \ldots\}$ are uniquely related to the moments $\langle s^k \rangle_L$ {or cumulants U_L, V_L, etc., see (11), (12)} of the distribution function. While the set $\{h'_L, r'_L, u'_L, v'_L \ldots\}$ could be obtained from $P_L(s)$ only via extensive least-square fitting procedure, quantities such as $\langle s^2 \rangle_L, U_L, V_L, \ldots$ are a direct output of the calculation, and thus more convenient to use.

Hence the variation of quantities such as U_L, V_L, etc. with block size can be interpreted in analogy with a renormalization group "flow diagram", Fig. 12. Upon change of length scale, U_L reaches the trivial fixed point values $U^* = 0$ for temperatures above T_c, as expected (14), and $U^* = 2/3$ for temperatures below T_c (20). At the critical point itself, nontrivial values of U^* are reached for $d = 2$ and $d = 3$, as expected from (37). From Fig. 12c it seems plausible that $U^* = 0$ for $T = T_c$ at $d = 4$, however, as expected for a critical behavior described by a gaussian fixed point. Of course, the smallness of N allowed us to study very small values of L only, and a study of larger systems would be very desirable to establish this point more convincingly.

Clearly, estimating the fixed point from the "flow diagram", Fig. 12, would not be a very accurate procedure. A more systematic method is to study the function $U_{L'} = U_{L'}(U_L)$, Figs. 13a, 14: estimates for the fixed point U^* result from the point where $U_{L'} = U_L$. If our values of L were large enough that corrections to the scaling forms (32), (37) are negligible, the estimates U_L^* for U^* should be independent of L. Due to corrections to scaling there is in fact, a slight dependence on L, which will be analyzed below. Here we draw attention to the fact that the function $U_{L'}(U_L)$ is a very smooth function throughout the critical region; in a wide environment of U_L^* it is well approximated by a straight line (for a global description of $U_{L'}(U_L)$ see Fig. 14a, left part, where all three fixed points are included). Thus rather accurate estimates for U_L^* can be obtained, without any influence of any ambiguity in locating T_c, estimating critical exponents, etc. In fact, plotting then the ratio $U_{L'}/U_L$ as a function of temperature, an estimate for T_c is obtained from the condition

$$(U_{L'}/U_L)_{T=T_c} = 1; \tag{52}$$

the estimate for T_c obtained in this way again is independent of any estimates for the critical exponents. Again we note that $U_{L'}/U_L$ is a rather smooth function of temperature, although one expects singularities similar to those described by (47a). In fact, if we perform the same procedure with free or periodic blocks, for which all moments $\langle s^k \rangle_L$ are analytic functions of temperature (47b), the variation of $U_{L'}/U_L$ with T is somewhat smoother (Fig. 13c).

Rather than studying $U_{L'}(U_L)$, one could study other moments such as $V_{L'}(V_L)$ as well: the scaling hypothesis implies analogous behavior for all moments. Of course, corrections to scaling would lead to systematic discrepancies between the estimates drawn from various moments, if L', L are finite. Actually the values of L', L chosen here are rather small, but Fig. 13c shows that the systematic discrepancy between the estimate $T_c^{(L)}$ for T_c drawn from $(U_{L'}/U_L)_{T_c} = 1$ and from $(V_{L'}/V_L)_{T_c} = 1$ is rather small – it is much smaller than the systematic dependence of $T_c^{(L)}$ on L itself. Thus the scaling form (32) is at least a reasonable approximation for all L down to $L = 3$ in the critical region, as expected from its direct analysis (Figs. 10, 11).

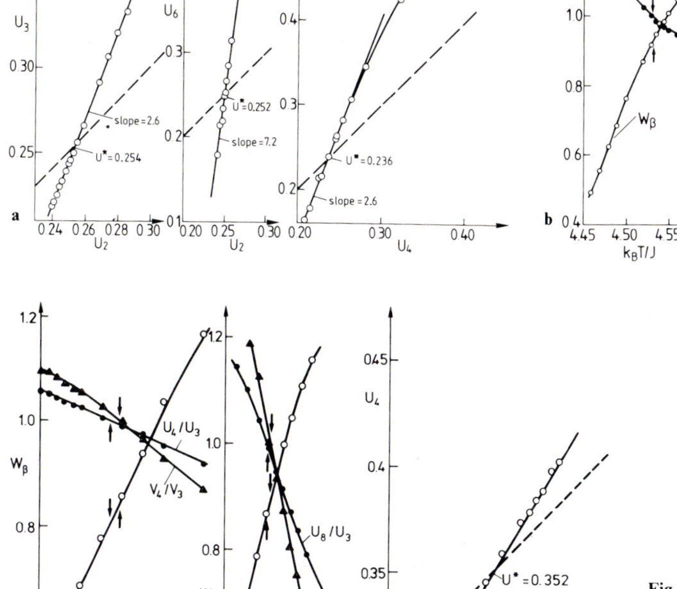

Fig. 13. a $U_{L'}$ plotted vs. U_L for three-dimensional subsystem blocks and various choices of L and L' **b** $U_{L'}/U_L$ and W_β plotted vs. temperature **c** $U_{L'}/U_L$, $V_{L'}/V_L$ and W_β plotted vs. temperature for three-dimensional blocks with periodic boundary conditions. Arrows indicate the resulting estimates for T_c and W_β^*

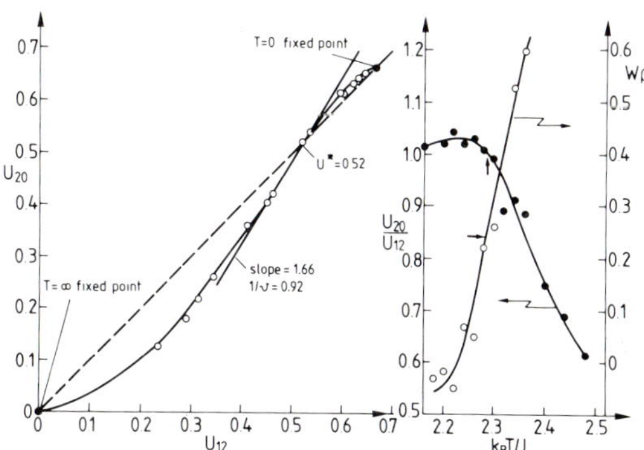

Fig. 14. U_{20} plotted vs. U_{12} for two-dimensional subsystem blocks (left part); U_{20}/U_{12} and W_β plotted vs. temperature (right part)

Estimating critical exponents is readily possible by a procedure very much analogous to Nightingale's phenomenological "finite size renormalization group" [11]. There one considers the scaling relation for the correlation length $\xi_L(K)$ of a strip of width L in one direction and infinite in all other directions $[K \equiv J/T]$

$$\xi_L(K) = b\,\xi_{L/b}(K'). \tag{53}$$

These correlation lengths for models with discrete degrees of freedom (Ising-, Potts models etc.) can be obtained from transfer matrix methods. Equation (53) then yields implicitly the relation $K' = K'(K)$, and from the behavior at the fixed point K^* one obtains exponent estimates (such as $\partial K'/\partial K|K^* = b^{1/\nu}$ [11]). A direct use of this method in conjunction with Monte Carlo is impractical, of course: neither is it convenient to calculate strips which are verly large in one direction, nor is it easy to obtain the correlation length with sufficient accuracy. Rather than using (53) to study $K'(K)$, one can equally well use the function $U_{L'}(U_L)$, however. From (47) we obtain readily

$$U_L = U^*[1 - c_1(\xi/L)^{-(1-\alpha)/\nu} + - \dots], \quad \text{(subsystem blocks)} \quad (54a)$$

and

$$U_L = U^*_{(f,p)}[1 - c_1^{(f,p)}(\xi/L)^{-1/\nu} + - \dots], \quad \text{(isolated blocks)} \quad (54b)$$

where c and $c_1^{(f,p)}$ are constants. Hence in the vicinity of U^* we have

$$\left.\frac{\partial U_{bl}}{\partial U_L}\right|_{U^*} \cong \frac{U_{bl} - U^*}{U_L - U^*} \cong \begin{cases} b^{(1-\alpha)/\nu} & \text{(subsystem blocks)} \\ b^{1/\nu} & \text{(isolated blocks)} \end{cases} \quad (55)$$

Thus the slope of the plot $U_{L'}(U_L)$ at U_L^* directly yields either the exponent $1/\nu$, as in the case of Nightingale's RG [11], or the exponent $(1-\alpha)/\nu = d - 1/\nu$. The latter value is a direct consequence of the singular temperature dependence of correlation functions of finite subsystem blocks near T_c, (47a).

From (35) we immediately can find the second critical exponent involved, by defining a function W_β:

$$W_\beta = -\ln[\langle s^2\rangle_{bL}/\langle s^2\rangle_L]/\ln b. \quad (56)$$

Since at T_c the dependence of $\langle s^2\rangle$ on L has a simple power-law form, $\langle s^2\rangle_L = L^{-2\beta/\nu}\frac{f_2(\infty)}{a C_0}$, the value W_β^* of W_β at the fixed point is an estimate for the exponent $2\beta/\nu$. Again the function W_β is a rather smooth function of U_L or temperature, Figs. 13, 14, and hence fairly accurate estimates for this exponent can be read off from the graphs. Note that no simultaneous fit to T_c, β and ν to the data is involved, as in the traditional finite size scaling analysis of Monte Carlo data [9, 24, 35].

From the examples shown in Figs. 13, 14 it is obvious that these exponent estimates are affected by some systematic dependences on L and scale factor $b = L'/L$, as expected since (32) should be true only asymptotically for $L \to \infty$. For obtaining accurate results, *corrections to scaling* in (32), (35), (47) etc. must be taken into account. Thus, (35) is replaced by

$$\langle s^k\rangle_L = \frac{L^{-k\beta/\nu}}{a^k C_0} f_k\left(\frac{\xi}{L}\right)\left\{1 + L^{-w} f_{k,c}\left(\frac{\xi}{L}\right) + \dots\right\}, \quad (57)$$

where w is an exponent describing corrections to scaling, and $f_{k,c}(\xi/L)$ is an associate scaling function. As a result, W_β^* becomes more complicated,

$$\begin{aligned} W_\beta^* &= \frac{2\beta}{\nu} - \frac{1}{\ln b}\ln\{[1 + (bL)^{-w} f_{k,c}(\infty)]/[1 + L^{-w} f_{k,c}(\infty)]\} \\ &\approx \frac{2\beta}{\nu} - \frac{L^{-w} f_{k,c}(\infty)}{\ln b}(b^{-w} - 1). \end{aligned} \quad (58)$$

In order that the exponent $2\beta/\nu$ is not biased by any assumption about the exponent w, it is advisable to extrapolate W_β^* as a function of $(\ln b)^{-1}$, treating L as a parameter: (58) implies then that W_β^* should be linear in $(\ln b)^{-1}$ for large b, and the intercept for $(\ln b)^{-1}$ should also be independent of L. Figure 15a shows that this method gives consistent results: a unique value for $(2\beta/\nu)$ is in fact obtained, within our accuracy the extrapolation yields a result independent of L,

$$2\beta/\nu \approx 1.03 \pm 0.01 \quad (59)$$

The systematic dependence on $(\ln b)$ and L can be described by (58) reasonably well, as the upper part of Fig. 15a shows: there a function $(2\beta/\nu)_{\text{corr}}$ is defined by adding a term $L^{-w} f_{k,c}(\infty)(b^{-w} - 1)/\ln b$ as a correction term to W_β^* and fitting the parameters $w, f_{k,c}(\infty)$ such that the points "collapse" on a horizontal straight line. Unfortunately the fitted parameters do not seem to have much physical significance, as we obtain $f_{2,c}(\infty) \approx 0.35$, $w \approx 1.8$, which is an unexpectedly large value for the correction exponent (corresponding to small correction effects, a fortunate fact for our analysis). What we think happens is that for such small L as used here there are several correction terms simultaneously coming into play, whose net effect fortunately cancels to some extent, and hence is fitted by an unreasonably large exponent. We feel, however, that this fact does not invalidate our estimate (59) above, as one expects W_β^* to behave rather smoothly as function of $(\ln b)^{-1}$ for large b. Equation (59) is in very good agreement with the estimates of field theoretic renormalization [8], while the conjectured rational value $2\beta/\nu = 1$ ($\beta = 5/16$, $\nu = 5/8$) seems less likely, and the traditional value due to high temperature series expansions, $2\beta/\nu \approx 0.98$ [39], seems to be clearly ruled out. Our findings thus provide further evidence that the discrepancy between field-theoretic renormalization [8] and high-temperature series [39] should not be taken as an evidence that the discrete Ising model and continuum S^4-field theory belong to different universality classes, as suggested by Baker [32, 33]. We think that the quoted high temperature series result is unreliable due to too short series; recent series reanalysis are not inconsistent with this conjecture, although the accuracy of the series expansion estimates is still somewhat controversial [48, 49].

From (58) we note that the correct exponent $2\beta/\nu$ can be obtained either by extrapolation $\ln b \to \infty$ (as done here) or by extrapolation $L \to \infty$, $\ln b \to 0$, since then $(b^{-w} - 1)/\ln b = (e^{-w \ln b} - 1)/\ln b \approx -w$. If we hence choose, following Nightingale [11], $b = (L+1)/L$,

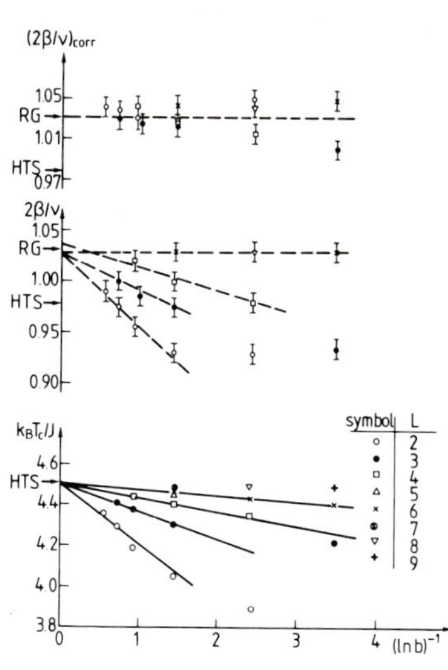

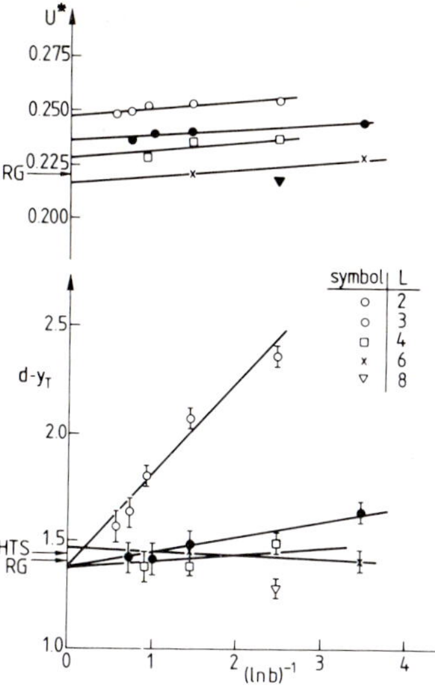

Fig. 15. a Estimates for $2\beta/\nu$ for three-dimensional subsystem blocks plotted vs. inverse logarithm of scale factor (middle part); estimates for $2\beta/\nu$ modified by fitting a correction term, cf. text (upper part); estimates for $k_B T_c/J$ for isolated blocks with free boundary conditions (lower part). **b** Estimates for U^* (upper part) and $(1-\alpha)/\nu = d - y_T$ for three-dimensional subsystem blocks (lower part)

one finds $W_\beta^* \approx \frac{2\beta}{\nu} + w f_{k,c}(\infty) L^{-w}$, i.e. the correction decreases in power-law form, and hence with increasing L the sequence of estimates W_β^* would be quickly convergent. We did not attempt to follow this method here, since very precise estimates for $\langle s_k \rangle_L$ are required to obtain W_β accurately enough.

From Figs. 15, 16, 17a it is seen that a systematic dependence on both scale factor b and linear dimension L not only affects the estimates for the exponents β/ν and $y_T = 1/\nu$ {or $d - y_T = (1-\alpha)/\nu$, respectively}, but the estimates for U^* and $k_B T_c/J$ as well, This fact is readily understood, generalizing (54) by including a correction term

$$U_L \approx U^*[1 - c_1(\xi/L)^{-(1-\alpha)/\nu} + L^{-w} f_{u,c}(\infty) + \dots], \quad \text{(subsystem blocks)}, \qquad (60\text{a})$$

$$U_L \approx U^*_{(f,p)}[1 - c_1^{(f,p)}(\xi/L)^{-1/\nu} + L^{-w} f_{u,c}^{(f,p)}(\infty) + \dots], \quad \text{(isolated blocks)} \qquad (60\text{b})$$

where the scaling function $f_{u,c}(\xi/L)$ involved in the correction term is approximated by $f_{u,c}(\infty)$ close to T_c. Now the fixed-point condition $U_L^* = U_{bL}^*$ is *not* satisfied for $\xi = \infty$, which would yield the bulk T_c, but rather the fixed-point condition is satisfied for a finite value of the correlation length ξ given by

$$\xi^{-(1-\alpha)\nu} = \frac{f_{u,c}(\infty)}{c_1} L^{-w-(1-\alpha)/\nu} \frac{b^{-1}-1}{\left(b^{\frac{1-\alpha}{\nu}} - 1\right)}, \quad \text{(subsystem blocks)}, \qquad (61\text{a})$$

or

$$\xi^{-1/\nu} = f_{u,c}^{(f,p)} L^{-w-1/\nu} \frac{b^{-w}-1}{b^{1/\nu}-1}, \quad \text{(isolated blocks)}. \qquad (61\text{b})$$

From (61) one easily obtains the expression for the shift ΔT_c between our estimate and the true T_c, remembering $\Delta T = T_c \xi_1^{1/\nu} \xi^{-1/\nu}$. From (61) we recognize the following facts: *(i)* $\Delta T_c \to 0$ for large b; for the values of b chosen in Figs. 15a, 16a one may, very roughly, approximate $b^{1/\nu} - 1$ by $\frac{1}{\nu} \ln b$, and hence the extrapolation of ΔT_c vs. $(\ln b)^{-1}$ is roughly linear; *(ii)* the shift decreases drastically with increasing L, this fact again being clearly borne out by the data.

By the same argument one can understand why the estimates for U^* (Figs. 15b, 16a) or V^* (Fig. 16b) do *not* converge to their asymptotic value independent of L, when being extrapolated vs. $(\ln b)^{-1}$. Equations (60), (61) imply instead

$$U_L^* = U^*\left[1 + L^{-w} f_{u,c}(\infty) \frac{1 - b^{-w-(1-\alpha)/\nu}}{1 - b^{-(1-\alpha)/\nu}}\right]_{b\to\infty} \to U^*[1 + L^{-w} f_{u,c}(\infty)], \quad \text{(subsystem blocks)}, \qquad (62\text{a})$$

or

$$U_L^* = U^*_{(f,p)}\left[1 + L^{-w} f_{u,c}^{(f,p)}(\infty) \frac{1 - b^{-w-1/\nu}}{1 - b^{-1/\nu}}\right]_{b\to\infty} \to U^*_{(f,p)}[1 + L^{-w} f_{u,c}^{(f,p)}(\infty)], \quad \text{(isolated blocks)}. \qquad (62\text{b})$$

Since a quantitative fit of the above expressions to data involving values of L as small as $L = 2, 3$ (and

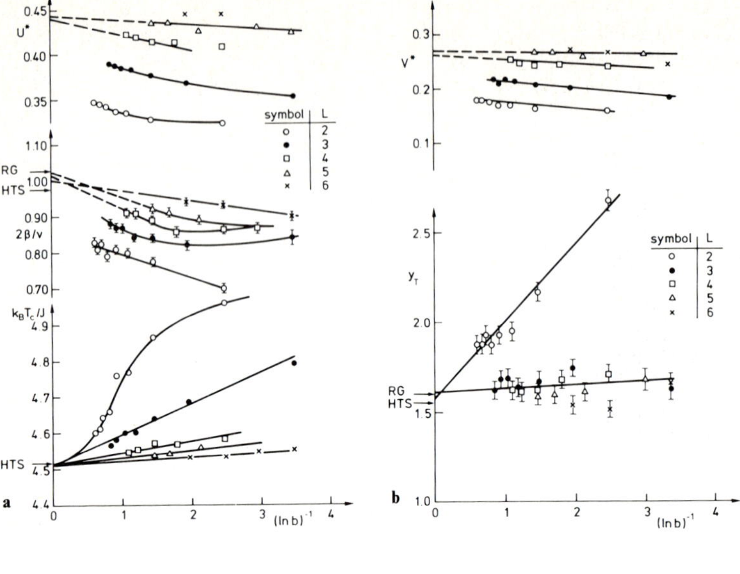

Fig. 16. a Estimates for U^* (upper part), $2\beta/\nu$ (middle part) and $k_B T_c/J$ (lower part) for isolated three-dimensional blocks with periodic boundary conditions: **b** Estimates for V^* (upper part) and $y_T = 1/\nu$ (lower part) for isolated three-dimensional blocks with periodic boundary conditions plotted vs. inverse logarithm of the scale factor

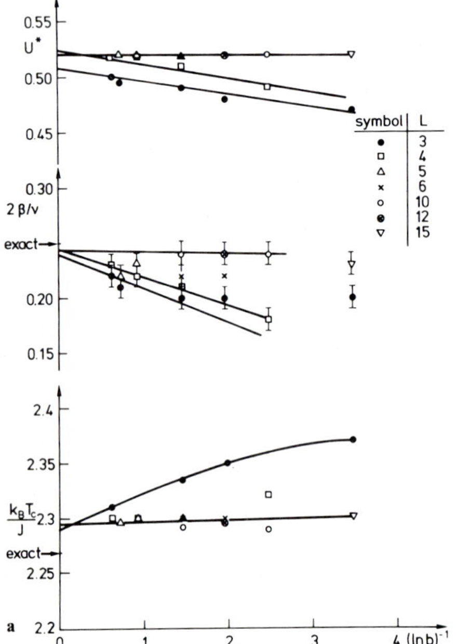

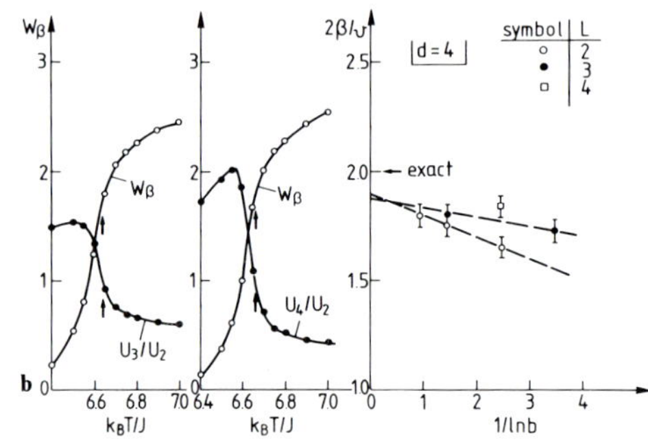

Fig. 17. a Estimates for U^* (upper part), $2\beta/\nu$ (middle part) and $k_B T_c/J$ (lower part) for two-dimensional subsystem blocks plotted vs. inverse logarithm of scale factor; **b** W_β and U_3/U_2 (or U_4/U_2, respectively) plotted vs. temperature for four-dimensional subsystem plots (left part); resulting estimates for $2\beta/\nu$ (right part)

temperatures not always very close to T_c, which may lead to additional correction effects not considered here) seemed premature, we took as a final estimate for U^* the value U_L^*, for larger values of L where the systematic effects seem to disappear in the scatter of our data due to their limited statistical accuracy. Hence we obtain for subsystem blocks

$$U^* \cong 0.52 \pm 0.01\,(d=2), \qquad U^* \cong 0.21 \pm 0.01\,(d=3) \quad (63)$$

while for isolated three-dimensional blocks we get

$$U^*_{(p)} \cong 0.44 \pm 0.02, \qquad U^*_{(f)} \cong 0.12 \pm 0.01. \quad (64)$$

Equation (63) is in reasonable agreement with a treatment of Bruce [10] yielding $U^* \cong 0.58$ $(d=2)$,

$U^* \cong 0.22$ ($d=3$), and are consistent with the direct scaling analysis of $P_L(s)$, see Sect. III.

While the estimates for T_c drawn from isolated blocks with either free (Fig. 15a) or periodic (Fig. 16a) boundary conditions are nicely convergent to the correct value for $(\ln b)^{-1} \to 0$, the estimates from subsystem blocks converge towards $k_B T_c/J \approx 4.55$ ($d=3$) rather than $k_B T_c/J \approx 4.51$ [39] and to $k_B T_c/J \approx 2.29$ ($d=2$) rather than to $k_B T_c/J \approx 2.269$ [37]. We interpret these 1%-discrepancies as finite-size effects due to the finiteness of our total lattice linear dimension, $N=60$ ($d=2$) and $N=24$ ($d=3$). One does in fact expect a shift of the "effective" T_c of the rounded transition in the finite system of the order of $N^{-1/\nu}$ [29], yielding shifts of the order of 1% in our case. Unfortunately, in our case the proportionality constant in the relation $\Delta T_c/T_c \propto N^{-1/\nu}$ seems to be rather close to unity, while much smaller prefactors seem to apply when $\Delta T_c/T_c$ is defined from the specific heat maximum [29].

In view of these finite-size effects, it may be worthwhile to study subsystem blocks also for other values of N, to make sure whether there is any appreciable effect on the exponent estimates. From this point of view, the (less convenient) study of isolated blocks is superior – but there for the same values of L the correction effects due to the finiteness of L (such as included in (58)) are somewhat larger, and for reaching good enough accuracy blocks distinctly larger than $L=10$ (the maximum block size included in Fig. 16) is indispensable (for blocks with free boundary conditions the situation is even worse). Such a more extensive study, though feasible in principle, is not attempted here.

The exponent estimates obtained from the subsystem blocks are summarized as

$$2\beta/\nu = 1.03 \pm 0.01, \quad 1/\nu = 1.60 \pm 0.05 \ (d=3), \tag{65a}$$

$$2\beta/\nu = 0.24 \pm 0.02, \quad 1/\nu = 0.9 \pm 0.1 \ (d=2). \tag{65b}$$

For $d=4$ the same analysis (Fig. 17b) would yield $2\beta/\nu \approx 1.9$ rather than the answer $2\beta/\nu = 1$ appropriate for a gaussian fixed point. Since for $d=4$ logarithmic correction terms to the leading mean field behavior are predicted [42], the simple scaling analysis upon which our method for estimating exponents is based, must be reconsidered with great care; we are not pursuing this topic further since the smallness of N and L accessible for $d=4$ do not warrant such an analysis. The values $U_L^* > 0$ seen for small L (Fig. 17b) are nonzero only due to correction terms: we except $U^*=0$, consistent with Fig. 12c, and then the exponents must be the mean-field exponents.

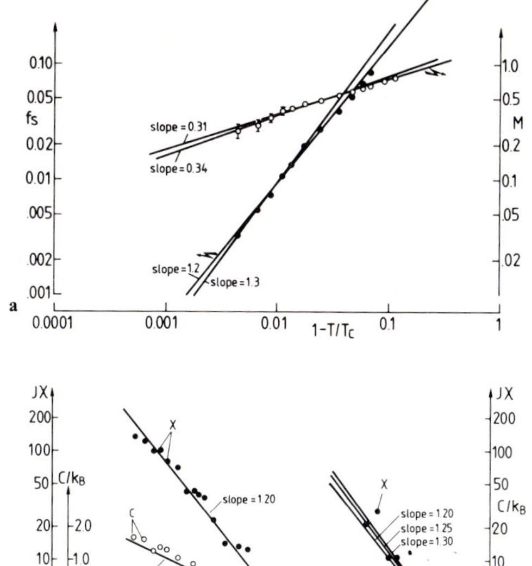

Fig. 18. a Magnetization M and "interface tension" $f_s = S_3(V_3/2)^{1/3} F_s/k_B T$ of the three-dimensional Ising model plotted vs. $(1-T/T_c)$. **b** Specific heat C and susceptibility χ plotted vs. $T/T_c - 1$ (above T_c, left part) and $1 - T/T_c$ (below T_c, right part). Points are calculated from fluctuations of energy and magnetization, respectively, using about 10^2 statistically independent observations, taken at "time" intervals of about 50–100 MCS/spin for a system of 24^3 spins. $k_B T_c/J = 4{,}51$ was taken from high-temperature series [39]

We note that the exponent estimates drawn from three-dimensional isolated blocks with periodic boundary conditions and $L \leq 10$ would be $2\beta/\nu \approx 1.02 \pm 0.03$, $1/\nu = 1.60 \pm 0.05$, consistent with (65a), though the value of $2\beta/\nu$ is less accurate. Of course, the direct finite size scaling analysis of such small systems in the usual form [10, 24, 35] would yield by far more imprecise estimates. Similarly, if we analyze the Monte Carlo data of the systems used for the subsystem analysis in the standard way [10], i.e. by estimating exponents from log-log plots of all quantities in the critical region, we obtain rather imprecise estimates either, Figs. 18, 19: Even implying knowledge of T_c, the scatter of the data allows to "fit" straight lines with an uncertainty in the slope of several percent, e.g. for $d=3$ we find $\gamma = 1.25 + 0.05$, $\beta = 0.32 + 0.02$, $2\nu = 1.25 + 0.05$ (Fig. 18). More important, *correction terms* which sometimes are important can be masked in the scatter of data points completely, and hence from the slope of the

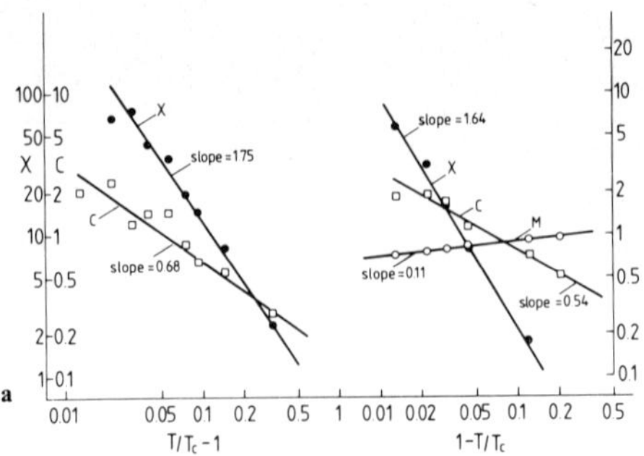

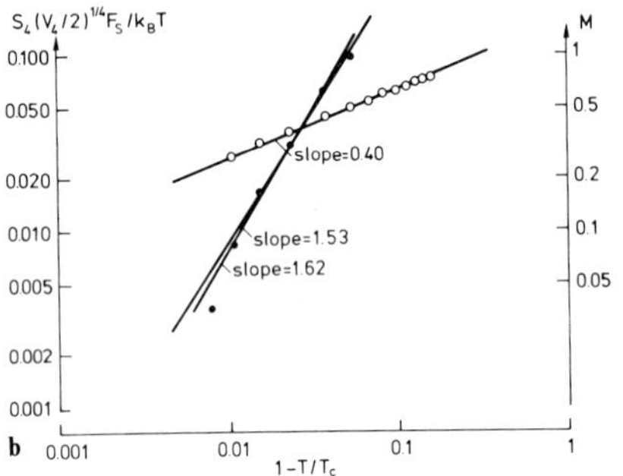

Fig. 19. a Magnetization M, susceptibility χ and specific heat C of the two-dimensional Ising model below T_c plotted vs. $(1-T/T_c)$ (right part); specific heat C and susceptibility χ above T_c plotted vs. T/T_c-1 (left part). Points are calculated from about 10^2 statistically independent observations taken at "time" intervals of about 50–100 MCS/spin for a system of 60^2 spins. $k_B T_c/J = 2.269$ was taken from the exact solution [37]; **b** "Interface tension" and magnetization of the four-dimensional Ising model plotted vs. $1-T/T_c$

resulting "straight line" an estimate is obtained, which is definitely off: this happens for the specific heat above T_c, where $\alpha \approx 0.45$ $(d=3)$ and $\alpha \approx 0.68$ $(d=2)$ would result from such a naive analysis; disregarding logarithmic corrections at $d=4$ leads to exponent estimates which are also systematically off, such as $\beta=0.40$ instead of $\beta=0.5$ (Fig. 19b). In contrast, the above analysis of data obtained from the same runs takes effects due to correction terms in a systematic way into account, and hence the error bars quoted are much more reliable than those of the direct estimates from log-log plots.

V. Conclusions

In this paper, the probability distribution function $P_L(s)$ for observing an order parameter value s in a finite block of linear dimension L was discussed in detail for d-dimensional Ising models. It was shown that for large L the properties of $P_L(s)$ can be interpreted in terms of the spontaneous magnetization M, susceptibility χ and interface tension F_s for temperatures T less than the critical temperature T_c, while above T_c $P_L(s)$ becomes simply a gaussian whose width determines the susceptibility χ. Right at T_c, $P_L(s)$ tends against a scaled universal form which is distinctly non-gaussian, and which depends strongly on the boundary conditions used (and presumably also somewhat on the shape of the block). For subsystem blocks we support the findings of Bruce, that $P_L(s)$ has a single peak at $s=0$ at criticality for $d=3$ while it is double-peaked for $d=2$. For $d=4$, $P_L(s)$ is also gaussian right at T_c.

A study of $P_L(s)$ in the context of Monte Carlo simulation can serve several purposes: first it was shown that reliable estimates of M, χ and F_s can be obtained; second, a Monte Carlo renormalization group analysis was proposed, from which one obtains exponent estimates much more accurately than from the standard fits to log-log plots; in fact, for $d=3$ we could confirm the value $2\beta/\nu \approx 1.03$ from field-theoretic renormalization and reject the high-temperature series expansion estimate $2\beta/\nu \approx 0.98$. Since our analysis is based on an extrapolation vs. the inverse logarithm of the scale factor for various L, the *uniqueness* of the extrapolated result (being independent of L) proves the underlying (hyper-) scaling assumptions to be fully consistent with our data. This method takes corrections to scaling in a systematic way into account, and finite-size effects are well understood: particularly for isolated blocks with periodic boundary conditions, finite-size effects are the ingredients of the finite-size scaling description of $P_L(s)$, on which the "renormalization group" analysis is based. This method, which is related in spirit to the Nightingale phenomenological renormalization of finite strips, is hence an alternative to the Swendsen Monte Carlo renormalization group and reaches a comparable accuracy, particularly for three-dimensional systems. We emphasize that our method neither needs a-priori knowledge of the critical point nor extremely precise computations right at the critical point (which are hard to perform because of "critical slowing down", see [9]).

It should also be interesting to generalize this analysis to other quantities, e.g. the distribution $P_L(u)$ of internal energies u in a block. This quantity will tend towards a gaussian (both above and below T_c) centered at the internal energy U of the bulk system, the width being given by the specific heat. At T_c, we again expect a scaled universal form. From a Monte Carlo renormalization group analysis of $P_L(u)$ similar to the above one should be able to get additional estimates for α/ν and $1/\nu$, and hence obtain a further check on the consistency of the analysis. $P_L(u)$ should also be useful for distinguishing second-order transitions from first-order transitions, where for large enough L a double-peaked structure of $P_L(u)$ must result which becomes sharper double-peaked with increasing L (while for second-order transitions near a tri-critical point a double-peak structure for smaller L must tend to a single-peak structure for large L).

Another possibility for future work along these lines would be the application to systems with continuous degrees of freedom. For example, for a liquid-gas system one would try to sample the block distribution function of density $P_L(\rho)$ in a grand-canonical simulation. Apart from the fact that $P_L(\rho)$ is no longer symmetric around $\rho=1/2$ as in the lattice gas and that the chemical potential at criticality is non-trivial, the extension of the above methods should be straightforward. One thus could relate the liquid-gas critical point to the interatomic potential. More complicated but nevertheless promising would be the application to systems with n order parameter components where $P_L(S_1, \ldots, S_n)$ needs to be sampled. This task should at least be easy in fully symmetric cases, where one can instead consider $P_L\left(\sum_{i=1}^{n} s_i^2\right)$, such as in usual XY or Heisenberg models. We hope to report on such and similar application in the future.

Thanks are due to A. Aharony, A.D. Bruce and M.P. Nightingale for stimulating discussions, and to G.A. Baker and R.B. Stinchcombe for drawing my attention to Refs. [50, 51].

References

1. Kampen, N.G. van: Phys. Rev. **135**, A362 (1964)
2. Kadanoff, L.P.: Physics **2**, 263 (1966)
3. Langer, J.S.: Ann. Phys. **65**, 53 (1971); Physica **73**, 61 (1974)
4. Langer, J.S., Baron, M., Miller, H.D.: Phys. Rev. A**11**, 1417 (1975)
   Billotet, C., Binder, K.: Z. Physik B – Condensed Matter **32**, 195 (1979)
5. Kawasaki, K., Imaeda, T., Gunton, J.D.: In: Studies in Statistical Mechanics. Raveche, H. (ed.). Amsterdam: North-Holland 1980
6. Wilson, K.G.: Phys. Rev. B**4**, 3174 (1971); B**4**, 3184 (1971)
7. Fisher, M.E.: Rev. Mod. Phys. **46**, 587 (1974)
   Domb, C., Green, M.S. (eds.): Phase Transitions and Critical Phenomena. Vol. 6. New York: Academic Press 1976
8. Le Guillou, J.C., Zinn-Justin, J.: Phys. Rev. B**21**, 3976 (1980)
9. Binder, K. (ed.): Monte Carlo Methods in Statistical Physics Berlin, Heidelberg, New York: Springer 1979
10. Bruce, A.D.: Preprint; see also Bruce, A.D., Schneider, T., Stoll, E.: Phys. Rev. Lett. **43**, 1284 (1979)
11. Nightingale, M.P.: Physica **83**A, 561 (1976); Proc. K. Ned. Acad. v. Wet. B**82** (3), 235 (1979)
12. Sneddon, L.: J. Phys. C**11**, 2823 (1978); C**12**, 3051 (1979)
    dos Santos, R.R., Sneddon, L.: Preprint
13. Racz, Z.: Phys. Rev. B**21**, 4012 (1980)
    Derrida, B., Vannimenus, J.: J. Phys. Lett. **41**, (1980)
    Derrida, B.: Preprint
    Nightingale, M.P., Blöte, H.W.J.: Physica **104**A, 352 (1980)
    Schick, M., Kinzel, W.: (unpublished)
    Blöte, H.W.J., Nightingale, M.P., Derrida. B.: Preprint
14. Ma, S.-K.: Phys. Rev. Lett. **37**, 461 (1976)
15. Friedman, Z., Felsteiner, J.: Phys. Rev. B**15**, 5317 (1977)
16. Reynolds, P.J., Stanley, H.E., Klein, W.: Phys. Rev. B**21**, 1223 (1980)
    Herrmann, H.J., Stauffer, D., Eschbach, P.D.: Phys. Rev. B**23**, 422 (1981)
17. Swendsen, R.H.: Phys. Rev. Lett. **42**, 859 (1979); Phys. Rev. B**20**, 2080 (1979)
18. Blöte, H.W.J., Swendsen, R.H.: Phys. Rev. B**20**, 2077 (1979)
19. Blöte, H.W.J., Swendsen, R.H.: Phys. Rev. B**22**, 4481 (1980)
20. Blöte, H.W.J., Swendsen, R.H.: Phys. Rev. Lett. **43**, 737 (1979)
    Swendsen, R.H., Krinsky, S.: Phys. Rev. Lett. **43**, 177 (1979)
    Rebbi, C., Swendsen, R.H.: preprint; Novotny, M.A., Landau, D.P., Swendsen, R.H.: Preprint
    Landau, D.P., Swendsen, R.H.: Preprint
21. Baumgärtner, A.: J. Phys. A**13**, L38 (1980)
    Kremer, K., Baumgärtner, A., Binder, K.: Z. Phys. B – Condensed Matter **40**, 331 (1981)
22. Redner, S., Reynolds, P.J.: Preprint
23. Shenker, S., Tobochnik, J.: Phys. Rev. B**22**, 4462 (1980)
24. Landau, D.P.: Phys. Rev. B**13**, 2297 (1976); B**14**, 255 (1976)
25. For connecting probability theory and renormalization group ideas, see Jona-Lasinio, G.: I. Nuovo Cimento **26**B, 99 (1975)
26. Binder, K., Rauch, H.: Z. Phys. **219**, 201 (1969)
27. Schulman, L.S.: J. Phys. A**13**, 237 (1980)
28. Binder, K., Kalos, M.H.: J. Stat. Phys. **22**, 363 (1980)
29. Fisher, M.E.: In: Critical Phenomena Green, M.S. (ed.), p. 1. New York: Academic Press 1971
    Suzuki, M.: Prog. Theor. Phys. **58**, 1142 (1977)
30. The correlation length ξ. which enters the second argument of $\tilde{P}$ could differ from the standard definition of ξ by a constant of order unity. For simplicity this constant is here absorbed in the definition of ξ
31. Patashinskii, A.Z.: Sov. Phys. JETP **26**, 1126 (1968)
32. Baker, G.A., Jr.: Phys. Rev. B**15**, 1552 (1977)
33. Baker, G.A., Jr., Kincaid, J.M.: J. Stat. Phys. **24**, 469 (1981)
34. Stell, G.: In: Critical Phenomena. Green, M.S. (ed.), p. 188. New York: Academic Press 1971
    Fisher, M.E.: In: Proceedings of the Twenty-Fourth Nobel Symposium on Collective Properties of Physical Systems. Lundquist, B., Lundquist, S. (eds.). p. 16. New York: Academic Press 1973
35. Binder, K.: Thin Solid Films **20**, 367 (1974)
36. Baumgärtner, A., Binder, K.: J. Chem. Phys. (in press)
37. Onsager, L.: Phys. Rev. **65**, 117 (1944)
38. Because of the symmetry $P_L(s)=P_L(-s)$ only data for $s>0$ are shown. The data in the figures arbitrarily are normalized to $\int_0^\infty P_L(s)\,ds=1$ rather than 1/2
39. Domb, C.: In: Phase Transitions and Critical Phenomena. Vol. 3. Domb, C., Green, M.S. (eds.), Vol. 3. New York: Academic Press 1974
40. Gaunt, D.S., Sykes, M.F., McKenzie, S.: J. Phys. A**12**, 871 (1979)
41. Mouritsen, O.G., Knak Jensen, S.J.: Phys. Rev. B**19**, 3663 (1979)
42. Brezin, E., Le Guillou, J.C., Zinn-Justin, J.: Phys. Rev. D**8**, 2418 (1973)
43. Binder, K.: (to be published)
44. Kinzel, W.: Phys. Rev. B**19**, 4584 (1979)

45. Racz, Z., Rujan, P.: Z. Phys. B – Condensed Matter **28**, 287 (1977)
Muto, S., Oguchi, T., Ono, I.: J. Phys. A**13**, 1799 (1980)
46. Leeuwen, J.M.J. van: In: Phase Transitions and Critical Phenomena. Vol. 6. Domb, C., Green, M.S. (eds.), Vol. 6. New York: Academic Press 1976
47. Müller-Krumbhaar, H.: Z. Phys. B – Condensed Matter (1980)
48. Nickel, B.G., Sharpe, B.: J. Phys. A**12**, 1819 (1979)
49. Gaunt, D.S., Sykes, M.F.: J. Phys. A**12**, L25 (1979)
Rehr, J.J.: J. Phys. A**12**, L179 (1979)
McKenzie, S.: J. Phys. A**12**, L185 (1979)
Zinn-Justin, J.: J. Phys. (Paris) **40**, 969 (1979)

Kurt Binder
Institut für Festkörperforschung
Kernforschungsanlage Jülich GmbH
Postfach 19 13
D-5170 Jülich 1
Federal Republic of Germany

Note Added in Proof

For Ising systems with finite susceptibility it has been proven rigorously that the fixed-point Hamiltonian for the block spin case is independent Gaussians [50]. This is consistent with the gaussian forms of the distribution chosen here off criticality. It should also be noted that the concept of obtaining a fixed-point probability distribution function in a real-space renormalization has also been used in the context of percolation studies [51]. Finally we mention that unintentionally it is not the order parameter distribution $P_L(s)$ as defined in (3), which was studied numerically in Sect. II, but a slightly different distribution $P_L'(s)$. In terms of the M states $\{s_i^{(L)}\}$ generated for the block variable these distributions can be expressed as $P_L(s)=(1/M)\sum_{i=1}^{M}\delta(s_i^{(L)}-s)$ and $P_L'(s)=(1/M)\sum_{i=1}^{M}\delta(s_i^{(L)}-s)/W(s_i^{(L)})$, where $W(s_i^{(L)})$ is the transition probability (for a single spin-flip) in state $s_i^{(L)}$. Since $W(s)$ is a very smooth function of its argument, the distributions $P_L(s)$ and $P_L'(s)$ are qualitatively similar, and tend towards the same distribution for large L. Since both distributions hence must have the same scaling behavior as $L\to\infty$, $T\to T_c$, they are equally well suited for the analysis presented in the paper.

50. Newman, C.M.: Commun. Math. Phys. **74**, 119 (1980); Baker, G.A., Krinsky, S.: J. Math. Phys. **18**, 590 (1977)
51. Stinchcome, R.B., Watson, B.P.: J. Phys. C**9**, 3221 (1976)

Helmholtz free energy of finite spin systems near criticality

E. Eisenriegler and R. Tomaschitz
Institut für Festkörperforschung, Kernforschungsanlage Jülich, D-5170 Jülich, West Germany
(Received 22 September 1986)

We present a quantitative comparison between $4-\epsilon$ expansion and Monte Carlo estimates of critical finite-size properties. Specifically we consider the Helmholtz free energy (where the overall order parameter rather than its conjugate field is kept fixed) at the bulk critical temperature for a cube with periodic boundary conditions within a $4-\epsilon$ expansion to one-loop order. An estimate for the complete asymptotic scaling function obtained from renormalization flow equations as well as systematic ϵ-expansion estimates for several amplitude ratios agree well with corresponding Monte Carlo results in three dimensions. The nonequivalence of thermodynamic ensembles for critical finite-size properties is discussed in some detail.

I. INTRODUCTION

Finite-size effects for critical behavior are inherent in the most powerful numerical methods for investigating critical phenomena, i.e., in Monte Carlo simulations and in transfer-matrix methods, see e.g., the recent review by Barber.[1] They are, of course, also of importance for the critical behavior of real systems.[1] While theoretical investigations of critical finite-size effects have a long tradition, see Onsager's original paper[2] and later work by Fisher and Ferdinand[3,4] on the two-dimensional Ising model, and while the general concept of "finite-size scaling" was already formulated some time ago by Fisher,[5,6] there is renewed interest recently. On the one hand, conformal invariance at a critical point has been exploited to obtain exact finite-size results, mainly for systems in space dimension $d=2$, see the review by Cardy.[7] On the other hand, expansions in the deviation $\epsilon=4-d$ from the upper critical dimension $d=d_>=4$ have been used to obtain finite-size universal quantities for periodic[8,9] and Dirichlet (or free)[10-12] boundary conditions. To test the reliability of the ϵ expansion when applied to finite-size quantities in $d=3$, there is an urgent need for quantitative comparison with corresponding results obtained by other means, either exact, see e.g., recent work on the spherical model,[13] or numerical.

In this paper we consider completely finite systems in thermal equilibrium where the total order parameter (and the volume) are not allowed to fluctuate but have sharp values mV (and V). An example is the Ising model with spins $S_i=\pm 1$ on a finite lattice (i; i in V) where the total magnetization $\sum_i S_i$ is fixed to mV. The corresponding Helmholtz total free energy is

$$F(m,T,L)=-k_BT\ln\sum_{\{S\}}\delta_{mV,\sum_i S_i}\exp[-\mathcal{H}_I/k_BT] , \quad (1.1)$$

with $\mathcal{H}_I$ the Ising Hamiltonian for vanishing magnetic field, compare e.g., Refs. 14 and 15. Another example is a model for a real gas, showing a liquid-gas critical point for $V\to\infty$ with critical density m_c, in a finite closed vessel with rigid walls (V fixed) which allow for thermal equilibrium but are impenetrable for the particles such that the total particle number is fixed to $(m_c+m)V$.

For the situation where the total order parameter is allowed to fluctuate in a magnetic field (or chemical potential) h, see the discussions of Privman and Fisher (Ref. 16) and Ref. 11. One expects also for the present situation of fixed total order parameter a universal finite size scaling form for the free energy near criticality. If the shape of the system has only one characteristic length L and $V\sim L^d$, as for a cube or a sphere, and if T denotes the temperature, then the leading singular behavior $F^{(s)}$ of the Helmholtz total free energy $F(m,T,L)$ with (m,T,L) close to $(0,T_{CB},\infty)$ should have the form

$$\frac{F^{(s)}(m,T,L)}{Vk_BT_{CB}}=L^{-d}B(y,w) . \quad (1.2)$$

Here B is a scaling function, k_B is Boltzmann's constant, T_{CB} denotes the bulk ($V\to\infty$) critical temperature and

$$y\sim mL^{\beta/\nu} , \quad (1.3)$$

$$w\sim(T-T_{CB})L^{1/\nu} . \quad (1.4)$$

If one fixes the scales in Eqs. (1.3) and (1.4) by

$$y=(L/\xi_m)^{\beta/\nu} , \quad (1.3')$$

$$w=(L/\xi_+)^{1/\nu} , \quad (1.4')$$

with ξ_m and ξ_+ *bulk* ($V=\infty$) correlation lengths for $T=T_{CB},m\neq 0$ and $T>T_{CB},m=0$ respectively, the function B turns out to be universal. Apart from the usual dependence on d (which is assumed smaller than $d_>$, see Ref. 8) and on dimensionality n of the order parameter, B depends in general also on the shape of the sample and the type of boundary condition on the surface.

We mention two general properties of the function B in Eq. (1.2): (i) For $(m,T)\neq(0,T_{CB})$, the free energy *per unit volume* approaches its L-independent bulk value f_b

$$\frac{1}{V}F(m,T,L)\to f_b(m,T) \quad (1.5)$$

as $L\to\infty$. Therefore, B in Eq. (1.2) should be proportional to $V\sim L^d$ in this case (where $L\gg$ bulk correlation length). For $T=T_{CB}$, in particular, this leads to the behavior

$$B(y,0)\approx b_m y^{\nu d/\beta}=b_m(L/\xi_m)^d, \quad y>>1 , \tag{1.6}$$

with the exponent $\nu d/\beta\equiv\delta+1$ reflecting the well-known[17] m dependence of the bulk free energy density $f_b^{(s)}$ at $T=T_{CB}$ and with the universal bulk amplitude

$$b_m=(\xi_m)^d f_b^{(s)}(m,T_{CB})/k_B T_{CB} . \tag{1.7}$$

Note that the limiting behavior (1.6) is independent of the boundary condition. (ii) In the opposite limit, where the system size L and thus wavelengths of possible fluctuations are much smaller than the bulk correlation length, there are no infrared divergencies possible [for $(m,T)\rightarrow(0,T_{CB})$ with L fixed] and the corresponding free energy should be *analytic* in m and $T-T_{CB}$, compare e.g., Wilson's[18] momentum shell integration procedure. Thus for $T=T_{CB}$ in particular,

$$B(y,0)\rightarrow B(0,0)+B_{2,0}y^2+B_{4,0}y^4+\cdots, \quad y<<1 , \tag{1.8}$$

if we assume $m\rightarrow-m$ symmetry in F. The universal finite-size amplitudes $B_{i,j}$ will depend on boundary conditions and shape.

Our interest in the quantity $F^{(s)}$ on the left-hand side (lhs) of Eq. (1.2) was stimulated by a Monte Carlo simulation of Binder[15] where he studied the "order parameter distribution function"

$$P_L(m)\sim\exp[-F(m,T_{CB},L)] , \tag{1.9}$$

for a $d=3$ Ising model on a finite lattice cube of $L\times L\times L$ spins with periodic as well as free boundary conditions, at $T=T_{CB}$. He found indeed evidence that P has scaling form for $(m,L)\rightarrow(0,\infty)$ as to be expected from Eqs. (1.2), (1.3), and (1.9), since F should be dominated by $F^{(s)}$ in this case. The corresponding asymptotic form of P will be denoted by $P^{(s)}$. Even without information on ξ_m in Eq. (1.3') he extracted from $P^{(s)}$ a universal function $p(\tilde{y})$ which is related to $\exp[-B(y,0)]$, with B from Eq. (1.2), simply by changes of scale such that

$$\int_{-\infty}^{\infty} d\tilde{y}p(\tilde{y})=\int_{-\infty}^{\infty} d\tilde{y}\,\tilde{y}^2p(\tilde{y})=1 \tag{1.10}$$

(see Sec. III C). Interestingly, the shape of p depends *qualitatively* on the boundary condition. For periodic boundary conditions, it shows a (symmetric) double-peak structure with a local minimum at $\tilde{y}=0$. This is in accord with the finding of Brézin and Zinn-Justin[8] that the coefficient $B_{2,0}$ in Eq. (1.8) is negative in this case. In the other case of free boundaries, p looks nearly like a Gaussian and $B_{2,0}$ should be positive.

The main goal of the present paper is to perform a *quantitative* comparison with other available information in $d=3$. The only attempt in this direction up to now is, to the best of the authors' knowledge, the calculation of the "renormalized coupling constant" in Ref. 8 for a periodic cube (PC) which is related to the fourth moment of the corresponding function $p(\tilde{y})$ which we denote by $p_{PC}(\tilde{y})$. The zeroth and second moments of p are unity by definition, see (1.10). The renormalized coupling constant was then compared with Monte Carlo data.[19] Here we calculate several other universal numbers contained in the double-peak structure of p_{PC} by systematic ϵ expansion and obtain an estimate for the whole universal function p_{PC} which shows a crossover between the two regimes characterized by Eqs. (1.6) and (1.8), respectively, by using a method similar to that in Ref. 9. The calculations are restricted to one-loop order. These estimates are compared with the Monte Carlo results of Binder.[15] Our explicit results for the quantities mentioned above are summarized in Sec. IV.

II. GINZBURG-LANDAU MODEL AND RENORMALIZATION

We consider the Ginzburg-Landau model for an n-component field $\boldsymbol{\phi}(r)$ with Hamiltonian

$$\mathcal{H}=\int_V d^dr\left[\tfrac{1}{2}(\nabla\boldsymbol{\phi})^2+\frac{t}{2}\boldsymbol{\phi}^2+\frac{g}{24}(\boldsymbol{\phi}^2)^2\right] , \tag{2.1}$$

where the r integration extends over a finite region with volume V in $d=4-\epsilon$ dimensional space. We will mainly discuss periodic boundary conditions where (for a cube of edge length L)

$$\boldsymbol{\phi}(\mathbf{r}+\mathbf{n}L)=\boldsymbol{\phi}(\mathbf{r}) , \tag{2.2}$$

where $\mathbf{n}$ is an integer vector, but will sometimes refer also to Dirichlet boundary conditions

$$\boldsymbol{\phi}(\mathbf{r})=0, \quad \mathbf{r}\in\text{surface of } V . \tag{2.3}$$

Renormalization of field theories of this type has been discussed by Symanzik[10] and Brézin[20] (see also the review by Diehl[21]). The deviation δF of the model's Helmholtz total free energy from the one at $(\mathbf{m},t)=(\mathbf{0},t_{CB})$, i.e., at vanishing order parameter $\mathbf{m}$ per unit volume and at the bulk critical temperature, is given in renormalized form by

$$\delta F^R(u,m^R,t^R,\Lambda)=-\ln[Z/(Z)_{m=0,t=t_{CB}}] , \tag{2.4a}$$

where

$$Z=\int D\boldsymbol{\phi}\,\delta\left[\mathbf{m}V-\int_V d^dr\boldsymbol{\phi}\right]\exp(-\mathcal{H}-\mathcal{H}_{ac}) , \tag{2.4b}$$

with $\mathcal{H}_{ac}$ defined in Eqs. (2.8) below. Note that δF depends only on the magnitude m of $\mathbf{m}$, due to $O(n)$ invariance. Here we use dimensional regularization,[22] for which $t_{CB}=0$, and define renormalized parameters u,m^R,t^R,Λ by

$$g=\mu^{\epsilon}16\pi^2 f Z_u u , \tag{2.5a}$$

$$m=\mu^{1-\epsilon/2}Z_{\phi}^{1/2}m^R , \tag{2.5b}$$

$$t=\mu^2 Z_\tau t^R , \tag{2.5c}$$

$$L=\mu^{-1}\Lambda , \tag{2.5d}$$

where μ is an arbitrary reference wave number and the Z factors contain only negative powers of ϵ, i.e.,

$$Z_{u,\phi,\tau}-1=\epsilon^{-1}b_{u,\phi,\tau}^{(1)}(u)+\epsilon^{-2}b_{u,\phi,\tau}^{(2)}(u)+\cdots , \tag{2.6}$$

with the $b^{(i)}$'s independent of ϵ. Different forms for the factor[23]

$$f=1+\epsilon f_1+\epsilon^2 f_2+\cdots \tag{2.7}$$

in Eq. (2.5a) have been used in the literature, e.g., $f\equiv 1$ in Refs. 24 and 25, $f=(4\pi)^{-\epsilon/2}\Gamma(2-\epsilon/2)$ in Ref. 22, $f=(4\pi)^{-\epsilon/2}(1-\epsilon/2)/\Gamma(1+\epsilon/2)$ in Ref. 26, and $f=(4\pi)^{-\epsilon/2}$ in Ref. 21. The form of the functions $b^{(i)}_{u,\phi,\tau}(u)$ in Eq. (6) (and thus of the corresponding coefficients in the renormalization group equation, see below) are independent of the particular choice of the coefficients f_i in Eq. (2.7).[27] With our choice $16\pi^2$ of the d-independent prefactor in Eq. (2.5a), the $Z_{u,\phi,\tau}$'s are as given in the review.[21] [The present variable u is obtained by multiplying the variables u used in Refs. 21, 22, 24, and 26 by factors of 1, $\frac{1}{2}$, $1/(16\pi^2)$, and 12, respectively.] The f_i's drop out also in systematic ϵ expansions of universal ratios. In order to allow for appropriate checks in explicit calculations in Sec. III, we will keep the f_i's arbitrary. In addition to the bulk counterterms with the structures $(\boldsymbol{\phi}^2)^2, (\nabla\boldsymbol{\phi})^2, \boldsymbol{\phi}^2$ (Ref. 22) which originate via (2.5a)–(2.5c) from the right-hand side (rhs) of Eq. (2.6), there are also $\boldsymbol{\phi}$-independent counterterms, represented by $\mathcal{H}_{\rm ac}$, in Eq. (2.4b). For periodic boundary conditions (BC's) as in Eq. (2.2), *bulk* counterterms alone are sufficient to remove the ultraviolet singularities in δF, i.e., to render δF^R in Eqs. (2.4) finite for $\epsilon\to 0$, see Ref. 20. This should hold also for Dirichlet BC's provided the surface is planar but may contain edges and corners as for a cube.[21] In these cases, therefore,

$$\mathcal{H}_{\rm ac}=\int_V d^d r \mu^d \tfrac{1}{2}(t^R)^2 a \ , \tag{2.8a}$$

which is the only ϕ-independent bulk counterterm[22] if only sources for ϕ and ϕ^2 are considered. Here

$$a(u,\epsilon)=\frac{a_1}{\epsilon}+\frac{a_2}{\epsilon^2}+\cdots \ , \tag{2.8b}$$

similar to the rhs of Eq. (2.6). For curved surfaces and Dirichlet BC's, there is in addition a *surface* contribution $\sim t^R$ to $\mathcal{H}_{\rm ac}$, see Refs. 10 and 11.

We briefly make contact with the ensemble considered in Refs. 8–11 where the total order parameter fluctuates in a magnetic field $\mathbf{h}$. The corresponding Gibbs total free energy deviation $\delta F_G^R(u,h^R,t^R,\Lambda)$ from $h=t=0$ is given by equations like (2.4a) and (2.4b) but with the δ function in (2.4b) replaced by $\exp(\mathbf{h}\int d^d r \boldsymbol{\phi})$ and the subscript $m=0$ on the rhs of (2.4a) by $h=0$. In addition, the renormalization (2.5b) for m is to be replaced by a corresponding one

$$h=\mu^{3-\epsilon/2}Z_\phi^{-1/2}h^R \tag{2.9}$$

for the magnetic field, compare Ref. 11 where δF_G^R was called δF^R. It is then easy to show that

$$\delta F_G^R(u,h^R,t^R,\Lambda)=-\ln[\delta Z_G^R/(\delta Z_G^R)_{h^R=t^R=0}] \ , \tag{2.10}$$

with

$$\delta Z_G^R=\int d^n m^R \exp[\mathbf{h}^R\mathbf{m}^R V\mu^d-\delta F^R(u,m^R,t^R,\Lambda)] \ . \tag{2.11}$$

The corresponding free energies

$$\delta f^R(u,m^R,t^R,\Lambda)=\delta F^R/(\mu^d V) \tag{2.12a}$$

and

$$\delta f_G^R(u,h^R,t^R,\Lambda)=\delta F_G^R/(\mu^d V) \tag{2.12b}$$

per unit volume are supposed to reach finite ("bulk") limits $\delta f_b^R(u,m^R,t^R)$ and $\delta f_{G,b}^R(u,h^R,t^R)$ as $\Lambda\to\infty$ for fixed $(m^R,t^R)\neq(0,0)$ and $(h^R,t^R)\neq(0,0)$, respectively (compare e.g., the explicit calculations in Sec. III). It is only in this limit that δf^R and δf_G^R are connected by the well-known Legendre-type relation[22,28]

$$\delta f_{G,b}^R(u,h^R,t^R)=\delta f_b^R(u,m_M,t^R)-h^R m_M \tag{2.13a}$$

with

$$\mathbf{m}_M=\mathbf{m}_M(u,\mathbf{h}^R,\mathbf{t}^R) \ , \tag{2.13b}$$

determined by

$$\frac{\partial}{\partial m_M}\delta f_b^R(u,m_M,t^R)=h^R \ . \tag{2.13c}$$

Equation (2.13c) determines the unique[29] minimum $m^R=m_M$ of the m^R-dependent function δf_b^R-$h^R m^R$ (with $u,\mathbf{h}^R,t^R$ fixed) since δf_b^R is convex from below.[14,28] The reason for Eqs. (2.13) is, of course, that the integrand on the rhs of Eq. (2.11) is sharply peaked at this minimum in the limit considered.

A renormalization group (RG) equation for δF^R follows from the observation that δF^R-$\mathcal{H}_{\rm ac}$ in Eqs. (2.4) is unchanged by changing the variables μ,u,m^R,t^R,Λ simultaneously in such a way that the bare parameters g,m,t,L remain fixed. The procedure is completely similar to the corresponding one for δF_G^R which was presented in Ref. 11, where δF_G^R was called δF^R. The only difference comes from the fact that Eq. (2.5b) has to be used for δF^R instead of Eq. (2.9). The RG property of δF^R then takes the form of Eqs. (A10)–(A15) in Appendix A of Ref. 11 with Q denoting the pair (m,t) and with the complete m "dimension"

$$d_m(u)\equiv d-d_h(u)=d_m^0+\eta_m(u) \tag{2.14a}$$

composed of its naive and anomalous parts

$$d_m^0=1-\epsilon/2, \quad \eta_m(u)=-\eta_h(u) \ . \tag{2.14b}$$

Here d_h,η_h are defined in Appendix A of Ref. 11. At the fixed point $\overset{*}{u}$ of u, see,[21]

$$d_m(\overset{*}{u})=\beta/\nu \ , \tag{2.14c}$$

with bulk exponents β and ν.[17] In the following we will discuss exclusively *asymptotic* behavior where $(m^R,t^R,\Lambda)\to(0,0,\infty)$. The corresponding asymptotic RG property for δf^R in Eq. (2.12a) is

$$\delta f^R(u,m^R,t^R,\Lambda)=l^d\delta f^R(\overset{*}{u},\tilde{m}^R,\tilde{t}^R,\tilde{\Lambda}) + \frac{1}{2}\frac{\overset{*}{a}_1}{d-2/\nu}[l^d(\tilde{t}^R)^2-(t^R)^2P(u)] \ , \tag{2.15}$$

with

$$\tilde{m}^R=D_m(u)m^R l^{-\beta/\nu} \ , \tag{2.16a}$$

$$\tilde{t}^R=D_t(u)t^R l^{-1/\nu} \ , \tag{2.16b}$$

$$\tilde{\Lambda}=\Lambda l \ , \tag{2.16c}$$

and holds only if the RG flow parameter l is sufficiently small. For curved surfaces, there is an additional term, $\sim t^R$, on the rhs of Eq. (2.15). $D_{m,t}$ and P are u-dependent, i.e., nonuniversal, amplitudes[30] and the quantity $\bar{a}_1^*$ is given by

$$\bar{a}_1^*=[(1+u\partial_u)a_1]_{u=u^*} \ , \tag{2.17a}$$

with a_1 from Eq. (2.8b). Equations (2.15)–(2.17a) remain valid if one uses a form of the counterterm a in Eq. (2.8b) where the u-dependent coefficients a_i on the rhs contain also an ϵ dependence from an arbitrary i-independent factor r with the form of f in Eq. (2.7), i.e.,

$$a_i=r(\epsilon)\hat{a}_i(u) \ , \tag{2.17b}$$

compare Ref. 11 where $r=(4\pi)^{\epsilon/2}$ was used. The two-loop results for the bulk specific heat, see e.g., Ref. 26, lead to

$$\hat{a}_1=\frac{1}{16\pi^2}\left[n-u\frac{n(n+2)}{3}(f_1+r_1)+O(u^2)\right] ,$$
$$\hat{a}_2=\frac{1}{16\pi^2}\left[u\frac{n(n+2)}{3}+O(u^2)\right] , \tag{2.17c}$$

with r_1 defined from r as f_1 from f in Eq. (2.7). Generally, $\hat{a}_i$ contains u powers of at least the order $i-1$. The first two terms on the rhs of Eq. (2.15) contain the flow parameter l and represent the (leading) *singular* contribution in the asymptotic regime. The last term which contains the amplitude P, is a *regular* contribution. The complete singular contribution is independent of the form of $r(\epsilon)$ in Eq. (2.17b) [while $\hat{a}_i$, see (2.17c), as well as the separate contributions of the first and second terms on the rhs of (2.15) do depend on $r(\epsilon)$]. *Corrections* to the rhs of Eq. (2.15) are smaller than the leading singular contribution by a factor l^{ω} when $l\to 0$ where ω is the correction to scaling exponent.[22] We note that for $u=\overset{*}{u}$, Eqs. (2.15) and (2.16) hold, without corrections, for arbitrary l with

$$D_{m,t}(\overset{*}{u})=P(\overset{*}{u})=1 \ . \tag{2.18}$$

The amplitudes $D_{m,t}(u)$ can be expressed by the longitudinal bulk correlation length $\xi\equiv\xi^R/\mu$ as defined e.g., via moments[31]

$$[\xi^R(u,m^R,t^R)]^2=\mu^2(2d)^{-1}M_2/M_0 \ ,$$
$$M_{2\lambda}=\int_\infty d^dr(\mathbf{r}^2)^\lambda G_{c,b}(r) \ , \tag{2.19}$$

of the longitudinal two point cumulant $G_{c,b}$ in the infinite bulk. The corresponding quantities $\xi_{m,+}$ in Eqs. (1.3′) and (1.4′) in the Introduction are then given by

$$\xi_m=\xi_m^R/\mu, \quad \xi_+=\xi_+^R/\mu \ ,$$
$$\xi_m^R=\xi^R(u,m^R,0), \quad \xi_+^R=\xi^R(u,0,t^R) \ , \tag{2.19a}$$

with $t^R>0$. As in Ref. 11, one finds

$$D_m(u)m^R=\psi_m(\xi_m^R)^{-\beta/\nu} \ ,$$
$$D_t(u)t^R=\psi_+(\xi_+)^{-1/\nu}, \quad t^R>0 \ , \tag{2.20a}$$

where

$$\psi_m=[\xi^R(\overset{*}{u},1,0)]^{\beta/\nu}, \quad \psi_+=[\xi^R(\overset{*}{u},0,1)]^{1/\nu} \ . \tag{2.20b}$$

Proceeding as in Ref. 32, one obtains the explicit expressions

$$\psi_m=\frac{1}{\sqrt{\overset{*}{u}}}\frac{2^{-3/2}}{\pi}\left[1+\frac{\epsilon}{4}\left[(\ln 4\pi)-C_E+\frac{n-1}{n+8}\ln 3+\frac{2}{3}\frac{n+5}{n+8}\right]+O(\epsilon^2)\right] , \tag{2.20c}$$

$$\psi_+=1+\frac{\epsilon}{2}\frac{n+2}{n+8}[1+2f_1+(\ln 4\pi)-C_E]+O(\epsilon^2) \ , \tag{2.20d}$$

where $C_E=0.577$ is Euler's constant. Since δF^R corresponds to the quantity $[F-F(0,T_{\text{CB}},L)]/k_BT$ from Sec. I, the deviation $\delta B(y,w)=B(y,w)-B(0,0)$ of B on the rhs of Eq. (1.2) can be expressed by means of Eqs. (2.12a), (2.15), (2.16), and (2.20a) in the form

$$\delta B([\Lambda/\xi_m^R]^{\beta/\nu},[\Lambda/\xi_+^R]^{1/\nu})=(\Lambda l)^d\left[\delta f^R[\overset{*}{u},\psi_m(\xi_m^R l)^{-\beta/\nu},\psi_+(\xi_+^R l)^{-1/\nu},\Lambda l]+\frac{\bar{a}_1^*}{d-2/\nu}\psi_+^2(\xi_+^R l)^{-2/\nu}\right] \tag{2.21}$$

valid for any l. Putting l^{-1} to a fixed multiple of either Λ, ξ_m^R, or ξ_+^R, one recognizes that the rhs of Eq. (21) is indeed a function of the two variables of δB on the lhs. This function is universal, i.e., it is independent of u and of the forms of f and r in Eqs. (2.7) and (2.17b), see below. This verifies the universal finite size scaling postulate given in Eqs. (1.2), (1.3′), and (1.4′). Note that $\Lambda/\xi_{m,+}^R=L/\xi_{m,+}$, due to Eqs. (2.5d) and (2.19a).

The asymptotic forms (2.15) or (2.21) for the Helmholtz free energy imply via Eqs. (2.10) and (2.11) a corresponding form for the Gibbs energy. One first notices that the regular term on the rhs of Eq. (2.15) is independent of m^R and can be taken out of the integral on the rhs of Eq. (2.11) and contributes additively to δF_G^R. The corresponding contribution from the singular part on the rhs of Eq. (2.15) [or equivalently from the rhs of Eq. (2.21)] has the form

$$-\ln\left[\int d^n y\exp\{\mathbf{y}\Lambda^{\Delta/\nu}[\mathbf{h}^R\psi_m/D_m(u)]-\delta B(y,w)\}\right]+\ln\{\ \}_{h^R=t^R=0}=\delta B_G[(\Lambda/\xi_h^R)^{\Delta/\nu},w] \tag{2.22}$$

as expected[16] for the asymptotic singular part of δF_G^R, see also Ref. 11 where δB_G was called δY. Here we have used the relation

$$\nu d-\beta=\beta\delta=\Delta \tag{2.23}$$

for bulk exponents and have denoted by ξ_h, the bulk correlation length ξ_m^R in Eq. (2.19a) with m^R expressed in terms of h^R via the asymptotic bulk equation of state for $t^R=0$. By means of Eq. (2.20a), the latter can be written as

$$(\xi_m^R)^{-\beta\delta/\nu}=(\xi_h^R)^{-\Delta/\nu}\equiv[h^R\psi_m/D_m(u)]b_m^{-1}\beta/(\nu d)\ , \tag{2.24}$$

with b_m from (1.6) and follows from Eq. (2.13c) by noticing that

$$\delta f_b^R(u,m^R,0)=b_m[\xi_m^R]^{-d}\ , \tag{2.25}$$

asymptotically. The relation (2.25) is a consequence of Eq. (2.12a), the remark below Eq. (2.20d) and Eqs. (1.2) and (1.6).

III. RESULTS FOR A CUBE WITH PERIODIC BOUNDARY CONDITIONS

The Ginzburg-Landau model for a cube with periodic boundary conditions has been discussed in Refs. 8 and 9. To evaluate δF^R in Eqs. (2.4), these authors split

$$\boldsymbol{\phi}(\mathbf{r})=\mathbf{m}+\boldsymbol{\varphi}(\mathbf{r}) \tag{3.1}$$

into the zeroth Fourier component which is clamped to $\mathbf{m}$ and the fluctuating part φ which contains the other Fourier components

$$\boldsymbol{\varphi}(\mathbf{r})=\sum_{\mathbf{q}(\neq 0)}\tilde{\boldsymbol{\phi}}_{\mathbf{q}}\exp(i\mathbf{q}\cdot\mathbf{r})\ ,\qquad \mathbf{q}=\frac{2\pi}{L}\mathbf{j}\ , \tag{3.2}$$

$\mathbf{j}$ integer vectors in d dimensions, i.e., on the rhs of Eq. (2.4b) one may remove the δ function, replace $D\phi$ by $D\varphi$ and $\mathcal{H}\equiv\mathcal{H}(\boldsymbol{\phi})$ by

$$\mathcal{H}(\mathbf{m}+\boldsymbol{\varphi})=\int_V d^dr\,(h_0+h_2+h_{3+4})\ , \tag{3.3}$$

where

$$h_0=\frac{t}{2}m^2+\frac{g}{4!}m^4\ , \tag{3.3a}$$

$$h_2=\frac{1}{2}\left[(\nabla\boldsymbol{\varphi})^2+\rho_{\parallel}\varphi_1^2+\rho_{\perp}\sum_{\alpha=2}^{n}\varphi_\alpha^2\right]\ , \tag{3.3b}$$

$$h_{3+4}=\frac{g}{6}m\varphi_1\boldsymbol{\varphi}^2+\frac{g}{4!}(\boldsymbol{\varphi}^2)^2\ . \tag{3.3c}$$

Here we assumed that $m_\alpha=m\delta_{\alpha 1}$ points into the 1 direction and

$$\rho_{\parallel,\perp}=t+(\tfrac{1}{2},\tfrac{1}{6})gm^2\ . \tag{3.3d}$$

Thus, an expansion of the Helmholtz free energy per unit volume, δf^R in (2.12a), with respect to the fluctuation nonlinearities h_{3+4} leads formally to precisely the same (connected and one-line-irreducible) diagrams as occur in the corresponding treatment[22] in infinite bulk. The only difference is that r integrations of vertices extend only over the region L^d and that the unperturbed propagator is

$$\langle\varphi_\alpha(r_1)\varphi_\beta(r_2)\rangle_2=\frac{\delta_{\alpha\beta}}{L^d}\sum_{\mathbf{q}(\neq 0)}e^{iq(r_1-r_2)}(q^2+\rho_{\parallel,\perp})^{-1}\ , \tag{3.4}$$

for $(\alpha=1,\alpha\neq 1)$. As pointed out by Brézin and Zinn-Justin,[8] this expansion makes sense [and all m,t derivatives are, in accord with (1.8), infrared finite, diagram by diagram] even at $m=t=0$ if $L<\infty$, due to the absence of the $q=0$ term on the rhs of Eq. (3.4). This is also the reason why one-line reducible diagrams (in which there is a propagator with r_1-r_2 freely integrated over L^d) do not contribute. Since the rhs of Eq. (3.4) tends for $L\to\infty$ into the corresponding propagator in infinite bulk, one obtains the bulk limit as $L\to\infty$, diagram by diagram, each of which is infrared finite if $\rho_{\parallel,\perp}>0$.

As in Refs. 8 and 9 we will not go beyond one-loop order up to which

$$\delta f^R=\tfrac{1}{2}(t^R)^2a+\mu^{-d}h_0+\frac{1}{2\Lambda^d}[\mathcal{L}(x_\parallel)+(n-1)\mathcal{L}(x_\perp)]\ , \tag{3.5}$$

where

$$x_{\parallel,\perp}=(L/2\pi)^2\rho_{\parallel,\perp} \tag{3.5a}$$

and

$$\mathcal{L}(x)=\sum_{\mathbf{j}(\neq 0)}\ln[1+x/j^2]\ , \tag{3.5b}$$

with $\mathbf{j}$ as in Eq. (3.2). The continuation of $\mathcal{L}$ to $d=4-\epsilon$ dimensions has been discussed in Refs. 8 and 9, see also the Appendix.

A. System size small compared to bulk correlation length ("strong finite size limit")

Consider the behavior for $\Lambda<\infty$ as $(m^R,t^R)\to(0,0)$. Due to Eqs. (2.5) this corresponds to small $x_{\parallel,\perp}$ in Eq. (3.5a) where one obtains a Taylor series

$$\mathcal{L}(x)=\sum_{\lambda=1,2,\ldots}A_\lambda x^\lambda\ . \tag{3.6}$$

The coefficients A_λ have an ϵ expansion

$$A_\lambda=-\frac{\pi^2}{\epsilon}\delta_{\lambda,2}+A_\lambda^0+O(\epsilon)\ . \tag{3.6a}$$

All A_λ^0's are known explicitly from work by Zucker,[33] compare the Appendix,

$$A_1^0=-8\ln 2=-5.545\ , \tag{3.6b}$$

$$A_2^0=\frac{\pi^2}{2}[(\ln\pi)-1-\tfrac{2}{3}(\ln 2)-6\zeta'(2)/\pi^2]=1.237\ , \tag{3.6c}$$

$$A_{\lambda>2}^0=-(-1)^\lambda\frac{8}{\lambda}(1-4^{1-\lambda})\zeta(\lambda-1)\zeta(\lambda)\ . \tag{3.6d}$$

Here ζ is the Riemann zeta function, in particular $\zeta'(2)=-0.9345$. Results for the universal amplitudes

$$B_{i,j}=\left[\frac{\partial^i}{\partial y^i}\frac{\partial^j}{\partial w^j}B\right]_{y=w=0}\Big/(i!j!)\ , \tag{3.7}$$

some of which have been introduced already in Eq. (1.8), now follow from Eq. (2.21) by using Eqs. (3.3a) and (3.5)–(3.6d) together with Eqs. (2.5), (2.7), (2.8), (2.17), and (2.20d) and (2.20e) from Sec. II. Consider, e.g., $B_{0,2}$ which determines the finite-size universal ratio

$$[(T/T_{\rm CB})-1]^2(\xi_+/L)^{2/\nu}C_m^{(s)}(m=0,T=T_{\rm CB},L)/k_B=-2B_{0,2}\ , \tag{3.8}$$

involving the singular part $C_m^{(s)}$ of the total specific heat $C_m=-T\partial_T^2F$ for the finite system with *clamped* total order parameter mV. Using the knowledge that

$$[\partial_{t^R}^2\delta f^R(\overset{*}{u},0,t^R,\Lambda)]_{t^R=0}=\frac{n}{16\pi^2}\left[r_1+\frac{A_2^0}{\pi^2}-\ln\Lambda+O(\epsilon)\right], \tag{3.9a}$$

and that

$$\bar{a}_1^*=\frac{n}{16\pi^2}\left[1+\epsilon\left[r_1-2\frac{n+2}{n+8}(f_1+r_1)\right]+O(\epsilon^2)\right], \tag{3.9b}$$

Eqs. (2.21) with $l^{-1}=\Lambda$ and (2.20d) give

$$B_{0,2}=-\frac{1}{\alpha}\frac{n}{64\pi^2}\left[1+\epsilon\left[\frac{n-4}{n+8}\frac{A_2^0}{\pi^2}+\frac{n+2}{n+8}[\tfrac{3}{2}+\ln(4\pi)-C_E]\right]+O(\epsilon^2)\right], \tag{3.10}$$

with α the specific-heat exponent. Note that the arbitrary numbers r_1 and f_1 which enter intermediate quantities such as (3.9a) or (3.9b) or (2.20d), drop out in the universal quantity (3.10) as they should. In a similar way one obtains

$$B_{4,0}=\frac{1}{\epsilon}\frac{n+8}{288\pi^2}\left[1+\epsilon\left[\frac{A_2^0}{\pi^2}+\ln(4\pi)-C_E+\frac{n-1}{n+8}\ln 3+\frac{2}{3}\frac{n+5}{n+8}-\frac{3(3n+14)}{(n+8)^2}\right]+O(\epsilon^2)\right] \tag{3.11}$$

and

$$B_{2,1}=\frac{1}{\epsilon}\frac{n+8}{48\pi^2}\left[1+\frac{\epsilon}{n+8}\left[(n+2)\frac{A_2^0}{\pi^2}+(n+5)(\tfrac{1}{3}+\ln(4\pi)-C_E)+\frac{n-1}{2}\ln 3-\frac{3(3n+14)}{n+8}\right]+O(\epsilon^2)\right]. \tag{3.12}$$

We note also the results

$$B_{2\lambda,0}=(A_\lambda^0/2)(4\pi^2)^{-\lambda}\left[1+\frac{n-1}{3^\lambda}\right]+O(\epsilon)\ , \tag{3.13}$$

$$B_{0,\lambda}=(A_\lambda^0/2)(4\pi^2)^{-\lambda}n+O(\epsilon)\ , \tag{3.14}$$

valid for $\lambda=1,3,4,5,\ldots$, i.e., $\lambda\neq 2$. Since $B_{2,0}<0$, due to Eqs. (3.13) and (3.6b) and since $\delta B(y,0)\to+\infty$ for $y\to\pm\infty$, see Eq. (1.6) and below, the function $\delta B(y,0)$ shows minima at $y=\pm y_M$ as well as nonvanishing zeros at $y=\pm y_0$. Using Eqs. (3.6), (3.11), and (3.13) one finds

$$y_M^2=\epsilon 6\frac{n+2}{n+8}|A_1^0|+O(\epsilon^2)\ , \tag{3.15a}$$

$$\delta B(y_M,0)=-\epsilon\frac{(n+2)^2}{n+8}(A_1^0)^2/8\pi^2+O(\epsilon^2)\ , \tag{3.15b}$$

$$y_0/y_M=\sqrt{2}[1+O(\epsilon^2)]\ . \tag{3.15c}$$

B. System size large compared to bulk correlation length (bulk limit)

Consider (m^R,t^R) inside the (bulk) one-phase region and let $\Lambda\to\infty$. Due to Eqs. (2.5), this corresponds to $\rho_{\parallel,1}>0$ and $x_{\parallel,1}\to+\infty$ in Eq. (3.5a). In this case,

$$\mathcal{L}(x)=-(x\pi)^{d/2}\Gamma(-d/2)-\ln(xC)+O(e^{-x})\ , \tag{3.16}$$

with C a d-dependent constant, compare Eqs. (A10) and (A11) in the Appendix. The first and second terms on the rhs of Eq. (3.16) lead to the bulk limit and its leading correction, respectively, of δf^R in Eq. (3.5), within one-loop order. Consider in particular the bulk limit and $t^R=0$. Then Eq. (2.21) for $l^{-1}=\xi_m^R$ can be used to exponentiate the logarithms occurring in $\delta\overset{*}{f}{}^R$ and one obtains with Eqs. (3.5), (3.16), and (2.20d) that the leading behavior of $\delta B(y,0)$ for $y\gg 1$ is indeed given by the rhs of Eq. (1.6) with

$$b_m=\frac{1}{\epsilon}\frac{n+8}{288\pi^2}\left[1+\frac{\epsilon}{2}\left[\ln(4\pi)-\tfrac{3}{2}-C_E+\frac{n-1}{n+8}\ln 3+\frac{4}{3}\frac{n+5}{n+8}-\frac{6(3n+14)}{(n+8)^2}\right]+O(\epsilon^2)\right]. \tag{3.17}$$

b_m is a universal bulk amplitude which we have calculated for later reference. The corresponding amplitude b_+ for $\delta B(0,w\gg 1)$ could be gleaned from Ref. 34.

C. Magnetization distribution for Ising symmetry and comparison with Monte Carlo results

The Ginzburg-Landau counterpart of the magnetization distribution $P_L(m)$ in an Ising model of finite size[15] is essentially the quantity Z in Eq. (2.4) [see also Eq. (1.9)], for spin dimensionality $n=1$, which enters as a distribution for m^R within the Gibbs ensemble, compare Eq. (2.11). By adjusting scales in the asymptotic scaling form of Z for $T=T_{\rm CB}$, one may define a universal function[15]

$$p(\tilde{y})=p_0\exp[-\delta B(y,0)],\ \ \tilde{y}=y/\kappa\ . \tag{3.18a}$$

Here, Eqs. (2.4a) and (1.2) have been used, see also the remark in front of Eq. (2.21). Note that $p_0 = p(0)$. The scale changes

$$p_0^2 = I_2/(I_0)^3, \quad \kappa^2 = I_2/I_0 , \tag{3.18b}$$

with

$$I_j = \int_{-\infty}^{\infty} dy\, y^j \exp[-\delta B(y,0)] \tag{3.18c}$$

follow from the requirement[15] that the zeroth and second moment of $p(\bar{y})$ should equal unity, see Eq. (1.10).

A systematic ϵ expansion for p_0 and κ follows from Eq. (3.18) and (1.8) by using the forms (3.11), (3.13) of the expansion coefficients $B_{j,0}$ for $n=1$. One finds

$$p_0^2 = \frac{1}{4}\frac{M_2}{M_0^3}\left[1 + \epsilon^{1/2} H \left[3\frac{M_2}{M_0} - \frac{M_4}{M_2}\right] + \epsilon H^2 \left[6\frac{M_2^2}{M_0^2} - \frac{9}{2}\frac{M_4}{M_0} + \frac{1}{2}\frac{M_6}{M_2}\right] + O(\epsilon^{3/2})\right], \tag{3.19}$$

with moments

$$M_j = \int_0^{\infty} dy\, y^j \exp(-y^4) = \Gamma\left[\frac{j+1}{4}\right] \Big/ 4 \tag{3.20}$$

and with

$$H \equiv [B_{2,0}^{(0)}(B_{4,0}^{(-1)})^{-1/2}]_{n=1} = -(4\sqrt{2}\ln 2)/\pi . \tag{3.21}$$

Here $B_{i,j}^{(l)}$ denotes the coefficient multiplying ϵ^l in the ϵ expansions of $B_{i,j}$ in Eqs. (3.11) and (3.13). This gives in terms of numbers

$$p_0 = 0.321[1 - 0.171\epsilon^{1/2} - 0.065\epsilon + O(\epsilon^{3/2})] . \tag{3.22}$$

Similarly,

$$\kappa^2 = \epsilon^{1/2}\frac{M_2}{M_0} 4\pi\sqrt{2}\left\{1 + \epsilon^{1/2} H \left[\frac{M_2}{M_0} - \frac{M_4}{M_2}\right] + \epsilon\left[(HM_2/M_0)^2 - \frac{1}{2}\left[\frac{A_2^0}{\pi^2} + \ln(4\pi) - C_E - \frac{5}{27}\right]\right] + O(\epsilon^{3/2})\right\} . \tag{3.23}$$

The positions $\pm\bar{y}_M$ of the symmetric maxima of p follow now from (3.18a) and the result (3.15a) for the corresponding minima of δB as

$$\bar{y}_M = 1.359\epsilon^{1/4}[1 - 0.250\epsilon^{1/2} + O(\epsilon)] . \tag{3.24}$$

The height of the maxima of p is determined by Eq. (3.15b) together with Eqs. (3.18a) and (3.22). The result is

$$p(\bar{y}_M) \equiv p_M = 0.321[1 - 0.171\epsilon^{1/2} + 0.324\epsilon + O(\epsilon^{3/2})] \tag{3.25}$$

and should be compared with (3.22). Table I shows a comparison of our estimates for $p_{0,M}$ and $\bar{y}_M$ with Monte Carlo results.

The universal scale κ^2 in (3.18b) has a more direct meaning in terms of the susceptibility in the Gibbs ensemble. The corresponding derivative of the Gibbs scaling function in Eq. (2.22) has the form

$$-[\partial_v^2 \delta B_G(v,0)]_{v=0} = (b_m \kappa v d/\beta)^2 , \tag{3.26}$$

where Eqs. (2.24) and (3.18b) have been used. Contrary to the fourth moment

$$r_2 \equiv \int_{-\infty}^{\infty} d\bar{y}\, \bar{y}^4 p(\bar{y}) = I_0 I_4/(I_2)^2 \tag{3.27}$$

of p, which can be expressed by fourth and second derivatives of δB_G in (2.22), with respect to its first argument and which was considered in Ref. 8, the quantity in Eq. (3.26) involves no fourth derivative and thus contains independent finite size information. Monte Carlo data for the finite size susceptibility from Refs. 19, 35, and 36 and information on the bulk correlation length ξ_h from Ref. 31 lead to the estimate

$$[(\xi_h/L)^{2\Delta/\nu} h^2](-\partial_h^2 F_G^{(s)}/k_B T_{CB})_{h=0} \approx (0.257)^{2\Delta/\nu} 1.55 \tag{3.26'}$$

for the quantity on the lhs of Eq. (3.26) for the $d=3$ simple cubic Ising model. This can be compared with the ϵ expansion

$$(b_m \kappa v d/\beta)^2 = 0.000\,963\,5\epsilon^{-3/2} \times [1 + 0.5013\epsilon^{1/2} - 0.1856\epsilon + O(\epsilon^{3/2})] \tag{3.26''}$$

of the rhs of Eq. (3.26) which follows from Eqs. (3.17) and (3.23), see Table I.

TABLE I. The first line quotes Monte Carlo results from Fig. 11 in Ref. 15 for the quantities $p_{0,M}$ and $\bar{y}_M$ introduced in Sec. III C. The second line shows our ϵ-expansion estimates. These follow from Eqs. (3.22), (3.24), and (3.25) by discarding contributions denoted by O and setting $\epsilon=1$. The agreement is improved significantly over that with the zero-loop expression $p = 0.321\exp(-0.114\bar{y}^4)$ where $p_0 = p_M = 0.321$ and $\bar{y}_M = 0$. The first and second lines in the last column show the rhs of Eqs. (3.26') and (3.26''), respectively.

p_0	p_M	$\bar{y}_M$	$-\partial_v^2 B_G$
0.25	0.35	1.1	0.0018
0.245	0.370	1.019	0.0013

Finally, we present an estimate for the complete scaling function p in Eq. (3.18a). To this end, we follow the strategy used by Rudnick, Guo, and Jasnow in Ref. 9 and use the (asymptotic) renormalization flow equation (2.15) for $t^R=0$. Since the scale D_m in Eq. (2.16a) drops out from the universal function p, one may set it equal to unity (or equivalently put $u=\overset{*}{u}$) for convenience. According to the form of the propagator given in Eqs. (3.4) and (3.3d), it is appropriate to introduce the variables

$$z=\overset{*}{u}^{1/2}m^R,\quad \tilde{z}=\overset{*}{u}^{1/2}\tilde{m}^R \tag{3.28}$$

instead of $m^R,\tilde{m}^R$. The corresponding function

$$S(z,\Lambda)=\delta f^R(\overset{*}{u},z/\overset{*}{u}^{1/2},0,\Lambda)\ , \tag{3.29}$$

which appears on both sides of Eq. (2.15) has to be considered now for arbitrary magnitudes of z and Λ. From Eqs. (3.5), (3.3), and (2.5) one finds that

$$S(z,\Lambda)=S_0(z,\Lambda)+S_1(z,\Lambda)+O(\epsilon)\ , \tag{3.30}$$

with

$$S_0(z,\Lambda)=z^4 2\pi^2/(3\overset{*}{u})\ , \tag{3.30a}$$

$$S_1(z,\Lambda)=z^4\pi^2(C_E+\ln(2\pi)-\tfrac{3}{2}+2\ln z)+\Lambda^{-4}\Delta(2\Lambda^2z^2)/2\ , \tag{3.30b}$$

for $n=1$. Here

$$\Delta(x)=\lim_{d\to 4}[\mathcal{L}(x)+(x\pi)^{d/2}\Gamma(-d/2)]\ . \tag{3.31}$$

As in Ref. 9 one may use the convenient representation

$$\Delta(x)=\vartheta(x)-\vartheta(0)\ , \tag{3.32}$$

where

$$\vartheta(x)=-\int_0^1 d\sigma[\Theta^4(\pi/\sigma)-1]\times(\sigma^{-3}e^{-x\pi\sigma}+\sigma^{-1}e^{-x\pi/\sigma})+\tfrac{1}{2}(1-x\pi)e^{-x\pi}+[1-(x\pi)^2/2]\mathrm{Ei}(-x\pi)-\ln(x\pi)\ , \tag{3.33}$$

with Ei the exponential integral and

$$\Theta(\sigma)=1+2\sum_{n=1,2,\ldots}^{\infty}e^{-\sigma n^2} \tag{3.34}$$

see the Appendix. For later reference, we note the limiting behavior

$$\Delta(x)=xA_1^0+x^2\left[-\frac{\pi^2}{2}\ln x+A_2^0+\frac{\pi^2}{2}(\tfrac{3}{2}-C_E-\ln\pi)\right]+O(x^3)\ , \tag{3.35}$$

for $x\ll 1$, which follows from Eqs. (3.6) and (3.31). Behavior for $x\gg 1$ is given by Eq. (3.16).

The evaluation of scaling functions by means of renormalization flow equations has been discussed by many authors, see e.g., Refs. 37 and 38. For a critical system, the "mass" M in the propagators—which provides a lower cutoff for wave-number integrations in a fluctuation expansion—becomes small. This makes the expansion useless since successive contributions contain increasing powers of $\ln M$. Using flow equations like (2.15), one thus maps the critical system onto a noncritical one where $M=O(1)$ and for which the expansion is meaningful. In the present case, the propagators are given by Eqs. (3.4) and the squared mass has the qualitative structure

$$M^2=\Lambda^{-2}+z^2\ . \tag{3.36}$$

In accord with the previous discussion, one may verify that the ratio S_1/S_0 of successive contributions to δf^R contains indeed a divergence of the type $\ln M$ as Λ^{-2} and z^2 become small: For $\Lambda^{-2}/z^2\ll 1$ or $=O(1)$, the second contribution on the rhs of Eq. (3.30b) can be neglected with respect to the first one and the $\ln M$ behavior arises from the $\ln z$ term. For $\Lambda^{-2}/z^2\gg 1$, the form (3.35) of Δ precisely cancels this $\ln z$ term and there arises a $\ln(\Lambda^{-1})$ term instead which is again of the form $\ln M$. We note that the two limits mentioned, i.e., $\Lambda\to\infty$ for fixed z and $z\to 0$ for fixed Λ, are the bulk and strong finite size limits, respectively, with the scaling variable

$$\hat{y}=\Lambda^{\beta/\nu}z\ , \tag{3.37}$$

being large or small, respectively.

Contrary to δf^R on the lhs of Eq. (2.15) with squared mass given by (3.36), δf^R on the rhs has a squared mass of magnitude

$$\tilde{M}^2=\tilde{\Lambda}^{-2}+\tilde{z}^2=\Lambda^{-2}l^{-2}+z^2l^{-2\beta/\nu}\ . \tag{3.38}$$

Choosing l such that

$$\tilde{M}^2=1\ , \tag{3.39}$$

δf^R on the rhs of (2.15) is noncritical and can, as a function of $\tilde{z}$ and $\tilde{\Lambda}$, be calculated perturbatively. Our approximation consists in terminating the series for S after the second term on the rhs of Eq. (3.30) and using the lowest-order estimate $\epsilon/3=\frac{1}{3}$ for $\overset{*}{u}$. With (2.12a) the approximate free energy δF_a^R reads

$$\delta F_a^R=\tilde{\Lambda}^d[\tilde{z}^4 2\pi^2+S_1(\tilde{z},\tilde{\Lambda})]\ . \tag{3.40}$$

For the exponents β,ν,d occurring in (3.38) and (3.40), we use their best estimates in $d=3$, see Refs. 25 and 39. Note that for $\hat{y}\ll 1$ and $\hat{y}\gg 1$ the matching condition (3.38), (3.39) reduces to the conditions $\tilde{\Lambda}^{-2}=1$ and $\tilde{z}^2=1$, respectively, which have been used previously in Secs. III A and III B. The solution of (3.38), (3.39) in the general case may be simplified by the observation that for $d=3$, $n=1$ (and $n=2,3,\ldots$), the replacement of $2\beta/\nu$ by 1 is in error only by about 3%. This leads to

$$\tilde{\Lambda}=\{[1+\hat{y}^4/4]^{1/2}-\hat{y}^2/2\}^{-1}\ ,\quad \tilde{z}=\hat{y}(\tilde{\Lambda})^{-1/2}\ . \tag{3.41}$$

Inserting (3.41) into Eq. (3.40) one obtains δF_a^R as a function of $\hat{y}$ and the desired estimate for p in Eq. (3.18a) follows from adjusting scales $\hat{p}_0,\hat{\kappa}$ in

$$p(\bar{y})=\hat{p}_0\exp(-\delta F_a^R),\quad \bar{y}=\hat{y}/\hat{\kappa}\ , \tag{3.42}$$

such that the zeroth and second moment of $p(\bar{y})$ equal unity. Figure 1 shows the Monte Carlo data for $p(\bar{y})$ from Ref. 15 together with our estimate from Eqs. (3.40)–(3.42) and (3.30b)–(3.34). The striking agreement

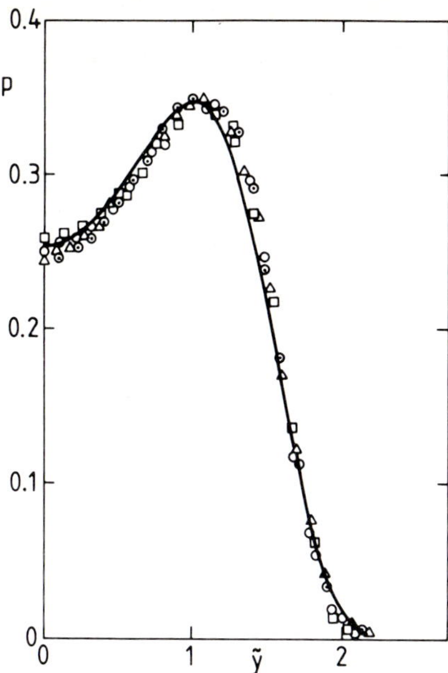

FIG. 1. Universal scaling form $p(\tilde{y})$ of the order parameter distribution function at $T=T_{\mathrm{CB}}$ as defined in Eqs. (3.18): The solid line represents the present estimate which follows from Eqs. (3.40), (3.41), and (3.42). The dotted circles, and the triangles, circles, and squares represent Monte Carlo data from Fig. 11 of Ref. 15.

should, however, be seen in the right perspective: Calculating scaling functions via flow equations suffers as usual from a certain degree of ambiguity. One can also argue in favor of approximate equations slightly different from those in (3.38) and (3.40). Our choice was guided by the requirement of matching closely the estimates from the *systematic* ϵ expansion in the second line of Table I, in particular the ratio p_M/p_0. Thus Fig. 1 demonstrates that the good agreement with Monte Carlo data as shown for the three amplitudes $p_0, p_M, \tilde{y}_M$ in Table I is maintained for the whole scaling function $p(\tilde{y})$.

IV. SUMMARY AND CONCLUDING REMARKS

In an *infinite* magnetic or liquid-gas (Ginzburg-Landau-model) system, the bulk thermodynamic state is—apart from the parameters u,d,n—completely specified by a pair of values for h,t [or equivalently for m,t via the bulk equation of state, Eq. (2.13b)]. Using either the Gibbs or the Helmholtz ensemble with correspondingly related variables (h,t) or (m,t) makes only a formal difference but actually describes the same bulk thermodynamic state ("equivalence of ensembles"). This follows, e.g., by recognizing that the entropy density is the same in both cases, i.e.,

$$-\partial_{t^R}\delta f^R_{G,b}(h^R,t^R)=-\partial_{t^R}\delta f^R_b(m^R,t^R)\ , \tag{4.1}$$

according to Eqs. (2.13). In the simplest situation $t>0$ and $h=0$, where also $m=0$, this implies the equality of specific heats per unit volume

$$-\partial^2_{t^R}\delta f^R_{G,b}(0,t^R)=-\partial^2_{t^R}\delta f^R_b(0,t^R),\quad t^R>0 \tag{4.1a}$$

for constant h and constant m, respectively.

The thermodynamic state of a *finite* macroscopic system close to its bulk critical point requires further specification. It depends firstly[1] on the shape of the system and the boundary conditions on its surface. The latter arise physically, e.g., for a real gas, from the attractive or repulsive forces between the molecules and the surface or from the change of effective intermolecular interactions near the surface. Secondly,[40,41] the state now actually depends on whether the total order parameter mV may fluctuate as, e.g., in a demixing alloy in equilibrium with the gas phase outside the crystal with a fixed chemical potential h (Gibbs ensemble) or whether mV is fixed as, e.g., in a real gas enclosed by rigid impenetrable walls (Helmholtz ensemble). Consider, e.g., the relative deviation

$$R(w)=[\partial^2_{t^R}f^R_G(0,t^R)/\partial^2_{t^R}f^R(0,t^R)]_{\mathrm{sing}}-1$$
$$\equiv[\partial^2_w\delta B_G(0,w)/\partial^2_w\delta B(0,w)]-1 \tag{4.2}$$

of the singular parts of the specific heats in the two ensembles (i.e., of the specific heats at constant h and constant mV, respectively), for $t\geq 0$, $h=0$, and $m=0$. Here $w=(L/\xi_+)^{1/\nu}$ has been introduced in Eq. (1.4′) and $\delta B, \delta B_G$ are scaling functions defined in Eqs. (1.2) and (2.22). According to Eq. (4.1a), R vanishes,

$$R(+\infty)=0\ , \tag{4.2a}$$

as the system becomes infinite, i.e., $L\to\infty$, for fixed $t>0$. However, R is nonvanishing for $w<\infty$, i.e., for a finite system. For a cube with periodic boundary conditions, one obtains[42]

$$R(0)=\frac{2(4-n)}{n}\left[\frac{M_{n+3}}{M_{n-1}}-\left[\frac{M_{n+1}}{M_{n-1}}\right]^2\right]+O(\epsilon^{1/2})\ , \tag{4.2b}$$

with the moments M from Eq. (3.20). For $n=1$, the ϵ^0 contribution to the rhs of Eq. (4.2b) leads to an estimate $R(0)\approx 0.81$. Thus the Gibbs and Helmholtz ensembles describe different physical situations and are nonequivalent, for finite-size properties.

In the present paper, we have considered the universal scaling form (1.2) for the asymptotic singular part of the Helmholtz free energy in case of a cube with periodic boundary conditions. For related amplitudes, defined in (3.7) or (1.8) and in (1.6), we have obtained systematic ϵ-expansion results in Eqs. (3.10)–(3.14) and in Eq. (3.17). Some of these determine the results given in Eqs. (3.15) for the minimum and the nontrivial zero of the scaling function $\delta B=B-B(0,0)$, see (1.2), at $T=T_{\mathrm{CB}}$. Estimates for corresponding amplitudes of the normalized "magnetization distribution" p for $n=1$ as defined in Eqs. (3.18) are given in Eqs. (3.22), (3.24), and (3.25) in terms of a systematic ϵ expansion and compare well with Monte Carlo results of Ref. 15, see Table I. This also holds true for our estimate for the complete scaling function p given in Fig. 1. Finally, a related susceptibility amplitude in the Gibbs ensemble, see Eqs. (3.26), is compared with Monte Carlo and (bulk-) series estimates in Table I. Generally, our one-loop estimates for asymptotic quantities compare surprisingly well with the numerical results which are mainly from systems[15] of modest size. Information about higher-loop contributions as well as simulations of larger sized systems would be clearly of interest.

We have used a rather general form of the renormalization scheme, compare Eqs. (2.7) and (2.17b), and have checked the independence of universal finite size amplitudes from the details in that scheme.

In conclusion, we would like to stress that despite recent progress,[8,9] the present understanding of finite size scaling is still incomplete, even for a cube with periodic boundary conditions. This refers to the possible importance of interface (or droplet-) configurations for $T < T_{CB}$ and $n=1$. As in the corresponding bulk problem, these cannot be included within a perturbative approach starting from a spatially homogeneous state. Phenomenological arguments[14,15] show that droplet configurations will contribute to the Helmholtz free energy of a finite system only in a region D in the m,t plane which is bounded from above by a borderline λ_D, situated below the borderline λ_c of the bulk coexistence region C, i.e., $D \subset C$. λ_D crosses the negative t axis at $tL^{1/v}=-O(1)$ and approaches λ_c asymptotically for $tL^{1/v}\to-\infty$. Thus in a Helmholtz ensemble with $m=0$, the fluctuation expansion developed in Refs. 8 and 9, see also Sec. III of the present paper, will break down for temperatures below T_c with $tL^{1/v}<-O(1)$, i.e., for a bulk correlation length smaller than the system size L. After all, the form (3.4) of the propagator indicates an instability of the homogeneous state for $m=0$ in about this regime, even for linear stability analysis.

The situation is more advantageous for the Gibbs ensemble (with $n=1$ as before) where one has to evaluate the integral in Eq. (2.11). Consider e.g., $tL^{1/v}\ll -1$ fixed. Then the two degenerate minima of δF^R (of depth $\sim L^d$) will be situated at $m=\pm|m_{\lambda_c}|$ right on the line λ_c and δF^R should increase by an amount $\sim(L|t|^v)^{d(d-1)/(d+1)}\gg 1$ on decreasing $|m|$ from $|m_{\lambda_c}|$ to $|m_{\lambda_D}|$ on the line λ_D.[14,15,43] Thus the integral in (2.11) is dominated by the small intervals of width

$$|m_{\lambda_c}|-|m_{\lambda_D}|\sim|t|^{\beta}(L|t|^{v})^{-d/(d+1)}$$

around $m=\pm|m_{\lambda_c}|$, where the droplet formation is still unimportant. Actually, Rudnick *et al.*[9] found consistent behavior for $tL^{1/v}\to-\infty$ in the case of the specific heat in the Gibbs ensemble.

For $n>1$, the propagators (3.4) contain singularities which are situated close to the bulk line λ_c and prevent a naive application of the fluctuation expansion even for the Gibbs ensemble, for $tL^{1/v}\ll-1$. These singularities are due to long-wavelength Goldstone modes and have to be carefully exponentiated even in bulk problems, see, e.g., Refs. 44 and 45. To obtain quantitative information about finite size properties in this case,[46] the use of a $d=2+\epsilon$ expansion seems preferable[8] in which case the complete renormalization flow with a critical as well as a zero-temperature (Goldstone) fixed point can be obtained in perturbation theory.

ACKNOWLEDGMENTS

We thank K. Binder, T. W. Burkhardt, and H. W. Diehl for useful discussions.

APPENDIX

We describe first how to find the coefficients A_λ in Eqs. (3.6). From (3.5b) we obtain

$$\mathcal{L}'(x)=\sum_j{}'\frac{1}{j^2+x}=\int_0^\infty d\sigma e^{-\sigma x}\{[\Theta(\sigma)]^d-1\} \tag{A1}$$

with Θ defined in (3.34), see also Ref. 8. Comparison with (3.6) leads to

$$\lambda!A_\lambda=(-1)^{\lambda-1}\int_0^\infty d\sigma\sigma^{\lambda-1}\{\Theta^d-1\}\ . \tag{A2}$$

Since[8]

$$\Theta(\sigma)\equiv(\pi/\sigma)^{1/2}\Theta(\pi^2/\sigma)$$
$$\approx(\pi/\sigma)^{1/2}[1+2e^{-\pi^2/\sigma}+\cdots]\ \text{ for } \sigma\to 0\ , \tag{A3}$$

the integral in (A2) is well defined (near its lower boundary) only for $d<2\lambda$. We will make use of the formula

$$\int_0^\infty d\sigma\sigma^{\lambda-1}[\Theta^4-1]=8[1-4^{1-\lambda}]\zeta(\lambda-1)\zeta(\lambda)\Gamma(\lambda)\ , \tag{A4}$$

which can be found as the "sum $S(0,4)$" in Table I of Ref. 33 and is valid for any real $\lambda>2$. Thus for $\lambda>2$, A_λ in (A2) tends to the finite value A_λ^0, given in Eq. (3.6d), as $d\to4$. To evaluate A_2, one may subtract and add $(\pi/\sigma)^{d/2}\exp(-\sigma)$ in the integrand of (A2) for $\lambda=2$ to obtain

$$-2A_2-\pi^{d/2}\Gamma(2-d/2)+O(\epsilon)=\int_0^\infty d\sigma\sigma[\Theta^4-1-(\pi/\sigma)^2e^{-\sigma}]\ . \tag{A5}$$

Writing the factor σ in front of the square bracket as $\lim\sigma^{1+a}$ with a tending to zero from positive values, the rhs of (A5) becomes

$$\lim_{a\to0}\left[-\pi^2\Gamma(a)+\int_0^\infty d\sigma\sigma^{1+a}(\Theta^4-1)\right]=\pi^2[1+\tfrac{2}{3}\ln2+6\zeta'(2)/\pi^2+C_E]\ , \tag{A6}$$

where again (A4) has been used for $\lambda-1=1+a>1$. Equations (A5) and (A6) lead to the result for A_2 given in Eqs. (6). Similarly, A_1 may be analytically continued from $d<2$ to $d\to4$ by subtracting and adding $(\pi/\sigma)^{d/2}(1+\sigma)\exp(-\sigma)$ to the integrand in (A2) for $\lambda=1$. This gives

$$A_1-\pi^{d/2}\{\Gamma(2-d/2)+\Gamma(1-d/2)\}+O(\epsilon)=\int_0^\infty d\sigma[\Theta^4-1-(\pi/\sigma)^2(1+\sigma)e^{-\sigma}] . \quad \text{(A7)}$$

The curly brackets on the lhs (and thus A_1) approach finite values (A_1^0) for $d\to 4$. To obtain A_1^0, the rhs of (A7) can be evaluated by inserting σ^b in front of the square bracket, using (A4) for $\lambda-1=b>1$ and analytic continuation to $b=0$. This leads to the result given in Eq. (3.6b) for A_1^0.

We conclude with a few remarks on how to obtain Eqs. (3.32) and (3.33). A corresponding derivation within a cutoff-regularized theory has been given already in Ref. 9. Within dimensional regularization one considers the quantity

$$\mathcal{L}(x)=\int_0^\infty \frac{d\sigma}{\sigma}(1-e^{-x\sigma})\{[\Theta(\sigma)]^d-1\} , \quad \text{(A8)}$$

with $\mathcal{L}(0)=0$ from (3.5b) and (A1) which is well defined for $d<2$. To continue to dimensions d near 4, one may use the identity

$$(x\pi)^{d/2}\Gamma(-d/2)=-\int_0^\infty \frac{d\sigma}{\sigma}[1-e^{-x\sigma}](\pi/\sigma)^{d/2} \quad \text{(A9)}$$

valid for $0<d<2$. According to (3.34) and (A3), the sum of the integrands on the rhs of (A8) and the right-hand side (A9) is proportional to σ^0 and $\sigma^{-1-d/2}$ for $\sigma\to 0$ and $\sigma\to+\infty$, respectively. This defines the sum $\tilde{\mathcal{L}}(x)$ of the lhs of (A8) and the left-hand side (A9) as an integral over σ which is well defined for all $d>0$. As in Ref. 9 one may separate the σ integration into two parts extending from 0 to τ and from τ to ∞, respectively, with corresponding contributions

$$\tilde{\mathcal{L}}_<(x)=\int_0^\tau \frac{d\sigma}{\sigma}(1-e^{-x\sigma})\{[\Theta(\sigma)]^d-(\pi/\sigma)^{d/2}\} -\int_0^\tau \frac{d\sigma}{\sigma}(1-e^{-x\sigma}) \quad \text{(A10a)}$$

and

$$\tilde{\mathcal{L}}_>(x)=\int_\tau^\infty \frac{d\sigma}{\sigma}(1-e^{-x\sigma})\{[\Theta(\sigma)]^d-1\} -\pi^{d/2}\int_\tau^\infty d\sigma\sigma^{-1-d/2}(1-e^{-x\sigma}) \quad \text{(A10b)}$$

to $\tilde{\mathcal{L}}$. Using the representations (A3) and (3.34) for Θ in Eqs. (A10a) and (A10b), respectively, and choosing for τ a value of $O(1)$, Eqs. (A10) serve for a convenient numerical evaluation of

$$\mathcal{L}(x)+(x\pi)^{d/2}\Gamma(-d/2)=\tilde{\mathcal{L}}_<(x)+\tilde{\mathcal{L}}_>(x) , \quad \text{(A11)}$$

for arbitrary $d>0$. For $d=4$, in particular, Eqs. (3.32) and (3.33) are obtained for $\tau=\pi$ by using Eqs. (5.1.1), (5.1.2), and (5.1.39) of Ref. 47 and partial integration for the last term on the rhs of (A10b). The form (3.16) for $\mathcal{L}(x\gg 1)$ follows also from Eqs. (A10) and (A11).

---

[1]M. N. Barber, in *Phase Transitions and Critical Phenomena,* edited by C. Domb and J. L. Lebowitz (Academic, New York, 1983), Vol. 8.

[2]L. Onsager, Phys. Rev. **65**, 117 (1944).

[3]M. E. Fisher and A. E. Ferdinand, Phys. Rev. Lett. **19**, 169 (1967).

[4]A. E. Ferdinand and M. E. Fisher, Phys. Rev. **185**, 832 (1969).

[5]M. E. Fisher, in *Critical Phenomena,* Proceedings of the International School of Physics, "Enrico Fermi," Vol. 51, edited by M. S. Green (Academic, New York, 1971); see also, Y. Imry and D. Bergman, Phys. Rev. A **3**, 1416 (1971).

[6]M. E. Fisher and M. N. Barber, Phys. Rev. Lett. **28**, 1516 (1972).

[7]J. Cardy, see Ref. 1, Vol. 11 (to be published).

[8]E. Brézin and J. Zinn-Justin, Nucl. Phys. B **257** (FS14), 867 (1985).

[9]J. Rudnick, H. Guo, and D. Jasnow, J. Stat. Phys. **41**, 353 (1985).

[10]K. Symanzik, Nucl. Phys. B **190** (FS3), 1 (1981).

[11]E. Eisenriegler, Z. Phys. B **61**, 299 (1985); Equation (A14) of Ref. 11 contains a misprint. The quantity $\overline{\mathcal{H}}_{I,\mathrm{ac}}[l]$ should correctly read $\overline{\mathcal{H}}_{I,\mathrm{ac}}[l']$.

[12]J. Rudnick, G. Gaspari, and V. Privman, Phys. Rev. B **32**, 7594 (1985).

[13]S. Singh and R. K. Pathria, Phys. Rev. B **33**, 672 (1985).

[14]L. S. Schulman, J. Phys. A **13**, 237 (1980); K. Binder, Phys. Rev. A **26**, 556 (1982).

[15]K. Binder, Z. Phys. B **43**, 119 (1981).

[16]V. Privman and M. E. Fisher, Phys. Rev. B **30**, 322 (1984).

[17]S. K. Ma, in *The Modern Theory of Critical Phenomena* (Benjamin, New York, 1976).

[18]K. G. Wilson, J. B. Kogut, Phys. Rep. **12C**, 75 (1974).

[19]M. N. Barber, R. B. Pearson, D. Toussaint, and J. Richarson, Phys. Rev. B **32**, 1720 (1985).

[20]E. Brézin, J. Phys. (Paris) **43**, 15 (1982).

[21]H. W. Diehl, see Ref. 1, Vol. 10.

[22]D. Amit, *Field Theory, the Renormalization Group, and Critical Phenomena* (McGraw-Hill, New York, 1978).

[23]Notice that also ϵ-independent prefactors other than $16\pi^2$ have been used in the renormalization corresponding to Eq. (2.5a). These lead only to a trivial redefinition of u in (2.5a) in $Z_{u,\phi,\tau}$ by the corresponding ϵ-independent numerical ratio.

[24]I. D. Lawrie, J. Phys. A **9**, 961 (1976).

[25]A. A. Vladimirov, D. I. Kazakov, and O. V. Tarasov, Zh. Eksp. Teor. Fiz. **77**, 1035 (1979) [Sov. Phys.—JETP **50**, 521 (1979)].

[26]V. Dohm, Z. Phys. B **60**, 61 (1985). This paper discusses in detail the role played by the factor f for bulk quantities.

[27]J. S. Kang, Phys. Rev. D **13**, 851 (1976).

[28]R. B. Griffiths, Phys. Rev. **152**, 240 (1966).

[29]m_M is unique in the h^R,t^R plane apart from the line $t^R<0$, $h^R=0$, compare Ref. 28.

[30] D_m is the inverse of D_h which was introduced in Appendix A of Ref. 11.

[31] ξ is the "effective range of correlation" or "second-moment correlation length" as defined in H. B. Tarko and M. E. Fisher, Phys. Rev. B **11**, 1217 (1975) in case of the Ising model.

[32] M. Combescot, M. Droz, and J. M. Kosterlitz, Phys. Rev. B **11**, 4661 (1975); see also, C. Bervillier, Phys. Rev. B **11**, 4964 (1976).

[33] I. J. Zucker, J. Phys. A **7**, 1568 (1974).

[34] P. C. Hohenberg, A. Aharony, B. I. Halperin, and E. D. Siggia, Phys. Rev. B **13**, 2986 (1976).

[35] G. Bhanot, D. Duke, and R. Salvador, Phys. Rev. B **33**, 7841 (1986).

[36] A. Hoogland, A. Compagner, and H. W. J. Blöte, Physica A (to be published).

[37] D. R. Nelson, Phys. Rev. B **11**, 3504 (1974).

[38] J. Rudnick and D. R. Nelson, Phys. Rev. B **13**, 2208 (1976).

[39] J. C. Le Guillou and J. Zinn-Justin, Phys. Rev. Lett. **39**, 95 (1977).

[40] T. L. Hill, *Thermodynamics of Small Systems* (Benjamin, New York, 1963).

[41] H. Furukawa and K. Binder, Phys. Rev. A **26**, 556 (1982).

[42] The contributions to $R(0)$ of order $\epsilon^{1/2}$ and ϵ^1 could be explicitly evaluated via Eqs. (4.2) and (2.22) by expanding δB for small y and w as in Sec. III C and by using the information contained in Eqs. (3.6) for the expansion coefficients (3.7) of δB. Note that $R(w)$ is a universal function which apart from shape and boundary condition depends only on the bulk universality class specified by d and n.

[43] K. Binder and D. P. Landau, Phys. Rev. B **30**, 1477 (1984).

[44] L. Schäfer and H. Horner, Z. Phys. B **29**, 251 (1978).

[45] I. D. Lawrie, J. Phys. A **14**, 2489 (1981).

[46] M. E. Fisher and V. Privman, Phys. Rev. B **32**, 447 (1985).

[47] *Handbook of Mathematical Functions,* edited by M. Abramowitz and I. A. Stegun (Dover, New York, 1965).

Nuclear Physics B280 [FS18] (1987) 340–354
North-Holland, Amsterdam

FINITE SIZE SCALING EFFECTS IN DYNAMICS

Yadin Y. GOLDSCHMIDT

Department of Physics and Astronomy, University of Pittsburgh, Pittsburgh, PA 15260, USA

Received 23 June 1986
(Revised 28 July 1986)

We calculate the linear relaxation time for a finite size system with a cubic geometry and "model A" dynamics both above T_c in a $4-\varepsilon$ expansion and below T_c in a $2+\varepsilon$ expansion, and express the results in a scaling form. The universal scaling functions are obtained to one-loop order. We use the method of the effective hamiltonian for the homogeneous modes. The quantum mechanical hamiltonian is supersymmetric. The large n limit (infinitely many spin components) is also considered both above and below T_c.

1. Introduction

In recent years there has been much interest in the theory of finite size scaling (FSS) [1]. The theory has been formulated for the first time by Fisher [2] and studied extensively in subsequent papers. It is very useful as an efficient extrapolation for numerical calculations which are typically performed on small systems. It is also the basis for a real space renormalization procedure [3]. Recently Brézin and Zinn-Justin [4] (BZ) developed a method to analytically perform a calculation of the size-dependent universal scaling functions in an ε expansion, which is singular about four dimensions [5]. This means that the series generated is in powers of $\varepsilon^{1/2}$ or $\varepsilon^{1/3}$ depending upon the geometry of the sample. The first case corresponds to a cubic geometry of volume L^d and periodic boundary conditions, the second case to a "cylindrical" geometry for which $d-1$ dimensions are of size L and one dimension is infinite. A calculation of the correlation length in powers of $2+\varepsilon$ has also been performed [4].

In this paper we extend their method for dynamics [6], and calculate the linear relaxation time for a sample with the cubic geometry. In sect. 2 the scaling form of the relaxation time is calculated in powers of $\varepsilon^{1/2}$ in $4-\varepsilon$ dimensions above and at T_c. The quantum mechanical hamiltonian for the homogeneous modes is supersymmetric. In sect. 3 we calculate the linear relaxation time below T_c in $2+\varepsilon$ dimensions using the nonlinear σ-model. We also discuss the behavior of the relaxation time below T_c for an Ising spin system. In sect. 4 we discuss the large n limit of infinitely many spin components and show that it is consistent with the ε

0619-6823/87/$03.50

expansion. We restrict ourselves in this paper to model A dynamics [6] – a nonconserved order parameter.

2. Expansion about the upper critical dimension (four)

Our aim is to calculate the linear relaxation time for an n-component spin system. The relaxation time is defined, for example, by Fisher and Racz [7]: Let $\psi(\tau)$ be a thermodynamic property which relaxes to zero with time τ. Let $\psi_0 = \psi(0)$ be its initial value. Then

$$\tau_{\mathrm{R}}^{(l)}(T) = \lim_{\psi_0 \to 0} \int_0^{\infty} \mathrm{d}\tau \, \frac{\psi(\tau)}{\psi_0} . \tag{2.1}$$

We will not consider here the nonlinear relaxation time which is defined for nonzero ψ_0 as $T \to T_{\mathrm{c}}$ [7, 8]. We will omit the superscript (l). We denote by $\tau_{\mathrm{R}}(\infty, T)$ the bulk relaxation time and by $\tau_{\mathrm{R}}(L, T)$ the relaxation time of the finite system. The relaxation time can also be defined [6] as $[\omega_{\psi}(\boldsymbol{k}=0)]^{-1}$ where

$$\psi_{\psi}(\boldsymbol{k}) = \frac{\Gamma_{\psi}(\boldsymbol{k})}{\chi_{\psi}(\boldsymbol{k})}, \qquad \frac{1}{\Gamma_{\psi}(\boldsymbol{k})} = i \left. \frac{\partial \chi_{\psi}^{-1}(\boldsymbol{k}, \omega)}{\partial \omega} \right|_{\omega=0} . \tag{2.2}$$

Here $\chi_{\psi}(\boldsymbol{k}, \omega)$ is the linear response function and $\chi_{\psi}(\boldsymbol{k}) = \chi_{\psi}(\boldsymbol{k}, \omega=0)$. Near T_{c} the bulk relaxation time diverges like $t^{-\nu z}$ where $t = (T - T_{\mathrm{c}})/T_{\mathrm{c}}$, ν is the correlation length exponent and z is the dynamical exponent. Model A dynamics is defined by the following markovian equations of motion

$$\partial_{\tau} \phi_i(\boldsymbol{x}, \tau) = -\lambda_0 \frac{\delta \mathbb{H}}{\delta \phi_i} + \zeta_i(\boldsymbol{x}, \tau),$$

$$\frac{H}{T} \equiv \mathbb{H} = \int \mathrm{d}^d x \left[\tfrac{1}{2} (\nabla \phi)^2 + \tfrac{1}{2} r_0 \phi^2 + \frac{1}{4!} u_0 (\phi^2)^2 \right],$$

$$\langle \zeta_i(\boldsymbol{x}, \tau) \rangle = 0, \qquad \langle \zeta_i(\boldsymbol{x}, \tau) \zeta_j(\boldsymbol{x}', \tau') \rangle = 2\lambda_0 \delta^{(d)}(\boldsymbol{x} - \boldsymbol{x}') \delta(\tau - \tau') \delta_{ij} . \tag{2.3}$$

The function $\zeta(x, \tau)$ is the gaussian white noise. All correlations and response functions can be calculated from the generating functional [9]

$$Z[J, \tilde{J}] = \int [\mathrm{d}\phi][\mathrm{d}\tilde{\phi}] \exp\Big\{ - \int \mathrm{d}^d x \, \mathrm{d}\tau \left[\lambda_0 \tilde{\phi}^2 - \tilde{\phi} \left[\partial_{\tau} \phi + \lambda_0 \left(-\nabla^2 \phi + r_0 \phi \right) \right] \right.$$
$$\left. - \tfrac{1}{6} \lambda_0 u_0 \tilde{\phi} \phi \phi^2 - \tfrac{1}{12} (n+2) \delta^{(d)}(0) \lambda_0 u_0 \phi^2 - J\phi - \tilde{J}\tilde{\phi} \right] \Big\} . \tag{2.4}$$

The term involving $\delta^{(d)}(x=0)$ originates from the jacobian [9] D:

$$D \equiv \det\left[\frac{\mathrm{d}}{\mathrm{d}\tau} + \lambda \frac{\delta^2 \mathbb{H}}{\delta\phi\delta\phi}\right]. \tag{2.5}$$

It serves as a counterterm and subtracts the contribution of the response function $\langle\tilde{\phi}(x,\tau)\phi(x,\tau)\rangle$ at equal points in space and time whenever it occurs. In (2.3) the average over the noise ζ has already been taken and vector indices have been suppressed. It is now useful to integrate out the auxiliary field $\tilde{\phi}$ and consider the generating functional

$$Z[J] = \int[\mathrm{d}\phi]\exp\left\{-\int \mathrm{d}^d x\,\mathrm{d}\tau\left[\frac{\dot{\phi}^2}{4\lambda_0} + \tfrac{1}{4}\lambda_0\left(-\nabla^2\phi + r_0\phi + \tfrac{1}{6}u_0\phi\phi^2\right)^2 - \tfrac{1}{12}(n+2)\delta^{(d)}(0)\lambda_0 u_0\phi^2 - J\phi\right]\right\}. \tag{2.6}$$

This functional was derived in a different manner by Munoz Sudupe and Alvarez-Estrada [10] who constructed the renormalized perturbation theory in $4-\varepsilon$. More generally it was used as a basis for renormalization of dynamical models by Zinn-Justin [11]. In passing from (2.4) to (2.6) we have dropped the term linear in $\dot{\phi} \equiv \partial_\tau\phi$ in the action which is a total derivative and contributes only to boundary terms.

We now expand the fields in Fourier modes in the definite dimensions of the box:

$$\phi(x,\tau) = \sum_q \mathrm{e}^{iq\cdot x}\phi_q(\tau), \tag{2.7}$$

where the components of q are quantized in units of $2\pi/L$. The modes with $q=0$ cannot be treated perturbatively. Defining

$$\phi(\tau) \equiv \phi_{q=0}(\tau),$$

these modes are an arbitrary function of τ and are not damped by the q-dependent terms in the action. The $q \neq 0$ modes can be treated perturbatively, and one can define an effective action $S_{\mathrm{eff}}[\phi(\tau)]$ by tracing out the $q \neq 0$ modes up to a given order in the loop expansion. We can consider first the simpler situation above the upper critical dimension which is four for the model described by (2.6).

In the case $d>4$ one can neglect loop contributions to the effective action for the $q=0$ modes. One has to be careful about the role of the jacobian in this case. This can be done by returning to eq. (2.3) and considering the equation of motion for the

$\boldsymbol{q}=0$ mode only

$$\partial_\tau \phi_i(\tau) = -\lambda_0 \frac{\delta \mathbb{H}}{\delta \phi_i} + \zeta_i(\tau),$$

$$\mathbb{H} = \tfrac{1}{2} r_0 \phi^2 + \frac{u_0}{4!}(\phi^2)^2,$$

$$\langle \zeta_i(\tau) \zeta_j(\tau') \rangle = 2\lambda_0 L^{-d} \delta(\tau - \tau') \delta_{ij}; \tag{2.8}$$

repeating the steps that led to eq. (2.6) one obtains:

$$Z[J] = \int [\mathrm{d}\phi] \exp[-S_{\mathrm{eff}}[\phi] + J\phi],$$

$$S_{\mathrm{eff}}[\phi] = L^d \int \mathrm{d}\tau \left\{ \frac{1}{4\lambda_0} (\dot{\phi})^2 + \tfrac{1}{4}\lambda_0 \left(r_0 \phi + \frac{u_0}{3!} \phi^2 \phi \right)^2 \right.$$

$$\left. - \tfrac{1}{2} L^{-d} \lambda_0 \left(n r_0 + \tfrac{1}{6}(n+2) u_0 \phi^2 \right) \right\}. \tag{2.9}$$

We have kept the constant term for reasons which will be explained below. It is obvious now that this effective action corresponds to a quantum mechanical problem in n dimensions as formulated by the Feynman path integral in imaginary time.

It is useful to realize that this action can alternatively be written in the form

$$S_{\mathrm{eff}}[\phi] = \frac{1}{\hbar} \int \mathrm{d}\tau \left\{ \tfrac{1}{2}\dot{\phi}^2 + \tfrac{1}{2} W^2(\phi) - \tfrac{1}{2}\hbar W'(\phi) \right\}, \tag{2.10}$$

with

$$\hbar \equiv 2\lambda_0 L^{-d}, \tag{2.11}$$

$$W(\phi) = \lambda_0 \phi \left(r_0 + \tfrac{1}{6} u_0 \phi^2 \right), \tag{2.12}$$

$$W'(\phi) \equiv \sum_i \mathrm{d}W_i(\phi)/\mathrm{d}\phi_i . \tag{2.13}$$

This action is related to the action of a supersymmetric quantum mechanical model

[17, 18] of the form

$$S_{\mathrm{SUSY}}(\phi)=\frac{1}{\hbar}\int \mathrm{d}\tau\left\{\tfrac{1}{2}\dot{\phi}^2-\psi^*\partial_\tau\psi+\tfrac{1}{2}W^2(x)+\tfrac{1}{2}\hbar[\psi^*,\psi]W'(\phi),\right. \quad (2.14)$$

where ψ is an anticommuting Grassmann variable. When the ψ field is integrated out the theory is equivalent to the bosonic action (2.10). This is provided supersymmetry is not broken as it is for the potential under consideration [17, 18]. We can associate with this action a quantum-mechanical hamiltonian in n dimensions by identifying $\phi(\tau)$ with the coordinates $\boldsymbol{q}(\tau)$

$$H(\boldsymbol{p},\boldsymbol{q})=\frac{\boldsymbol{p}^2}{2(L^d/2\lambda_0)}+\tfrac{1}{4}L^d\lambda_0\boldsymbol{q}^2\left(r_0+\frac{u_0}{3!}\boldsymbol{q}^2\right)^2-\tfrac{1}{12}(n+2)\lambda_0u_0\boldsymbol{q}^2-\tfrac{1}{2}n\lambda_0r_0. \quad (2.15)$$

In particular the relaxation time, governing the exponential decay of correlations with time, is related to the gap of the hamiltonian (2.15), i.e.

$$\tau_{\mathrm{R}}(L)=(E_1-E_0)^{-1}, \quad (2.16)$$

where E_0, E_1 are the ground state and first excited state of the hamiltonian respectively. Since the potential is O(n) symmetric, and so are the lowest lying states, the gap can be found by solving a one-dimensional radial problem. Because of the underlying supersymmetry the ground state energy E_0 vanishes. It is useful to make the following changes of variables

$$\boldsymbol{q}\to L^{-d/4}u_0^{-1/4}\boldsymbol{q},$$
$$\boldsymbol{p}\to L^{d/4}u_0^{1/4}\boldsymbol{p}, \quad (2.17)$$

and express the hamiltonian in the new variables

$$H(p,q)=2\lambda_0u_0^{1/2}L^{-d/2}\Big[\tfrac{1}{2}\boldsymbol{p}^2+\tfrac{1}{2}\boldsymbol{q}^2\left(\tfrac{1}{2}u_0^{-1/2}L^{d/2}t+\tfrac{1}{12}\boldsymbol{q}^2\right)^2$$
$$-\tfrac{1}{24}(n+2)\boldsymbol{q}^2-\tfrac{1}{2}n\left(\tfrac{1}{2}u_0^{-1/2}L^{d/2}t\right)\Big], \quad (2.18)$$

where we have replaced r_0 by $t(\propto T-T_c)$ which is legitimate at the tree approximation (loop corrections are irrelevant for $d>4$). Thus the relaxation time $\tau_{\mathrm{R}}(L)$ must be of the scaling form

$$\tau_R(L) = \frac{1}{2\lambda_0} u_0^{-1/2} L^{d/2} f\left(\tfrac{1}{2} u_0^{-1/2} L^{d/2} t\right), \tag{2.19}$$

where $f(x)$ is the inverse gap of the hamiltonian

$$h(\boldsymbol{p},\boldsymbol{q}) = \tfrac{1}{2}\boldsymbol{p}^2 + \tfrac{1}{2}\boldsymbol{q}^2\left(x + \tfrac{1}{12}\boldsymbol{q}^2\right)^2 - \tfrac{1}{24}(n+2)\boldsymbol{q}^2 - \tfrac{1}{2}nx. \tag{2.20}$$

Relation (2.19) can be also cast in the form

$$\tau_R(L,t) = \tau_R(\infty,t)\, g\left(\frac{L}{\xi_\infty(t)} L^{(d-4)/4}\right), \tag{2.21}$$

where

$$\tau_R(\infty,t) = \frac{1}{\lambda_0 t} \tag{2.22}$$

is the mean-field relaxation time for the infinite system, and $\xi_\infty(t) \propto t^{-1/2}$ is the mean-field correlation length. Eq. (2.21) is consistent with the breaking of the usual form of finite size scaling above four dimensions, which is well known [5].

Loop corrections which result from integrating out the $q \neq 0$ modes are irrelevant above four dimensions. All they cause is a shift of T_c, a finite renormalization of u_0 and λ_0, and they also generate terms in the action that lead to contributions which are down by powers of L compared to the leading contribution to the scaling function calculated above.

We now turn to the case $d < 4$. In this case loop corrections are important since they are responsible for long-distance singularities. Their effect can be taken into account using the renormalization group and ε expansion. To one-loop order an infinite renormalization as well as a finite renormalization of λ_0 is not required (this occurs only at the two-loop order) but a renormalization of u_0 and a shift of t is required. We realize that the renormalization of these parameters can be obtained from the *statics* properties of the system. (These statements can be verified by performing the loop expansion associated with the generating functional (2.6).) The shifts in t and g have been calculated previously [4] for the system with the cubic geometry. Introducing the dimensionless coupling constant

$$g = \mu^{\varepsilon} u_R \tag{2.23}$$

and the scaling variable

$$y = tL^{1/\nu}, \tag{2.24}$$

the quantity $x=\frac{1}{2}t_R g^{-1/2}L^{d/2}$ becomes at the fixed point

$$g^*=8\pi^2\frac{6\varepsilon}{n+8}+o(\varepsilon^2), \tag{2.25}$$

$$\begin{aligned} x &= \tfrac{1}{2}t_R L^{d/2} g^{-1/2}|_{\text{fixed point}} \\ &= \tfrac{1}{2}(g^*)^{-1/2}\Bigg\{ y-\tfrac{1}{4}\varepsilon y+\frac{n-4}{4(n+8)}\varepsilon y\ln y \\ &\quad +\frac{\varepsilon}{4\pi^2}y\int_0^\infty du\, u\, e^{-uy/4\pi^2}\left[A^4(u)-1-\frac{\pi^2}{u^2}\right] \\ &\quad +\varepsilon\frac{2(n+2)}{n+8}\int_0^\infty du\, e^{-uy/4\pi^2}\left[A^4(u)-1-\frac{\pi^2}{u^2}\right]+o(\varepsilon^2)\Bigg\}, \end{aligned} \tag{2.26}$$

where

$$A(u)=\sum_{n=-\infty}^{\infty} e^{-un^2}. \tag{2.27}$$

Hence, below four dimensions we find that

$$\frac{\tau_R(L)}{L^2}=\frac{1}{2\lambda}(g^*)^{-1/2}f(x)(1+o(\varepsilon)), \tag{2.28}$$

where $f(x)$ is the inverse gap of the hamiltonian (2.20), x is a function of the variable $y=tL^{1/\nu}$ as expressed in eq. (2.26) and g^* is given in (2.25). At $T=T_c$ $(y=0)$ one obtains

$$\frac{\tau_R(L)}{L^2}=\frac{1}{2\lambda}(g^*)^{-1/2}f\left(\varepsilon^{1/2}\cdot\frac{1}{4}\frac{n+2}{\sqrt{3(n+8)}}\frac{I}{\pi}\right)(1+o(\varepsilon)), \tag{2.29}$$

with

$$\frac{I}{\pi}=\frac{1}{\pi}\int_0^\infty du\left(A^4(u)-1-\frac{\pi^2}{u}\right)=-1.7650\ldots. \tag{2.30}$$

In higher loop order the denominator L^2 on the l.h.s. of (2.28) and (2.29) will be replaced by L^z where z is the dynamical critical exponent, since it is well known [12, 8] that at T_c, $\tau_R(L)\sim L^z$. z has been calculated in $4-\varepsilon$ and found to be [13]

$$z=2+0.7261\cdot\frac{n+2}{2(n+8)}\varepsilon^2+o(\varepsilon^3). \tag{2.31}$$

Dividing (2.28) by (2.29) we find

$$\frac{\tau_{\mathrm{R}}(L,t)}{\tau_{\mathrm{R}}(L,T=T_{\mathrm{c}})}=\frac{f(x)}{f\left(\varepsilon^{1/2}\cdot\frac{1}{4}(n+2)I/\pi\sqrt{3(n+8)}\right)}, \tag{2.32}$$

which is a universal function of $tL^{1/\nu}$. Eq. (2.28) can also be written in the form

$$\frac{\tau_{\mathrm{R}}(L)}{\tau_{\mathrm{R}}(\infty)}=\tfrac{1}{2}(g^{*})^{-1/2}y^{\nu z}f(x)(1+\mathrm{o}(\varepsilon)), \tag{2.33}$$

where to order one-loop

$$\nu z=1+\frac{(n+2)}{2(n+8)}\varepsilon, \tag{2.34}$$

and y was defined in (2.24).

3. Expansion about the lower critical dimension

In order to investigate the behavior of the relaxation time below the critical temperature it is useful to investigate the nonlinear σ-model [14,15] in a $2+\varepsilon$ expansion. Considering the case $n>2$ near two dimensions the critical temperature is very small ($\mathrm{o}(\varepsilon)$) and hence the expansion parameter for $T<T_{\mathrm{c}}$ is the temperature divided by the spin stiffness constant, which we will denote in this section by t. (This is *not* the temperature difference from T_{c} as it was in the previous section.)

The dynamics of the non-linear σ-model has been first investigated by de Dominicis, Ma and Peliti [16]. It was shown later by Bausch, Janssen and Yamuzaki [19], that the same dynamics can be derived from the generating functional

$$Z(J,\tilde{J})=\int[\mathrm{d}\phi][\mathrm{d}\tilde{\phi}]\prod_{x\tau}[\delta(\phi^{2}-1)\delta(\tilde{\phi}\cdot\phi)]\exp\{-S(\phi,\tilde{\phi})-J\phi-\tilde{J}\tilde{\phi}\},$$

$$S(\phi,\tilde{\phi})=\frac{1}{t}\int\mathrm{d}^{d}x\,\mathrm{d}\tau\left\{\lambda_{0}\tilde{\phi}^{2}+i\tilde{\phi}\left[\partial_{\tau}\phi+\lambda_{0}\frac{\delta H}{\delta\phi}\right]\right\},$$

$$H=\tfrac{1}{2}\int\mathrm{d}^{d}x\,(\nabla\phi)^{2}. \tag{3.1}$$

It is now possible to use the identity

$$\prod_{x\tau}\delta(\tilde{\phi}\cdot\phi)=\int[\mathrm{d}\omega]\exp\left(i\int\mathrm{d}^{d}x\,\mathrm{d}\tau\,\omega\tilde{\phi}\cdot\phi\right), \tag{3.2}$$

and integrate out the field $\tilde{\phi}$ for the case $\tilde{J}=0$ to obtain

$$Z[J]=\int[\mathrm{d}\phi][\mathrm{d}\omega]\prod_{x\tau}\delta(\phi^2-1)$$

$$\times\exp\left\{-\frac{1}{4\lambda_0 t}\int\mathrm{d}^dx\,\mathrm{d}\tau\left[\dot{\phi}^2-\lambda_0^2(\nabla^2\phi)^2-2\lambda_0 t\omega\phi\cdot\nabla^2\phi+t^2\omega^2\right]+J\phi\right\},\tag{3.3}$$

where we again dropped a total time derivative. Note that because $\phi^2=1$ one has $\phi\cdot\nabla^2\phi=-(\nabla\phi)^2$. The field ω can now be integrated over to yield

$$Z[J]=\int[\mathrm{d}\phi]\prod_{x\tau}\delta(\phi^2-1)\exp\{-S_{\mathrm{eff}}(\phi)+J\phi\},$$

$$S_{\mathrm{eff}}(\phi)=\frac{1}{4\lambda_0 t}\int\mathrm{d}^dx\,\mathrm{d}\tau\left\{(\partial_\tau\phi)^2+\lambda_0^2\left[(\nabla^2\phi)^2-(\nabla\phi)^2\cdot(\nabla\phi)^2\right]\right\}.\tag{3.4}$$

This form of the action coincides with the form derived by Lebedev [20]. We will now restrict the spatial integration to a box of volume L^d, and consider the action

$$S(\phi)=\frac{1}{4\lambda_0 t}\int_{L^d}\mathrm{d}^dx\int_0^T\mathrm{d}\tau\left\{(\dot{\phi})^2+\lambda_0^2\left[(\nabla^2\phi)^2-(\nabla\phi)^2(\nabla\phi)^2\right]\right.,\tag{3.5}$$

with the constraint $\phi^2=1$. Here we denoted by T some finite time cutoff, and it should not be confused with the temperature that in our present formalism appears in the form of the variable t (in units of the spin stiffness constant). We can now follow similar steps used by BZ [4], the difference being the appearance of the laplacian operator in (3.5) instead of the gradient operator in their treatment of the cylindrical geometry. Defining

$$\phi(\boldsymbol{x},\tau=0)=\boldsymbol{u}_1,\qquad\phi(\boldsymbol{x},\tau=T)=\boldsymbol{u}_2,\tag{3.6}$$

with

$$\boldsymbol{u}_1\cdot\boldsymbol{u}_2=\cos\theta,\qquad|\boldsymbol{u}_1|=|\boldsymbol{u}_2|=1,\tag{3.7}$$

the generating functional $Z(J=0)$ defined in (3.5) has to be of the form

$$Z(T,\theta)=\langle\boldsymbol{u}_2|\mathrm{e}^{-TH}|\boldsymbol{u}_1\rangle=\sum_l\delta_l P_l(\cos\theta)\mathrm{e}^{-T\varepsilon_l}.\tag{3.8}$$

Here ε_l are the eigenvalues of the quantum hamiltonian H, $P_l(\cos\theta)$ are Gegenbauer polynomials, orthogonal with respect to the measure $\sin^{n-2}\theta$ and δ_l are degeneracy factors. Parametrizing the fields $\boldsymbol{\phi}$ in an appropriate way [4] that takes into account the condition $\boldsymbol{\phi}^2 = 1$ and the boundary conditions (3.6), we can re-write the action (3.5) keeping only quadratic terms in the fields, which is legitimate for a calculation carried out to one-loop order. We can thus integrate out over the fields $\boldsymbol{\phi}$ in the expression for $Z(J=0)$ in (3.5) and obtain, after some labor:

$$\ln\frac{Z(T,\theta)}{Z(T,0)} = -\frac{\theta^2}{4\lambda_0 t}L^d - \tfrac{1}{2}(n-2)\left[\operatorname{tr}\ln\left(-\frac{\theta^2}{T^2} + \lambda_0^2(\nabla^2)^2 - \frac{\mathrm{d}^2}{\mathrm{d}\tau^2}\right)\right.$$
$$\left. - \operatorname{tr}\ln\left(\lambda_0^2(\nabla^2)^2 - \frac{\mathrm{d}^2}{\mathrm{d}\tau^2}\right)\right]. \quad (3.9)$$

The eigenvalues of the operator $\lambda_0^2(\nabla^2)^2 - \mathrm{d}^2/\mathrm{d}\tau^2$ with the appropriate boundary conditions for the eigenfunctions (periodic boundary conditions of $\boldsymbol{x}$ and vanishing eigenfunctions for $\tau = 0, T$) are:

$$\lambda_0^2\left(\left(\frac{2\pi}{L}\boldsymbol{p}\right)^2\right)^2 + \frac{m^2\pi^2}{T^2}, \quad (3.10)$$

where $p_1, \ldots, p_d$ are integers and m are positive integers. Thus (3.9) takes the form

$$\ln\frac{Z(T,\theta)}{Z(T,0)} = \theta^2\left\{-\frac{L^d}{4\lambda_0 tT} + \frac{n-2}{2T^2}\sum_{\boldsymbol{p}}\sum_{m=1}^{\infty}\frac{1}{\lambda_0^2(4\pi^2\boldsymbol{p}^2/L^2)^2 + \pi^2 m^2/T^2}\right\} + \mathrm{O}(\theta^4). \quad (3.11)$$

Here an expansion in θ about $\theta = 0$ has been performed which is the relevant region of θ for small t. For large T one has

$$\sum_{\boldsymbol{p}}\sum_{m}\frac{1}{\lambda_0^2(4\pi^2\boldsymbol{p}^2/L^2)^2 + \pi^2 m^2/T^2} = \tfrac{1}{6}T^2 + \sum_{\boldsymbol{p}}{}'\frac{1}{\lambda_0}\frac{TL^2}{8\pi^2\boldsymbol{p}^2}, \quad (3.12)$$

where the prime denotes all values of $\boldsymbol{p} \neq 0$. The sum over $\boldsymbol{p}$ is dimensionally regularized.

$$\sum_{\boldsymbol{p}}{}'\frac{1}{\boldsymbol{p}^2} = \int_0^\infty \mathrm{d}u \sum_{\boldsymbol{p}}{}'\mathrm{e}^{-u\boldsymbol{p}^2} = \int_0^\infty \mathrm{d}u\left[A^d(u) - 1\right], \quad (3.13)$$

where $A(u)$ has been defined in eq. (2.27). For small u, $A(u) \sim (\pi/u)^{1/2}$, hence the integral converges for $d<2$. Expanding in $\varepsilon=d-2$ we obtain

$$\sum_p{}' \frac{1}{p^2} = \pi^{d/2}\int_0^\infty \frac{\mathrm{d}u}{(\mathrm{e}^u-1)^{d/2}} + \int_0^\infty \mathrm{d}u\left(A^d(u)-1-\frac{\pi^{d/2}}{(\mathrm{e}^u-1)^{d/2}}\right)$$

$$= -\frac{2\pi}{\varepsilon} - \pi\ln\pi + \mathrm{o}(\varepsilon) + \int_0^\infty \mathrm{d}u\left(A^2(u)-1-\frac{\pi}{\mathrm{e}^u-1}\right) + \mathrm{O}(\varepsilon)$$

$$= -\frac{2\pi}{\varepsilon} + A + \mathrm{O}(\varepsilon), \tag{3.14}$$

$$A \equiv -\pi\ln\pi + \int_0^\infty\left(A^2(u)-1-\frac{\pi}{\mathrm{e}^u-1}\right)\mathrm{d}u = -2.8247\dots. \tag{3.15}$$

Collecting together the previous results we obtain

$$\ln\frac{Z(T,\theta)}{Z(T,0)} = \theta^2\left\{\tfrac{1}{12}(n-2) - \frac{L^d}{4\lambda_0 tT} - \frac{1}{\varepsilon}\frac{n-2}{8\pi}\frac{1}{\lambda_0}\frac{L^2}{T} + \frac{n-2}{16\pi^2}\frac{L^2A}{\lambda_0 T} + \mathrm{O}(\varepsilon)\right\}$$

$$+\mathrm{O}\left(\theta^4; t, \frac{1}{T^2}\right). \tag{3.16}$$

To project the two lowest eigenvalues we have to multiply (3.16) by 1 and $\cos\theta$ (the first Gegenbauer polynomials) and integrate with the measure $\sin^{n-2}\theta$. The integrals can be evaluated by the steepest descent technique. We finally obtain:

$$\tau_{\mathrm{R}}^{-1}(L) = (n-1)\frac{\lambda_0 t}{L^d}\left[1 + \frac{n-2}{4\pi^2}L^{-\varepsilon}\left(-\frac{2\pi}{\varepsilon} + A\right) + \mathrm{O}(\varepsilon)\right]. \tag{3.17}$$

The ε pole is cancelled by introducing the renormalized temperature parameter [15]

$$t = t_{\mathrm{R}} + \frac{n-2}{2\pi\varepsilon}t_{\mathrm{R}}^2 + \mathrm{O}(t_{\mathrm{R}}^3) \tag{3.18}$$

to one-loop order. λ_0 is not renormalized and one can set $\lambda=\lambda_0$. Introducing the running coupling constant $t_{\mathrm{R}}(L)$ [15] which is, at this order, given by the relation

$$\frac{1}{t_{\mathrm{R}}} = \frac{1}{t_{\mathrm{R}}^*} + L^{-\varepsilon}\left(\frac{1}{t_{\mathrm{R}}(L)} - \frac{1}{t_{\mathrm{R}}^*}\right),$$

$$t_{\mathrm{R}}^* - \frac{2\pi\varepsilon}{n-2} + \mathrm{O}(\varepsilon^2), \tag{3.19}$$

we can write (3.17) in the form

$$\frac{\tau_R(L)}{L^2} = \frac{1}{(n-1)\lambda t_R(L)}\left[1 - A\frac{n-2}{4\pi^2}t_R(L) + O(t_R^2)\right]. \tag{3.20}$$

For large L, $t_R(L)$ vanishes. It is related to the correlation length by

$$t_R(L) = \left(\frac{\xi(t_R)}{L}\right)^{d-2}\left[1 - \frac{1}{2\pi}\frac{n-2}{d-2}\left(\frac{\xi}{L}\right)^{d-2} + O\left(\frac{1}{L^{2(d-2)}}\right)\right]. \tag{3.21}$$

Hence for large L, $T < T_c$ we can rewrite (3.20) in the form

$$\frac{\tau_R(L)}{L^2} = \frac{1}{\lambda}\frac{1}{n-1}\left(\frac{L}{\xi}\right)^{d-2}\left[1 + \frac{n-2}{2\pi}\left(\frac{\xi}{L}\right)^{d-2}\left(\frac{1}{d-2} - \frac{1}{2\pi}A\right) + O\left(\frac{1}{L^{2(d-2)}}\right)\right]. \tag{3.22}$$

For an Ising system below T_c the situation is different. Returning back to the hamiltonian (2.20) with $n = 1$ we show in fig. 1 the qualitative shapes of the potential for various values of the parameter x. We know, because of the underlying supersymmetry, that the ground state is at $E = 0$. For $x < 0$, i.e. $T < T_c$ we see that there is a barrier of the potential rising above $E = 0$, and thus the ground state is a linear combination of states lying in the two outermost wells. This indicates the existence of an exponentially small gap between the ground state and first excited state which can be calculated using instanton calculus. (Notice that had we failed to include the terms originating from the jacobian in the hamiltonian (2.20), the potential for $x < 0$ would have had three minima of equal depth. But the frequency of oscillation in the two outer wells would have been twice as large as the frequency in the middle well. Thus the ground state would originate from the state in the middle well which is contrary to the correct situation below T_c.)

Thus we observe, that as opposed to the $O(N)$ case for which the relaxation time grows as a power of L (see eq. (3.22)), for the Ising case the relaxation time grows exponentially with increasing L.

A detailed instanton calculation to determine the relaxation time is beyond the scope of this article. It ought to include also the contribution of the $q \neq 0$ modes, since the reversal of the magnetization of the sample occurs via the development of a bubble of the opposite phase, thus one is forced to consider inhomogeneous configurations*.

* See the note added at the end of the paper.

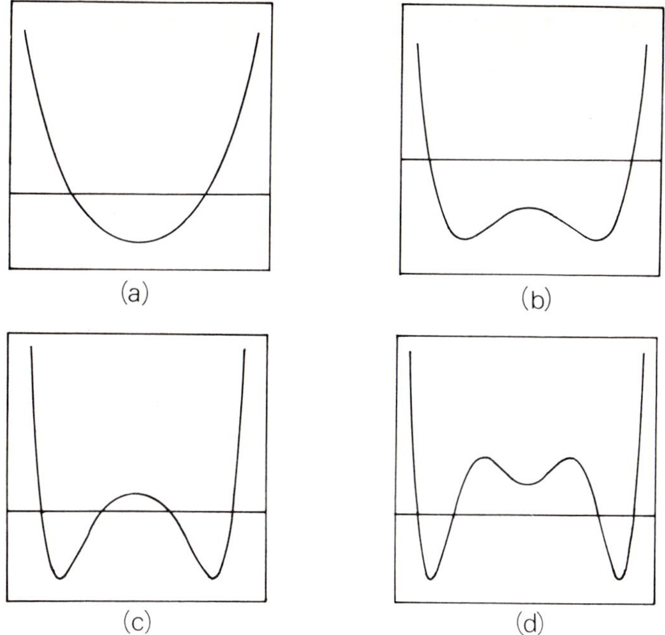

Fig. 1. Qualitative shape of the potential of the quantum hamiltonian for the case (a) $x > \frac{1}{2}$, (b) $0 < x < \frac{1}{2}$, (c) $-\frac{1}{2} < x < 0$, (d) $x < -\frac{1}{2}$. The horizontal line denotes the zero energy ground state.

4. The relaxation time in large *n* limit

In the large n limit there is a simple relation between the relaxation time for the cubic geometry and the correlation length both above and below T_c. We will follow the notation of Brézin [5]. In the large n limit we can show that the time dependent correlation function for the model defined by eq. (2.3) is given by the expression

$$C(\tau) = \frac{1}{L^d} \sum_{\boldsymbol{q}} \int \frac{d\omega}{2\pi} \frac{e^{i\omega\tau} \cdot 2\lambda}{\omega^2 + \lambda_0^2 (q^2 + m_L^2)^2} \tag{4.1}$$

where m_L depends on r_0 and u_0 but not on momentum and frequency. It is the same self-energy part as that appearing in the static calculation. As $\tau \to \infty$ the leading contribution to $C(\boldsymbol{q}, \omega)$ comes from $\boldsymbol{q} = 0$. Hence

$$C(\tau) \sim \frac{1}{L^d} \int \frac{d\omega}{2\pi} \frac{e^{i\omega\tau}}{\omega^2 + \lambda_0^2 (m_L^2)^2} \sim e^{-\lambda_0 m_L^2 |\tau|}, \tag{4.2}$$

from which we can infer that

$$\tau_R(L) = \frac{1}{\lambda_0} m_L^{-2}(t). \tag{4.3}$$

The quantity $m_L^{-1}(T_c)$ has been evaluated in ref. [5] and identified with the correlation length. Hence we obtain

$$\tau_L(T_c) \propto \begin{cases} \dfrac{1}{\lambda_0}\varepsilon^{-1/2}L^2, & \varepsilon = 4-d \to 0^+ \\ \dfrac{1}{\lambda_0}L^2(\ln L)^{1/2}, & d=4 \\ \dfrac{1}{\lambda_0}L^2L^{(d-4)/2}, & d>4 \end{cases} \tag{4.4}$$

These results are consistent with our previous results above and below four dimensions.

Below T_c, relation (4.3) still holds, but m_L has not been evaluated for the cubic geometry. Brézin evaluated m_L for the cylindrical geometry with one infinite dimension [5]. It is possible to modify his equation (60) to our case. The second term on the r.h.s. has to be replaced by

$$\sum_n{}' \int \frac{d^d q}{(2\pi)^d} e^{iq\cdot nL}\left[q^2+m_L^2\right]^{-1} = L^{2-d}\frac{1}{(4\pi)^{d/2}}\int_0^\infty e^{-ty^2}\tilde{g}(t)t^{-d/2}\,dt, \tag{4.5}$$

where

$$\tilde{g}(t) = \sum_n{}' e^{-n^2/4t} \tag{4.6}$$

and $y = Lm_L$. For $y_L \ll 1$ one can replace $\tilde{g}(t)$ by its asymptotic behavior $\sim (4\pi t)^{d/2}$ and hence one finds

$$\beta - \beta_c = L^{-d}m_L^{-2}, \tag{4.7}$$

from which we obtain at $T < T_c$

$$\tau_R(L) = \frac{1}{\lambda_0}L^d(\beta-\beta_c). \tag{4.8}$$

From eq. (3.19) and the fact that $n\beta \sim 1/t$ we see that for large L

$$\beta - \beta_c \sim \frac{1}{n}L^{-\varepsilon}t_R^{-1}(L) \tag{4.9}$$

and thus eq. (4.8) agrees with our result (3.20).

5. Conclusions

We have carried here a computation of the scaling behavior of the relaxation time for a system with a cubic geometry and periodic boundary conditions. We have used model A dynamics. It will be useful in the future to consider other geometries and also other types of dynamics. Our results should be useful for an estimation of the equilibration time in Monte Carlo calculations as a function of the system size.

This work was supported by the National Science Foundation under grant #DMR-8302323.

Note added

After this work had been submitted for publication we learnt from Dr. J. Zinn-Justin (private communication), that a similar work has just been completed by himself together with J.C. Niel. They obtained many of the results of this paper, and also carried out a detailed instanton calculation to evaluate the relaxation time of the Ising model below T_c. They found $\tau_R \sim \exp(L^{d-1} \cdot 2\sigma(T))$.

References

[1] M.N. Barber, *in* Phase transitions and critical phenomena, vol. VIII, eds. C. Domb and J. Lebowitz (Academic Press, NY, 1984) and references therein

[2] M.E. Fisher, *in* Critical phenomena, Proc. 51st Enrico Fermi Summer School, Varena, ed. M.S. Green (Academic Press, NY, 1972);
M.E. Fisher and M.N. Barber, Phys. Rev. Lett. 28 (1972) 1516

[3] M.P. Nightingale, Physica 83 (1976) A561

[4] E. Brézin and J. Zinn-Justin, Nucl. Phys. B257 [FS14] (1985) 868

[5] E. Brézin, J. de Phys. 43 (1982) 15

[6] P.C. Hoenberg and B.I. Halperin, Rev. Mod. Phys. 49 (1977) 435

[7] M.E. Fisher and Z. Racz, Phys. Rev. B13 (1976) 5039

[8] J.M. Sancho et al., J. Phys. A13 (1980) L443

[9] R. Bausch et al., Z. Phys. B24 (1976) 113

[10] A. Munoz Sudupe and R.F. Alvarez-Estrada, J. Phys. A16 (1983) 3049

[11] J. Zinn-Justin, Nucl. Phys. B275 [FS17] (1986) 135

[12] M. Suzuki, Prog. Theor. Phys. 58 (1977) 1142

[13] B. Halperin et al., Phys. Rev. Lett. 29 (1972) 1548;
C. de Dominicis et al., Phys. Rev. B12 (1975) 4945

[14] A.M. Polyakov, Phys. Lett. 59B (1975) 79

[15] E. Brézin and J. Zinn-Justin, Phys. Rev. Lett. 36 (1976) 691; Rev. B14 (1976) 3110

[16] C. de Dominicis et al., Phys. Rev. B15 (1977) 4313

[17] E. Witten, Nucl. Phys. B188 (1981) 513

[18] F. Cooper and B. Freedman, Ann. of Phys. 146 (1983) 262

[19] R. Bausch, H.K. Janssen and Y. Yamazaki, Z. Phys. B37 (1980) 163

[20] V. Lebedev, Phys. Lett. 105A (1984) 173

Nuclear Physics B285 [FS19] (1987) 519–534
North-Holland, Amsterdam

DYNAMICAL RELAXATION IN FINITE SIZE SYSTEMS
Non-linear and linear decay of the magnetization

Yadin Y. GOLDSCHMIDT

Department of Physics and Astronomy, University of Pittsburgh, Pittsburgh, PA 15260, USA

Received 15 January 1987

We calculate analytically the non-linear time-dependent equation of state for the magnetization in a finite size system. The calculation is done to one-loop order in the $q \neq 0$ modes. The equation is obtained in an expansion in powers of $\varepsilon^{1/2}$ where $\varepsilon = 4 - d$ up to $o(\varepsilon)$. The stochastic equation of state is solved numerically at the critical point and the results are compared with a recent Monte Carlo simulation. The solution displays the full crossover from the bulk power law decay to the finite size exponential relaxation. We also present more explicitly the results obtained for the linear relaxation time in an earlier paper, in a form which enables a direct comparison with simulations. To simplify the calculation we exploit the supersymmetry of the corresponding quantum mechanical hamiltonian.

1. Introduction

A quantitative understanding of finite size effects is very useful for an efficient interpretation of data obtained by numerical Monte Carlo simulations, which are typically performed on small systems. Finite size scaling (FSS) theory has been initiated by Fisher [1] and has been the subject of many subsequent investigations [2]. Brézin and Zinn-Justin [3] showed how to calculate the size-dependent universal functions in an ε expansion, which is singular about four dimensions [4]. Thus the series generated are in powers of $\varepsilon^{1/2}$ or $\varepsilon^{1/3}$ depending upon the geometry of the sample. Rudnick et al. [5] reported on theoretical and numerical results for the free energy.

Recently I showed [6] how to extend the finite size calculation to the dynamics of Ising and vector spin systems, and I calculated the universal scaling function associated with the linear relaxation time. This relaxation time governs the exponential decay of the autocorrelation function for the magnetization. The calculation was performed for the cubic geometry with periodic boundary conditions in $4 - \varepsilon$ dimensions in powers of $\varepsilon^{1/2}$, and in $2 + \varepsilon$ dimensions using the non-linear σ-model. Related work has been reported by Niel and Zinn-Justin [7] and more recently by Diehl [8].

0619-6823/87/$03.50

The essence of the method which I have used in ref. [6] (to be denoted by (I) in this sequel) is as follows:

In a finite size scaling calculation, it is necessary to treat the $q=0$ mode of the order parameter, say the magnetization separately from the $q\neq 0$ modes [3]. The $q\neq 0$ modes (q is discrete for a system in a finite box) can be treated perturbatively using the loop expansion. They can be traced over to yield an effective equation of motion, or an effective action for the $q=0$ mode:

$$e^{-S_{\mathrm{eff}}(\phi_{q=0})} = \operatorname*{Tr}_{\phi_{q\neq 0}} e^{-S(\phi_{q=0},\phi_{q\neq 0})}. \tag{1.1}$$

The action appearing on the r.h.s. of eq. (1.1) is that derived from the generating functional for the dynamical correlation functions [9]. The effective dynamics for the $q=0$ mode is then reduced to a solution of a quantum-mechanical problem with $\hbar \propto L^{-d}$, where L is the size of the box. The quantum mechanical hamiltonian associated with the dynamics of the $q=0$ mode is shown in (I) to be supersymmetric. A non-perturbative treatment of the $q=0$ mode is essential to avoid infrared divergences, as opposed to the perturbative treatment of the $q\neq 0$ modes. Thus in (I) and in sect. 2 of this paper, we proceed by solving for the gap of the effective hamiltonian.

In this paper our aim is twofold:

First, we present the results of the theory for the linear relaxation time, based on the formalism developed in (I), in a more explicit way which enables a direct comparison with Monte Carlo simulations. We calculate explicitly the linear relaxation time in $d=3$ ($\varepsilon=1$ in $4-\varepsilon$ expansion) as a function of the variable $y=tL^{1/\nu}$ (where $t=(T-T_c)/T_c$, and L is the system size). The supersymmetry of the effective hamiltonian is used to simplify the calculation. This is discussed in detail in sect. 2.

Second, we derive for the first time an expression for the finite-size time-dependent non-linear equation of state for the magnetization to one loop order. The "surprise" is that for $d<4$, unlike the linear case, it is not enough, in the one-loop calculation, to just replace the coupling constants by their L dependent counterparts, but the equation of state turns out to be of a form which allows both for the correct non-classical behavior in the non-linear regime where the magnetization is large, and in the linear regime where the magnetization is small. We continue to solve the equation numerically. For $\varepsilon=1$ (three dimensions), the results are compared with recent Monte Carlo simulations by Kikuchi and Okabe [10] who simulated the critical relaxation of the magnetization of an Ising system. Our results display the full crossover from the bulk non-linear relaxation behavior $M\sim\tau^{-\beta/\nu z}$ (where M is the magnetization and τ is the time) to a finite size relaxation behavior $M\sim\exp\{-\tau/\tau_{\mathrm{R}}(L)\}$ with $\tau_{\mathrm{R}}(L)\sim L^z$ being the linear relaxation time. Our numerical solution of the non-linear equation for the magnetization uses only a very small fraction of the computer time that is needed for the MC simulation.

2. The linear relaxation time

The linear relaxation time, governs the long time behavior of the autocorrelation function of the magnetization

$$\langle M(0)M(\tau)\rangle \sim \exp(-\tau/\tau_R(L)). \tag{2.1}$$

For a system of the cubic geometry with linear size L and periodic boundary conditions, I have shown in ref. [6] that for $T \geqslant T_c$, $d = 4-\varepsilon$

$$\frac{\tau_R(L)}{L^2} = \frac{1}{2\lambda}\frac{\sqrt{3}}{4\pi}\varepsilon^{-1/2}f(x(y))(1+o(\varepsilon)), \tag{2.2}$$

with $f(x)$ being the inverse gap of the quantum-mechanical hamiltonian

$$\mathscr{H} = \tfrac{1}{2}p^2 + \tfrac{1}{2}q^2\left(x + \tfrac{1}{12}q^2\right)^2 - \tfrac{1}{8}q^2 - \tfrac{1}{2}x \tag{2.3}$$

and x is a function of the scaling variable $y = tL^{1/\nu}$ (with $t = (T-T_c)/T_c$) through the relation [3,6,7]:

$$x(y) = \frac{1}{8\pi}\sqrt{\frac{3}{\varepsilon}}\left(y - \tfrac{1}{4}\varepsilon y - \tfrac{1}{12}\varepsilon y \ln y + 4\pi^2\varepsilon y F_2(y) + \tfrac{8}{3}\pi^2\varepsilon F_1(y)\right), \tag{2.4}$$

with

$$F_n(y) = \frac{1}{(4\pi^2)^n}\int_0^\infty \mathrm{d}u\, u^{n-1}\left(A^4(u) - 1 - \frac{\pi^2}{u^2}\right)\exp\left\{-\frac{uy}{4\pi^2}\right\}, \qquad n = 1,2,$$

$$A(u) = \sum_{n=-\infty}^{\infty} \exp(-n^2u). \tag{2.5}$$

The gap of the hamiltonian (3) for a few discrete values of x can be extracted from ref. [11] (one has to divide their parameters K and the gap μ, by $2\sqrt{6}$ in order to compare with our conventions). Nevertheless in order to obtain the value of the gap for many more values of x which is needed in particular for plotting the relaxation time as a function of $y = tL^{1/\nu}$ we found it useful to use the supersymmetry of the hamiltonian (2.3) which as discussed in ref. [6] can be written in the form

$$\mathscr{H} = \tfrac{1}{2}p^2 + \tfrac{1}{2}W^2(q) - \tfrac{1}{2}W'(q), \tag{2.6}$$

$$p^2 = -\frac{\mathrm{d}^2}{\mathrm{d}q^2}, \qquad W(q) = q\left(x + \tfrac{1}{12}q^2\right). \tag{2.7}$$

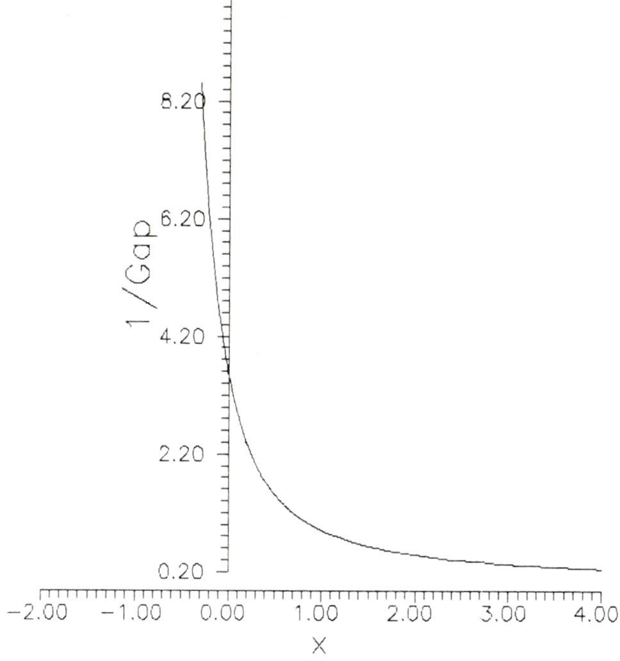

Fig. 1. A plot of the inverse gap of the hamiltonian versus x.

The supersymmetry implies [12] that the ground state of the hamiltonian has an energy $E_0 = 0$ and the wave function is given exactly by

$$\psi_0(q) = N_0 \exp\left\{ -\tfrac{1}{48}\left(q^2 + 12x\right)^2 \right\}. \tag{2.8}$$

we have chosen a variational wave function for the first excited state of the form

$$\psi_1(q) = N_1 q\, e^{-\alpha(q^2+\beta)^2}, \tag{2.9}$$

where α, β are variational parameters. This function has one node and is automatically orthogonal to the exact ground state. Calculating $\langle\psi_1|H|\psi_1\rangle/\langle\psi_1|\psi_1\rangle$ and minimizing it by varying the parameters α and β using a two variable minimization routine, we obtained a very good estimate of the first excited state energy E_1, which is the value of the gap. For the few discrete values calculated in ref. [11] the results agree very well. The inverse gap is plotted versus the variable x in fig. 1. For large values of x it is clear from eq. (2.3) that the term $\frac{1}{2}x^2q^2$ in the potential dominates and hence the gap of the hamiltonian is proportional to x and $f(x) \sim x^{-1}$. Inspection of eq. (2.4) shows that for large y, $x(y) \sim y^{1-\varepsilon/12}$ (since $F_2(y) \sim -y^{-2}$ and $F_1(y) \sim -y^{-1}$ for large y). But reexamining eq. (2.2) reveals that for large y,

$\tau_R \sim y^{-1+\varepsilon/12}$ which differs from the correct asymptotic behavior $y^{-z\nu}$. We can obtain the correct behavior by including the o(ε) terms originating from the shift in the coupling constant to order one loop. Eq. (2.2) originated from the expression [6]

$$\tau_R(L)/L^2 = (2\lambda)^{-1}\tilde{g}^{-1/2}L^{-\varepsilon/2}f(x) \tag{2.10}$$

evaluated at the fixed point, where $\tilde{g}$ is the shifted dimensionless coupling constant to order one loop:

$$\tilde{g} = g + \frac{3}{32\pi^2}g^2(1+\ln t) - \tfrac{3}{2}g^2F_2(tL^2), \tag{2.11}$$

with $F_2(y)$ defined by eq. (2.5). Hence at the fixed point $g^* = \frac{16}{3}\pi^2\varepsilon$

$$\tilde{g}^{-1/2}L^{-\varepsilon/2}\big|_{\text{fixed point}} = (g^*)^{-1/2}\left(1 - \tfrac{1}{4}\varepsilon - \tfrac{1}{4}\varepsilon\ln y + \varepsilon 4\pi^2F_2(y)\right). \tag{2.12}$$

Thus eq. (2.10) reads to order one loop

$$\frac{\tau_R(L)}{L^2} = \frac{1}{2\lambda}\frac{\sqrt{3}}{4\pi}\varepsilon^{-1/2}\left(1 - \tfrac{1}{4}\varepsilon - \tfrac{1}{4}\varepsilon\ln y + \varepsilon 4\pi^2F_2(y)\right)f(x(y)) \tag{2.13}$$

and now for large y, $\tau_R \sim y^{-1-\varepsilon/6}$ which is the desired behavior to o(ε).

Note that corrections to the effective potential involving higher order operators, like a contribution proportional to q^5 to $W(q)$ of eq. (2.7), would lead to corrections of o($\varepsilon^{3/2}$) which are higher than those included in eq. (2.13). The same is true for contributions to the effective action involving derivatives of the field, which also enter only at o($\varepsilon^{3/2}$). Note also that eq. (2.13) has a finite limit as $y \to 0$, i.e. at the bulk critical temperature. Certainly $f(x)$ is finite at the point

$$x_c \equiv x(y=0) = \sqrt{\tfrac{1}{3}}\,\pi F_1(0)\varepsilon^{1/2}, \qquad F_1(0) = -0.14045. \tag{2.14}$$

By substituting

$$\ln y = \int_0^\infty \frac{du}{u}\left(\exp\left\{-\frac{u}{4\pi^2}\right\} - \exp\left\{-\frac{uy}{4\pi^2}\right\}\right) \tag{2.15}$$

and using the expression (2.5) for $F_2(y)$, it is readily found that as $y \to 0$

$$\frac{\tau_R(L)}{L^2} = \frac{1}{2\lambda}\frac{\sqrt{3}}{4\pi}\varepsilon^{-1/2}\left(1 - \tfrac{1}{4}\varepsilon - \tfrac{1}{4}\varepsilon C\right)f\left(\sqrt{\tfrac{1}{3}}\,\pi F_1(0)\varepsilon^{1/2}\right), \tag{2.16}$$

with

$$C = \int_0^\infty \frac{du}{u}\left[\exp\left\{-\frac{u}{4\pi^2}\right\} - \frac{u^2}{\pi^2}A^4(u) + \frac{u^2}{\pi^2}\right]. \tag{2.17}$$

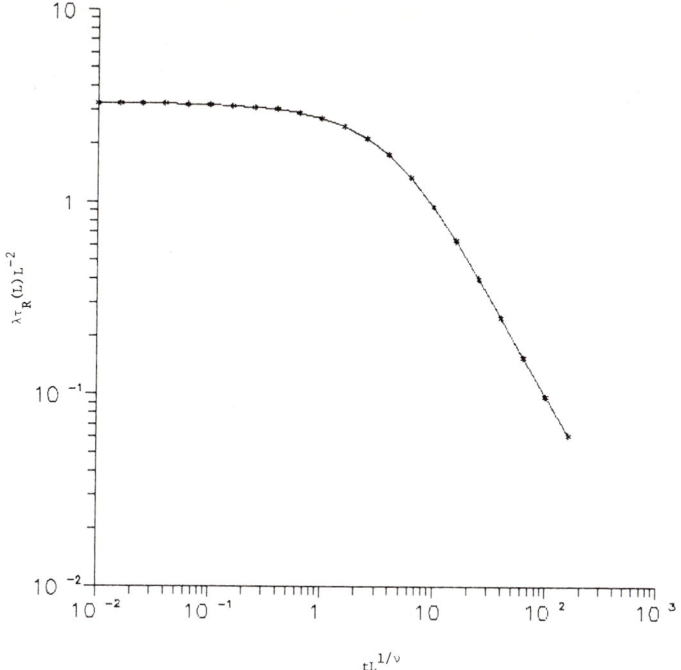

Fig. 2. A plot of the linear relaxation time versus reduced temperature in scaling units.

In fig. 2 we display a log-log plot of $\lambda \cdot \tau_R(L)/L^2$ versus the variable $y = tL^{1/\nu}$ as obtained from eq. (2.11) with $\varepsilon = 1$ using our variational results for the inverse gap and the function $x(y)$ as given by (2.4) and (2.5). In plotting eq. (2.11) we have omitted the coefficient $\sqrt{3}\,\varepsilon^{-1/2}/(8\pi)$ which contributes only to a vertical shift of the graph in the log-log plot. We have also replaced the other prefactor by the expression

$$\exp\left\{-\tfrac{1}{4}\varepsilon - \tfrac{1}{4}\varepsilon \ln y + 4\pi^2 \varepsilon F_2(y)\right\},$$

which is the same to $o(\varepsilon)$. The linear part of the graph corresponds to a bulk-like behavior $\tau_R \sim t^{-z\nu}$, whereas when $y \to 0$, $\tau_R(L) \to AL^z$ with $z = 2$ to first order in ε. So far we are aware only of results of MC simulations in two dimensions by Miyoshita and Takano [13] who obtained the linear relaxation time away from T_c and plot it against the variable $tL^{1/\nu}$. Our results in three dimensions behave qualitatively as theirs, but a direct comparison is impossible, since $\varepsilon = 2$ is too big to substitute in the $4 - \varepsilon$ expansion. In three dimensions we are aware only of simulations at T_c [14]. Thus the situation calls for more MC results in three dimensions for the behavior of the linear relaxation time in the vicinity of T_c.

3. The non-linear relaxation time

We now turn to the non-linear relaxation of the magnetization at T_c. In this case it is preferable to consider directly the equation of motion for the $q=0$ mode, rather than the corresponding effective hamiltonian. Starting with Langevin equation of motions for a field $\phi(x,\tau)$ for model A dynamics (a non-conserved order parameter), we expand the field in Fourier modes as discussed in ref. [6]:

$$\phi(x,\tau)=\sum_{q} e^{i q\cdot x}\phi_q(\tau), \tag{3.1}$$

where the components of $\boldsymbol{q}$ are quantized in units of $2\pi/L$. We denote $\phi(\tau)=\phi_{q=0}(\tau)$. I consider first the case $d>4$, the contribution of the $q\neq 0$ modes can be neglected [6] and the equation for ϕ reads

$$\partial_\tau\phi=-\lambda_0\left(r_0\phi+\tfrac{1}{6}u_0\phi^3\right)+\eta(\tau), \tag{3.2}$$

$$\langle\eta(\tau)\eta(\tau')\rangle=2\lambda_0 L^{-d}\delta(\tau-\tau'). \tag{3.3}$$

Upon making the rescaling transformation

$$\phi\to u_0^{-1/4}L^{-d/4}\phi,\qquad \tau\to\frac{1}{2\lambda_0}u_0^{-1/2}L^{d/2}\tau,\qquad \eta\to 2\lambda_0 u_0^{1/4}L^{-3d/4}\eta, \tag{3.4}$$

we obtain the equation

$$\partial_\tau\phi=-x\phi-\tfrac{1}{12}\phi^3+\eta(\tau), \tag{3.5}$$

with

$$\langle\eta(\tau)\eta(\tau')\rangle=\delta(\tau-\tau'), \tag{3.6}$$

$$x=\tfrac{1}{2}u_0^{-1/2}L^{d/2}r_0. \tag{3.7}$$

We have solved eq. (3.5) numerically by the following procedure: We divided the interval $(0,T)$ into K subintervals of length Δ. Starting with an initial value ϕ_0 we iterated the discrete difference equation

$$(\phi_{i+1}-\phi_i)/\Delta=-x\phi_i-\tfrac{1}{12}\phi_i^3+\eta_i, \tag{3.8}$$

where η_i are random variables with a normal (gaussian) distribution satisfying

$$\langle\eta_i\rangle=0,\qquad \langle\eta_i\eta_j\rangle=\frac{1}{\Delta}\delta_{ij}. \tag{3.9}$$

The solution vector $\{\eta_i\}$ was stored, and the procedure was repeated N times starting from the same initial value ϕ_0. The average solution

$$\frac{1}{N}\sum_{k=1}^{N}\{\phi_i\}^{(k)}$$

was then calculated, and consisted the numerical solution of eq. (3.5) in the interval $(0, T)$ which is a function $g_x(\phi_0, \tau)$. The values we have chosen for the different parameters will be discussed later. For $d>4$ we see (using the inverse of the transformation (3.4)) that the time dependent magnetization will behave like

$$M(\tau)=u_0^{-1/4}L^{-d/4}g_x\left(M_0L^{d/4}u_0^{1/4}, 2\lambda_0 u_0^{1/2}L^{-d/2}\tau\right). \tag{3.10}$$

We will discuss below the dependence of g_x on the initial value M_0. Eq. (3.10) is consistent with the breakdown of finite size scaling above four dimensions. From the renormalization group point of view [15, 4] the magnetization in dimension $d<4$ ought to scale according to the formula

$$M(h, t, \tau, L)=L^{-\beta/\nu}G\left(hL^{\beta\delta/\nu}, tL^{1/\nu}, \tau L^{-z}\right), \tag{3.11}$$

where h is the value of the magnetic field which induces the initial magnetization, and is then turned off to zero. [16] At $T=T_c$ $(t=0)$ one has $h=M_0^{\delta}$, hence

$$M(M_0, \tau, L)=L^{-\beta/\nu}G_1\left(M_0L^{\beta/\nu}, \tau L^{-z}\right)=\tau^{-\beta/\nu z}G_2\left(M_0L^{\beta/\nu}, \tau L^{-z}\right). \tag{3.12}$$

Notice that formally our expression (3.12) differs from eq. (3.2) of ref. [10] by the appearance of the argument $M_0L^{\beta/\nu}$ in G_2. We would like to comment here on the dependence of G_2 (or G_1) on the parameter M_0. If $M_0L^{\beta/\nu}\gg 1$ there exists a region for which

$$M_0^{-z\nu/\beta}\ll\tau<L^z. \tag{3.13}$$

In this region we expect G_2 to behave like a constant independent of the initial magnetization M_0, and

$$M\sim A\tau^{-\beta/\nu z}, \tag{3.14}$$

which is the bulk non-linear relaxation at T_c. On the other hand if $M_0L^{\beta/\nu}\sim 1$, then condition (3.13) never holds and by the time that $\tau\gg M_0^{-z\nu/\beta}$ it turns out that $\tau\gg L^z$, hence one expects to see only the *linear* relaxation

$$M\sim B\exp\left\{-\tau/\tau_R^{(l)}(L)\right\}. \tag{3.15}$$

Thus the existence of a universal scaling function

$$M(\tau, L)=L^{-\beta/\nu}\tilde{G}_1(\tau L^{-z})=\tau^{-\beta/(\nu z)}\tilde{G}_2(\tau L^{-z}), \tag{3.16}$$

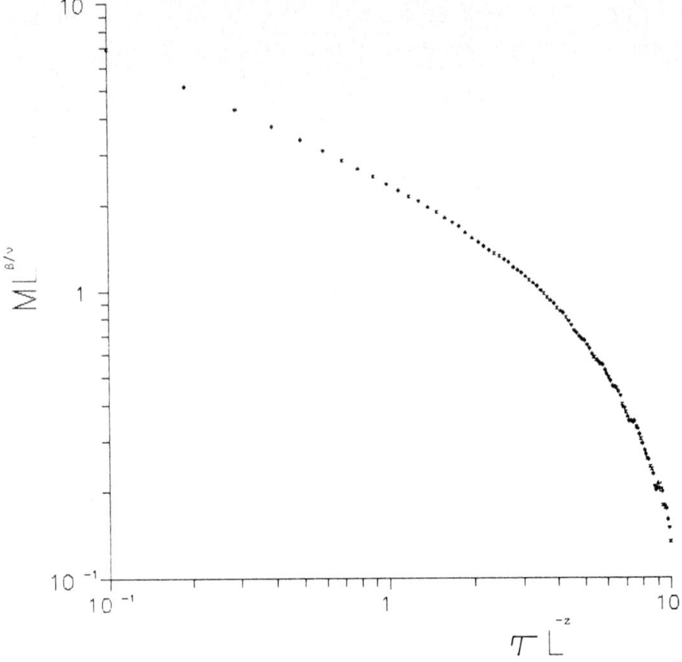

Fig. 3. A plot of the magnetization versus time in scaling units for $d \geqslant 4$ (mean field).

which do not depend on M_0 and crosses over from the non-linear dependence (3.14) to the linear dependence (3.15) can be achieved provided the condition

$$M_0 L^{\beta/\nu} \gg 1 \tag{3.17}$$

holds. This can be accomplished by taking $M_0 = 1$ and $L \sim 20$–60 as was used practically in the Monte Carlo simulation [10].

Before we turn to the discussion of the stochastic equation in the case of $d < 4$ let us first display the mean-field results for $d > 4$, as obtained from the numerical solution of eqs. (3.8) and (3.9). The function $g_{x=0}(\phi_0, \tau)$, is plotted versus τ in a log-log plot, in fig. 3, using an initial value $\phi_0 = 20$. One can see clearly the initial power law decay $g \sim \tau^{-1/2}$ which appears as a straight line with slope $-\frac{1}{2}$ in the figure. The function then crosses over to an exponential decay as can be seen more clearly from fig. 4, where a semi-log plot of the same function is depicted. The time interval $(0, 10)$ has been divided into 1000 segments, thus $\Delta = 0.01$. The number of times that the solution vector has been obtained for averaging has been $N = 1600$. When we repeated the calculation using an initial value $\phi_0 = 1$ only the exponential decay regime has been observed as is clear from the discussion above. When starting with $\phi_0 = 10$ the function nearly overlapped with that obtained with $\phi_0 = 20$ when the time interval was slightly shifted.

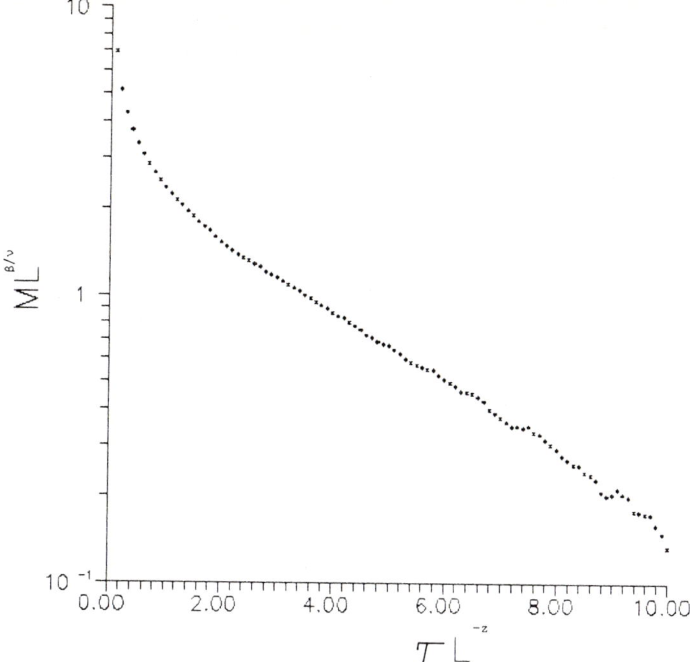

Fig. 4. A plot of the magnetization versus time on a semilog scale which emphasizes the region of linear relaxation. Again $d \geqslant 4$ (mean field).

We turn now to the case $d < 4$. It turns out that in order to obtain the correct non-linear relaxation when $d < 4$ it is *not* enough to simply fix the value of x in eq. (3.5) or (3.8) to the value x_c as given by eq. (2.14). The form of the stochastic equation of motion or the $q = 0$ mode should be such that it will describe correctly the non-linear regime where M is large, and where the bulk relaxation behavior $M \sim \tau^{-\beta/(\nu z)}$ should emerge with the correct non-mean-field exponent. The non-linear time-dependent equation of state for the *bulk*, to one-loop order, has been derived by Bausch et al. [17]. Their equation is deterministic, whereas our equation for the $q = 0$ mode of the order parameter in the finite size system is stochastic and the solution has yet to be averaged over the random noise. We only outline here the steps of the calculation, the details being presented in the appendix. We write down the action appearing in the generating functional for the dynamical correlation function, using the Martin-Siggia-Rose field [9]. We separate the action into three parts and distinguish the $q = 0$ component of the field, denoted by $\phi(\tau)$ from the $q = 0$ components $\phi_q(\tau)$. Thus

$$S = S_0 + S_1 + S_2 , \tag{3.18}$$

where

$$S_0 = \int d\tau \tilde{\phi}(\tau)\left(\partial_\tau + \lambda_0\left(r_0 + \tfrac{1}{6}u_0\phi^2(\tau)\right)\phi(\tau) - \int d\tau \tilde{\phi}(\tau)\eta(\tau)\right.$$
$$+ \int d\tau \sum_q{}' \left[-\lambda_0\tilde{\phi}_q(\tau)\tilde{\phi}_{-q}(\tau) + \tilde{\phi}_{-q}(\tau)\right.$$
$$\left.\times\left(\partial_\tau + \lambda_0\left(q^2 + r_0 + \tfrac{1}{2}u_0\phi^2(\tau)\right)\phi_q(\tau)\right)\right], \tag{3.19}$$

$$S_1 = \tfrac{1}{2}\lambda_0 u_0 \int d\tau \tilde{\phi}(\tau)\left[\phi(\tau)\sum_q{}' \phi_q(\tau)\phi_{-q}(\tau) + \tfrac{1}{3}\sum_{q_i q_2}{}' \phi_{q_1}(\tau)\phi_{q_2}(\tau)\phi_{-q_1-q_2}(\tau)\right],$$

$$S_2 = \tfrac{1}{2}\lambda_0 u_0 \int d\tau \sum_q{}' \tilde{\phi}_{-q}(\tau)$$
$$\times\left[\phi(\tau)\sum_{q_1}{}' \phi_{q_1}(\tau)\phi_{q-q_1}(\tau) + \tfrac{1}{3}\sum_{q_1 q_2}{}' \phi_{q_1}(\tau)\phi_{q_2}(\tau)\phi_{q-q_1-q_2}(\tau)\right].$$

The prime denotes summation over $q \neq 0$. $\eta(\tau)$ is the $q=0$ Fourier component of the noise $\eta(x,\tau)$. The $q \neq 0$ components of the noise have been integrated out. Factors of L^{-d} associated with summations over momenta have been suppressed. From S_0 we obtain the free response and correlation functions for the $q \neq 0$ modes *in the background of the* $q=0$ *mode*:

$$G_q(\tau,\tau') = \theta(\tau-\tau')\exp\left\{-\lambda_0\int_{\tau'}^{\tau} d\tau''\left[q^2 + r_0 + \tfrac{1}{2}u_0\phi(\tau'')\right]\right\}, \tag{3.20}$$

$$C_q(\tau,\tau') = 2\lambda_0\int_{-\infty}^{\infty} d\tau''\, G_q(\tau,\tau'')G_{-q}(\tau,\tau''). \tag{3.21}$$

The equation of motion for the magnetization is given by

$$\left\langle \frac{\delta S}{\delta\tilde{\phi}(\tau)} \right\rangle = 0, \tag{3.22}$$

where the average is taken only with respect to the $q \neq 0$ modes. To one-loop order we obtain

$$0 = \partial_\tau\phi(\tau) + \lambda_0\left[r_0 + \tfrac{1}{6}u_0\phi^2(\tau)\right]\phi(\tau) - \underset{q\neq 0}{\leftarrow\!\bigcirc} - \eta(\tau), \tag{3.23}$$

with

$$\underset{q\neq 0}{\leftarrow\!\bigcirc} = -\tfrac{1}{2}\lambda_0 u_0\phi(\tau)2\lambda_0\int_{-\infty}^{\tau} d\tau'\left(\frac{1}{L^d}\sum_q{}' \exp(-2\lambda_0(\tau-\tau')q^2)\right)$$
$$\times\exp\left(-2\lambda_0\int_{\tau'}^{\tau} d\tau''\left[r_0 + \phi^2(\tau'')\right]\right); \tag{3.24}$$

the rest of the details are given in the appendix.

Upon the introduction of renormalized parameters g instead of u_0 and t instead of r_0 and making the change of variables

$$\tau \to \frac{1}{2\lambda} g^{-1/2} L^{z} \tau \qquad \text{with } z = 2 + \mathrm{o}(\varepsilon^2), \tag{3.25}$$

$$\phi \to g^{-1/4} L^{-\beta/\nu} \phi \qquad \text{with } \beta/\nu = 1 - \tfrac{1}{2}\varepsilon + \mathrm{o}(\varepsilon^2), \tag{3.26}$$

$$\eta \to 2\lambda g^{1/4} L^{-2-\beta/\nu} \eta, \tag{3.27}$$

we can express the equation of motion at the fixed point $g^* = 16\pi^2\varepsilon/3$ in terms of variable $y = tL^{1/\nu}$ in the form:

$$\begin{aligned} 0 = {} & \partial_\tau \phi + \tfrac{1}{2}(g^*)^{-1/2} y\phi + \tfrac{1}{12}\phi^3 - \eta(\tau) + \tfrac{1}{4}\sqrt{g^*}\,\phi F_1\left(y + \tfrac{1}{2}\sqrt{g^*}\,\phi^2\right) \\ & + \frac{\sqrt{g^*}}{64\pi^2}\phi\left(y + \tfrac{1}{2}\sqrt{g^*}\,\phi^2\right)\ln\left(y + \tfrac{1}{2}\sqrt{g^*}\,\phi^2\right) \\ & + \tfrac{1}{4}\sqrt{g^*}\,\phi(\tau)\int_0^\infty \mathrm{d}\tau'\left[A^4(4\pi^2\tau') - 1\right] \\ & \times\left[\exp\left\{-\int_0^{\tau'} \mathrm{d}\tau''\left(y + \tfrac{1}{2}\sqrt{g^*}\,\phi^2(\tau - \sqrt{g^*}\,\tau'')\right)\right\}\right. \\ & \left. - \exp\left\{-\tau'\left(y + \tfrac{1}{2}\sqrt{g^*}\,\phi^2(\tau)\right)\right\}\right], \end{aligned} \tag{3.28}$$

with the random noise satisfying $\langle\eta(\tau)\eta(\tau')\rangle = \delta(\tau - \tau')$.

The function $F_1(y)$ has been defined in eq. (2.5). The last term in (3.28) is a non-markovian (memory) term. At the bulk critical point $y = 0$, we finally obtain:

$$\begin{aligned} 0 = {} & \partial_\tau \phi + \tfrac{1}{12}\phi^3 - \eta(\tau) + \sqrt{\tfrac{1}{3}}\,\pi\varepsilon^{1/2}\phi F_1\left(2\sqrt{\tfrac{1}{3}}\,\pi\varepsilon^{1/2}\phi^2\right) \\ & + \tfrac{1}{24}\varepsilon\phi^3\ln\left(2\sqrt{\tfrac{1}{3}}\,\pi\varepsilon^{1/2}\phi^2\right) + \tfrac{1}{4}\sqrt{\varepsilon}\,\phi(\tau)\int_0^\infty \mathrm{d}\tau'\left[A^4(\sqrt{3}\,\pi\tau') - 1\right] \\ & \times\left[\exp\left\{-\tfrac{1}{2}\sqrt{\varepsilon}\int_0^{\tau'} \mathrm{d}\tau''\,\phi^2(\tau - \sqrt{\varepsilon}\,\tau'')\right\} - \exp\left\{-\tfrac{1}{2}\sqrt{\varepsilon}\,\tau'\phi^2(\tau)\right\}\right]. \end{aligned} \tag{3.29}$$

Let us discuss briefly some limiting situations associated with eq. (3.29).

First if the initial value ϕ_0 is very large, there is a region where the bulk relaxation is observed. This is because for large argument $F_1(y) \sim -y^{-1}$ and hence the term including $F_1(\phi^2)$ is negligible compared to the other terms. The random noise and the non-markovian term are also small compared to the other terms. It is then clear that $\phi \sim \tau^{-\beta/(\nu z)}$ is an approximate solution to the equation in this region with $\beta/\nu z = \frac{1}{2} - \frac{1}{4}\varepsilon + o(\varepsilon^2)$.

On the other hand at much later times when ϕ is already very small we can use the expansion

$$F_1(y) = F_1(0) - \frac{1}{16\pi^2} y \ln y - \frac{1}{16\pi^2}(1+C)y + o(y^2), \tag{3.30}$$

where C is the constant defined in eq. (2.17), to expand the term containing F_1 in eq. (3.29). It is easy also to verify that the non-markovian term is equal to

$$\frac{\sqrt{3}\,\varepsilon^{3/2}}{144\pi^3}\phi(\tau)\frac{d}{d\tau}\phi^2(\tau)\int_0^\infty d\tau'\left[A^4(\tau')-1\right]\tau'^2 + o(\varepsilon^2), \tag{3.31}$$

and thus this term is of higher order and can be discarded in a calculation to $o(\varepsilon)$. We thus find in that region eq. (3.29) becomes

$$0 = \partial_\tau\phi + \sqrt{\tfrac{1}{3}}\,\pi F_1(0)\varepsilon^{1/2}\phi + \tfrac{1}{12}\left(1 - \tfrac{1}{2}(1+C)\varepsilon\right)\phi^3 - \eta(\tau) + o(\varepsilon^{3/2}). \tag{3.31}$$

Eq. (3.31) is just the one expected for linear relaxation from the formalism developed in (I), and the linear relaxation time is given by the inverse gap of the corresponding quantum mechanical hamiltonian. (Notice that our rescaling of the

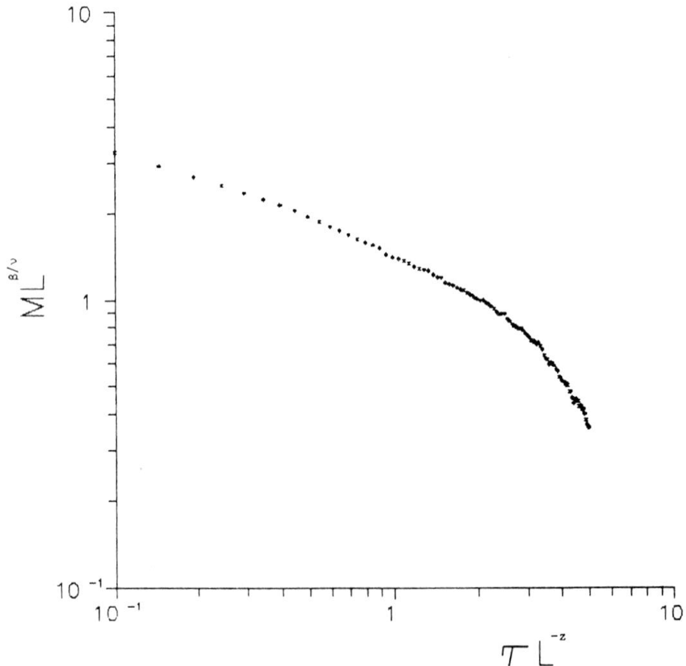

Fig. 5. A plot of the magnetization versus time for $d = 3$ ($\varepsilon = 1$).

field ϕ in (3.26) differs from that used in (I) by terms of $o(\varepsilon)$; that's the reason for the appearance of the $o(\varepsilon)$ correction to the coefficient of ϕ^3 in eq. (3.31)). Thus for a large initial value ϕ_0 the solution of eq. (3.29) is expected to display the full crossover from the bulk power law non-linear relaxation to the finite size linear relaxation which is exponential with characteristic time $\tau_R(L)$.

After finding the solution of eq. (3.29) in the form

$$\overline{\phi(\tau)} = F(\phi_0, \tau), \tag{3.32}$$

the final result for the non-linear relaxation of the magnetization, taking into account the change of variables (3.25) and (3.26) is

$$M(M_0, \tau, L) = (g^*)^{-1/4} L^{-\beta/\nu} F\big((g^*)^{1/4} M_0 L^{\beta/\nu}, 2\lambda (g^*)^{1/2} \tau L^{-z}\big). \tag{3.33}$$

The function $F(\phi_0, \tau)$ has been calculated numerically from eq. (3.29) using the method outlined in the beginning of the section. We verified that the memory term (last term in eq. (3.29)) is negligible even in the non-linear regime compared to other terms and thus it was not retained in the numerical calculation. The function $F(\phi_0, \tau)$ is plotted versus τ in fig. 5 using a log-log scale and in fig. 6 using a semilog scale. The initial value chosen was $\phi_0 = 20$. One can observe clearly the

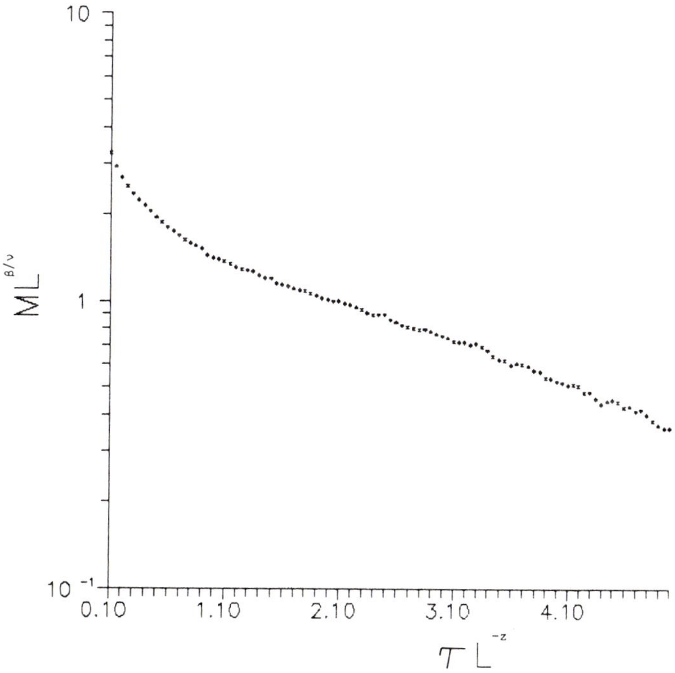

Fig. 6. A plot of the magnetization versus time for $d=3$, which emphasizes the region of linear relaxation.

crossover from the power law decay to an exponential decay at large time. This figure should be compared with fig. 5 of the MC simulation by Kikuchi and Okabe [10]. For this calculation we have used $\Delta = 0.005$ and $N = 400$.

4. Summary and discussion

In this paper we have shown how to obtain analytically the stochastic equation of state for the non-linear relaxation in a finite size system. Below four dimensions the equation is written in a form which fulfills the correct finite size scaling behavior. The solution at the bulk critical point displays the full crossover from the bulk power law decay to the finite size exponential relaxation as a function of time. Thus at long times the decay is governed by the linear relaxation time. We also solve explicitly for the linear relaxation time by solving for the gap of the associated quantum mechanical hamiltonian. The results are obtained in the vicinity of T_c and show a crossover from the bulk power law increase of the relaxation time to the finite size saturation at T_c. The results are cast in a form which enables comparison with Monte Carlo simulation. Our results agree nicely with results of non-linear critical relaxation in three dimensions. More simulations in the vicinity of T_c are needed for comparison with the shape of the linear relaxation time. Using our results it is straightforward to obtain the relaxation of a system of any size, starting with different values of the initial magnetization. Besides obtaining the linear relaxation time it is also possible to use our methods to obtain an approximate expression for the full autocorrelation function of the magnetization. This will require solving for higher energy levels and eigenfunctions of the associated supersymmetric hamiltonian.

This work has been supported in part by the National Science Foundation.

Appendix

In this appendix we explain some of the intermediate steps used in the derivation of eq. (3.28). We start with eqs. (3.23) and (3.24). Using the function $A(u)$ defined in eq. (2.5) we find

$$\underset{q\neq 0}{\leftarrow\!\bigcirc} = -\lambda_0^2 u_0 L^{-d}\phi(\tau)\frac{L^2}{8\pi^2\lambda_0}\int_0^\infty \mathrm{d}\tau'\left[A^d(\tau')-1\right]$$

$$\times\exp\left\{-\frac{L^2}{4\pi^2}\int_0^{\tau'}\mathrm{d}\tau''\left[r_0+\tfrac{1}{2}u_0\phi^2\left(\tau-\frac{L^2}{8\pi^2\lambda_0}\tau''\right)\right]\right\}. \tag{A.1}$$

From the last integral subtract and add the term

$$\int_0^\infty \mathrm{d}\tau'\frac{\pi^{d/2}}{(\tau')^{d/2}}\exp\left\{-\frac{L^2}{4\pi^2}\tau'\left(r_0+\tfrac{1}{2}u_0\phi^2(\tau)\right)\right\}. \tag{A.2}$$

We substitute the result in eq. (3.23). We then use the identity

$$\int_0^{\infty} \frac{\mathrm{d}u}{u^{2-\varepsilon/2}} \exp\left\{-\frac{us}{4\pi^2}\right\} = \frac{1}{\varepsilon} \frac{\Gamma\left(1+\frac{1}{2}\varepsilon\right)}{1-\frac{1}{2}\varepsilon} \left(s/4\pi^2\right)^{1-\varepsilon/2} \tag{A.3}$$

and proceed to carry out the usual bulk renormalizations within the framework of minimal subtraction. Upon introducing renormalized coupling constants t instead of r_0 and g (dimensionless) instead of u_0, and carrying out the rescaling transformations (3.25) through (3.27) we obtain eq. (3.28).

References

[1] M.E. Fisher, *in* Critical phenomena, Proc. 51st Enrico Fermi Summer School, Varena, ed. M.S. Green (Academic Press, NY, 1972);
M.E. Fisher and M.N. Barber, Phys. Rev. Lett. 28 (1972) 1516.
[2] M.N. Barber, *in* Phase transitions and critical phenomena, vol. VIII, ed. C. Domb and J. Lebowitz (Academic Press, NY, 1984) and references therein
[3] E. Brézin and J. Zinn-Justin, Nucl. Phys. B257 (1985) 868
[4] E. Brézin, J. de Phys. 43 (1982) 15
[5] J. Rudnick, H. Guo and D. Jasnow, J. Stat. Phys. 41 (1985) 353
[6] Y.Y. Goldschmidt, Nucl. Phys. B280 [FS18] (1987) 340
[7] J. Niel and J. Zinn-Justin, Nucl. Phys. B280 [FS18] (1987) 355
[8] H. Diehl, to be published
[9] P.C. Martin, E.D. Siggia, H.A. Rose, Phys. Rev. A8 (1973) 423;
R. Bausch et al., Z. Phys. B24 (1976) 113;
A. Munoz Sudupe and R.F. Alvarez-Estrada, J. Phys. A16 (1983) 3049
[10] M. Kikuchi and Y. Okabe, J. Phys. Soc. Japan, 55 (1986) 1359
[11] H. Dekker and N.G. Van Kampen, Phys. Lett. 73A (1979) 374
[12] E. Witten, Nucl. Phys. B188 (1981) 513;
F. Cooper and B. Freedman, Ann. of Phys. 146 (1983) 262
[13] S. Miyoshita and H. Takano, Prog. Theor. Phys. (Japan) 73 (1985) 1122
[14] M.C. Yalabik and J.D. Gunton, Phys. Rev. B25 (1982) 534;
N. Jan et al., J. Stat. Phys. 33 (1983) 1;
C. Kalle, J. Phys. A17 (1984) L801;
R.B. Pearson et al., Phys. Rev. B31 (1985) 4472;
S. Wansleben and D.P. Landau, Univ. of Georgia preprint (1986)
[15] M. Suzuki, Prog. Theor. Phys. 58 (1977) 1142
[16] J.M. Sancho et al., J. Phys. A13 (1980) L443
[17] R. Bauch and H.K. Janssen, Z. Phys. B25 (1976) 275;
R. Bauch et al., Z. Phys. B36 (1979) 179

Finite-size tests of hyperscaling

K. Binder
Institüt für Physik, Universität Mainz, D-6500 Mainz, Postfach 3980, Germany and Institut für Festkörperforschung der Kernforschungsanlage Jülich, D5170 Jülich, Germany

M. Nauenberg
Natural Sciences, University of California, Santa Cruz, California 95064

V. Privman
Baker Laboratory, Cornell University, Ithaca, New York 14853

A. P. Young
Mathematics Department, Imperial College, London, England
(Received 10 April 1984)

The possible form of hyperscaling violations in finite-size scaling theory is discussed. The implications for recent tests in Monte Carlo simulations of the $d=3$ Ising model are examined, and new results for the $d=5$ Ising model are presented.

Recently two extensive Monte Carlo calculations[1,2] have been carried out which test the validity of hyperscaling in the three-dimensional Ising model. In the calculation by Freedman and Baker,[1] they "observe a systematic downward trend by more than twice the statistical error for a quantity which should be constant if hyperscaling is satisfied," while Barber, Pearson, Toussaint, and Richardson[2] conclude that their results require "the anomalous dimension to be small, and are consistent with hyperscaling." The purpose of this paper is to resolve this apparent conflict by showing that these two collaborations evaluated different anomalous exponents, and that the exponent evaluated by Barber *et al.*, which they erroneously identified with Fisher's anomalous dimension,[3] actually vanishes on theoretical grounds. New Monte Carlo results are also presented for the five-dimensional Ising model which confirm the theoretical predictions for hyperscaling violation in this case.

To start, we investigate the consequence of the existence of a "dangerous irrelevant variable" on the finite-size scaling form of the free energy, and on the hyperscaling relations for critical exponents. We then show that consistency conditions require that the anomalous exponent introduced by Barber *et al.*[2] must be zero. Finally, we discuss the finite-size scaling form of the renormalized coupling constant for the Ising model,[4] and the results of a Monte Carlo simulation for the dimensionality $d=5$.

According to the renormalization-group derivations[5,6] of finite-size scaling, the singular part of the free energy f_L and the correlation length[7] ξ_L have the form

$$f_L=L^{-d}f(tL^{y_T},hL^{y_H},uL^{y_U}) , \tag{1}$$

$$\xi_L=L\xi(tL^{y_T},hL^{y_H},uL^{y_U}) , \tag{2}$$

where t is the reduced temperature $t=(T-T_c)/T_c$ (T_c is the transition temperature of the infinite system), h is the magnetic field, and u is an irrelevant variable, while $y_T>0$, $y_H>0$, and $y_U<0$ are renormalization-group exponents. We use equality signs to indicate here asymptotic scaling relations. If the free-energy scaling function $f(x,y,z)$ is singular in the limit $z\to 0$, then u is called a dangerous irrelevant variable.[3,6] For simplicity we assume here that the correlation scaling function $\xi(x,y,z)$ is regular in this limit, but later we consider the possibility that it is singular. We assume that for small z,

$$f(x,y,z)=z^{p_1}\bar{f}(xz^{p_2},yz^{p_3}) . \tag{3}$$

The choice of this particular pattern, multiplicative singular powers of z, is motivated by the known[3] mechanism for the *bulk* scaling at $d>d_c=4$. Equation (3) implies

$$f_L=L^{-d^*}F(tL^{y_T^*},hL^{y_H^*}) , \tag{4}$$

where scale factors have been absorbed in t and h, and $d^*=d-p_1y_U$, $y_T^*=y_T+p_2y_U$, and $y_H^*=y_H+p_3y_U$ are the effective exponents. Note that the anomalous exponent introduced by Barber *et al.*[2] corresponds to $d-d^*$, while Fisher's exponent[3] ω^* is given by

$$\omega^*=d-d^*/(y_T^*\nu) . \tag{5}$$

These two exponents would be the same if $y_T^*\nu=1$, which was implicitly assumed by Barber *et al.*,[2] but has not been proven. We will show instead that $d^*=d$ for cubic or similarly shaped systems,[6] as usually employed in Monte Carlo simulations.[1,2]

The existence of the limit $L\to\infty$ of f_L corresponding to the bulk free energy f_∞ implies that for $x\to\pm\infty$ and $y|x|^{-\Delta}$ fixed,

$$F(x,y)=|x|^{d^*/y_T^*}Y_\pm(y|x|^{-\Delta}) , \tag{6}$$

where $\Delta = y_H^*/y_T^*$, and $Y_\pm$ is the scaling function for $t \gtrless 0$. Likewise the existence of the bulk correlation length ξ_∞ in this limit leads to the asymptotic form

$$\xi(x,y,0) = |x|^{-\nu} X_\pm(y|x|^{-\Delta'}) , \tag{7}$$

where

$$\nu = 1/y_T \text{ and } \Delta' = y_H/y_T , \tag{8}$$

and we assumed for simplicity that the scaling function ξ in Eq. (2) is nonsingular as $z \to 0$. Taking suitable derivatives of the free energy, one finds the following scaling relations $\alpha = 2 - d^*/y_T^*$, $\beta = d^*/y_T^* - \Delta$, and $\gamma = 2\Delta - d^*/y_T^*$, where α, β, and γ are the thermodynamic exponents for the specific heat, magnetization, and susceptibility, respectively. Consequently y_T^* and y_H^* can be written as

$$y_T^* = d^*/(\gamma + 2\beta) , \tag{9}$$

$$y_H^* = d^*(\gamma + \beta)/(\gamma + 2\beta) . \tag{10}$$

Hyperscaling violations occur if y_T^* and/or Δ differ from the exponents $y_T(=1/\nu)$ and $\Delta'(=y_H/y_T)$ appearing in the scaling form of the correlation length. However, we will show that even in this case one must have $d^* = d$.

We now give three arguments which support $d^* = d$. First, consider the finite-size magnetization m_L and susceptibility χ_L:

$$m_L = \langle s \rangle_L , \tag{11}$$

$$\chi_L = L^d(\langle s^2 \rangle_L - \langle s \rangle_L^2) , \tag{12}$$

where $s = (1/L^d)\sum s_i$, s_i is the spin at the ith site, and $\langle\ \rangle_L$ denotes the thermal average. According to Eq. (4), m_L and χ_L have the scaling form

$$m_L = \frac{\partial f_L}{\partial h} = L^{y_H^* - d^*} V(tL^{y_T^*}, hL^{y_H^*}) , \tag{13}$$

$$\chi_L = \frac{\partial^2 f_L}{\partial h^2} = L^{2y_H^* - d^*} W(tL^{y_T^*}, hL^{y_H^*}) , \tag{14}$$

where the scaling functions V and W are obtained from F.

For the bulk magnetization, m_b, as well as the susceptibility, χ_b, special care must be taken in order of the limits $L \to \infty$ and $h \to 0^\pm$ below the critical temperature. Taking first the limit $L \to \infty$, and afterwards the limit $h \to 0^\pm$, one obtains the conventional bulk values,

$$m_b = a|t|^\beta \text{ and } \chi_b = b|t|^{-\gamma} \text{ as } T \to T_c^- , \tag{15}$$

so that, by Eq. (12),

$$\langle s^2 \rangle_L = m_b^2 + L^{-d}\chi_b \quad (t<0,\ L \to \infty) . \tag{16}$$

However, if we first set $h = 0$ and then take the limit $L \to \infty$, then $\langle s \rangle_L \equiv 0$ but we expect (see Refs. 4 and 6 for details)

$$\lim_{L \to \infty} [\langle s^2 \rangle_{L,h=0}] = m_b^2 \tag{17}$$

[in fact, Eq. (16) applies in this limit as well[4,6]]. By Eqs. (12) and (17), the zero-field susceptibility behaves according to

$$\chi_L \propto L^d |t|^{2\beta} \text{ as } T \to T_c^- . \tag{18}$$

To obtain this result from the scaling form, Eq. (14), we must require[6]

$$\lim_{x \to \pm\infty} W(x,0) \propto |x|^{2\beta} \tag{19}$$

and

$$d^* = 2(y_H^* + \beta y_T^*) - d . \tag{20}$$

Substituting Eqs. (9) and (10) in Eq. (20), we obtain the relation

$$d^* = d , \tag{21}$$

which also implies $p_1 = 0$ in Eq. (3).

Second, we show that this condition also follows from the finite-size scaling form of f_L at the ferromagnetic phase boundary due to the existence of a discontinuity fixed point.[8] In this case the finite-size magnetization m_L below the critical temperature takes the form[6]

$$m_L = m_b \tanh(m_b h L^d) \tag{22}$$

for $L \gg \xi_b$, and $|h| \ll (m_b \xi_b)^{-1} L^{1-d}$, where ξ_b is the bulk correlation length, and m_b is the bulk magnetization, Eq. (15). Hence, χ_L is given by

$$\chi_L = L^d m_b^2 \cosh^{-2}(m_b h L^d) . \tag{23}$$

For $h = 0$, χ_L reduces to Eq. (18) when $T \to T_c^-$, and therefore implies $d^* = d$, Eq. (21).

Finally, we consider the finite-size scaling properties of the zero-field probability distribution $P(s)$ of the magnetization below the critical temperature. Binder[4] has shown that for large L and s near $\pm m_b$, $P_L(s)$ can be written approximately in the form

$$P_L(s) = \frac{L^{d/2}}{2(2\pi\chi_b)^{1/2}} \left(e^{-(s-m_b)^2 L^d/2\chi_b} + e^{-(s+m_b)^2 L^d/2\chi_b} \right) . \tag{24}$$

Hence the arguments of the exponential functions have the form

$$(s|t|^{-\beta} + a)^2 \frac{(|t|L^{y_T^*})^{\gamma+2\beta}}{2b} ,$$

which demonstrates the occurrence of the scaling combination $|t|L^{y_T^*}$ in $P_L(s)$, where $y_T^* = d/(\gamma + 2\beta)$. Likewise, by including a magnetic field h, the arguments of the exponential become $(s \pm m_b - \chi_b h)^2 L^d/2\chi_b$ and the term linear in h can be written as

$$(b|t|L^{d/(\gamma+2\beta)})^{-(\gamma+\beta)} h L^{y_H^*} .$$

This shows the dependence of $P_L(s)$ on the scaling variable $L^{y_H^*}$, where $y_H^* = d(\gamma+\beta)/(\gamma+2\beta)$. These expressions for y_T^* and y_H^* are precisely Eqs. (9) and (10), but with $d^* = d$. If we assume that no other scaling variables occur, the scaling form Eq. (4) follows, with $d^* = d$.

Next, we discuss some recent Monte Carlo calculations[1,2] which test possible hyperscaling violation in the

$d=3$ Ising model. These calculations evaluated the $h=0$ finite-size renormalized coupling g_L introduced by Binder,[4]

$$g_L=\frac{\langle s^4\rangle_L}{\langle s^2\rangle_L^2}-3=\left[\frac{\chi_L^{(4)}}{L^d\chi_L^2}\right]_{h=0}, \tag{25}$$

where χ_L is given by Eq. (14), and

$$\chi_L^{(4)}=\frac{\partial^4 f_L}{\partial h^4}\approx L^{4y_H^*-d^*}W^{(4)}(tL^{y_T^*}) . \tag{26}$$

Substituting Eqs. (14) and (28) in Eq. (27), we find that for $t=0$,

$$g_L=L^{d^*-d}G(tL^{y_T^*}) , \tag{27}$$

where G is a scaling function. Since $d^*=d$, it follows that g_L is a constant (nonzero in general) at $T=T_c$.

The finite-size scaling analysis of the Monte Carlo data by Barber *et al.*[2] implies that indeed $d-d^*=0$, with an error of ± 0.04 due to the uncertainty in their determination of the critical temperature. This was interpreted by them as evidence for the validity of hyperscaling, but we have shown here that $d^*=d$ even when hyperscaling is violated.

For $t>0$ and $L\to\infty$, χ_L and $\chi_L^{(4)}$ take their bulk values $\chi_L\propto t^{-\gamma}$ and $\chi_L^{(4)}\propto t^{-\gamma-2\Delta}$ so that

$$g_L\propto(t^{d^*/y_T^*}L^d)^{-1} \text{ for } L\gg t^{-1/y_T^*} , \tag{28}$$

where we have used $\gamma-2\Delta=-d^*/y_T^*$. Baker and Freedman[1] evaluated g_L for values of $t>0$ such that the correlation length $\xi_L=cL$, where c is a constant. After some algebra, this yields

$$g_L\propto L^{-\omega^*} , \tag{29}$$

where ω^*, Eq. (5), is Fisher's anomalous dimension,[3] and is positive. For lattices of dimensions $L=3$ to 60 they find $\omega^*\simeq 0.2$. Note, however, that their result depends on the assumption, Eq. (6), that the scaling function ξ in Eq. (2) is not singular as $uL^{y_U}\to 0$.

We now briefly turn our attention to the possibility that $\xi(x,y,z)$ is singular in the limit $z\to 0$. In analogy with Eq. (3) we assume that

$$\xi(x,y,z)=z^{q_1}\overline{\xi}(xz^{q_2},yz^{q_3}) , \tag{30}$$

which implies

$$\xi_L\sim L^{1+q_1y_U}Z(tL^{y_T^{**}},hL^{y_H^{**}}) , \tag{31}$$

where $y_T^{**}=y_T+q_2y_U$ and $y_H^{**}=y_H+q_3y_U$. Likewise in the limit $x\to\pm\infty$ and $y\,|x|^{-\Delta^{**}}$ fixed,

$$Z(x,y)\to|x|^{-\nu}\widetilde{Z}_\pm(y\,|x|^{-\Delta^{**}}) , \tag{32}$$

where $\nu=(1+q_1y_U)/y_T^{**}$ and $\Delta^{**}=y_H^{**}/y_T^{**}$. Since the finite-size correlation length ξ_L is bounded by L, we require $q_1y_U\le 0$. Even if one adopts a plausible assumption that for $t=h=0$, the correlation length increases up to the linear dimensions of the lattice, which implies that $q_1=0$ and $\nu=1/y_T^{**}$, y_T^{**} need not be equal to y_T leading to a new possible source for hyperscaling violation.

Finally, we comment on the effect of boundary conditions on our analysis. The predictions presented here are unchanged in the thermodynamic limit ($L\to\infty$) if the reduced temperature t in Eq. (4) is replaced by a "shifted" variable t_L, where

$$t_L=[T-T_c(L)]/T_c , \tag{33}$$

in which $T_c(L)$ is a "pseudo T_c" for the finite system. This could, for example, be defined as the temperature where the probability distribution of the magnetization starts to develop a two-peak structure. One then defines a shift exponent ψ by

$$[T_c(L)-T_c]/T_c=AL^{-1/\psi} , \tag{34}$$

where A is a constant. For systems with a surface it is expected that $\psi=\nu$, because properties of the finite system should differ from their bulk values when ξ_L approaches the system size. In this case the shift in T_c is greater than the range of the finite-size rounding if hyperscaling is violated, because $y_T^*>\frac{1}{2}$. However, we shall argue below that $\psi=1/y_T^*$ for periodic boundary conditions.

A well-known example which violates hyperscaling is the Ising model (more generally the n-component vector model) for $d>4$, where critical exponents stick at their mean field values. It is straightforward to show[6] that the free energy of a finite system scales as in Eq. (4) with $y_T^*=d/2$, $y_H^*=d/4$ compared with $y_T=1/\nu=2$, and $y_H=(d+2)/2$. However, these[6] renormalization-group arguments do not give information on the size of the shift in T_c, for which an explicit calculation must be performed. This is possible for the n-component model with periodic boundary conditions in the limit $n\to\infty$. By a straightforward extension of the work of Brézin[9] one can evaluate the scaling function $W(x,0)$ for the susceptibility, and see that the scaling form of Eq. (4) is valid, so the coefficient of the $L^{-1/\nu}$ term in Eq. (34) must vanish for this model. We observe that this is a general result for periodic boundary conditions *provided that* $tL^{y_T^*}$ is the only temperature variable which enters thermodynamic scaling, and then $\psi=1/y_T^*$ follows by a standard argument.[10]

We have tested our claim that $d^*=d$ by Monte Carlo simulations on the five-dimensional Ising model on a cubic lattice with nearest-neighbor interaction, J, for sizes between $L=3$ and 7. Figure 1 shows that the data for g_L intersect at $T/J\simeq 8.77$, $g_L\simeq -1.00$. A nonzero value of g_L is just what is expected from Eq. (27) with $d^*=d$. Furthermore, the temperature agrees precisely with the estimate of T_c from high-temperature series by Fisher and Gaunt.[11] In Fig. 2 we show that the data scale well with the predicted exponent $y_T^*=\frac{5}{2}$ instead of $y_T=1/\nu=2$. It is possible to evaluate the scaling function $G(x)$ in Eq. (27) exactly for $d>4$ because, in the critical region, the probability distribution for the magnetization per spin s has the mean-field form

$$P_L(s)\propto\exp[-L^d(ct_Ls^2+us^4)] , \tag{35}$$

where c and u are constants. By the substitution $\phi=(uL^d)^{1/4}s$, one finds that

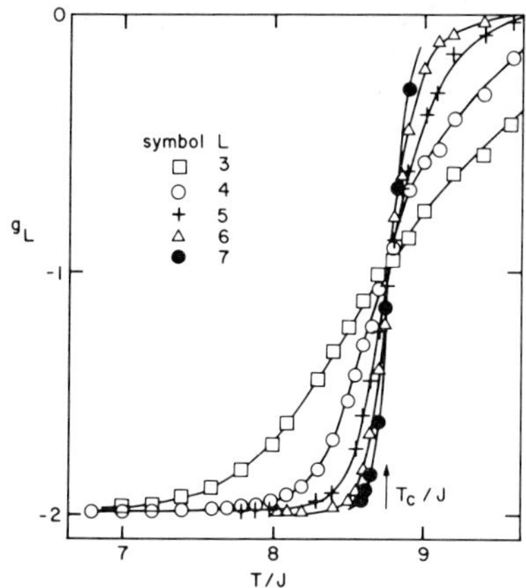

FIG. 1. Monte Carlo results for g_L as a function of T/J for sizes $L=3$ to 7, where 16 000 iterations per spin have been performed for each data point. The arrow marks the value $T_c/J=8.77$ from Ref. 11.

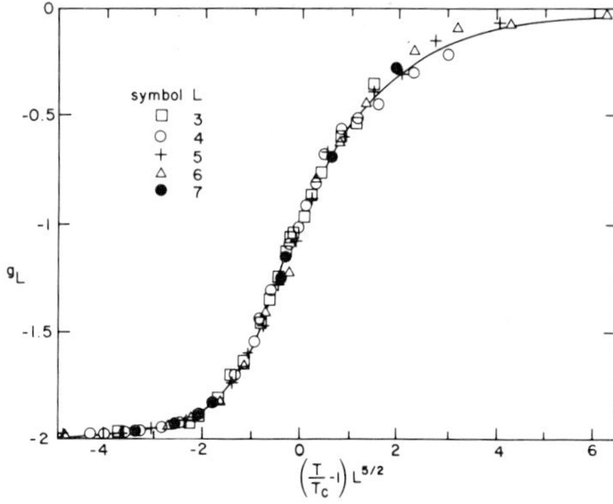

FIG. 2. The data for g_L shown in Fig. 1 plotted vs the scaling variable $(T/T_c-1)L^{5/2}$, where $T_c=8.77J$. The solid line is obtained from the mean-field form of the free energy, as described in the text.

$$G(x)=\frac{\langle\phi^4\rangle}{\langle\phi^2\rangle^2}-3\ , \tag{36}$$

where the averages are obtained by integrating ϕ from $-\infty$ to $+\infty$ with the weight

$$P_L(\phi)\propto\exp[-a(x-b)\phi^2-\phi^4]\ , \tag{37}$$

where $a=c/u^{1/2}$ and $ab=A$ is the coefficient of the shift relation, Eq. (34) (with $\psi^{-1}=y_T^*=d/2$). The solid curve in Fig. 2 is obtained from Eqs. (36) and (37) with $b=0.37$, $a=0.56$, and fits the data very well.

To conclude, we propose that the free energy of a finite system with periodic boundary conditions scales as in Eq. (4) with y_T^* and y_H^* given by Eqs. (9) and (10) and $d^*=d$. For other boundary conditions, where the system has a surface, it is probably necessary to use both $tL^{y_T^*}$ and $tL^{1/\nu}$ for a complete asymptotic description. When hyperscaling is violated it is not possible to determine the correlation length exponent ν from the free energy, and its derivatives, and it is necessary to carry out a *separate* calculation, using either renormalization-group techniques or explicitly looking at the spatial dependence of the correlations.

We would like to thank M. E. Fisher and J. Rudnick for helpful comments. M. Nauenberg and V. Privman acknowledge support for this research from the National Science Foundation under Grants Nos. PHY-81-15541 and DMR-81-17011.

[1]B. A. Freedman and G. A. Baker Jr., J. Phys. A 15, L715 (1982).

[2]M. N. Barber, R. B. Pearson, D. Toussaint, and J. L. Richardson, NSF Report No. NSF-ITP83-144 (unpublished).

[3]M. E. Fisher, in *Renormalization Group in Critical Phenomena and Quantum Field Theory,* proceedings of a conference, edited by J. D. Gunton and M. S. Green (Temple University, Philadelphia, 1974); and in *Critical Phenomena,* Vol. 186 of *Lecture Notes in Physics,* edited by F. J. W. Hahne (Springer, Berlin, 1983).

[4]K. Binder, Z. Phys. B 43, 119 (1981).

[5]A recent review is given by M. N. Barber, in *Phase Transitions and Critical Phenomena,* edited by C. Domb and J. L. Lebowitz (Academic, New York, 1983), Vol. 8.

[6]V. Privman and M. E. Fisher, J. Stat. Phys. 33, 385 (1983).

[7]For a finite system of linear dimension L some care must be taken in the definition of the correlation length, particularly for temperatures T below the critical temperature T_c. For periodic boundary conditions we adopt the relation

$$2d\xi_L^2=\sum_{i,j}(\vec{r}_i-\vec{r}_j)^2(\langle s_is_j\rangle-c_L)\Big/\sum_{i,j}(\langle s_is_j\rangle-c_L)\ ,$$

where $\vec{r}_i$ is the position of lattice site i,

$$c_L=\frac{1}{L^d}\sum_i\langle s_is_{i'}\rangle\ ,$$

and i' is the site with $\vec{r}_{i'}=\vec{r}_i+\frac{1}{2}(1,1,\ldots,1)L$. Note that for $L\to\infty$, $c_L\to m_b^2$, where m_b is the bulk magnetization, so

$\xi_L^2 \rightarrow \xi_b^2$, the square of the bulk correlation length, defined as the second moment of the connected correlation function $\langle s_i s_j \rangle - \langle s_i \rangle \langle s_j \rangle$.

[8]B. Nienhuis and M. Nauenberg, Phys. Rev. Lett. **35**, 477 (1975); T. Niemeiyer and J. M. J. van Leeuwen, in *Phase Transitions and Critical Phenomena*, edited by C. Domb and M. S. Green (Academic, New York, 1976), Vol. 6.

[9]E. Brézin, J. Phys. (Paris) **43**, 15 (1982).

[10]M. E. Fisher, in *Critical Phenomena*, Vol. 51 of *Proceedings of the "Enrico Fermi" International School of Physics*, edited by M. S. Green (Academic, New York, 1971).

[11]M. E. Fisher and D. S. Gaunt, Phys. Rev. **133**, A224 (1964).

Journal of Statistical Physics, Vol. 33, No. 2, 1983

Finite-Size Effects at First-Order Transitions

Vladimir Privman[1] **and Michael E. Fisher**[1]

Received May 24, 1983

Finite-size rounding of a first-order phase transition is studied in "block"- and "cylinder"-shaped ferromagnetic scalar spin systems. Crossover in shape is investigated and the universal form of the rounded susceptibility peak is obtained. Scaling forms on the low-temperature side of the critical point are considered both above and below the borderline dimensionality, $d_{>} = 4$. A method of phenomenological renormalization, applicable to both odd and even field derivatives, is suggested and used to estimate universal amplitudes for two-dimensional Ising models at $T = T_c$.

KEY WORDS: First-order transitions; finite-size effects; scaling theory; Ising models; phenomenological renormalization; borderline dimensionality.

1. INTRODUCTION

In this paper we study various finite-size effects arising at the zero-field phase boundary of a ferromagnetic scalar (or Ising-like) spin system, with an emphasis on the rounding of the first order transition for $T < T_c$. Finite-size scaling theory[1,2] for behavior close to a critical point has attracted appreciable theoretical and experimental effort (see Ref. 3 for a recent review). Transfer matrix calculations utilizing finite-size scaling ideas have led to some remarkably precise and accurate estimates for the critical exponents of several $(d = 2)$-dimensional models (see the overview presented by Nightingale[4]). More recently, a study of finite size effects at *first-order* phase transitions has been initiated. A scaling theory was developed,[5] and the properties of renormalization groups close to the associated "discontinuity" fixed points[6] were invoked. It was found[5,7–9] that

[1] Baker Laboratory, Cornell University, Ithaca, New York 14853.

0022-4715/83/1100-0385$03.00/0

the rounding of a first-order transition as a function of the ordering field, say, H, takes place on a scale proportional to the inverse of the total volume, V, of the system. However, if the system is already infinite in one direction, so forming a cylinder of cross-sectional area A, the rounding in field is known to be *exponentially* narrow in A; this property has been derived analytically[9-11] for the field-driven transitions in Ising model cylinders and numerically[9,12] for temperature-driven first-order transitions in several $(d = 2)$-dimensional Potts models with $q > 4$ states per site. One of the motivations for our present work is to understand the crossover in the sharpness of a first-order transition that must evidently take place when the shape of a system goes over from a totally finite "block" geometry to an elongated, cylindrical geometry with, ultimately, one infinite dimension. In addition we hoped to obtain more explicit forms for the scaling functions describing the characteristic rounding behavior.

In Section 2 we combine an introductory discussion with a derivation of the rounding form for the "block" situation and we present a scaling formulation which allows a crossover to cylindrical geometry. In Section 3 we call on the transfer matrix approach to obtain an explicit form for the rounding which interpolates between the block and cylinder geometries. We compare our results with an approximate scaling analysis for low temperatures presented recently by Cardy and Nightingale[13] (see also Appendix B).

Section 4 addresses various further questions including the nature of the "corrections-to-scaling" for different types of boundary condition. We also exhibit the rounding behavior in the solvable infinite-range model which displays mean field critical behavior. Cylindrical geometry is taken up again and two regimes of the exponential rounding are identified: one close to the $d = 1$, $T \equiv 0$, limiting phase transition, and the other arising when the d-dimensional phase transition is approached at fixed temperature in the range $0 < T < T_c$. The interplay between the two forms of behavior is related to the detailed nature of the asymptotic degeneracy of the two largest eigenvalues of the transfer matrix. For the standard square lattice Ising model, we report (in Appendix A) some exact results, based on Onsager's solution.[14,15]

Recently Brézin[16] has argued on the basis of exact calculations in the infinite-component, $n \to \infty$ limit, that the usual finite-size scaling theory for the *critical* region[1,2] is valid only for d less than the borderline dimensionality $d_> = 4$. In Section 5 we consider the approach to criticality and derive the large argument asymptotics of the scaling functions for both block and cylinder geometries. We present a generalization of the finite-size scaling hypothesis which is applicable for $d > d_>$ and which agrees with Brézin's explicit results: a "dangerous irrelevant variable" plays a central role.

Finally, in Section 6, we argue that the so-called "single-phase functions" introduced by Schulman and Privman[10,11] satisfy a finite-size scaling postulate near criticality. This property allows phenomenological renormalization calculations[4] to be performed for both even and *odd* field derivatives of the free energy below and at T_c. We illustrate the application of the method and report numerical estimates for the first few critical exponents and universal critical amplitudes for the square and the triangular lattice Ising models. Some open problems are mentioned briefly in Section 7.

2. SCALING FORMS FOR FIRST-ORDER TRANSITIONS

For definiteness we consider d-dimensional hypercubic lattices with lattice spacing a and cell volume a^d. A rectangular or *block* lattice geometry will be specified by sides of lengths $L_1, L_2, \ldots L_d$ parallel to the lattice axes. Frequently, we will consider the simple, finite geometry

$$L_1 = L_{\parallel} \equiv L, \qquad L_2 = L_3 = \cdots L_d = L_{\perp} \tag{2.1}$$

which represents a *rod* if $L_{\parallel} > L_{\perp}$, as we will usually consider, or a *slab* if $L_{\parallel} < L_{\perp}$. The total volume is

$$V = \prod_{i=1}^{d} L_i = L_{\perp}^{d-1} L_{\parallel} \tag{2.2}$$

and the cross-sectional areas are

$$A_j = V/L_j = \prod_{i \neq j}^{d} L_i \qquad \text{with} \quad A \equiv A_1 = L_{\perp}^{d-1} \tag{2.3}$$

Unless explicitly stated otherwise, *periodic boundary conditions* will be assumed in each direction. An infinite *cylinder* geometry is specified by $L_1 = L_{\parallel} \equiv L \to \infty$.

At each site i, with position vector $\mathbf{r}_i$, we suppose a *scalar* spin variable, s_i, is located which interacts with an external magnetic field

$$H = hk_B T \tag{2.4}$$

and, via couplings of finite range, with spins on other sites, j. Various explicit calculations will be discussed and presented for the nearest neighbor spin-1/2 ferromagnetic Ising model with Hamiltonian

$$\mathscr{H} = -J \sum_{\langle ij \rangle} s_i s_j - Ha^d \sum_i s_i \qquad (s_i = \pm 1) \tag{2.5}$$

the unorthodox factor a^d is introduced here for dimensional reasons that will become more evident below. For simplicity we will use the language of this model even when our considerations can be readily generalized.

The scaling ansatz for first-order transitions advanced by Fisher and Berker (FB)[5] can be written for the singular part of the reduced free energy density of a simple ferromagnet below criticality, $T < T_c$, in the form

$$f_s(H, T; L_j) \equiv F_s/k_B TV \approx A_0(T) L_0^{-d} W[B(T) H L_0^d; l_j] \tag{2.6}$$

where the shape ratios $l_j = L_j/L_0$ are assumed fixed and of order unity: a convenient normalization is provided by

$$\prod_{j=1}^{d} l_j = 1 \quad \text{so that} \quad L_0^d = V \tag{2.7}$$

Technically the temperature is an irrelevant variable when $T < T_c$ and the coefficients $A_0(T)$ and $B(T)$ then represent nonuniversal amplitudes. The scaling function, $W(y; l_j)$, must, when conveniently normalized, satisfy

$$W(y; l_j) \approx -|y|, \qquad \text{as} \quad y \to \pm\infty \tag{2.8}$$

in order to reproduce the first-order transition in the infinite-size limit, $L_0 \to \infty$. Indeed, the magnetization density in that limit becomes

$$m \equiv -\frac{1}{V}\left(\frac{\partial F_s}{\partial H}\right)_T \approx \pm k_B T A_0(T) B(T) = \pm m_0(T) \tag{2.9}$$

for small $H \gtrless 0$, where $m_0(T)$ is the bulk spontaneous magnetization which vanishes near criticality as

$$m_0(T) \sim |t|^\beta \qquad \text{with} \quad t = (T - T_c)/T_c \tag{2.10}$$

The zero-field susceptibility in the finite system is then given by

$$\chi_0(T, L_0) = \left(\frac{\partial m}{\partial H}\right)_{H=0} \approx -k_B T A_0 B^2 W_0''(l_j) L_0^d \equiv C(T; l_j) V \tag{2.11}$$

where $W_0''(l_j) = (d^2W/dy^2)_{y=0}$. Now, as discussed in FB, the zero-field susceptibility for large L_0 can be expressed in terms of the correlation functions, $\langle s_i s_j \rangle$, for large spin separations. Specifically, the fluctuation relation yields

$$\chi_0 = (a^{2d}/Vk_B T) \sum_i \sum_j \langle s_i s_j \rangle \tag{2.12}$$

The zero-field correlation functions, $\langle s_i s_j \rangle$, depend here, of course, on the size and shape of the system but for $T < T_c$ we may argue that the predominant configurations of the system correspond to the spins being spontaneously magnetized "up" or spontaneously magnetized "down." This encapsulates the usual argument[5,17–19] leading to the identification of the (*short*) long-range order as

$$\lim_{|\mathbf{r}_i - \mathbf{r}_j| \to \infty} \lim_{L_0 \to \infty} \langle s_i s_j \rangle = m_0^2(T) \tag{2.13}$$

and to the conclusion that when $L_0 \to \infty$ the sum in (2.12) may be correctly estimated by the corresponding replacement $\langle s_i s_j \rangle \Rightarrow m_0^2$.[17–19] If, at least provisionally, we accept this heuristic argument we find, for $L_0 \to \infty$,

$$\chi_0 \approx (m_0 V)^2 / k_B T V \quad \text{or} \quad C(T; l_j) = m_0^2 / k_B T \tag{2.14}$$

which, although nonrigorous, would seem to be rather generally valid.

Combining (2.9), (2.11), and (2.14) yields the identifications

$$A_0 = -W_0''(l_j), \qquad B(T) = m_0(T)/k_B T A_0 \tag{2.15}$$

from which we conclude that A_0 is independent of T and that $W_0''(l_j)$ is actually independent of the shape ratios, l_j (see also below). More importantly, we see from (2.6) that the finite size of the system at the first-order transition enters the FB scaling ansatz *only* through the natural combination

$$y_V = E_V(T)/k_B T = m_0(T) H V / k_B T \tag{2.16}$$

which represents the dimensionless ratio of the total bulk ordering energy of the transition to the thermal energy! This constitutes a most appealing and, as we will see, rather general way of restating the main scaling conclusion of FB.

If one takes to heart, more seriously still, the predominance of the two spontaneously magnetized configurations, the partition function for the system near the transition should be approximated well by

$$Z(H; T, L_j) \sim e^{+m_0 h V} + e^{-m_0 h V} \tag{2.17}$$

This leads immediately to the explicit results

$$f_s(H, T; L_j) \approx -V^{-1} \ln\{2 \cosh[m_0(T) H V / k_B T]\} \tag{2.18}$$

$$m_s(H, T; L_j) \approx m_0(T) \tanh[m_0(T) H V / k_B T] \tag{2.19}$$

which, clearly, are in the expected scaling forms, the scaling function in (2.6) being $W(y_V) = -\ln(2 \cosh y_V)$ while $A_0 = 1$.

The simple scaling picture thus obtained is intuitive and, seemingly, quite satisfactory. It cannot, however, be regarded as complete: in particular, it indicates that the rounding of the transition is always on the scale

$$H_V \simeq k_B T / m_0(T) V \tag{2.20}$$

which is *shape-independent*, whereas we know[9–11] that in the case of a long block or rod with $L_\parallel \gg L_\perp$, which approaches a cylinder if $L_\parallel \to \infty$, the rounding should become *exponentially* small in $L_\perp$. Likewise, χ_0 always diverges linearly with $V = \prod_j L_j$ according to (2.11) [or (2.19)], whereas for a cylinder χ_0 diverges exponentially rapidly with $L_\perp$. We aim to understand this deficiency of the simple scaling analysis and to repair it so that the crossover between block and cylinder geometry can be described more satisfactorily.

One unstated assumption leading to (2.16)–(2.20) is immediately obvious, namely, here (as in FB) it has been implicitly assumed that the dimensions satisfy

$$L_j \gg \xi_\infty(T) \qquad (\text{all } j) \tag{2.21}$$

where ξ_∞ represents the *bulk* ($L_j \to \infty$) *single-phase correlation length* measured by the decay, or by the second spatial moment of the *net* correlation function

$$G(\mathbf{r}_i - \mathbf{r}_j) = \langle s_i s_j \rangle - \langle s_i \rangle \langle s_j \rangle$$

This condition is needed in order that the long-distance behavior of the correlation functions, $\langle s_i s_j \rangle$, dominates in (2.12). However, it should cause no problems except near the critical point where ξ_∞ diverges as $|t|^{-\nu}$.

More serious is the fact that in asserting the predominance of the states of "up" or "down" with total magnetization $\pm Vm_0(T)$, we have overlooked all configurations in which some regions of the system are magnetized "up" while others are magnetized "down," as illustrated in Fig. 1. Such configurations are, of course, suppressed by a Boltzmann factor representing the excess free energy associated with the interface (or domain wall) between the oppositely magnetized regions: Including them does, however, increase the entropy. For *block* geometry, configurations of this sort may thus yield corrections which, relative to the bulk, will be of order $A/V \sim 1/L_0$. This already suggests corrections to (2.18) [or (2.6)] of order $1/V^{1/d}$ which remain undisplayed. (See, however, Sections 3 and 4 below.)

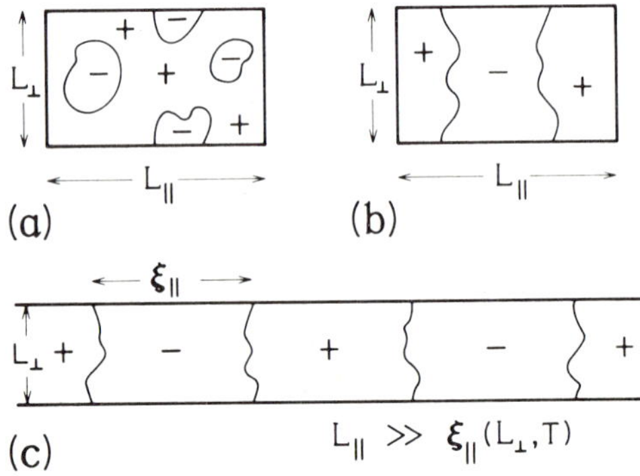

Fig. 1. Some typical configurations of a system of finite size exhibiting nonuniform ordering (or magnetization) forming "up" (or +) and "down" (or −) domains in a block geometry, (a) and (b), and in a cylinder geometry, (c), where $L_\parallel \to \infty$.

In the case of a *cylinder* geometry the effects are more serious. As illustrated in Fig. 1c, whenever $L_{\parallel} \gg L_{\perp}$ we expect the dominant configurations to involve domain walls of area $A = L_{\perp}^{d-1}$ which reach *across* the system so resulting in fluctuations which break the system into successive regions of "up" and "down" magnetization of a characteristic length which we will call $\xi_{\parallel}(T; L_{\perp})$. This length may be expressed in terms of the ratio of the two largest eigenvalues of the transfer matrix below T_c (see Section 3); but, more generally, it is related to the interfacial tension (or domain wall free energy) $\Sigma(T)$ via[15]

$$\xi_{\parallel}(T; L_{\perp}) \sim \exp[A\Sigma(T)/k_B T] \tag{2.22}$$

so diverging exponentially fast as $L_{\perp} \to \infty$. (There is a more slowly varying prefactor present here which will be discussed in Sections 4 and 5.)

These observations suggest that for a cylinder (with $L_{\parallel} = \infty$) scaling should involve the ratio

$$y_A = E_A(T)/k_B T = m_0(T) H A \xi_{\parallel}(T; A)/k_B T \tag{2.23}$$

which would replace (2.16). Thus we might expect the scaling relation

$$m_s(H, T; L_{\parallel} = \infty, L_{\perp}) \approx m_0(T) Y_{\infty}[m_0 h L_{\perp}^{d-1} \xi_{\parallel}(L_{\perp})] \tag{2.24}$$

to hold for a cylinder. This then predicts rounding on the scale

$$H_A = k_B T/m_0(T)\xi_{\parallel}(T; L_{\perp})A \sim \exp(-\sigma L_{\perp}^{d-1}) \tag{2.25}$$

where the reduced interfacial tension is

$$\sigma(T) = \Sigma(T)/k_B T \tag{2.26}$$

Thus we obtain the anticipated exponentially small rounding and an exponential divergence of $\chi_0(L_{\perp})$. The surmise (2.24) will be justified analytically in Section 3 and an explicit form for the scaling function $Y_{\infty}(y)$ will be obtained.

The presence of a diverging correlation length, $\xi_{\parallel}$, suggests that when $L_{\parallel}$ is finite it should enter specifically through the combination $x = L_{\parallel}/\xi_{\parallel}$ which is actually the same as the scaling ratio y_V/y_A. Thus we may extend (2.6) [or (2.19) and (2.24)] to obtain the combined scaling forms for the magnetization

$$\begin{aligned} m(H, T; L_j) &\approx m_0(T) Y(y_V, y_A) \\ &\approx m_0(T) Y_0[m_0 h V; L_{\parallel}/\xi_{\parallel}(L_{\perp})] \end{aligned} \tag{2.27}$$

which should describe *both block* and *cylinder* geometries and the crossover between them. Thus when $L_{\parallel}/\xi_{\parallel}$ becomes large we should have $Y(y_V \to \infty, y_A) \to Y_{\infty}(y_A)$ so that (2.24) applies and the rounding is exponentially small. Conversely, if the shape ratios l_j remain fixed as $L_{\parallel}, L_{\perp}$

$\to\infty$ we have $L_{\parallel}/\xi_{\parallel}\to 0$ and expect

$$Y_0(y_V; x\to 0)\to \tanh y_V \tag{2.28}$$

so reproducing (2.19) and yielding rounding of order $1/V$.

Some caution is necessary, however, since if $L_{\parallel}$ becomes significantly smaller than $L_{\perp}$ one goes over to a slab geometry (or, for $d=2$, simply to a rod oriented along the second axis). Since a slab of finite thickness $L_{\parallel}$ may exhibit a transition in the limit $L_{\perp}\to\infty$ a different crossover must arise: even in the absence of a transition, however, we should expect a different scaling combination, say, $L_{\perp}/\tilde{\xi}(T; L_{\parallel}, L_{\perp})$, to enter in a nontrivial way. In the limit $L_{\perp}\to\infty$ there will be finite-size corrections of order $\exp(-L_{\parallel}/\xi_{\infty})$ (or some power thereof) and, in the case of a first-order transition, interface corrections of order $\exp[-L_{\parallel}S\Sigma(T)/k_BT]$ where S is some characteristic $(d-2)$-dimensional perimeter of a heterophase fluctuation. For $d=2$ we clearly have, by symmetry, corrections of the form $\exp(-L_{\parallel}\Sigma/k_BT)$ [which, via the hyperscaling relation $\Sigma\sim 1/\xi_{\infty}$, are also of the form $\exp(-L_{\parallel}/\xi_{\infty})$, at least, in the critical region!] and evidently $\tilde{\xi}\sim\exp(L_{\parallel}\Sigma/k_BT)$. Thus it is plausible generally that the new length $\tilde{\xi}$ diverges exponentially fast with $L_{\parallel}$. If so it would suggest that the crossover to slab behavior occurs when $L_{\parallel}$ is of order $\ln L_{\perp}$; we will find that the same criterion arises in the transfer matrix analysis presented in the next section.

In summary, scaling at a first-order transition in a block geometry should be controlled by the "bulk" combination, $y_V = m_0HV/k_BT$; but if $L_{\parallel}$ becomes much larger than $L_{\perp}$, the further, "cross-sectional area" combination, $y_A = m_0HA\xi_{\parallel}/k_BT$ enters, with $\xi_{\parallel}(L_{\perp})$ diverging in accord with (2.22); in the cylinder limit, $L_{\parallel}=\infty$, only y_A determines the rounding of the first-order transition. We now test these conclusions in various ways: we will find that they even extend *into* the critical region.

3. TRANSFER MATRIX ANALYSIS

The transfer matrix method for Ising spin systems is well known[20] but it is not always recalled that it is of considerable generality, applying to continuous variables and extending to continuum systems.[21,22] If $\Lambda_r = \Lambda_r(H, T; L_{\perp})$ with $r = 0, 1, 2, \ldots, R(L_{\perp})$ are the eigenvalues of the transfer matrix for a system of cross-sectional dimensions $L_{\perp}$, arranged in order of decreasing magnitude, the reduced free energy density is given by

$$f(H, T; L_{\parallel}, L_{\perp}) = -V^{-1}\ln\left(\sum_{r=0}^{R}\Lambda_r^{L/a}\right) \tag{3.1}$$

with $L_{\parallel}\equiv L$. The upper limit $R(L_{\perp})$ in general diverges as $\exp(cA/a^{d-1})$, where c is a constant and $A = L_{\perp}^{d-1}$. In the cylinder limit, $L_{\perp}\to\infty$, the first

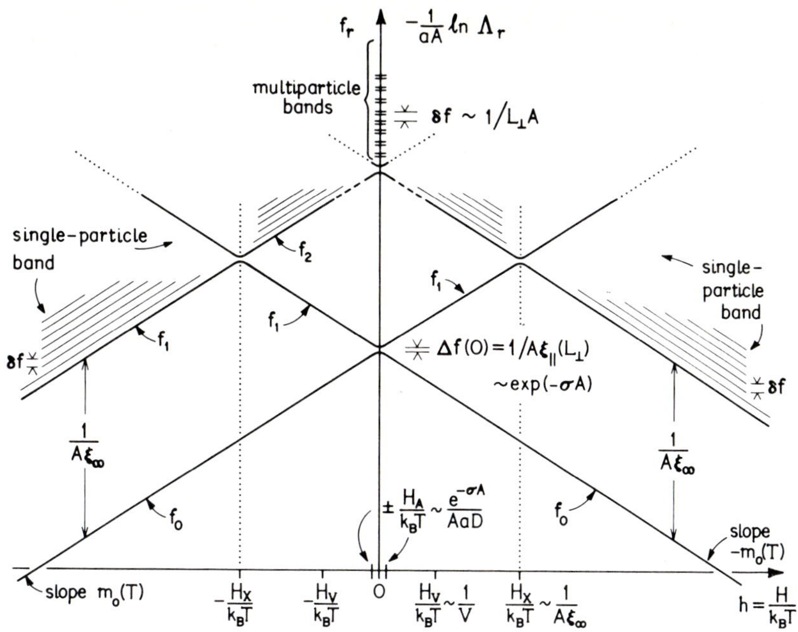

Fig. 2. Depiction of the transfer matrix spectrum for a system of cross-sectional area $A = L_\perp^{d-1}$ as a function of field, H, through a first-order transition in terms of the "free-energy levels" $f_r = -(\ln \Lambda_r)/Aa$. Details of the spectrum above f_2 are incomplete.

term in the sum gives the complete result. For $T < T_c$ and H not too large we will see, further, that the *first two terms* give a satisfactory representation[2] for a rather wide range of $L_\parallel$. To show this consider the behavior of the "free energy levels"

$$f_r(H, T; L_\perp) = -(Aa)^{-1}\ln \Lambda_r(H, T; L_\perp) \tag{3.2}$$

Our knowledge of the structure of the low-lying levels as a function of $h = H/k_BT$ is summarized schematically in Fig. 2. This figure synthesizes what is known rigorously in zero field for the two-dimensional Ising model,[(14)] what can be concluded by general analyses,[(21,22,26)] what can be surmised by studying behavior at large fields, H, and in general field for

[2] A similar conclusion regarding the adequacy of only the first two eigenvalues has been reached independently in a somewhat different context by Kleban and Akinci,[(23)] who have studied the shape dependence of the specific heat of a two-dimensional Ising model on an $m \times n$ torus in the critical region in the scaling limit, $m, n \to \infty$ with m/n fixed as analyzed originally by Ferdinand and Fisher.[(24)] Two-eigenvalue dominance in yet another context has been found by Bruce.[(25)]

$T = 0$, what can be checked in certain simple models,[27] and what can reasonably be conjectured about the nature of the asymptotic degeneracy of Λ_0 and Λ_1 and tested numerically.[10,11] The first point is that the lowest level, f_0, determines the bulk free energy and hence, for h of order unity but small, has a slope $\mp m_0(T)$. The next crucial feature concerns the asymptotic degeneracy in zero field which fixes the gap

$$\Delta f(H=0, T; L_\perp) \equiv f_1 - f_0 = 1/A\xi_\parallel(T; L_\perp) \tag{3.3}$$

in terms of the longitudinal (zero-field) correlation length,

$$\xi_\parallel(T; L_\perp)/a \approx D(T; L_\perp)\exp[A\Sigma(T)/k_B T], \quad \text{as} \quad L_\perp \to \infty \tag{3.4}$$

and of the interfacial tension $\Sigma(T) \equiv k_B T\sigma(T)$.[15] Note we include the amplitude factor which is slowly varying in the sense that

$$A^{-1}\ln D(T; L_\perp) \to 0, \quad \text{as} \quad L_\perp \to \infty \tag{3.5}$$

In Appendix A we outline an explicit evaluation of $D(T; L_\perp)$ for the two-dimensional Ising model and find, as might have been guessed, that it varies as a power of $L_\perp$ (see further below).

The third point is that the gap, $\Delta f = f_1 - f_0$, for $h = O(1)$ small, is equal to $1/A\xi_\infty(T)$ where, as in Section 2, ξ_∞ denotes the bulk single-phase correlation length. Above the gap lies a quasicontinuum of "single-particle" levels[21,22,26] with a spacing δf of magnitude $1/AL_\perp$ forming a band containing of order A/a^{d-1} levels. In a zeroth-order approximation (which is exact at $T = 0$) the lowest levels cross linearly in zero field and again at a field H_X which, it now follows, must be given by

$$2m_0 H_X/k_B T = 1/A\xi_\infty \tag{3.6}$$

Furthermore, it appears[10,11,27] that the "avoided crossings" that appear for $T > 0$ are, to leading order in $L_\perp$, still located at H_X. From this we can conclude that under the condition

$$T < T_c \quad \text{and} \quad |H| \ll H_X \propto k_B T/m_0\xi_\infty L_\perp^{d-1} \tag{3.7}$$

the two lowest free energy levels are separated from the higher levels by a gap of order $1/A\xi_\infty$. This in turn means that under the same conditions we have

$$\Lambda_2/\Lambda_1 < \exp(-a/\xi_\infty) \tag{3.8}$$

and hence can write the total free energy density as

$$f = -V^{-1}\ln[\Lambda_0^{L/a} + \Lambda_1^{L/a}] + \mathscr{E}(L_\parallel) \tag{3.9}$$

where $\mathscr{E}(L_\parallel)$ denotes terms exponentially small as $L_\parallel \to \infty$. To be more explicit, note that the sum $\sum_{r=2}^{R}(\Lambda_r/\Lambda_0)^{L/a}$ is, by (3.8), certainly bounded by $R\exp(-L/\xi_\infty) \approx \exp(cA/a^{d-1} - L/\xi_\infty)$ and so is exponentially small when $L \equiv L_\parallel \gg L_\perp^{d-1}\xi_\infty(T)/a^{d-1}$. More realistically, however, only the first, single-particle band of order $(L_\perp/a)^{d-1}$ levels need be counted to estimate the leading correction so that the error term in (3.9) is negligible when

$$L_\parallel \gg (d-1)\xi_\infty(T)\ln(L_\perp/a) \tag{3.10}$$

which is a much milder condition! (Compare with the penultimate paragraph in Section 2.)

Now one can argue, and substantiate by various detailed checks,[10,11,27] that the "avoided crossing" of the two largest eigenvalues as a function of the field, h, or, better, of the two lowest free energy levels, $f_0(H)$ and $f_1(H)$, may, for large $L_\perp/a$, be described correctly by the roots of a quadratic equation that represents the characteristic determinant of a real symmetric 2×2 matrix whose diagonal matrix elements are smoothly crossing, symmetrically related functions, $f_+(H)$ and $f_-(H) = f_+(-H)$, while the off-diagonal elements serve to produce the splitting $\Delta f(0)$ given by (3.3). The "single-phase free energies" thus introduced are then given by

$$f_\pm(H,T;L_\perp) = \tfrac{1}{2}(f_0 + f_1) \mp \tfrac{1}{2}h\left\{(f_1 - f_0)^2/h^2 - \left[\Delta f(0)/h\right]^2\right\}^{1/2} \tag{3.11}$$

Further, one can show,[10,11] as is to be anticipated, that as $L_\perp/a \to \infty$ one has

$$f_\pm \approx f_\infty(0,T) \mp m_0(T)h - \tfrac{1}{2}k_BT\chi_\infty(T)h^2 \mp O(h^3) \tag{3.12}$$

where $f_\infty(0,T) = f(H=0, T; L_j \to \infty)$ is the bulk zero-field free energy, $m_0(T)$ is, as before, the bulk spontaneous magnetization and, similarly, $\chi_\infty(T)$ is the initial susceptibility of the infinite system. The residual errors in (3.12) arise from the finite transverse dimensions of the layers: for periodic boundary conditions, we thus anticipate[1] that they are of order $\exp(-L_\perp/\xi_\infty)$, provided only that the avoided crossings between $f_1(H)$ and $f_2(H)$ at $H \simeq H_X$ are not reached: this is ensured if (3.7) is respected. Conversely, then, for the two lowest free energy levels we expect the representation

$$f_0, f_1 \approx f_\infty(T) - \tfrac{1}{2}k_BT\chi_\infty(T)h^2 \mp \left[m_0^2h^2 + 1/4A^2\xi_\parallel^2(L_\perp)\right]^{1/2} \tag{3.13}$$

to be accurate to order h^3 and $\exp(-L_\perp/\xi_\infty)$ as regards $f_0(H,T;L_\perp)$ although similar accuracy for $f_1(H,T;L_\perp)$ should apply only up to $|H| \lesssim H_X$ [given by (3.6)]. This expression is clearly useful in that all the rapid dependence on H has been isolated explicitly in terms of $\xi_\parallel(T;L_\perp)$.

If we accept (3.13) and, in order to obtain the leading crossover behavior in h, neglect the $\chi_\infty h^2$ term, we obtain from (3.2) and (3.9) the final result

$$f_s \equiv f(H,T;L_\parallel,L_\perp) - f_\infty(T)$$
$$\approx -V^{-1}\ln 2\cosh\left\{V\left[m_0^2(T)h^2 + 1/4A^2\xi_\parallel^2(T,L_\perp)\right]^{1/2}\right\} \quad (3.14)$$

this should be valid for $T < T_c$ up to corrections in $L_\parallel$ and $L_\perp$ which are exponentially small provided that (3.10) is met and that the further condition

$$L_\perp/\xi_\infty(T) \gg 1 \quad (3.15)$$

is satisfied. Note that on combining this with (3.10) the shape ratio $L_\parallel/L_\perp$ is required to be bounded below only by $\xi_\infty \ln(L_\perp/a)/L_\perp$ which becomes arbitrarily small for large $L_\perp$; i.e., flat slabs are allowed! For finite $L_\perp$, accuracy can be improved, if desired, by replacing f_∞ in (3.14) by $\frac{1}{2}[f_+(L_\perp)+f_-(L_\perp)]$ and $m_0(T)$ by $[f_-(L_\perp)-f_+(L_\perp)]/2h$.[10,11] For systems with anisotropic interactions (3.15) must, naturally, be replaced by $L_j/\xi_\infty^{(j)} \gg 1$ where $\xi_\infty^{(j)}$ is the correlation length in the direction j.

To bring out the main features of the conclusion (3.14) consider, first, the limit of large H/k_BT: the result then reduces simply to

$$f_s(H,T;L_j) \approx -V^{-1}\ln 2 - m_0|h| \quad (3.16)$$

which reproduces correctly the appearance of the bulk first-order transition. Next consider the *block limit*

$$x = L_\parallel/\xi_\parallel(L_\perp) \approx L_\parallel \exp(-\sigma L_\perp^{d-1})/D(L_\perp) \to 0 \quad (3.17)$$

where we have invoked (3.4) and (3.5). Evidently one now recaptures *precisely* the naive "two-peak" result (2.18), which scales solely in terms of the bulk ratio $y_V = m_0HV/k_BT$ with, therefore, rounding on the scale H_V [see (2.20)]. Note that the ratio of H_V to H_X is given by $2\xi_\infty/L_\parallel \ll 1/(d-1)\ln(L_\perp/a)$, where (3.10) has been recalled: thus the condition (3.7) is amply met in the region of interest.

On departure from the block limit a cursory inspection of (3.14) suggests that sharp structure in the magnetization as a function of H might set in on the scale $H_A = k_BT/m_0A\xi_\parallel \ll H_V$ [see (2.25)]. In fact, however, this is *not* the case since when $x = L_\parallel/\xi_\parallel$ is small *and* $|H| \lesssim H_A$ the whole argument of the cosh in (3.14) is then likewise small and one may expand to obtain

$$Vf_s \approx -\ln 2 - \tfrac{1}{8}x^2 - \tfrac{1}{2}y_V^2\left(1 - \tfrac{1}{12}x^2 + \cdots\right)$$
$$+ \tfrac{1}{12}y_V^4\left(1 - \tfrac{1}{5}x^2 + \cdots\right) + O\left(x^4, y_V^6\right) \quad (3.18)$$

Consequently the deviations in magnetization, susceptibility, etc. enter merely as multiplicative factors which depart from unity as x^2.

More generally, we see that (3.14) has exactly the two-variable crossover scaling form anticipated in (2.27): explicitly we can rewrite (3.14) as

$$f_s(H,T;L_j) \approx -V^{-1}\ln\left[2\cosh\left(y_V^2+\tfrac{1}{4}x^2\right)^{1/2}\right] \tag{3.19}$$

so that the crossover scaling function for the magnetization in (2.27) is

$$Y_0(y;x) = \frac{y}{\left(y^2+\frac{1}{4}x^2\right)^{1/2}}\tanh\left[\left(y^2+\tfrac{1}{4}x^2\right)^{1/2}\right] \tag{3.20}$$

which satisfies the limit relation (2.28). A similar result, with $m_0(T)$ replaced by 1 in the limit $T \to 0$ can be derived using the low-temperature approximate renormalization-cum-rescaling approach of Blöte, Nightingale and Cardy[9,13] but the form of the slowly varying prefactor, $D(T, L_\perp)$, in the expression (3.4) for $\xi_\parallel$ is not reproduced correctly: see the next section and Appendix B for details.

Lastly, notice that in the *cylinder limit* $x = L_\parallel/\xi_\parallel \to \infty$ one obtains simply

$$f_s(H,T;L_j) \approx -\left[m_0^2(T)h^2 + 1/4A^2\xi_\parallel^2(T;L_\perp)\right]^{1/2} \tag{3.21}$$

so that the scaling function for the magnetization in (2.24) becomes

$$Y_\infty(y_A) = 2y_A/\left(1+4y_A^2\right)^{1/2} \tag{3.22}$$

and rounding is now only on the scale $H_A \ll H_V \ll H_X$. The zero-field susceptibility hence diverges as

$$\chi_0(T;L_\perp) \approx \frac{2m_0^2(T)}{k_B T} L_\perp^{d-1}\xi_\parallel(T,L_\perp) \approx \frac{2m_0^2}{k_B T} L_\perp^{d-1} a D(L_\perp) e^{A\Sigma/k_B T} \tag{3.23}$$

where the full coefficient has now been identified.

In summary, the transfer matrix calculations bear out all the general scaling features anticipated in Section 2 and, furthermore, provide the explicit scaling functions and amplitudes describing the cylinder limit *and* the crossover from the block limit.

4. SOME FURTHER ASPECTS

The considerations of Section 2 leading to the scaling form (2.27) are, of course, not much more than heuristic; similarly, although we believe the transfer matrix analysis of Section 3 leading to the explicit result (3.14) is rather convincing, it is certainly not rigorous. The main points open to question will, we trust, have been evident to the reader. In this section we comment further on a few more detailed issues.

First, recall the restriction to periodic boundary conditions in all finite directions. Under these conditions no matrix elements are needed in (3.1) and one may assert[1-3] that corrections to the leading scaling results are of order $\exp(-L_j/\xi_\infty^{(j)})$. This conclusion should hold equally for *anti*periodic or helical boundary conditions in one or more of the transverse directions since these, also, respect the symmetry $H \Leftrightarrow -H$ and so do not shift the location of the susceptibility peak. For free or open boundary conditions, however, changes in the free energy of relative magnitude $a/L_\perp$ and $a/L_\parallel$ occur and, if there are also surface magnetic fields the position of the susceptibility peak may shift[5] to a field $H_\sigma(L_j)$ of order $(aA_i/V) \sim (a/L_i)$ which, asymptotically, is much larger than the rounding field $H_V \sim a^d/V$. One may reasonably conjecture that the *same* forms of rounding and scaling will be valid with H simply replaced by the shifted field[1] $\dot{H} = H - H_\sigma(L_j)$ but we have not investigated the more detailed arguments required to substantiate this. (Such shifts have been studied in the critical region by scaling hypotheses[28] and local mean-field theory.[29])

When the finite-width corrections are exponentially small one may, as remarked, improve the accuracy of (3.14) by using the terms in (3.12) of higher order in the field. Thus, for the susceptibility the leading correction to the scaling peak in the block limit is given, for $|H| \lesssim H_X$, by

$$\chi(H, T; L_j) \approx \frac{Vm_0^2(T)}{k_B T \cosh^2(Vm_0 h)} + \chi_\infty(T) + O(H) \tag{4.1}$$

This formula might be useful in analyzing well-equilibrated Monte Carlo data on a finite system in small fields below T_c.

It is instructive, in passing, to examine the rounding of the first-order transition in the *infinite range* or Husimi–Temperley model which yields mean-field theory in the thermodynamic limit. The Hamiltonian for N spins may be written

$$\mathscr{H} = -\frac{1}{2}\frac{J}{N}\left(\sum_{i=1}^{N} s_i^2\right) - \overline{H}\sum_{i=1}^{N} s_i \tag{4.2}$$

The thermodynamics of the model (and references to the literature) are given by Thompson.[20] Via a Kac–Hubbard transformation one sees that the partition function is proportional to

$$\mathscr{Z}_N(H, T) = \int_{-\infty}^{\infty} \exp\left[Ng(\mu; K, \bar{h})\right] d\mu \tag{4.3}$$

with $K = J/k_B T$, $\bar{h} = \overline{H}/k_B T$, and

$$g(\mu) = -\tfrac{1}{2}K\mu^2 + \ln\cosh(K\mu + \bar{h}) \tag{4.4}$$

For $T < T_c$ (given by $K_c = 1$) and $\overline{H} \geqslant 0+$ the magnetization, $m(\overline{H}, T)$, is the positive root of

$$m = \tanh(Km + \bar{h}) \tag{4.5}$$

while for $\bar{H} \leqslant 0-$ the negative root is to be chosen. For large N one easily sees that the transition is rounded on the scale $\bar{H}_N \sim 1/N$ which corresponds to the previous block limit since N fills the role of V even though the model has no proper spatial geometry. Indeed, for small $\bar{h}$ one finds the singular behavior of the free energy per spin is

$$\left[f(\bar{H}, T) - f_\infty(0, T)\right]_s = -N^{-1}\ln\left[2\cosh(m\bar{h}N)\right] \tag{4.6}$$

with $f_\infty(\bar{H}, T) = g[m(\bar{H}, T); \bar{H}, T]$; this is precisely analogous to the previous "two-peak" scaling results, (2.18) and (3.14), in the block limit. The corrections to this leading behavior follow by using the method of steepest descents to evaluate (4.3) and are thus of relative order $1/N$, $1/N^2$, etc. By way of example one finds that the initial susceptibility is

$$\chi_0(T, N) = m_0^2(T)N/k_BT + \chi_\infty\left[1 - 2K + 2J(1-K)\chi_\infty\right] + O(N^{-1}) \tag{4.7}$$

where the limiting bulk susceptibility is $\chi_\infty = (k_BT\cosh^2 Km - J)^{-1}$.

Although the two-peak result is confirmed by the infinite range model and, more pertinently, by the transfer matrix analysis, it cannot be regarded as beyond question except, perhaps, in the low-temperature region, $\exp(-2dJ/k_BT) \ll 1$, where only few excitations are present. Indeed, the question of the probability distribution of the magnetization (especially in Heisenberg spin systems) is a matter of some subtlety.[25,27,30-32]

In the cylinder limit $L_\parallel/\xi_\parallel \gg 1$, the rounding is on the scale $H_A \simeq k_BT/m_0A\xi_\parallel$ so that the behavior of $\xi_\parallel(T; L_\perp)$ is of interest. In the low-temperature limit one easily shows that the prefactor in (3.4) behaves for all d as

$$D(T; L_\perp) \to \tfrac{1}{2}, \qquad \text{for} \quad T \to 0, \quad L_\perp \text{ fixed} \tag{4.8}$$

Since $m_0(T)$ and $\Sigma(T)$ also approach constants when $T \to 0$ the rounding thus varies as

$$H_A(T; L_\perp) \propto T\exp(-\Sigma_0 A/k_BT)/A \tag{4.9}$$

when $T \to 0$ at fixed $A = L_\perp^{d-1}$. This result is implicit also in Refs. 10 and 11; however, the arguments of Blöte, Nightingale, and Cardy[9,13] yield rounding on the scale $T\exp(-\Sigma_0 A/k_BT)/L_\perp^d$ which is smaller than (4.9) by a factor $1/L_\perp$. The reason for this discrepancy is explained in Appendix B: qualitatively, the inadequacy of their argument is the use of renormalization group flows *linearized* around the discontinuity fixed point at $T, H = 0$, $L_\perp = \infty$, to renormalize down to finite $L_\perp$.

The low-temperature limiting form (4.9) is valid only while the total probability, $A\exp[-2(d-1)J/k_BT]$, of a layer excitation is small: see region (i) of Fig. 3. In the opposite limit, namely, $L_\perp \to \infty$ at fixed $T > 0$ (with $T < T_c$), (4.8) fails. The analysis of Appendix A shows, for the two-dimensional Ising model, that one may then write

$$\frac{\xi_\parallel(T; L_\perp)}{\xi_\infty(T)} \approx P(T)\left(\frac{A}{\xi_\infty^{d-1}}\right)^\omega \exp\left[L_\perp^{d-1}\sigma(T)\right] \qquad (4.10)$$

where $P(T)$ is a slowly varying function with a finite limit at $T = T_c -$, while $\omega(d=2) = \frac{1}{2}$. It is reasonable to conjecture that this expression with an appropriate exponent $\omega(d)$ or, possibly, different functional form, holds generally for $L_\perp \to \infty$ at T fixed for $d \leqslant d_> = 4$ (see Section 5 for $d > d_>$): see region (ii) in Fig. 3 which is bounded, following (3.15), also by $L_\perp \sim \xi_\infty(T)$. The rounding is then smaller than (4.9) by a factor $1/A^\omega$ and is also "enhanced" by a factor $\xi_\infty^{\omega(d-1)-1}$ which diverges as $T \to T_c -$ if $\omega(d-1) > 1$. Note, further, that if one invokes the hyperscaling relation,[33-35] $\mu = (d-1)\nu$, for the surface tension exponent defined via

$$\sigma(T) = \Sigma(T)/k_BT \approx B_\sigma\left[(T_c - T)/T_c\right]^\mu, \qquad \text{for} \quad T \to T_c - \qquad (4.11)$$

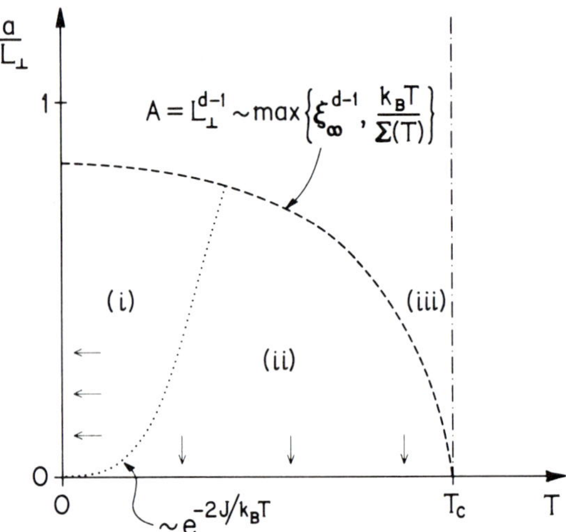

Fig. 3. Sketch of the region, (i) with (ii), of validity of asymptotic scaling at the first-order transition in the cylinder limit $L_\parallel \gg \xi_\parallel$ for cross-section area $A = L_\perp^{d-1}$. The arrows indicate the order of limits appropriate in the two distinct regimes (i) and (ii). The domain of critical behavior of $\xi_\parallel(T, L_\perp)$ is marked (iii).

the ratio $\xi_{\parallel}/\xi_{\infty}$ becomes, near criticality, a function only of the combination $L_{\perp}/\xi_{\infty}$. This accords with the natural scaling hypothesis for the critical region advanced in the following section. Hyperscaling is valid only for $d \leqslant 4$, although (3.4) should hold above $d = 4$: the modifications of naive finite-size scaling needed when hyperscaling fails are also discussed in the next section.

5. FIRST-ORDER SCALING NEAR CRITICALITY

We consider now how the finite-size scaling description of a first-order transition in the block and cylinder limits goes over, as it should, into standard finite-size scaling theory[1-3,16] for critical points when $T \to T_c -$. We also address the situation above $d = 4$ dimensions where naive finite-size scaling and hyperscaling breakdown[16] because, as we will demonstrate, a *dangerous irrelevant variable*[36] must be taken into account.

The finite-size scaling hypothesis for the critical region asserts that all unbounded lengths, say, L, should be scaled by the correlation length $\xi_{\infty}(T)$ which diverges as $|t|^{-\nu}$ with $t = (T - T_c)/T_c$. With the notation of Section 2, namely, $V = L_0^d$ and $l_j = L_j/L_0$, we may thus write the hypothesis

$$f_c(H, T; L_j) \approx |t|^{2-\alpha} W_c\left(h/|t|^{\Delta}, L_0|t|^{\nu}; l_j\right) \tag{5.1}$$

with $\Delta = \beta + \gamma$. Note that, as usual, analytic background terms have been subtracted from $f(H, T; L_j)$ to define f_c: for simplicity we have also dropped the metrical or amplitude factors A, B, etc. which permit one to normalize the scaling function, $W_c(y, z; l_j)$ [compare with (2.6)].

Now, if the neglect of irrelevant variables is permissible, (5.1) should be valid in the full domain $H, t, L_0^{-1} \to 0$. But this critical scaling domain overlaps the region of validity of the first-order scaling results (2.27), (3.14) and (3.19) for $f_s = f - f_{\infty}(H = 0, T)$. We recall that the first-order scaling involved the two combinations

$$y_V = m_0 h V \approx B h L_0^d |t|^{\beta}, \qquad \text{as} \quad T \to T_c - \tag{5.2}$$

where (2.7), (2.10) and (2.16) have been used, and

$$x = \frac{L_{\parallel}}{\xi_{\parallel}(L_{\perp})} \approx p_c \frac{L_{\parallel}|t|^{\nu-(d-1)\omega\nu}}{A^{\omega}} \exp\left[-\frac{\sigma(T) L_0^{d-1}}{l_1}\right] \tag{5.3}$$

where p_c is a constant and we have used (2.1) and (4.10). At first sight these scaling variables are not in the scaling form (5.1): however, if *hyperscaling* holds,[32-35] so that $d\nu = \Delta + \beta$ and $\mu = (d-1)\nu$, it is easy to check the relation

$$y_V \approx Byz^d \quad \text{with} \quad y = h/|t|^\Delta \quad \text{and} \quad z = L_0|t|^\nu \tag{5.4}$$

as $T \to T_c -$ and, likewise, using (4.11),

$$x \approx X(z; l_1) = p_c l_1^{1+\omega} z^{1-(d-1)\omega} \exp(-B_\sigma z^{d-1}/l_1) \tag{5.5}$$

where we recall that $\omega = \frac{1}{2}$ for $d = 2$ (by Appendix A) but is not determined for $d > 2$. (One might, indeed, have $z^{(d-1)\omega}$ replaced by a different, but still slowly varying function of z for $d = 3$.)

Having checked that, indeed, the first-order scaling forms for $T < T_c$ do respect critical scaling when $d < d_> = 4$ we can use (3.19) to conclude that the critical scaling function, $W_c(y, z; l_j)$, behaves in accord with

$$W_c(y, z; l_j) - W_c(0, z; l_j) \approx -\ln 2\cosh\left[B^2 y^2 z^{2d} + \tfrac{1}{4}X^2(z; l_1)\right]^{1/2} \tag{5.6}$$

as $z \to \infty$ (with $T < T_c$); this expression encompasses both the block and cylinder limits and also the crossover between them i.e., the full range $L_\| \equiv L_1 \gtrsim L_j$ for $j = 2, 3, \ldots$. It is remarkable that such an explicit result can be found for the finite-size scaling function in the critical region!

An interesting point raised by Brézin[16] for the cylinder limit is relevant to phenomenological renormalization group calculations[4]; this concerns the finite-size scaling properties of the spectral gap $\Delta f(0, T; L_\perp)$ or, equivalently, of $\xi_\|(T; L_\perp)$ in the critical region and, in particular, at the critical point. By the general finite-size scaling hypothesis one anticipates

$$\xi_\|(H, T; L_\perp) \approx |t|^{-\nu} Z_\|(h/|t|^\Delta, L_\perp |t|^\nu) \tag{5.7}$$

where we have included a dependence on the field but, for simplicity, supposed $L_2 = \cdots = L_d = L_\perp$ (while $L_\| \to \infty$). A check on this scaling ansatz is also provided by (4.10) which, if hyperscaling holds, yields

$$Z_\|(0, z) \approx p_\| z^{(d-1)\omega} \exp(B_\sigma z^{d-1}) \tag{5.8}$$

as $z \to \infty$, where $p_\|$ is a constant. At the critical point itself, (5.7) then gives

$$\xi_{\|,c}(L_\perp) \equiv \xi_\|(0, T_c; L_\perp) \sim L_\perp = A^{1/(d-1)} \tag{5.9}$$

which, of course, also follows directly from the finite-size scaling principle. However, if hyperscaling fails, as it does for $d > d_> = 4$, the general relation (3.4) for $\xi_\|$ is *inconsistent* with the scaling ansatz (5.7) which must, thus, be restricted to $d < 4$. This conclusion is, indeed, confirmed by Brézin's calculations[16] for the multicomponent limit $n \to \infty$. (Note that for finite systems, especially with nonperiodic boundary conditions, this limit must be distinguished from the standard spherical model.[1,37])

1988

Note: The authors have pointed out that in the originally published version a factor of z^{-d} is missing on the right-hand side of eq. (5.6), so that it should read:

$$W_c(y, z; l_j) - W_c(0, z; l_j) \approx -z^{-d} \ln 2\cosh\left[B^2 y^2 z^{2d} + \tfrac{1}{4}X^2(z; l_1)\right]^{1/2} \tag{5.6}$$

One may also define an *overall* size-dependent correlation length,[16] say, $\xi_V(H, T; L_j)$, in the block limit via the general scaling relation

$$\xi_V/a = (k_B T\chi/a^d)^{\nu/\gamma} \tag{5.10}$$

where $\chi \equiv \chi(H, T; L_j)$ is the finite-size susceptibility. The thermodynamic scaling relation (5.1) then leads to the analog of (5.7), namely,

$$\xi_V(H, T, L_j) \approx |t|^{-\nu} Z_V(h/|t|^\Delta, L_0|t|^\nu; l_j) \tag{5.11}$$

and hence, formally, as $y = h/|t|^\Delta$ and $z = L_0|t|^\nu \to 0$, to

$$\xi_{V,c}(L_j) \equiv \xi_V(0, T_c; L_j) \sim L_0 = V^{1/d} \tag{5.12}$$

which parallels (5.9). [If one defines ξ_V via $(f_c)^{-1/d}$, which is dimensionally appropriate, one would obtain (5.11) and, thence, (5.12) only when the hyperscaling relation $2 - \alpha = d\nu$ holds.]

Now Brézin concluded[16] by direct calculation for $n \to \infty$ that both (5.9) and (5.12) fail for $d > 4$, the length exponents no longer being unity but, rather, depending on d. Since mean field theory holds for the bulk system when $d > d_>$, one might hope to obtain the correct results by appropriate scaling arguments without the need for explicit calculation. To that end, we appeal to our observation that in the first-order cylinder limit ($L_\| \to \infty$) the finite dimensions enter principally through the fluctuation energy necessary to create an interface across the system and hence through the dimensionless combination

$$y_\Sigma = \frac{A\Sigma(T)}{k_B T} \approx B_\sigma L_\perp^{d-1} |t|^\mu = B_\sigma (L_\perp |t|^{3/2(d-1)})^{d-1} \tag{5.13}$$

in which, for $d > 4$, we have used the mean-field result $\mu = 3/2$. This suggests that for $d > 4$, where $\nu = 1/2$, and $\Delta = 3/2$, the scaling ansatz (5.7) should be replaced by

$$\xi_\|(H, T; L_\perp) \approx |t|^{-1/2} Z_\|(h/|t|^{3/2}, L_\perp |t|^{3/2(d-1)}) \tag{5.14}$$

At criticality this yields

$$\xi_{\|,c}(L_\perp) \sim a(L_\perp/a)^{(d-1)/3} \sim A^{1/3} \tag{5.15}$$

providing the scaling function is well behaved. This surmise is, in fact, confirmed precisely by Brézin's $n \to \infty$ calculations![16] [On the borderline $d = 4$ Brézin obtains $\xi_{\|,c} \sim L_\perp [\ln(L_\perp/a)]^{1/3}$.]

The parallel argument for the block limit suggests that only the bulk combination

$$y_V = m_0 h V \quad \text{scaling as} \quad t^{\beta+\Delta} L_0^d = (L_0 t^{2/d})^d \tag{5.16}$$

with $\beta+\Delta=2$ for $d>4$, should be important so that for $d>4$, one should replace (5.11) by

$$\xi_V(H,T;L_j)\approx|t|^{-1/2}Z_V(h/|t|^{3/2},L_0|t|^{2/d};l_j) \tag{5.17}$$

where the l_j are fixed. At criticality this leads to

$$\xi_{V,c}(L_j)\sim a(L_0/a)^{d/4}\sim V^{1/4} \tag{5.18}$$

which also agrees precisely with Brézin's result for $n\to\infty$.

From these considerations we learn that correlation lengths all scale in the usual way with $|t|^{-\nu}$ but that the overall dimensions of a system enter, in the block limit, through the total bulk ordering free energy ratio $y_V=F_V/k_BT$ and, in the cylinder limit, through the interfacial free energy ratio $y_\Sigma=F_\Sigma/k_BT$. Below the borderline $d_>=4$ hyperscaling prevails and *all* lengths and distances, including the correlation lengths ξ_∞, $\xi_\|$, etc. scale in the same way with $t^{-\nu}$; above $d=4$, however, the dimensions L_j scale with new powers of t.

It is instructive to place these results within a renormalization group context[38,39] in which *all* lengths and distances should renormalize with the spatial rescaling factor b as

$$f_c(h,t,u;R)\approx b^{-d}f_c(b^{\lambda_h}h,b^{\lambda_t}t,b^{\lambda_u}u;R/b) \tag{5.19}$$

Here R denotes collectively the various lengths and distances and we have included the parameter u which derives from the strength of the fourth-order term in the Landau–Ginzburg–Wilson effective Hamiltonian

$$\mathscr{H}[s]/k_BT=\int d^dx\left[\tfrac{1}{2}(\nabla s)^2-h_0s+\tfrac{1}{2}r_0s^2+u_0s^4\right] \tag{5.20}$$

with $r_0\sim t$. Further we have chosen to renormalize in the standard way which keeps the coefficient of $(\nabla s)^2$ constant.

For $d<4$ one anticipates a nontrivial fixed point[38,39] and u in (5.19) represents the deviation from the fixed point value, $u^*>0$, the corresponding eigenvalue, λ_u being negative. The choice $b=|t|^{-1/\lambda_t}$, as $t\to0$, leads to the standard scaling form where f_c scales as $|t|^{d/\lambda_t}$, h scales as $|t|^{\lambda_h/\lambda_t}$, and R scales as $|t|^{-1/\lambda_t}$. If we introduce the correction-to-scaling exponent $\theta=-\lambda_u/\lambda_t>0$ we see that the field u enters only in the combination $u|t|^\theta$ which *vanishes* as $t\to0$, so that u is an *irrelevant variable*. One thus obtains formally the standard exponent identifications, $2-\alpha=d\nu=d/\lambda_t$ and $\Delta=\lambda_h/\lambda_t$, which entail hyperscaling *and* the scaling of all lengths as $|t|^{-\nu}$. The main tacit assumption is that u is a *harmless* irrelevant variable which may be set equal to zero (corresponding to u_0 at the fixed point value u^*) without resulting in any singular or anomalous behavior.

When the dimensionality exceeds $d_>=4$ the critical behavior is controlled by the Gaussian fixed point[38,39] and one has the renormalization group eigenvalues

$$\lambda_h = \tfrac{1}{2}d + 1, \quad \lambda_t = 2, \quad \text{and} \quad \lambda_u = 4 - d \tag{5.21}$$

The last of these is still negative so that u is again an irrelevant variable. However, one now has $u^* = 0$ so that coefficient $u \cong u_0$ cannot be allowed to vanish for obvious reasons of stability (at and below criticality). If one modifies the renormalization group transformation so that the spin is rescaled (with b) so as to keep u fixed at u_0 one finds that the coefficient of the $(\nabla s)^2$ term now rescales rapidly to $+\infty$ as b grows. This justifies the classical saddle point approximation, equivalent to mean field theory, which, for a homogeneous bulk system, yields

$$f_c \approx \min_s \left\{ -h_0 s + \tfrac{1}{2} r_0 s^2 + u s^4 \right\} \tag{5.22}$$

Rescaling by putting $s = \tilde{s}/u^{1/2}$ shows that u enters the free energy in the form

$$f_c(h, t, u) = u^{-1} \tilde{f}(hu^{1/2}, t) \tag{5.23}$$

and, thus the magnetization, m, has a prefactor $u^{-1/2}$. The divergence of f and m as $u \to 0$ confirms the *dangerous* character[36] of the variable u. One may again use the general renormalization group relation (5.19) with the choice $b = |t|^{-1/\lambda_t} = |t|^{-1/2}$ but must recognize that f_c and h will entail factors of u as in (5.23). One thus anticipates

$$f_c \approx \frac{|t|^{d/2}}{u|t|^{\theta}} W_0 \left[\frac{h\left(u|t|^{\theta}\right)^{1/2}}{|t|^{(d+2)/4}}, u|t|^{\theta}; R|t|^{1/2} \right] \tag{5.24}$$

where the correction-to-scaling exponent is now $\theta = \tfrac{1}{2}(d-4) > 0$. If u is fixed at its initial physical value, u_0, it is readily checked that (5.24) represents the standard, classical scaling prediction with $\alpha = 0$, $\Delta = 3/2$ and $\nu = 1/2$. From this one likewise finds the bulk, zero-field behavior,

$$m_0 \approx B_0 |t|^{1/2}/u^{1/2} \quad \text{and} \quad \chi = C_0^{\pm}/|t| \qquad (t \gtrless 0) \tag{5.25}$$

where all the u dependence has been displayed; however, for *finite systems* the variable u may still enter in combination with the various dimensions, as we now indicate.

The bulk or block ratio can be written

$$y_V = f_c V \approx W_{00} \frac{|t|^2}{u} L_0^d = W_{00} \frac{\left(L_0 |t|^{1/2}\right)^d}{\left(u|t|^{\theta}\right)} = W_{00} \frac{\left(L_0 |t|^{2/d}\right)^d}{u} \tag{5.26}$$

where $W_{00} = W_0(0,0)$ and the dangerous irrelevant scaling combination $u|t|^{\theta}$ has been explicitly isolated. This likewise exhibits length scaling as $|t|^{-\nu}$ but shows that L_0 scales with the "anomalous" power $|t|^{-2/d}$ when

$u \equiv u_0$ is fixed, as proposed in (5.16). By the same token $u|t|^{\theta}$ must now enter (5.11) and (5.24) in a singular fashion i.e., as a divisor of $L_0|t|^{\nu}$. For the corresponding interfacial or cylinder ratio note, first, that $\Sigma(T)$ is quite generally,[33,34] proportional to $m_0^2\xi_\infty/\chi$ when $t \to 0$. Thus we have

$$y_\Sigma = \frac{\Sigma A}{k_B T} \approx \frac{B_0^2|t|}{u} \frac{|t|}{C_0^-} \frac{a_0 L_\perp^{d-1}}{|t|^{1/2}} = D_0 \frac{\left(L_\perp |t|^{1/2}\right)^{d-1}}{\left(u|t|^{\theta}\right)}$$

$$= \frac{D_0}{u}\left(L_\perp |t|^{3/2(d-1)}\right)^{d-1} \tag{5.27}$$

where $D_0 = a_0 B_0^2 / C_0^-$ and a_0 is a constant. Again we see standard renormalization group scaling with the dangerous irrelevant combination, $u|t|^{\theta}$, leading to the anomalous cross-sectional exponent $3/2(d-1)$, as advanced in (5.14).

In summary, our discussion of finite size scaling for $d > d_> = 4$ suggests that the *full* scaling form, encompassing single-phase behavior and the block and cylinder first-order transition limits will entail the *three* scaled variables

$$L_j|t|^{1/2}, \quad L_0^d t^2 \equiv V t^2, \quad \text{and} \quad L_\perp^{d-1}|t|^{3/2} \equiv A|t|^{3/2} \tag{5.28}$$

Thus all, for example, should enter as arguments in (5.7), (5.11), (5.14), (5.17), and (5.24) for a complete asymptotic description. By contrast, when $d < 4$ all limits are covered by the single combination $L_j|t|^{\nu}$.

6. PHENOMENOLOGICAL RENORMALIZATION USING SINGLE-PHASE FUNCTIONS

The phenomenological renormalization technique is a numerical method for studying bulk criticality which has been applied successfully to several two-dimensional problems. (See Ref. 4 for a recent review.) Various versions of the method have been used since the approach was originally introduced by Nightingale.[40] For simplicity we consider here only $d = 2$ and the $L \times \infty$ geometry and recall one of the most frequently used techniques. Thus finite-size scaling for, say, the zero-field susceptibility implies that the relation

$$\frac{\ln\left[\chi_L(T_0)/\chi_{L'}(T_0)\right]}{\ln\left[\chi_{L''}(T_0)/\chi_{L'}(T_0)\right]} \approx \frac{\ln(L/L')}{\ln(L''/L')} \tag{6.1}$$

is valid asymptotically as L, L' and $L'' \to \infty$ when $T_0 = T_c$. Conversely, the solution, $T_0(L, L', L'')$, of this relation regarded as an equality approximates the true critical point, T_c. Given a reliable estimate for T_c one may likewise estimate the exponent ratio γ/ν from

$$\gamma/\nu \simeq \ln[\chi_L(T_c)/\chi_{L'}(T_c)]/\ln(L/L') \tag{6.2}$$

Similar calculations have been performed using the specific heat and using $\xi_{\parallel}(T)$; in the latter case ν may be estimated by linearizing around the phenomenological fixed point.(40) More recently, Hamer(41) has suggested a variant using a function which approaches the spontaneous magnetization.

Now the "single-phase" free energy functions,(10,11) $f_{\pm}(H, T; L)$, defined in (3.11) should obey finite-size scaling because they are obtained by algebraic manipulation of quantities, namely, $f_0(H, T; L)$ and $f_1(H, T; L)$, which should also obey scaling. This entails the new assumption that the first excited free energy level, $f_1(H, T; L)$, obeys finite-size scaling; however, this is certainly plausible since f_1 and f_0 are branches of the same analytic function(27) of H. Notice that the single-phase free energies $f_{\pm}(H)$, are not even functions of H, unlike $f_0(H)$. Thus the corresponding scaling functions for *all* the field derivatives of $f_{\pm}(H)$ have nonvanishing values at $H = 0$. Therefore the $f_{\pm}(H)$ can be used in phenomenological renormalization calculations designed to study *odd* derivatives of the free energy *at* the first-order boundary when $T \to T_c -$ (as well as higher-order even derivatives).

To illustrate the potentialities of this observation, we report here numerical calculations for the nearest-neighbor square and triangular lattice Ising models (with isotropic interactions). We use periodic boundary conditions and consider the expansion

$$m_+^{(L)}(T_c, H) = -\left(\frac{\partial f_+^{(L)}}{\partial h}\right)_c = \sum_{k=0}^{\infty} c_k^{(L)} H^k \tag{6.3}$$

From the scaling ansatz we have, for $T \simeq T_c -$ and small H,

$$m_+^{(L)}(T, H) \approx D_1 L^{-\beta/\nu} Y_+(D_2 t L^{1/\nu}, D_3 H L^{\Delta/\nu}) \tag{6.4}$$

where the scaling function $Y_+(x, y)$ should be universal and, hence, independent of lattice structure with a proper assignment of the nonuniversal amplitudes, or metrical factors, D_i. Then the critical point expansion coefficients should vary as

$$c_k^{(L)} \approx D_1 D_3^k L^{(k\Delta - \beta)/\nu} (\partial^k Y_+/\partial y^k)_0 / k! \tag{6.5}$$

when $L \to \infty$, where the subscript 0 denotes evaluation at $x = y = 0$. With this observation in mind we have calculated numerically the approximants

$$(\gamma_k/\nu)^{(L)} = \ln[c_k^{(L)}/c_k^{(L-a)}]/\ln[L/(L-a)] \tag{6.6}$$

which, as $L \to \infty$, should converge to γ_k/ν, where

$$\gamma_k = k\Delta - \beta = \gamma + (k-1)\Delta \tag{6.7}$$

so that $\gamma_0 = -\beta$, $\gamma_1 = \gamma$, $\gamma_2 = \gamma + \Delta = \beta + 2\gamma$, etc.

For planar Ising models we have,[14,42,43] $\beta = \frac{1}{8}$, $\gamma = 1\frac{3}{4}$, and $\nu = 1$ and the validity of the relations for γ_k with $k = 2, 3, \ldots$ is well established in series extrapolation studies.[44–46] As a test of the approach, therefore, we list in Table I data for $(\gamma_k/\nu)^{(L)}$ for $L/a = 8$, 9 and 10 and several values of k, for the square and the triangular lattices. The calculations yielding these data are standard and will not be discussed here.[4,11] The expansion coefficients $c_k^{(L)}$ are obtained by numerical differentiation of $f_+(H, T_c; L)$ which, as a result of roundoff errors, restricts k to 8 for the triangular lattice and 6 for the square lattice. The values of $(\gamma_k/\nu)^{(L)}$ have been truncated to a sufficient number of places to display the nature of the convergence to γ_k/ν, and are otherwise accurate to the order displayed or higher. (Only in the last few entries for $L = 10a$ do roundoff errors manifest themselves.)

Frequently such phenomenological renormalization data for critical exponents are fitted well by the form $|(\gamma_k/\nu)^{(L)} - \gamma_k/\nu| \propto L^{-p_k}$ with an effective power law convergence exponent p_k of magnitude[4,47] around 2. Here, however, an analysis of $(\gamma_k/\nu)^{(L)}$ for $L/a = 6, 7, \ldots, 10$ does not reveal any regular pattern of convergence: indeed the $k = 0$ sequence, $(\beta/\nu)^{(L)}$, for the triangular lattice is not even monotonic! [Note that the restriction to $L/a \leqslant 10$ is the price paid for studying high-order derivatives so that $f_+(H)$ must be computed to high accuracy; when only low-order derivatives are of interest it is feasible[4] to go up to $L/a \lesssim 16$.] Extrapolation of exponent data using special methods[9,48] to accelerate convergence may in some cases[4] improve agreement with conjectured or exact values (if known) by an order of magnitude. However, we have not attempted to perform any such extrapolations because the data here are clearly not in a regime of asymptotic convergence.[47] Instead, let us discuss further those features which may be of interest in applications to other models.

First, note that the accuracy of critical exponents estimates for the higher-order derivatives is no worse (perhaps even somewhat better) than for the second derivative. This is not necessarily surprising: the higher-order derivatives of the free energy are more difficult to calculate accurately but the rate of their convergence to the $L \to \infty$ limit, which is determined by corrections to the leading finite-size scaling behavior, should not be qualitatively different. However, the values of $(\beta/\nu)^{(L)}$ are relatively closer to β/ν than are the other exponents estimates. (See also Hamer.[41]) There is also a striking difference in the accuracy of the estimates found for the two different lattices, which is probably related to the fact that the triangular lattice is more closely packed.

Table I. Values of $(\gamma_k/\nu)^{(L)} - \gamma_k/\nu$ for the Triangular and the Square Lattices[a]

	Triangular				Square			
k	$L = 8a$	$L = 9a$	$L = 10a$		$L = 8a$	$L = 9a$	$L = 10a$	
0	-2.02	-2.12	-1.94	$\times 10^{-6}$	6.10	4.62	3.63	$\times 10^{-4}$
1	4.56	3.63	2.97	10^{-2}	8.16	6.61	5.48	10^{-2}
2	-2.64	-1.93	-1.46	10^{-3}	-2.42	-2.00	-1.68	10^{-2}
3	-8.95	-7.98	-6.74	10^{-4}	-7.39	-5.91	-4.82	10^{-2}
4	13.6	4.99	1.30	10^{-4}	-13.6	-10.9	$-8.83(\pm 1)$	10^{-2}
5	5.62	2.98	1.69	10^{-3}	-2.27	-1.82	-1.48	10^{-1}
6	18.2	10.4	6.37	10^{-3}	-4.13	-3.33	$-2.8(\pm 1)$	10^{-1}
7	-4.82	-3.01	$-1.97(\pm 1)$	10^{-1}				
8	-21.8	-13.4	$-8.6(\pm 1)$	10^{-3}				

[a] The values are rounded to three figures or roundoff errors are indicated. The powers of 10 represent constant factors for each entry in the row.

We may also use our data to estimate the scaling function $Y_+(x, y)$ in (6.4). There is some arbitrariness in fixing the metrical factors D_i but a convenient approach[49] is to specify them by requiring

$$Y_+(x, y) = 1 + x + y + O(xy, x^2, y^2) \tag{6.8}$$

Then the expansion coefficients, Y_k, in

$$Y_+(0, y) = 1 + y + Y_2 y^2 + Y_3 y^3 + \cdots \tag{6.9}$$

should be universal and are approximated by

$$Y_k^{(L)} = \frac{c_k^{(L)}}{c_0^{(L)}} \left(\frac{c_0^{(L)}}{c_1^{(L)}} \right)^k \tag{6.10}$$

Our numerical results for both the square and triangular lattices are summarized by

$$\begin{gathered} Y_2 = -8.35 \pm 0.30, \qquad Y_3 = 104 \pm 7 \\ Y_4 = -(1.53 \pm 0.16) \times 10^3, \qquad Y_5 = (2.35 \pm 0.35) \times 10^4, \\ Y_6 = -(3.0 \pm 0.7) \times 10^5 \end{gathered} \tag{6.11}$$

The coefficients alternate in sign regularly to this order but some preliminary calculations suggest that this does not continue in higher orders.

As regards verifying the expected universality of $Y_+(0, y)$, the coefficients Y_k do not provide an optimal test. The problem is that parameters calculated from the second free-energy derivative data ($k = 1$) seem, as mentioned, to converge more slowly than for other derivatives. Owing to the factor $(c_1^{(L)})^k$ in (6.10), the associated uncertainties are amplified in the estimation of the Y_k. As an alternative test of universality it is better to study the ratios

$$R_k^{(L)} = c_k^{(L)} c_{k-2}^{(L)} / \left(c_{k-1}^{(L)}\right)^2 \tag{6.12}$$

which involve only low powers of the $c_j^{(L)}$, thus avoiding error accumulation. These ratios serve to approximate the universal scaling function coefficient ratios

$$R_k = Y_k Y_{k-2} / (Y_{k-1})^2, \qquad k = 2, 3, \ldots \quad \text{with} \quad Y_0 = Y_1 = 1 \tag{6.13}$$

In Table II we list several $R_k^{(L)}$ for $L/a = 9$ and 10. It appears again that the $R_k^{(L)}$ for $L/a = 6, 7, \ldots, 10$ do not yet exhibit any simple pattern of convergence although the triangular lattice values appear to be more rapidly convergent. The values for $L = 10a$ for the two lattices agree to within 4% for $k = 2$ and 6 but to better than 1% for $k = 3$, 4, and 5: this represents a gratifying confirmation of the anticipated universality.

Table II. The Ratios $R_k^{(L)}$ for the Triangular and Square Ising Lattices[a]

	Triangular		Square	
k	$L = 9a$	$L = 10a$	$L = 9a$	$L = 10a$
2	− 4.20	− 4.17	− 4.39	− 4.33
3	0.988	0.991	0.979	0.983
4	0.893823	0.893825	0.898477	0.897653
5	0.82861	0.82868	0.83817	0.83647
6	0.6922	0.6925	0.7211	0.7159

[a] The values have been rounded to the last place displayed.

As a final comment on numerical methods using the cylinder geometry, note that the position of a first-order phase boundary might be located[8,9] by searching for points at which the susceptibility (or the specific heat) diverges exponentially with the cross-sectional area $A = L_\perp^{d-1}$. Similarly, one could examine the spectral gap $1/\xi_\parallel(H, T; L) = A(f_1 - f_0)$ as a function of H. By the analysis of Section 3 this has the scaling form

$$1/\xi_\parallel^2(H, T; L_\perp) \approx (2Am_0 h)^2 + 1/\xi_\parallel^2(0, T; L_\perp) \tag{6.14}$$

for small $h = H/k_B T$. If one uses the standard phenomenological rescaling approach (following, e.g., Rikvold *et al.*[50]) and solves the relation

$$\xi_\parallel(H, T; L_\perp)/L_\perp = \xi_\parallel(H, T; L'_\perp)/L'_\perp \tag{6.15}$$

for H one then finds that the phase boundary is given correctly up to errors of order $1/\xi_\parallel(L_\perp)$ and $1/\xi_\parallel(L'_\perp)$, which are exponentially small in $L_\perp$ and $L'_\perp$. (At a continuous transition one expects errors decaying asymptotically[47] as $L_\perp^{-(1+\theta)/\nu}$ where θ is the leading singular correction-to-scaling exponent.) One can see similarly that the effective renormalization group eigenvalue, $\lambda^{(L)} = 1/\nu^{(L)}$, as computed formally by the usual rule,[4] diverges like $A = L_\perp^{d-1}$ at a first-order transition. This sort of behavior has, indeed, been observed[9] in a study of the first-order thermal transition in planar q-state Potts models with $q > 4$.

7. CONCLUDING REMARKS

Finite-size effects at a first-order phase boundary evidently provide a rich panorama of phenomena, involving a profound dependence on the shape of a system. We have focused mainly on just two geometries, namely, block and cylinder, and have studied the crossover between them. However, there are other geometries of significant theoretical and experimental interest, for example, the slab geometry where finite-size and, especially, surface phenomena have been a subject of intensive but by no means exhaustive study.[51]

It should also be emphasized that we have considered only boundary conditions which do not break the $H \Leftrightarrow -H$ symmetry. Since the shift in the first-order transition due to symmetry breaking surface effects is asymptotically larger than the rounding,[5] further interesting effects may arise with more realistic, free or pinned boundary conditions.[28,29] In particular the surface contributions themselves become important.[1,28,29]

Finally, recall that we have discussed only scalar (and discrete) spin systems. For systems with vector (and/or continuous) spins less information on the rounding of the first-order transitions is available but important qualitative differences arise[13,16] from the different dependence of $\xi_\parallel$ on $L_\perp$ which reflects the replacement of relatively sharp domain walls by indefinitely diffuse Bloch walls.

ACKNOWLEDGMENTS

We have enjoyed useful discussions with P. Nightingale and wish to thank M. Barma, E. Brézin, D. A. Huse, R. Pandit, and S. Trugman for instructive interactions. The financial support of the Rothschild Fellowship Foundation (for V.P.) is appreciated. The researches reported have also received support from the National Science Foundation (under Grant No. DMR-81-17011) and, in part, through the Materials Science Center at Cornell University.

APPENDIX A: SPECTRAL GAP FOR THE SQUARE LATTICE ISING MODEL

We outline the calculation of the asymptotic form of the spectral gap, $a/\xi_\parallel(T)$, for the square lattice Ising model with couplings $J_1 \equiv J_\parallel$ along the cylinder axis and J_2 between spins in the same layer. In terms of the variables $K_i = J_i/k_B T$ and K_i^* related by

$$\tanh K_i^* = \exp(-2K_i), \qquad i = 1, 2 \tag{A1}$$

one has[14] $K_1 = K_2^*$ at $T = T_c$ and $K_1 > K_2^*$ for $T < T_c$. The surface tension for an interface normal to the axis is given by[14,15]

$$\bar{\sigma} \equiv a\sigma(T) = a\Sigma(T)/k_B T = 2(K_1 - K_2^*) \tag{A2}$$

while the spectral gap for a lattice of width $M = L_\perp/a$ sites is[14]

$$\frac{a}{\xi_\parallel} \equiv \ln \frac{\Lambda_0}{\Lambda_1} = \frac{1}{2} \sum_{k=0}^{M-1} \left[\gamma\left(\frac{2k+1}{M} \right) - \gamma\left(\frac{2k}{M} \right) \right] \tag{A3}$$

where for $T < T_c$ the function $\gamma(x)$ is analytic on the real axis, with period 2, being given by

$$\gamma(x) = \cosh^{-1}(\operatorname{ch} 2K_1^* \operatorname{ch} 2K_2 - \operatorname{sh} 2K_1^* \operatorname{sh} 2K_2 \cos \pi x) > 0 \tag{A4}$$

in which we have used the convenient abbreviations

$$\operatorname{ch} z \equiv \cosh z \quad \text{and} \quad \operatorname{sh} z \equiv \sinh z \tag{A.5}$$

Then if $c_r(T)$ is the rth Fourier coefficient of $\gamma(x; T)$ one has

$$a/\xi_{\parallel}(T, L_{\perp}) = -M \sum_{j=0}^{\infty} c_{(2j+1)M}(T) \tag{A6}$$

Because $\gamma(x)$ is analytic on the real axis the Fourier coefficients decrease exponentially with r and thus we have

$$a/\xi_{\parallel} \approx -Mc_M = -M \int_0^2 \gamma(x) \cos(M\pi x)\, dx \tag{A7}$$

where the corrections are of relative order c_{3M}/c_M which is of magnitude $\exp(-2\sigma L_{\perp})$ as may be seen from the explicit result below.

After an integration by parts, some algebra, and a shift of the contour of integration into the complex plane one obtains

$$a/\xi_{\parallel} \approx \pi^{-1}(e^{\bar{\sigma}} I_{M-1} - e^{-\bar{\sigma}} I_{M+1}) e^{-\bar{\sigma} M} \tag{A8}$$

up to exponential corrections, where

$$I_M = \int_0^{4K_2^*} \frac{e^{-M\tau}\, d\tau}{\left[\operatorname{ch}(\bar{\sigma} + \tau) - \operatorname{ch} \bar{\sigma}\right]^{1/2} \left[\operatorname{ch} 2(K_1 + K_2^*) - \operatorname{ch}(\bar{\sigma} + \tau)\right]^{1/2}} \tag{A9}$$

When $M = L_{\perp}/a$ is fixed, the limiting behavior of this integral as $T \to 0$ is obtained when $e^{-M\tau}$ varies slowly over $[0, 4K_2^*]$, that is, for $K_2^* \ll a/L_{\perp}$. One then finds that I_M approaches $\pi/\operatorname{sh} 2K_1$, which leads to

$$\xi_{\parallel}(T, L_{\perp}) \approx \tfrac{1}{2} a e^{2K_1 L_{\perp}/a} \tag{A10}$$

for $T \to 0$ with $L_{\perp}$ fixed.

By contrast, the behavior for large $L_{\perp}$ at fixed T in $(0, T_c)$ is obtained when $e^{-M\tau}$ varies rapidly or $L_{\perp}/a \gg 1/K_2^*$. In this limit we find

$$I_M \approx \left[\pi \operatorname{sh} 2K_1^* \operatorname{sh} 2K_2 / 2M \operatorname{sh} \bar{\sigma}\right]^{1/2} \tag{A11}$$

which finally leads to

$$\xi_{\parallel}(T; L_{\perp}) \approx \left(\frac{\pi}{2 \operatorname{sh} 2K_1^* \operatorname{sh} 2K_2}\right)^{1/2} \left(\frac{L_{\perp}/a}{\operatorname{sh} \bar{\sigma}(T)}\right)^{1/2} e^{L_{\perp}\sigma} \left[1 + O\left(\frac{1}{L_{\perp}}\right)\right] \tag{A12}$$

which is valid as $L_\perp/a \to \infty$ at fixed nonzero $T < T_c$. The crossover between the two limiting forms (A10) and (A12) occurs for $L_\perp K_2^*/a$ of order unity: see the dotted curve in Fig. 3 which thus corresponds to $L_\perp/a \approx \exp(2J_2/k_BT)$. The exponential corrections in (A7) lead to crossover to critical behavior as $A\Sigma(T)/k_BT$ becomes small where, in this case, $d=2$ and so $A \equiv L_\perp$: see the dashed curve in Fig. 3. Via hyperscaling[35] this condition is equivalent to $L_\perp/\xi_\infty(T)$ of order unity as implied by the scaling criterion (3.15). To discuss the behavior of $\xi_\parallel(T, L_\perp)$ in the critical region, marked (iii) in Fig. 3, a more elaborate analysis is required which is not considered here.

APPENDIX B: RENORMALIZATION DOWN TO ONE DIMENSION

Blöte, Nightingale, and Cardy (BNC)[9,13] have discussed the effects of finite-size on the first-order transition with the aid of a low-temperature renormalization group rescaling approach. Here we present a brief critique of their arguments which demonstrates why they obtain results which, we believe, are not fully correct.

Let us first reformulate the BNC technique in a form suitable for comparing with our analysis in Sections 2–4. The basic idea of BNC is to use for finite-size systems the accepted *bulk* renormalization group recursion relations, as linearized about the $T, H = 0$ discontinuity fixed point, namely,[6,52]

$$h' = b^d h \quad \text{and} \quad T' = b^{1-d}T, \quad \text{so} \quad H' = bH \tag{B1}$$

where b is the standard spatial rescaling factor. For an $L_\parallel \times L_\perp^{d-1}$ geometry with $L_\parallel \geqslant L_\perp$ and periodic boundary conditions, these bulk relations are then supplemented with

$$L'_\parallel = L_\parallel/b \quad \text{and} \quad L'_\perp = L_\perp/b \tag{B2}$$

and the usual bulk flow equation for the free energy is extended by postulating

$$f_s(H, T; L_\parallel, L_\perp) \approx b^{-d} f_s(H', T'; L'_\parallel, L'_\perp) \tag{B3}$$

This framework is, of course, consistent with the standard asymptotic finite-size scaling hypothesis[1-3,5] and should thus be valid in the vicinity of the bulk first-order transition for *block* geometry. However, (B3) neglects, in particular, corrections due to nonlinearities of the renormalization group away from the fixed point (which can, at least in leading orders, be embodied in nonlinear scaling fields) as well as singular corrections to scaling. Nevertheless BNC *assume* that one may validly neglect all these corrections even as $L_\perp$ is renormalized down to $L'_\perp = a$, which corresponds to a *one*-dimensional chain. Thus they choose $b = L_\perp / a$ and, using (B3), assert

$$f_s(H,T;L_\parallel,L_\perp) \approx \left(\frac{a}{L_\perp}\right)^d f_s\left[H\frac{L_\perp}{a}, T\left(\frac{a}{L_\perp}\right)^{d-1}; \frac{L_\parallel}{L_\perp}a, a\right] \quad \text{(B4)}$$

where, it is argued, the right-hand side may be evaluated in terms of the free energy, $f_a(H',T';L')$, of a one-dimensional Ising model of length L'.

Now the two transfer matrix eigenvalues for a simple Ising chain with $\bar{h}' = H'a^d/k_BT$ and $K' = J/k_BT'$ are

$$\Lambda_0, \Lambda_1 = e^{K'}\left[\cosh\bar{h}' \pm \left(\sinh^2\bar{h}' + e^{-4K'}\right)^{1/2}\right] \quad \text{(B5)}$$

In leading order for small T' and H' (effective control over higher orders having, in any case, been lost) the singular part of the free energy is thus given by

$$f_a \approx -\frac{1}{L'}\ln 2\cosh\left[(L'/a)\left(\bar{h}'^2 + e^{-4K'}\right)^{1/2}\right] \quad \text{(B6)}$$

On appealing to (B4) with $L' \equiv L'_\parallel = L_\parallel a/L_\perp$, etc. one obtains a result which may be written as

$$f_s(H,T;L_\parallel,L_\perp) \approx -\frac{1}{V}\ln 2\cosh\left[\left(\frac{HV}{k_BT}\right)^2 + \left(\frac{L_\parallel}{L_\perp}e^{-\tilde{\sigma}_0}\right)^2\right]^{1/2} \quad \text{(B7)}$$

where the $T = 0$ surface tension enters, correctly, through the identification

$$\tilde{\sigma}_0 \equiv 2KL_\perp^{d-1}/a^{d-1} = \Sigma(T=0)A/k_BT \quad \text{(B8)}$$

in which the last part of this equation follows as in Appendix A.

Comparison with our results, as embodied in (3.14) or (3.19), shows that there is agreement only if one makes the correct correspondence $m_0(T=0) = 1$ and further accepts the identification

$$\xi_\parallel(T,L_\perp) \Rightarrow \tfrac{1}{2}L_\perp \exp(\Sigma A/k_BT) \quad \text{(B9)}$$

for $T \to 0$. However, the exact evaluation of $\xi_\parallel$ for the $d = 2$ Ising model (Appendix A) shows that the BNC result is too large by a factor $(L_\perp/a)$. [See (A10), which gives the result for $T \to 0$; when $L_\perp \to \infty$ at fixed $T > 0$ the error is of order $(L_\perp/a)^{1/2}$ but comparison in this limit is not really justifiable.] More generally, by (4.8) there are no grounds for accepting (B9) for other dimensionalities either!

The basic flaw in the BNC argument is that one is not justified in using the linearized flow equation (B3) when $L_\parallel/\xi_\parallel \gg 1$. This may be understood heuristically by noting that the nature of the renormalization group transformation is to smear out microscopic details. However, in the cylinder geometry the most probable configurations entail homogeneous domains of mean length $\xi_{\|}$ separated by roughly $L_{\|}/\xi_{\|}$ distinct interfaces: but an interface at low temperatures is a *micro*scopic structure, on the scale $\xi_{\infty} = O(a)$. Consequently the flow equation (B3), linearized about the bulk discontinuity fixed point, misses some of the needed information.

To demonstrate the deficiency of the BNC approach more concretely, consider $d = 2$ and suppose one renormalizes only down to $L'_{\perp} = 2a$ which describes an $L'_{\|} \times 2a$ "ladder" (or double-chain) with periodic boundary conditions. This system can again be solved exactly, and in place of (B6) one finds, for small T' and H',

$$f_{2a}(H',T';L') \approx -\frac{1}{2L'} \ln 2\cosh\left[(L'/a)(4\bar{h}'^2 + e^{-8K'})^{1/2}\right] \quad \text{(B10)}$$

As a matter of fact this expression can be derived from (B6) merely by recognizing the scaling properties of an Ising strip for low field and temperature! On using (B3) as before but with $b = L_{\perp}/2a$ one finds, instead of (B7) the *different* result

$$f_s \approx -\frac{1}{V} \ln 2\cosh\left[\left(\frac{HV}{k_B T}\right)^2 + \left(\frac{2L_{\|}}{L_{\perp}} e^{-\tilde{\sigma}_0}\right)^2\right]^{1/2} \quad \text{(B11)}$$

The necessary identification of $\xi_{\|}$ again differs from the correct result for $d = 2$ by a power of $L_{\perp}$ but now a further factor of 2 is also present! One may rationalize the different answers by noting that, in effect, the BNC approach fails to differentiate factors of 2 which are true numerical constants from those equal to $L'_{\perp}/a$ which rescale nontrivially. Thus wrong factors of $L_{\perp}$ and wrong numerical factors are not surprising. In any event, an adequate treatment along the BNC lines would have to go beyond the simple linearized renormalization group relation (B3).

REFERENCES

1. M. E. Fisher, in *Critical Phenomena*, Proceedings of the Enrico Fermi International School of Physics, Vol. 51, M. S. Green, ed. (Academic, New York, 1971).
2. M. E. Fisher and M. N. Barber, *Phys. Rev. Lett.* **28**:1516 (1972).
3. M. N. Barber, in *Phase Transitions and Critical Phenomena*, Vol. 7, C. Domb and J. L. Lebowitz, eds. (Academic, New York). [in press].
4. M. P. Nightingale, *J. Appl. Phys.* **53**:7927 (1982).
5. M. E. Fisher and A. N. Berker, *Phys. Rev. B* **26**:2507 (1982).
6. B. Nienhuis and M. Nauenberg, *Phys. Rev. Lett.* **35**:477 (1975).
7. Y. Imry, *Phys. Rev. B* **21**:2042 (1980).
8. P. Kleban and C.-K. Hu, preprint, University of Maine, 1982.
9. H. W. J. Blöte and M. P. Nightingale, *Physica* **112A**:405 (1982).
10. V. Privman and L. S. Schulman, *J. Phys. A* **15**:L231 (1982).
11. V. Privman and L. S. Schulman, *J. Stat. Phys.* **29**:205 (1982).
12. H. W. J. Blöte, M. P. Nightingale, and B. Derrida, *J. Phys. A* **14**:L45 (1981).

13. J. L. Cardy and P. Nightingale, *Phys. Rev. B* **27**:4256 (1983).
14. L. Onsager, *Phys. Rev.* **65**:117 (1944).
15. M. E. Fisher, *J. Phys. Soc. Japan Suppl.* **26**:87 (1969).
16. E. Brézin, *J. Phys. (Paris)* **43**:15 (1982).
17. M. E. Fisher, in *Essays in Physics*, Vol. 4, G. K. T. Conn and G. N. Fowler, eds. (Academic Press, London, 1972), p. 43.
18. D. Jasnow and M. E. Fisher, *Phys. Rev. B* **3**:895 (1971).
19. T. D. Schultz, D. C. Mattis, and E. W. Lieb, *Rev. Mod. Phys.* **36**:856 (1964).
20. C. J. Thompson, *Mathematical Statistical Mechanics* (MacMillan, New York, 1972), Chaps. 5, etc.
21. W. J. Camp and M. E. Fisher, *Phys. Rev. B* **6**:946 (1972).
22. W. J. Camp, *Phys. Rev. B* **6**:960 (1972).
23. P. H. Kleban and G. Akinci, preprint, University of Maine, 1983.
24. A. E. Ferdinand and M. E. Fisher, *Phys. Rev.* **185**:832 (1969).
25. A. D. Bruce, *J. Phys. C* **14**:3667 (1981).
26. M. E. Fisher and W. J. Camp, *Phys. Rev. Lett.* **26**:565 (1971).
27. C. M. Newman and L. S. Schulman, *J. Math. Phys.* **18**:23 (1977).
28. M. E. Fisher and H. Nakanishi, *J. Chem. Phys.* **75**:5857 (1981).
29. H. Nakanishi and M. E. Fisher, *J. Chem. Phys.* **78**:3279 (1983).
30. R. B. Griffiths, *Phys. Rev.* **152**:240 (1966).
31. L. S. Schulman, *J. Phys. A* **13**:237 (1980).
32. K. Binder, *Phys. Rev. Lett.* **47**:693 (1981).
33. B. Widom, *J. Chem. Phys.* **43**:3982 (1965).
34. B. Widom, in *Phase Transitions and Critical Phenomena*, Vol. 3, C. Domb and M. S. Green, eds. (Academic Press, New York, 1972), p. 79.
35. M. E. Fisher, in *Collective Properties of Physical Systems*, Proc. Nobel Symp., Vol. 24, B. Lundqvist and S. Lundqvist, eds. (Academic Press, New York, 1974), p. 16.
36. M. E. Fisher, in *Renormalization Group in Critical Phenomena and Quantum Field Theory —Proceedings of a Conference*, J. D. Gunton and M. S. Green, eds. (Temple University, Philadelphia, 1974), p. 65.
37. M. N. Barber and M. E. Fisher, *Ann. Phys. (N.Y.)* **77**:1 (1973).
38. K. G. Wilson and J. Kogut, *Phys. Rep.* **12C**:74 (1974).
39. M. E. Fisher, *Rev. Mod. Phys.* **46**:597 (1974).
40. P. Nightingale, *Proc. Kon. Ned. Akad. Wet. B* **82**:235 (1979).
41. C. J. Hamer, *J. Phys. A* **15**:L675 (1982).
42. C. N. Yang, *Phys. Rev.* **85**:809 (1952).
43. M. E. Fisher, *Physica* **25**:521 (1959).
44. J. W. Essam and D. L. Hunter, *J. Phys. C* **1**:392 (1968).
45. D. S. Gaunt and G. A. Baker Jr., *Phys. Rev. B* **1**:1184 (1970).
46. G. A. Baker Jr. and D. Kim, *J. Phys. A* **13**:L103 (1980).
47. V. Privman and M. E. Fisher, *J. Phys. A* **16**:L295 (1983).
48. C. J. Hamer and M. N. Barber, *J. Phys. A* **14**:2009 (1981).
49. P. Pfeuty, D. Jasnow, and M. E. Fisher, *Phys. Rev. B* **10**:2088 (1974).
50. P. A. Rikvold, W. Kinzel, J. D. Gunton, and K. Kaski, *Phys. Rev. B* [in press].
51. K. Binder, in *Phase Transitions and Critical Phenomena*, Vol. 7, C. Domb and J. L. Lebowitz, eds. (Academic, New York) [in press].
52. T. Niemeijer and J. M. J. van Leeuwen, in *Phase Transitions and Critical Phenomena*, Vol. 6, C. Domb and M. S. Green, eds. (Academic, New York, 1976).

2. Applications

Introduction

Finite-size scaling really became an important tool with the development by Nightingale of the so-called phenomenological renormalization method. Although in principle it can be used in any number of dimensions for which finite-size scaling is valid, for practical purposes its usefulness has largely been restricted to two dimensions.

Consider then some two-dimensional lattice model with a single adjustable parameter which we can think of as the temperature T, defined on an infinite strip of width L with periodic boundary conditions. The transfer matrix $\hat{t}$ for this strip is usually a finite matrix, and we can find its largest eigenvalues $\lambda_0, \lambda_1, \ldots$. As is well known, $-\ln \lambda_0$ gives the free energy per unit length, $\ln(\lambda_0/\lambda_1)$ gives the inverse correlation length $\xi(T, L)^{-1}$, and so on. The simplest way to calculate these eigenvalues is often to calculate $\hat{t}^n$ acting on a suitable state for n large.

Finite-size scaling predicts that

$$\xi(T, L) = L\phi\big((T-T_c)^{1/\nu}L\big), \tag{2.1}$$

which suggests that we define a mapping $T \to T'$ by

$$L^{-1}\xi(T, L) = (L-1)^{-1}\xi(T', L-1). \tag{2.2}$$

From (2.1) we see that this mapping has a fixed point at $T = T_c$, and that

$$\frac{T'-T_c}{T-T_c} = \left(\frac{L-1}{L}\right)^{-\nu}. \tag{2.3}$$

In practice, then, one numerically calculates $\xi(T, L)$ and defines an effective $T_c(L)$ by

$$L^{-1}\xi(T_c(L), L) = (L-1)^{-1}\xi(T_c(L), L-1). \tag{2.4}$$

Equation (2.3), with T_c replaced by $T_c(L)$, then defines an approximant $\nu(L)$. To get reliable values of T_c and ν, it is necessary to perform an extrapolation for large L. It is possible to also calculate other exponents by adding suitable terms of the Hamiltonian.

That these ideas work for the Ising model, where, thanks to Onsager, we know exact results in finite-width strips, was shown by Nightingale (paper 2.1). In paper 2.2 we see these methods applied to the Potts model. One should remark that, despite its name, phenomenological renormalization is not a true RG transformation, in that it does not preserve a whole probability distribution, or set of correlation functions, but rather just one quantity ($L^{-1}\xi$ in the above example). Barber [1] has shown how, by considering more quantities, one can gain access to the projection of the scaling variables into the space of physical parameters, but this process becomes increasingly complicated as more information is required.

The methods described above are limited by the size of the matrices with which one can calculate. Larger values of L can be treated by going to the so-called quantum Hamiltonian limit. This involves considering anisotropic couplings, parallel to and across the strip. Call them $K_{\|}$ and $K_{\perp}$. Then in the limit in which $K_{\|} \to \infty$ and $K_{\perp} \to 0$, in such a way that we stay near the critical curve, the transfer matrix simplifies: it can be written $\hat{t} \approx \hat{1} - K_{\perp} \hat{H}$, where $\hat{H}$ is a sum of local matrices, and can be thought of as the Hamiltonian of a quantum system in one less dimension. For many models, this limit of extreme anisotropy does not modify the universality class, and the advantage is that powerful methods are available to calculate the lowest few eigenvalues of $\hat{H}$. Paper 2.3, by Hamer and Barber, illustrates the method.

For some systems, however, anisotropy is an essential feature. For the model of directed percolation, for example, it is known that there is a preferred direction, and that the correlation lengths parallel and perpendicular to this direction, $\xi_{\|}$ and $\xi_{\perp}$, diverge with different exponents $\nu_{\|}$ and $\nu_{\perp}$ in the infinite system. In the finite-width strip, finite-size scaling is thus modified. If we align the preferred direction along the strip, we would expect finite-size effects to occur when $\xi_{\perp} \sim L$. At this point $\xi_{\|} \sim L^{\nu_{\|}/\nu_{\perp}}$. Thus the factors of L^{-1} and $(L-1)^{-1}$ in (2.2) should be replaced by $L^{-\nu_{\|}/\nu_{\perp}}$ and $(L-1)^{-\nu_{\|}/\nu_{\perp}}$ respectively. Kinzel and Yeomans (paper 2.4) showed that $\nu_{\|}$ and $\nu_{\perp}$ can then be extracted by comparing the values of $\xi_{\|}$ on strips of three different widths.

Many modern problems of critical behavior in statistical physics are of a geometrical nature, for example percolation and self-avoiding random walks. In that case there are no Boltzmann weights, and the concept of a transfer matrix requires rethinking. Nevertheless, it does make sense, and finite-size scaling has had some spectacular successes for these types of model, as shown by papers 2.5–2.7, by Derrida et al. The question of the rate of convergence of all the above methods is interesting. RG ideas would suggest that corrections should be of the form L^{-y}, where $-y$ is the leading irrelevant RG eigenvalue. Numerically this does not seem to be the case. Privman and Fisher (paper 2.8) show that such terms can however be seen by looking at very large values of L. Why these terms have such small amplitudes that, for practical purposes, they are swamped by L^{-2} terms is not understood.

We continue the selection of papers in this section with a different subject: how do finite-size effects enter a massive, asymptotically free theory (papers 2.9, 2.10)? This question is important for lattice gauge theories. Finally we include a study (paper 2.11) of the zeroes of the partition function of a finite Ising model. Such studies are important because they may give us clues to the eventual solution of such models.

Reference

[1] M.N. Barber, Phys. Rev. B 27 (1983) 5879; Physica A 130 (1985) 171.

Physica **83A** (1976) 561–572 © *North-Holland Publishing Co.*

SCALING THEORY AND FINITE SYSTEMS

M.P. NIGHTINGALE

Instituut voor Theoretische Fysica, Universiteit van Amsterdam, Amsterdam, The Netherlands

Received 1 December 1975

A renormalization group transformation is introduced with the help of which critical properties of infinite systems can be related to finite systems. As a numerical example the method is applied to the two-dimensional Ising model. The critical point and critical point exponent are computed in addition to the amplitude of the logarithmic singularity in the specific heat.

1. Introduction

In a previous publication[1]) it was shown how logarithmic singularities in magnetic lattice models can be explained within the Kadanoff theory[2]). In particular an expression was derived for the amplitude of the divergent contribution to the specific heat in case of a logarithmic divergency. Renormalisation group theory, a generalisation of Kadanoff's theory, was shown to give rise to a simple correction.

We shall presently illustrate these theoretical results for the two-dimensional Ising model. For that purpose we shall use approximate renormalisation group equations which on a phenomenological level map the Ising model into itself. Besides the above mentioned amplitude we find approximate values for the critical point and the critical point exponents.

2. Renormalisation group theory

Schematically the renormalisation group approach to critical phenomena can be summarised as follows.

Consider a d-dimensional lattice of $N = n \times n \times \cdots$ (d times) sites. At each lattice site i there is a spin which can be in either of the two states: up, denoted by $s_i = +1$ or down, denoted by $s_i = -1$. There is nearest neighbour interaction only, each pair i and j contributing Js_is_j to the total energy. The partition function

Z_N, which is a function of $K = J/kT = \beta J$ only, is given by

$$Z_N(K) = \sum_{\{s_i\}_N} \exp\left(-K \sum_{\text{n.n.}} s_i s_j\right) = \sum_{\{s_i\}_N} \exp - \mathscr{H}(K; \{s_i\}_N), \tag{1}$$

where $\{s_i\}_N$ denotes a configuration of spins on the $n \times n \times \cdots$ lattice and $\sum_{\text{n.n.}}$ is a sum over nearest neighbours. We introduce a dimensionless free energy per site $f_N(K)$ by means of

$$f_N(K) = \frac{1}{NJ} F_N(J, \beta) = -\frac{1}{NK} \ln Z_N(K). \tag{2}$$

The function $f_N(K)$ introduced in this way is a function of $K = \beta J$ only as indicated; this in contrast with the free energy as it is usually defined which is a function of J and β separately.

In order to obtain a scaling relation block spins of L^d spins per block are introduced. On a lattice of $N/L^d = (n/L) \times (n/L) \times \cdots$ (d times) sites a block spin configuration is defined as a constrained set of the original configurations. The summation in eq. (1) can then be carried out in two steps

$$Z_N(K) = \sum_{\{\sigma\}_{N/L^d}} \sum_{\{s\}_N} \chi(\{s\}_N \mid \{\sigma\}_{N/L^d}) \exp - \mathscr{H}(K; \{s\}_N),$$

where

$$\chi(\{s\}_N \mid \{\sigma\}_{N/L^d}) = \begin{cases} 1 & \text{if } \{s\}_N \text{ is compatible with } \{\sigma\}_{N/L^d}, \\ 0 & \text{otherwise.} \end{cases}$$

The function χ has the property that

$$\sum_{\{\sigma\}_{N/L^d}} \chi(\{s\}_N \mid \{\sigma\}_{N/L^d}) = 1.$$

Any function with this property may be introduced[7]).

In the spirit of Kadanoff's conjecture we assume that the assembly of block spins can be considered as an Ising model. Each spin of the scaled model has an internal free energy $f^{(i)}_{L,N}(K)$ while the scaled interaction parameter is given by $K_{L,N}(K)$. For the partition function this leads to the expression

$$\begin{aligned} Z_N(K) &= \sum_{\{\sigma\}_{N/L^d}} \exp\left\{-\frac{N}{L^d} f^{(i)}_{L,N}(K) - \mathscr{H}(K_{L,N}; \{\sigma\}_{N/L^d})\right\} \\ &= \exp\left\{-\frac{N}{L^d} f^{(i)}_{L,N}(K)\right\} Z_{N/L^d}(K_{L,N}). \end{aligned} \tag{3}$$

Both $f^{(i)}_{L,N}$ and $K_{L,N}$ depend upon K and upon L and N as indicated; the dependence upon N is expected to be a weak one though. From the definition of the free energy (2) and eq. (3) follows the scaling relation

$$L^d K f_N(K) = f_{L,N}^{(1)}(K) + K_{L,N} f_{N/L^d}(K_{L,N}). \tag{4a}$$

Taking the thermodynamic limit we obtain

$$L^d K f(K) = f_L^{(1)}(K) + K_L f(K_L), \tag{4b}$$

where the limiting functions have been denoted by dropping the N indices.

The equation

$$K_L = K_L(K) \tag{5}$$

is called the renormalisation group equation.

The essential assumption in the procedure is that both $K_L(K)$ and $f_L^{(1)}(K)$ are analytic functions of their argument K even at the critical point K_c. This assumption has a number of important implications for the critical behaviour. For our purpose the following is relevant.

I. The critical point K_c is a fixed point of the transformation (6) *i.e.*

$$K_c = K_L(K_c). \tag{6}$$

To see this suppose that $K = K_c$ is a critical point of the system, in other words that the function $f(K)$ has a singularity at $K = K_c$. Then the function $f(K_L)$ is singular at $K_L(K) = K_c$. Let us further assume that there is only one critical point in the system. Then it follows from eq. (5) that K_c has to be a solution of eq. (6).

II. Critical properties are determined by the coefficients of the Taylor series $K_L(K)$ and $f_L^{(1)}(K)$ at $K = K_c$.

Because of the assumed analyticity of these functions series expansions of the following form exist

$$\begin{aligned} K_L(K) - K_c &= \sum_{p=1}^{\infty} \lambda_p^{(L)} (K - K_c)^p, \\ f_L^{(1)}(K) &= \sum_{p=0}^{\infty} \varphi_p^{(L)} (K - K_c)^p. \end{aligned} \tag{7}$$

Several critical exponents can be expressed in the quantity y introduced by means of

$$y = {}^L\!\log \lambda_1^{(L)} \qquad \text{or} \qquad \lambda_1^{(L)} = L^y. \tag{8}$$

In the case that $y = d/2$ we have shown[1]), using the series expansions given above, that the free energy in the neighbourhood of the critical point is given by [see ref. 1 eq. (32)]

$$Kf(K) = -\left(\varphi_2^{(L)} + \lambda_2^{(L)} \frac{\varphi_1^{(L)}}{L^d - \lambda_1^{(L)}}\right) \frac{(K - K_c)^2 \ln |K - K_c|}{L^d \ln \lambda_1^{(L)}} + \cdots.$$

The remaining terms on the right-hand side of this equation are of no current interest: they do not contribute to the logarithmic singularity in the specific heat.

For the specific heat per spin $c(K)$ we have the thermodynamic relation

$$c(K)/k_{\mathrm{B}} = -(1/K)\,\partial^2 f/\partial\,(1/K)^2 .$$

Applying this relation to the equation for the free energy gives an expression for the amplitude A of the logarithmic singularity in the specific heat for $K \simeq K_c$ $c(K)/k_{\mathrm{B}} \simeq A \ln |K - K_c|$

$$A = -2K_c^2 \left(\varphi_2^{(L)} + \lambda_2^{(L)} \frac{\varphi_1^{(L)}}{L^d - \lambda_1^{(L)}} \right) \frac{1}{L^d \ln \lambda_1^{(L)}} . \tag{9}$$

We note that A is determined by the fixed point of $K_L(K)$ together with expansion coefficients of this function and those of the function $f_L^{(i)}(K)$ up to second order in $K - K_c$.

3. Renormalisation group equations and finite systems

As we saw in the previous section the functions $f_L^{(i)}$ and K_L play a central role in the theory. We now shall discuss eq. (4) which contains $f_{L,N}^{(i)}$ and $K_{L,N}$, the corresponding functions for finite systems.

The functions $f_{L,N}^{(i)}$ and $K_{L,N}$ occurring in eq. (4) are analytic functions of their argument K, for they refer to finite systems in which no singularities are present. Because of the assumed analyticity of $f_L^{(i)}$ and K_L, therefore, all relevant properties of these functions can be approximated by the corresponding properties of $f_{L,N}^{(i)}$ and $K_{L,N}$. In particular from the scaling relation (4) for finite systems increasingly accurate approximate values for K_c, y and A can be obtained taking larger values of N.

Up to this point we did not go into the question how to find the functions $f_{L,N}^{(i)}$ and $K_{L,N}$ which were introduced in eq. (3). The procedure we use to this end is a phenomenological one in the following sense. We assume the thermodynamics of both the N-spin and the N/L^d-block-spin system, in other words the functions f_N and f_{N/L^d}, to be known. The functions $f_{L,N}^{(i)}$ and $K_{L,N}$ are then defined so as to relate both systems according to eq. (3) or eq. (4) equivalently. It is clear that we need a second scaling relation for $f_{L,N}^{(i)}$ and $K_{L,N}$ to be determined.

In the Kadanoff picture one has a scaling relation for the correlation length ξ measured in units of the lattice spacing

$$\xi(K_L) = \frac{1}{L}\,\xi(K) . \tag{10}$$

We saw that the critical point is a fixed point of the renormalisation group equation. From eq. (10) the converse follows. If K^* satisfies

$$K^* = K_L(K^*)$$

then

$$\xi(K^*) = \frac{1}{L}\,\xi(K^*)$$

so that $\xi(K^*) = \infty$ and therefore $K^* = K_c$ (unless K^* is a trivial fixed point). If ξ_N is the correlation length of a finite system of N sites the corresponding relation which connects two finite systems reads

$$\xi_{N/L^d}\,(K_{L,N}) = \frac{1}{L}\,\xi_N(K). \tag{11}$$

As opposed to eq. (10) this equation does not imply an infinite correlation length at the fixed point, different functions appearing on the left- and right-hand side of the equation.

From the thermodynamics of finite systems, that is from knowledge of the functions ξ_N, ξ_{N/L^d}, f_N and f_{N/L^d} in particular, one can now in principle compute the functions $K_{L,N}$ and $f^{(i)}_{L,N}$ from eqs. (4) and (11).

4. Two-dimensional Ising model

Our present aim is twofold. First we show that the properties as implied by the simple Kadanoff-like picture as outlined above are borne out by the Onsager solution of the finite two-dimensional Ising model. Second we will show that the procedure as outlined above, that is deriving properties of the phase transition in the infinite system by studying the renormalisation group equations of a finite system, is a good one in the sense that a) even for small systems one gets nice results, b) the convergence to the values for the infinite system is a a quick one.

We shall first give a survey of some equations which play a central role in the Onsager solution.

Consider a two-dimensional Ising model of m rows of n spins each. Denoting the partition function of this system as $Z_{n,m}$ one has[4])

$$Z_{n,m} = \mathrm{Tr}\; V_n^m,$$

i.e. the partition function can be expressed as the trace of the mth power of a $2^n \times 2^n$ matrix V^n (assuming suitable boundary conditions). The transfer matrix V

is symmetric and has eigenvalues‡

$$\omega_1^{(n)} > \omega_2^{(n)} \cdots > \omega_{2^n}^{(n)}.$$

Therefore

$$Z_{n,m} = \sum_{s=1}^{2^n} (\omega_s^{(n)})^m \simeq (\omega_1^{(n)})^m \qquad \text{for} \qquad m \gg 1.$$

If $f_n(K)$ is the dimensionless free energy per spin of an infinite strip n spins wide,

$$f_n(K) = -\lim_{m\to\infty} \frac{1}{nmK} \ln Z_{n,m}(K) = -\frac{1}{nK} \ln \omega_1^{(n)}(K). \tag{12}$$

For the correlation length ξ_n of such a strip one can derive[5])

$$\{\xi_n(K)\}^{-1} = \lim_{m\to\infty} \{\xi_{n,m}(K)\}^{-1} = \ln \{\omega_1^{(n)}(K)/\omega_2^{(n)}(K)\}. \tag{13}$$

Diagonalising V_n Onsager found all its eigenvalues, the largest of which satisfy[6])

$$\left.\begin{aligned} \ln \omega_1^{(n)} - \tfrac{1}{2}n \ln (2 \sinh 2K) &= \tfrac{1}{2} (\gamma_1^{(n)} + \gamma_3^{(n)} + \cdots + \gamma_{2n-1}^{(n)}) \\ \ln \omega_2^{(n)} - \tfrac{1}{2}n \ln (2 \sinh 2K) &= \tfrac{1}{2} (\gamma_2^{(n)} + \gamma_4^{(n)} + \cdots + \gamma_{2n}^{(n)}) \end{aligned}\right\} \tag{14}$$

where the $\gamma_r^{(n)}$'s are implicitely given by

$$\cosh \gamma_r^{(n)}(K) = \cosh 2K \coth 2K - \cos (r\pi/n).$$

We now apply the renormalization group procedure on a 2-dimensional lattice of $N = n \times m$ sites. The sacling relation (4a) becomes

$$L^2 K f_{n,m}(K) = f_{L,n,m}^{(\mathrm{i})}(K) + K_{L,n,m} f_{n/L,m/L}(K_{L,n,m}) \tag{15}$$

and the scaling relation (11) for the correlation length

$$\xi_{n,m}(K) = L\xi_{n/L,m/L}(K_{L,n,m}). \tag{16}$$

In the $m \to \infty$ limit (4b) becomes

$$L^2 K f_n(K) = f_{L,n}^{(\mathrm{i})}(K) + K_{L,n} f_{n/L}(K_{L,n}) \tag{17}$$

and eq. (16):

$$\xi_n(K) = L\xi_{n/L}(K_{L,n}). \tag{18}$$

‡ $\omega_s^{(n)}$ has been used instead of $\lambda_s^{(n)}$ in order to avoid confusion with the coefficients $\lambda_p^{(L)}$ in eq. (7).

We note that with the help of eqs. (12) and (13) we can transform eqs. (17) and (18) into two identical relations for the largest two eigenvalues of the transfer matrix

$$\{\omega_s^{(n)}(K)\}^L = \exp\left\{-\frac{n}{L} f_{L,n}^{(i)}(K)\right\} \omega_s^{(n/L)}(K_{L,n}), \qquad s = 1, 2. \tag{19}$$

5. Numerical results

The results shown in tables 1 through 4 were obtained by substituting Onsager's expressions for $\omega_1^{(n)}$ and $\omega_2^{(n)}$ in eq. (19) and subsequently solving (18) for $K_{L,n}$ and (17) for $f_{L,n}^{(i)}$ as functions of K. From these functions $K_{L,n}(K)$ and $f_{L,n}^{(i)}(K)$ one obtains for each pair of values L, n the fixed point K^* determined by $K^* = K_{L,n}(K^*)$ and also the coefficients $\lambda_1^{(L,n)}$, $\lambda_2^{(L,n)}$, $\varphi_1^{(L,n)}$ and $\varphi_2^{(L,n)}$ of the expansion of $K_{L,n}(K) - K^*$ and $f_{n,L}^{(i)}(K)$ in powers of $K - K^*$. These are needed to obtain an approximate value for the critical exponent y, given by (8) and for the amplitude A given by (9). The so obtained values of K^*, y and A are given in tables 1, 2 and 3 for various pairs of values n and n/L.

TABLE I

n	n/L	K^*	y	A
16	2	0.437807	1.0331	−0.449
16	3	0.439030	1.0223	−0.454
16	4	0.439730	1.0145	−0.464
16	5	0.440080	1.0101	−0.472
16	6	0.440264	1.0076	−0.476
16	7	0.440370	1.0061	−0.479
16	8	0.440438	1.0051	−0.481
16	9	0.440484	1.0043	−0.482
16	10	0.440517	1.0038	−0.483
16	11	0.440542	1.0034	−0.484
16	12	0.440561	1.0031	−0.485
16	13	0.440576	1.0028	−0.486
16	14	0.440588	1.0026	−0.486
16	15	0.440598	1.0024	−0.487
Extrapolation		0.440670	1.0002	−0.493
Exact		0.440687	1.0000	−0.495

Table I shows, as expected, that the results improve as the smallest system – a strip of n/L block spins – entering into the calculation becomes larger by increasing n/L, keeping n constant. In table 2 n/L is kept constant and n increases. It is

TABLE II

n	n/L	K^*	y	A
4	3	0.4309	1.059	−0.386
5	3	0.4334	1.049	−0.404
6	3	0.4350	1.043	−0.416
7	3	0.4361	1.038	−0.425
8	3	0.4368	1.034	−0.432
9	3	0.4374	1.032	−0.437
10	3	0.4378	1.029	−0.441
11	3	0.4381	1.028	−0.444
12	3	0.4384	1.026	−0.447
13	3	0.4386	1.025	−0.449
14	3	0.4387	1.024	−0.451
15	3	0.4389	1.023	−0.453
16	3	0.4390	1.022	−0.454
Exact		0.4407	1.000	−0.495

TABLE III

n	n/L	K^*	y	A
3	2	0.422361	1.0792	−0.382
4	3	0.430884	1.0591	−0.386
5	4	0.435953	1.0369	−0.417
6	5	0.438258	1.0233	−0.441
7	6	0.439310	1.0157	−0.456
8	7	0.439831	1.0113	−0.465
9	8	0.440116	1.0085	−0.471
10	9	0.440286	1.0067	−0.476
11	10	0.440394	1.0054	−0.479
12	11	0.440466	1.0045	−0.481
13	12	0.440516	1.0038	−0.483
14	13	0.440552	1.0032	−0.484
15	14	0.440578	1.0028	−0.486
16	15	0.440598	1.0024	−0.487
Extrapolation		0.440676	1.0004	−0.493
Exact		0.440687	1.0000	−0.495

clear that even if the results may grow better for still larger values of n one may not expect them to converge towards the exact critical values, because one of the dimensions of the system remains finite. Table 3 shows that the results converge towards their exact limiting values if both n and n/L increase, which is illustrated

TABLE IV

n	n/L	K^*	y	A
20	19	0.44064266	1.001517	−0.4891
25	24	0.44066465	1.000957	−0.4909
30	29	0.44067414	1.000658	−0.4919
35	34	0.44067890	1.000480	−0.4926
40	39	0.44068154	1.000367	−0.4930
45	44	0.44068312	1.000289	−0.4933
50	49	0.44068412	1.000233	−0.4935
Extrapolation		0.44068675	1.000001	−0.4946
Exact		0.44068679	1.000000	−0.4945

again in table 4. Qualitatively one may expect that the value of a parameter like $p = 1/n + L/n$ may be a measure for the deviation from the exact value. Quantitatively one might use this observation to device an exterpolation scheme.

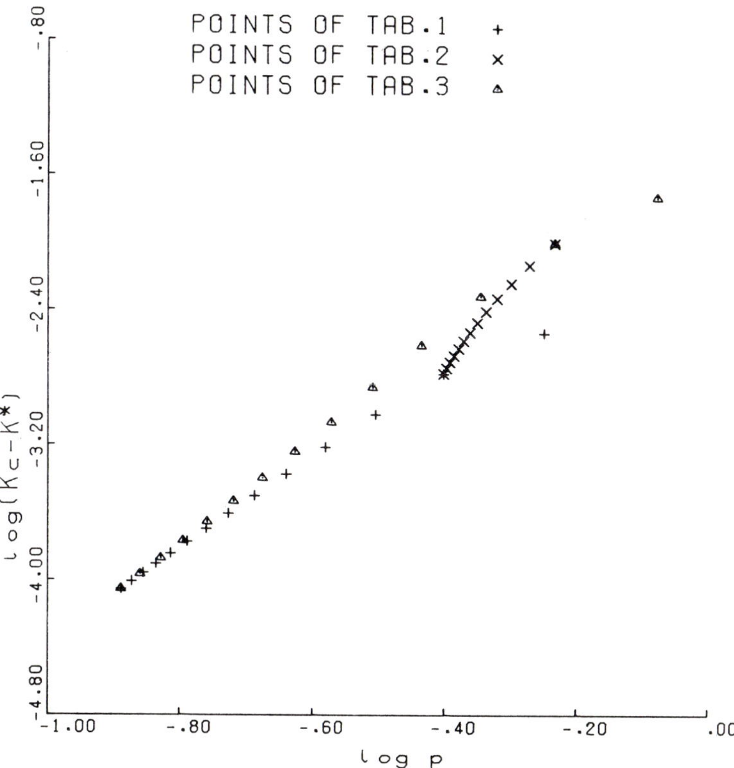

Fig. 1. Double logarithmic plot of the deviation from exact critical point as a function of p.

In figure 1 through 3 the values of $^{10}\log(K_c - K^*)$, $^{10}\log(y - y_c)$ and $^{10}\log \times (A - A_c)$ have been plotted as functions of $\log p$, where K_c, y_c and A_c denote the exact values. The figures showing the data from tables 1 and 3 lead to the linear plot as a function of $\log p$ in the range of small values of p. Conversely on this fact an exterpolation procedure can be based: the best values for K^*, y and A are those for which the linearity of this logarithmic plot becomes optimal.

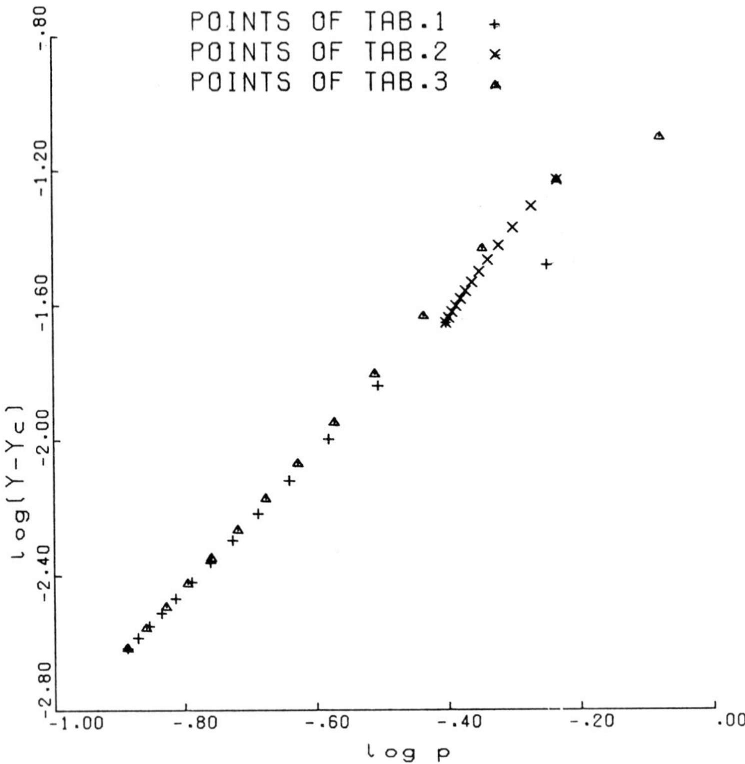

Fig. 2. Double logarithmic plot of the deviation from exact critical exponent as a function of p.

The results show that one obtains reasonable values for the critical properties for a relatively small system. The systems involved are small enough to render the calculation feasable for other two dimensional models for which no exact solution is available.

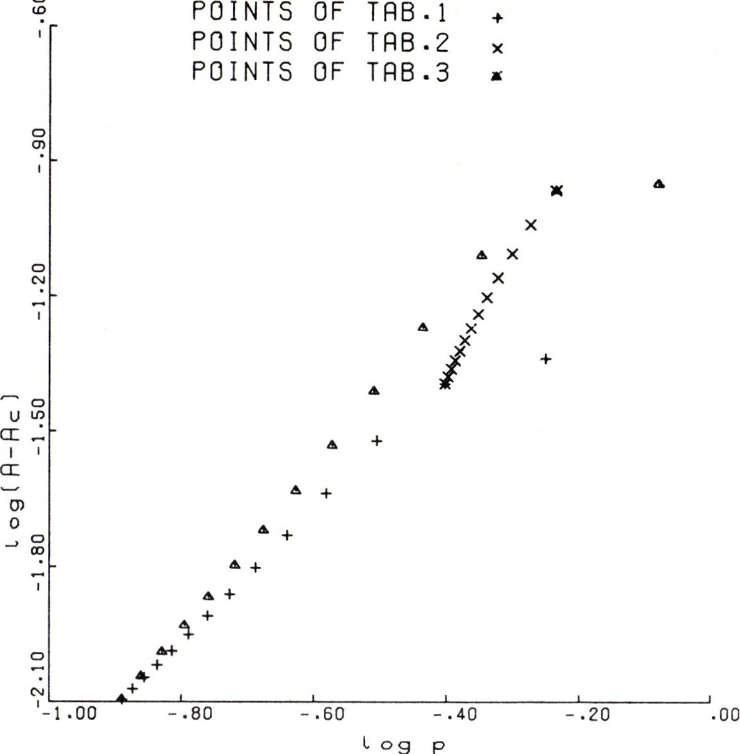

Fig. 3. Double logarithmic plot of the deviation from exact amplitude of logarithmic singularity as a function of p.

Acknowledgement

The author wishes to thank Professor J. de Boer for encouraging remarks. Furthermore he would like to thank Mr. A.H. 't Hooft for stimulating discussions and critically reading the manuscript prior to publication.

These investigations are part of the research programme of the "Stichting fundamenteel onderzoek der materie (F.O.M.)" which is financially supported by the "Nederlandse organisatie voor zuiver wetenschappelijk onderzoek (Z.W.O.)".

References

1) M. P. Nightingale and A. H. 't Hooft, Physica **77** (1974) 390.
2) L. P. Kadanoff, Physics **2** (1966) 263.
3) K. G. Wilson, Phys. Rev. **B4** (1971) 3174.
4) C. Domb, Adv. in Phys. **9** (1960) §§ 3.1 and 3.2.2.
5) Ref. 4, § 3.3.1.
6) Ref. 4, § 3.5.1.
7) L. P. Kadanoff and A. Houghton, Phys. Rev. **B11** (1975) 377.

Physica **112A** (1982) 405–465 *North-Holland Publishing Co.*

CRITICAL BEHAVIOUR OF THE TWO-DIMENSIONAL POTTS MODEL WITH A CONTINUOUS NUMBER OF STATES; A FINITE SIZE SCALING ANALYSIS

H.W.J. BLÖTE and M.P. NIGHTINGALE*

Laboratorium voor Technische Natuurkunde, Technische Hogeschool Delft, Delft, The Netherlands

Received 18 December 1981

We investigate the critical behaviour of the two-dimensional, q-state Potts model, using finite-size scaling and transfer matrix methods. For the continuous transition range ($0 < q \leq 4$), we present accurate values of the thermal and magnetic exponents. These are in excellent agreement with the conjecture of den Nijs, and that of Nienhuis et al. and Pearson, respectively. Finite size scaling is extended for the description of the first order region ($q > 4$). For completely finite systems, we recover the power law behaviour described by discontinuity fixed point exponents; however, for systems that are infinite in one direction, exponential behaviour occurs. This is illustrated numerically by the exponential divergences of the susceptibility and specific heat with increasing system size for $q \gg 4$.

These results for continuous q were obtained from a transfer matrix constructed for a generalized Whitney polynomial representing the Potts models. An effective algorithm to compute the dominant eigenvalues of this essentially nonsymmetric transfer matrix is developed.

1. Introduction

The Potts model[1]) has been the subject of many publications in recent years. It has been related to percolation[2]), absorption of rare gases on graphite[3]), cubic ferromagnets[4]), crystallographic transitions[5]), spin glasses[6,7]) and other models. For a theoretical description of the two-dimensional ferromagnetic model, a number of different techniques has been applied, including series expansions[8–12]), Monte-Carlo simulations[13]), analytic[14–19]) and Monte-Carlo renormalization[20,21]), finite size scaling[22,23]) and exact analytical treatments[24,25]). We refer the reader to a recent review article on the Potts model by Wu[26]). The picture that has emerged implies a phase transition between the ordered and disordered phases at a coupling K_c which is exactly $\log(\sqrt{q}+1)$ for the simple quadratic, q-state Potts model. It was shown by Baxter[24]) that the phase transition is continuous for $0 < q \leq 4$, and first order

* Present address: Department of Physics, University of Washington, Seattle WA 98195, USA.

for $q > 4$, with an exactly known latent heat. This result could, at first qualitatively[15]), and later more accurately[14,16,17]) be reproduced by real space renormalization treatments.

In the Potts model, the symmetry of the order parameter depends on the number of states. Thus we may expect non-universal behaviour as a function of q. In the context of scaling theory there are two exponents, viz. the temperature and magnetic exponents y_T and y_h, in terms of which the standard critical exponents α, β, γ, δ,... can be expressed. An expression for y_T as a function of $q (0 < q \leq 4)$ was proposed by den Nijs[27]) and one for y_h by Nienhuis et al.[17]) and by Pearson[28]) independently. At $q = 2$ these expressions reproduce the exact results $y_T = 1$ and $y_h = 15/8$ for the Ising model. The values for $q = 3$ and $q = 4$ are those of the hard hexagon model[29]) and the Baxter–Wu model[30,31]) which are believed to belong to the same universality classes, respectively[32,33]). In the case $q = 3$, this belief is corroborated by a variety of numerical treatments[12,20–23]). Recently, the conjecture of den Nijs[27]) for the temperature exponent has been given a more solid analytical basis by treatments of den Nijs[34]) and of Black and Emery[35]).

It has of late become clear that very accurate estimates of critical exponents may be obtained by combining finite size scaling with transfer matrix techniques. Exact numerical results for finite systems are derived by means of the transfer matrix technique and a comparison with scaling laws yields the required exponents. This circumstance, together with the fact that a transfer matrix can be constructed for non-integral values of q, opens the possibility to compare the aforementioned conjectures to accurate numerical results. This is the purpose of the present paper.

In section 2, we will describe the technique of finite scaling in some detail from the point of view of the renormalization group theory. Some special problems occur when finite-size scaling is applied at and near the critical value $q = 4$. These will be discussed in section 3, where we also go over the finite size behaviour in the region of the first-order transition: $q > 4$. The construction of the transfer matrix for non-integral q will be shown in section 4. A rather technical description of the correspondence between integer transfer matrix subscripts and lattice graphs will be given in Appendix I. In section 4.4 we discuss the numerical method employed for the calculation of the dominant eigenvalues of the non-symmetric transfer matrices of very large order (up to about 6×10^4). Section 5 is devoted to the numerical techniques used to extract the desired exponents from the finite size data. In section 6 we give our numerical results; these are compared to existing theory and conjectures in section 7.

A short report of the present work has already been published[36]). We feel that, in the present article, it is desirable to present, apart of more numerical data with an improved accuracy, also more detailed descriptions of the underlying theory and of our algorithms.

2. Finite size scaling theory

Finite size scaling[37]) can be regarded as an application of renormalization group theory. Consider a system described by variables situated on a d-dimensional lattice. For an infinite system one introduces scaled interaction parameters and derives scaling relations for e.g. the free energy per site via a renormalization transformation with Kadanoff-blocks of linear dimension L. Clearly, the linear dimension n of a finite system is rescaled to n/L under renormalization. Finite size scaling may therefore be formulated as follows: the inverse finite size $1/n$ should be considered a scaling field, in addition to the usual temperature and symmetry breaking fields; moreover this finite size scaling field has an exponent equal to one. Using this formulation we derive the scaling relations employed in our calculations.

Suppose that p is a quantity which scales with an exponent x; i.e. if u_1, $u_2, \ldots$ are the relevant scaling fields, with positive exponents $y_1, y_2, \ldots$, then

$$p(u_1, u_2, \ldots) = L^x p(L^{y_1}u_1, L^{y_2}u_2, \ldots). \tag{2.1}$$

The derivative $p^{(k_1,k_2,\ldots)}$ of p of order k_1 with respect to u_1, k_2 with respect to $u_2, \ldots$, then satisfies

$$p^{(k_1,k_2,\ldots)} \sim |u_j|^{-(k_1y_1+k_2y_2+\cdots+x)/y_j},$$

for $u_i = 0$ with $i \neq j$. One obtains this expression after taking the appropriate derivative in eq. (2.1) and choosing L such that $L^{y_j}|u_j| = 1$. According to the formulation above, one finds in particular

$$p^{(k_1,k_2,\ldots)} \sim n^{(k_1y_1+k_2y_2+\cdots+x)} \tag{2.2}$$

at the infinite system critical point $u_1 = u_2 = \cdots = 0$.

In the calculations described below, we use eq. (2.2) for the free energy and the inverse correlation length, and their derivatives. These scale with $x = -d$ (d is dimensionality) and $x = -1$, respectively. In particular, we find the following eqs. which we shall use to analyse data on finite systems in order to obtain the temperature and magnetic exponents y_T and y_h. We denote the specific heat by C, susceptibility by χ, and the derivatives of the inverse correlation length with respect to temperature and magnetic field by κ_T and κ_h, respectively. At the infinite system critical point these quantities satisfy (possibly up to a constant as explained below):

$$\left.\begin{aligned} &C \sim n^{2y_T-d} && \text{(a)} \\ &\chi \sim n^{2y_h-d} && \text{(b)} \\ &\kappa_T \sim n^{y_T-1} && \text{(c)} \\ &\kappa_h \sim n^{y_h-1} && \text{(d)} \end{aligned}\right\} . \tag{2.3}$$

Making use of eqs (2.3.c) or (2.3.d) to determine y_T or y_h is essentially equivalent to the phenomenologic renormalization method[38,39]) for calculating these exponents.

At this point the following remarks are to be made. Firstly, less singular terms modify the pure law behaviour (2.3). As usual, these corrections to scaling[40,41]) have various sources. These are: (i) the presence of irrelevant scaling fields ignored sofar, (ii) in the case of higher order derivatives the non-linearity in e.g. temperature and magnetic field of the corresponding scaling fields, (iii) the expected $1/n$ dependence of the scaling fields. Secondly, we remark that eq. (2.1), when applied to the free energy, holds for the singular part thereof only. Consequently, the expressions for C and χ hold only up to a constant term: the regular contribution at the critical point.

In particular, the consequences of regular contributions to the free energy cannot be neglected in the description of the finite size behaviour of quantities which do not diverge at the critical point. Examples thereof (to be discussed below) for the Potts model are the critical free energy and the specific heat for small q. Since finite size scaling theory is not usually applied in these cases, let us review the role of the regular part of the free energy in some detail. This may also serve for the purpose of future reference.

For simplicity, we shall consider just one scaling field u with exponent y, in addition to the finite size n. The standard renormalization group arguments[40,42,43]) lead to the following scaling relation for the free energy

$$f\left(u, \frac{1}{n}\right) = g_L\left(u, \frac{1}{n}\right) + L^{-d} f\left(L^y u, \frac{L}{n}\right). \tag{2.4}$$

From the fact that scaling with lengths L_1 and L_2 in succession is equivalent to scaling with a length L_1L_2 at once, it follows that

$$g_L\left(u, \frac{1}{n}\right) = \int_1^L l^{-(d+1)} \dot{g}\left(l^y u, \frac{l}{n}\right) \mathrm{d}l, \tag{2.5}$$

where $\dot{g} = (\partial g_L/\partial L)_{L=1}$.

Assuming an expansion

$$\dot{g}\left(u, \frac{1}{n}\right) = \sum_{i,j=0}^{\infty} \gamma_{ij} u^i n^{-j},$$

one immediately obtains, setting $L = n$,

$$f\left(u, \frac{1}{n}\right) = \sum_{i,j=0}^{\infty} \gamma_{ij} \frac{n^{iy+j-d}}{iy+j-d} u^i n^{-j} + n^{-d} f(n^y u, 1), \tag{2.6}$$

where $(n^{iy+j-d} - 1)/(iy+j-d)$ is to be replaced by $\log n$ if $iy+j-d = 0$. Note in particular that these logarithmic anomalies have indeed been found in the two-dimensional Ising model[25,44]) and the three-dimensional ideal Bose gas[45]). Finally, we remark that the logarithmic singularities in the specific heat in the cases $T \to T_c$, $n = \infty$ and $T = T_c$. $n \to \infty$, respectively, have the same amplitudes[37,25]).

3. Finite size behaviour for q ⩾ 4

In this section we consider the finite size behaviour of the Potts model for the larger values of q, i.e. q near the critical value $q = 4$ and beyond. First we discuss the region $q \approx 4$. As starting point we take the renormalization equations of Nauenberg and Scalapino[46]). Up to an analytical deformation, these equations are precisely those emerging from the renormalization treatment of the Potts model of Nienhuis et al.[15]). Denoting the temperature and dilution fields arising in this context by φ and ψ, respectively, the differential renormalization equations for a fractional, infinitesimal change of scale dx take the form

$$\frac{\mathrm{d}\psi}{\mathrm{d}x} = a(\psi^2 + \mu^2),$$

$$\frac{\mathrm{d}\varphi}{\mathrm{d}x} = (y_\mathrm{T} + b\psi)\varphi, \tag{3.1}$$

where $\mu^2 = q - 4$; a, b and y_T are constants which may be determined[46]) from Baxter's expression for the latent heat[24]) and the conjecture of den Nijs[27]) for y_T: $a = 1/\pi$, $b = 3/4\pi$ and $y_\mathrm{T} = 3/2$. In addition to these eqs. (3.1), there is the renormalization equation for the finite size

$$\frac{\mathrm{d}n^{-1}}{\mathrm{d}x} = n^{-1}.$$

The singular part of the free energy f_s satisfies

$$f_\mathrm{s}(\varphi, \psi, n^{-1}) = \mathrm{e}^{-xd} f_\mathrm{s}(\varphi(x), \psi(x), n^{-1}(x)). \tag{3.2}$$

For $q = 4$ one finds, by integration of eqs. (3.1) and choosing $n(x) = 1$,

$$f_\mathrm{s}(\varphi, \psi, n^{\;1}) = n^{\;d} f_\mathrm{s}(n^y z^{b/a} \varphi, z\psi, 1), \tag{3.3}$$

where $z \equiv (1-\psi a \log n)^{-1}$. Differentiation of (3.3) twice with respect to φ yields the specific heat. By expansion of the analytic scaling function we find

$$C(n) = n^{2y_T-d} z^{2b/a} \sum_{i=0}^{\infty} a_i(\psi z)^i. \tag{3.4}$$

Note that the leading singular behaviour $a_0 n^{2y_T-d} z^{2b/a}$ differs from (2.3.a). Since z decreases only very slowly as a function of n, correction terms are expected to modify this behaviour over a wide range of finite system sizes.

From the differential equations (3.1) it follows that the expressions (3.3) and (3.4) are also valid in the neighbourhood of $q=4$: in the region $|\psi(x)|^2 \gg \mu^2$, $n(x)=1$ i.e.

$$|q-4| \ll (\log n)^{-2}$$

for large n. Clearly, similar results can be derived for the susceptibility.

In the remainder of this section we consider the finite size behaviour in the region $q>4$, where the Potts model displays a first order transition. First we deal with the critical free energy, and finally we treat the way in which δ-functions in susceptibility and specific heat develop with increasing size.

First order transitions may be characterized by the absence of critical fluctuations. Therefore, due to the exponential decay of correlations one expects the difference of the free energy of a finite system of size n with periodic boundary conditions and an infinite system to be exponentially small in n. The non-critical two-dimensional Ising system does indeed show this behaviour, as was already pointed out by Onsager[25]). Combining this with the result of the previous section, one arrives at the following relation (for all q):

$$f\left(\frac{1}{n}\right) = f(0) + an^{-d}\, e^{-n/\xi} \tag{3.5}$$

at the infinite system critical point. Here a is an amplitude depending on q; ξ is of the order of the critical correlation length

$$\xi \begin{cases} = \infty & \text{for } q \leq 4 \\ \sim e^{\pi^2/\sqrt{q-q_c}} & \text{for } q>4. \end{cases} \tag{3.6}$$

The expression for $q>4$ is obtained by using the renormalization eqs. (3.1) in combination with the scaling relation for the correlation length

$$\xi(\varphi, \psi) = e^x \xi(\varphi(x), \psi(x)). \tag{3.7}$$

First order transitions are signalled by a latent heat and a discontinuity in the spontaneous magnetization. Equivalently, δ-function singularities occur in specific heat and susceptibility. We shall discuss the finite size behaviour of these quantities in terms of the zero temperature discontinuity fixed point. As an example we work out the case of the susceptibility of the Ising model below the critical temperature.

Consider a system with spins $s_i = \pm 1$ on a d-dimensional lattice with reduced hamiltonian $\mathscr{H}$ (i.e. conventional hamiltonian times $-1/k_B T$)

$$\mathscr{H} = K \sum_{\langle i,j \rangle} s_i s_j + h \sum_i s_i,$$

where $\langle i, j\rangle$ runs through all nearest neighbour sites. From the Migdal–Kadanoff recursion relations one obtains the following scaling relation for f_s, the singular part of the free energy, close to the zero temperature fixed point:

$$f_s\left(K, h, \frac{1}{n}\right) = L^{-d} f_s\left(K', h', \frac{L}{n}\right), \tag{3.8}$$

where

$$K' = L^{d-1}K, \quad h' = L^d h.$$

We assume that the renormalization can be carried out until the finite size is scaled down to one. Although this is an oversimplification, we expect thus to obtain the leading singular behaviour. We consider two special finite systems.

Case I: a hypercube of n^d spins. Choosing $L = n$, and substituting the free energy of one spin in a magnetic field, we obtain

$$f_s\left(K, h, \frac{1}{n}\right) = n^{-d} \log(e^{h'} + e^{-h'}). \tag{3.9}$$

Consequently, the susceptibility per particle diverges as follows

$$\chi\left(K, 0, \frac{1}{n}\right) \sim n^d \tag{3.10}$$

for any $K > K_c$, the critical coupling. This result has also been obtained by Imry[47]) using different arguments (which do not apply to case II). In other words, χ is proportional to the number of spins: a result which may, for very low temperatures, also be obtained from a spin–spin correlation function which approaches a nonzero constant for large distances.

Case II: as above, except infinite in one direction. Instead of (3.9) we find, substituting the free energy of an infinite one-dimensional Ising chain,

$$f_s\left(K, h, \frac{1}{n}\right) = n^{-d} \log\{e^{K'}(\cosh h') + [e^{2K'}(\cosh h')^2 - 2 \sinh 2K']^{1/2}\}. \tag{3.11}$$

From this expression we find an exponentially diverging susceptibility

$$\chi\left(K, 0, \frac{1}{n}\right) \sim n^{d} \exp(2n^{d-1} K). \tag{3.12}$$

The essential assumptions leading to (3.10) and (3.12) are: (i) the behaviour of the system can be analysed in terms of the zero temperature fixed point; (ii) at this fixed point, the symmetry breaking field has a discontinuity exponent d; (iii) the temperature is a (dangerous) irrelevant variable with exponent $-(d-1)$. Therefore, the analysis such as given above seems equally applicable both to the susceptibility and to the specific heat of the Potts model at the critical temperature. Note that in the renormalization description of Nienhuis et al.[15]) of the first order transition in the Potts model the temperature couples to a symmetry breaking field: the chemical potential of vacancies.

4. The transfer matrix

4.1. *Mapping the Potts model with a field onto an extended Whitney polynomial*

The most straightforward way to construct a transfer matrix for a Potts model on a strip with linear size n is realized on basis of the degrees of freedom of a column of n q-state Potts variables. This leads to a transfer matrix with dimensionality q^n, or somewhat smaller if symmetry properties are used. Such procedures have already been described[22,23,48]) and, in conjunction with finite size scaling, have yielded accurate results for the critical exponents in the cases $q=2$ and $q=3$. However, for a systematic study of the Potts model there are some problems. In the first place, the transfer matrix becomes unpleasantly large for high q, so that only very small system sizes can be treated. Secondly, the bond percolation problem, which is contained in the behaviour linear in δq of the model[2]) with $q = 1 + \delta q$, cannot be investigated by these means. Further, for an investigation of the validity of the conjectures of den Nijs[27]) and of Nienhuis et al.[17]) and Pearson[28]), it is desirable to calculate the exponents also for nonintegral values of q.

Under these circumstances, a transfer matrix formulation for general q is very useful. For this purpose, we may map the Potts model on other models in which q enters as a continuous variable, and make use of the transfer matrix of that model. Two obvious possibilities arise: the transformation to a Whitney polynomial[49,24]) and to the ice model[50]). We have chosen the first possibility; a straightforward construction of the transfer matrix for the ice model would yield a matrix of higher order. Besides, it is possible to generalize the first method such that a magnetic field can be included[2]).

Consider a simple quadratic Potts lattice with (for the sake of simplicity) equal horizontal and vertical coupling K. The q-state Potts variables $\sigma_i = 1, \ldots, q$ (i is the site label) are further coupled to an external field by a magnetic coupling h. The reduced hamiltonian is written as

$$\mathcal{H} = K \sum_{\langle ij \rangle} \delta_{\sigma_i \sigma_j} + h \sum_{\langle i0 \rangle} \delta_{\sigma_i \sigma_0}, \tag{4.1}$$

where the first summation runs over all nearest neighbour pairs in the quadratic lattice. The second term couples each lattice site to an external or "ghost" site labelled 0, which has a fixed value, e.g. $\sigma_0 = 1$. The partition function can be written as

$$Z = \sum_{\{\sigma_i\}} \prod_{\langle ij \rangle} \exp(K\delta_{\sigma_i \sigma_j}) \prod_{\langle k0 \rangle} \exp(h\delta_{\sigma_k \sigma_0}). \tag{4.2}$$

Following the method of Kasteleyn and Fortuin[51]), it is possible to transform the hamiltonian (4.1) which contains site variables, into one which contains bond variables. We define these bond variables on the lattice $\mathcal{L}$ that connects all nearest neighbouring Potts sites. Each bond variable b_{ij} can assume two values: "absent" or "present". Likewise, we define bond variables b_{k0} on the lattice $\mathcal{L}^0$ that connects the ghost site to all sites on $\mathcal{L}$. Further, we define $u \equiv \exp(K) - 1$ and $v \equiv \exp(h) - 1$ so that

$$\exp(K\delta_{\sigma_i \sigma_j}) = 1 + u\delta_{\sigma_i \sigma_j}, \tag{4.3a}$$

$$\exp(h\delta_{\sigma_k \sigma_0}) = 1 + v\delta_{\sigma_k \sigma_0}. \tag{4.3b}$$

Substituting this into (4.2) we obtain a polynomial in u en v. Working out the two products, we obtain 2^M terms where M is the total number of lattice edges on the combined lattice $\mathcal{L}' = \mathcal{L} \cup \mathcal{L}^0$. It is possible to give a one-to-one corresponding between these terms and configurations of bond variables. Whether or not a factor $u\delta_{\sigma_i \sigma_j}$ appears in a particular term may be considered to depend on the value of the bond variable b_{ij}; if b_{ij} is "present" we take the term $u\delta_{\sigma_i \sigma_j}$, and if it is "absent" we take the term 1 from (4.3a). Similarly, we let the bond variable b_{k0} decide whether or not a factor $v\sigma_{\sigma_k \sigma_0}$ appears in the second product in eq. (4.2).

It is convenient to represent the set of bond variables b_{ij} which have the value "present" by a graph G on $\mathcal{L}$: similarly, the bond variables b_{k0} define a graph G^0 on $\mathcal{L}^0$. Expression (4.2) can now be written as

$$Z = \sum_{\{\sigma_i\}} \sum_{G \subset \mathcal{L}} \sum_{G^0 \subset \mathcal{L}^0} \prod_{\langle ij \rangle \in G} u\delta_{\sigma_i \sigma_j} \prod_{\langle k0 \rangle \in G^0} v\delta_{\sigma_k \sigma_0}. \tag{4.4}$$

The new summations are over all graphs on lattices $\mathcal{L}$ and $\mathcal{L}^0$ (i.e. over all possible configurations of bond variables on these lattices), and the products are only over those bonds appearing in the graphs G and G^0. After interchanging the summations in eq. (4.4) it is possible to execute the summation over the Potts variables $\{\sigma_i\}$ for a given G and G^0. For instance, if no bond appears in $G' \equiv G \cup G^0$, we find a term q^N where N is the number of Potts sites on $\mathscr{L}$. If one bond appears on G, we obtain an extra factor u/q so that the term is uq^{N-1}; if the bond appears on G^0 instead, we find an extra factor v/q. An additional factor u/q or v/q appears for each new bond b_{ij} that appears on G', unless the bond closes a loop on G'. In the latter case, sites i and j were already connected by some path of bonds on G', so that $\delta_{\sigma_i\sigma_j} = 1$ necessarily in all surviving terms of Z, and an additional factor u or v appears instead of u/q or v/q. Thus we find

$$Z = q^N \sum_{G \subset \mathscr{L}} \sum_{G^0 \subset \mathscr{L}^0} (u/q)^{b(G)} (v/q)^{b(G^0)} q^{l(G')}, \tag{4.5}$$

where $b(G)$ is the number of bonds present on G, and $l(G')$ is the number of independent loops on G'. Expression (4.5) may be considered as the partition function of a system of $3N$ bond variables with a hamiltonian

$$\mathscr{H}' = N \log q + b(G) \log(u/q) + b(G^0) \log(v/q) + l(G') \log(q), \tag{4.6}$$

i.e. there are chemical potentials log (u/q) and log (v/q) for the bonds on G and G^0, and there is an additional energy log (q) for each independent loop on G'. We have now transformed the Potts model into a model in which q enters as a continuously variable parameter. Further, the number of degrees of freedom of the new model is smaller than that of the old model for large q. The price that we had to pay for these advantages is the introduction of a nonlocal interaction, the last term in (4.6).

For $h = v = 0$, expression (4.5) simplifies because only one term in the summation over G^0 survives. It reduces to a Whitney polynomial on a planar lattice as described by Baxter[24]) for the Potts model in the absence of a magnetic field. The simplified expression will be referred to as "simple Whitney polynomial" in order to distinguish it from the "extended Whitney polynomial" given by eq. (4.5). The underlying hamiltonians may be considered as to describe "Whitney models". It should be kept in mind that the interactions appearing in (4.6) are real only for $u > 0$ and $v > 0$, corresponding to ferromagnetic coupling and positive field.

4.2. *The construction of the transfer matrix*

Let us imagine a rectangular Potts lattice $\mathscr{L}_m$ consisting of n horizontal rows and m vertical columns; n, the system width, will be treated as fixed. The boundary conditions in the vertical direction do not yet have to be specified; the system is chosen to be open-ended in the horizontal direction.

The extended Whitney polynomial is written as

$$Z_m = q^{nm} \sum_{G' \subset \mathscr{L}'_m} (u/q)^{b(G)} (v/q)^{b(G^0)} q^{l(G')}, \tag{4.7}$$

The nonlocal interaction that enters into this expression by the number of loops $l(G')$ poses a problems for the construction of a transfer matrix for (4.7). To clarify this, we first consider the "usual" case of local interactions, such as e.g. in the original Potts hamiltonian (4.1). In that case, $\mathscr{H}$ can be written as a sum over column–column contributions directly in terms of the Potts variables σ_i of the columns involved. The values of the indices α and β of the transfer matrix $T_{\alpha\beta}$ may conveniently be chosen to correspond to the internal degrees of freedom (i.e. spin configurations) of the respective columns. The column–column contribution to $\mathscr{H}$ is completely determined by α and β. However, in the present model the hamiltonian (4.6) cannot be written as a sum over column–column contributions in terms of the bond variables b_{ij}, and we have to construct a transfer matrix in a different way.

In order to describe what happens to the partition function when a new column $m+1$ is added to the right hand side of the lattice $\mathscr{L}_m$, we define an extended lattice

$$\left.\begin{aligned} \mathscr{L}'_{m+1} &\equiv \mathscr{L}'_m \cup \ell'_{m+1} \\ \ell'_{m+1} &= \ell_{m+1} \cup \ell^0_{m+1} \end{aligned}\right\}, \tag{4.8}$$

where (i) ℓ_{m+1} consists of n horizontal bonds between the sites of columns m and $m+1$ and – in the case of periodic boundary conditions – n vertical bonds in column $m+1$, and (ii) ℓ^0_{m+1} consists of n bonds between the ghost site and column $m+1$. Further, let g' denote the graph on ℓ'_{m+1} that describes which of the $3n$ new bonds are present. It is convenient to split g' according to $g' \equiv g \cup g^0$, where g and g^0 are graphs on ℓ_m and ℓ^0_m, respectively. From eq. (4.6) one infers the increase of energy $\Delta\mathscr{H}'$, when the new column is added:

$$\Delta\mathscr{H}' = n \log q + b(g)\log(u/q) + b(g^0)\log(v/q) + l'(G', g')\log(q), \tag{4.9}$$

where l' is the number of loops closed upon appending the new subgraph g' to G'.

The numbers $b(g)$ and $b(g^0)$ depend on the newly appended subgraph only, but l' requires some information about the graph G' on $\mathscr{L}'_m$. This information comprises: firstly, which of the sites of the rightmost column of $\mathscr{L}'_m$ are connected to the ghost site via some path of bonds in G'; secondly, how are the remaining sites on the rightmost column interconnected via some path of bonds in G. This information is referred to by the word "connectivity". Fig. 1 gives some examples for a strip of width 4, and illustrates the use of these

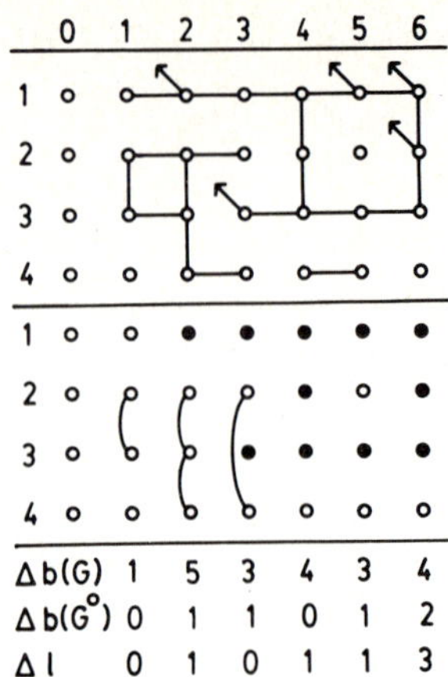

Fig. 1. Illustration of the definition of the general connectivities on a horizontal strip of width 4. The upper part of the figure shows a possible lattice bond configuration; 45° arrows stand for bonds to the ghost site. The lower part of the picture gives the connectivity at column $i (i = 1, 2, \ldots)$, where all bonds up to and including this column are taken into account; the lattice is considered to be built up from the left to the right. Full dots represent sites connected to the ghost site (therefore they are always mutually connected). The numbers below the connectivities are the increases of the numbers of bonds and loops as defined in the text.

connectivities for the determination of l'. It is shown in Appendix I that the number of possible connectivities d_n on a n-site column is asymptotically proportional to 5^n. Each connectivity may be represented by an integer α such that $1 \leqslant \alpha \leqslant d_n$. A particular correspondence between the connectivities and these integers is also given in Appendix I. We will make use of this numbering implicitly when we refer to a "connectivity α" by which we mean the connectivity which corresponds to α. The connectivity α of the rightmost column of $\mathscr{L}'_m$ is determined by the graph G' on $\mathscr{L}'_m$; $\alpha = \mu(G')$. Contributions to the partition function Z_m from all graphs G' on $\mathscr{L}'_m$ can now be sorted according to the state of connectivity β of column m (the rightmost one), leading to the restricted partition function $Z_m(\beta)$, which we consider as a vector with d_n components:

$$Z_m(\beta) \equiv \sum_{G'|\beta} (u/q)^{b(G)} (v/q)^{b(G^0)} q^{l(G')} . \tag{4.10}$$

The summation is only over those graphs G' that produce the specified connectivity β on column m; $\mu(G') = \beta$.

We proceed to investigate what happens when the interactions with and within the $(m+1)$th column are taken into account. We have observed that the increase of energy, as given by eq. (4.9), is determined by g' and the old connectivity β of column m. Also the new connectivity γ of the rightmost column $m+1$ depends only on β and g'. Thus we may write

$$\mu(G' \cup g') \equiv \mu'(\beta, g') \tag{4.11}$$

and

$$\mathcal{H}'(G' \cup g') = \mathcal{H}'(G') + \Delta\mathcal{H}'(\beta, g'), \tag{4.12}$$

with $\beta = \mu(G')$ in these equations. These relations allow us to write down the elements of Z_{m+1} as a function of those of Z_m; using $H' \equiv G' \cup g'$:

$$\begin{aligned} Z_{m+1}(\gamma) &\equiv \sum_{H'|\gamma} \exp(\mathcal{H}'_{m+1}(H') \\ &= \sum_{\beta} \sum_{g'|\gamma,\beta} \sum_{G'|\beta} \exp(\Delta\mathcal{H}'(\beta, g') \exp(\mathcal{H}'(G')). \end{aligned} \tag{4.13}$$

The second summation on the right hand side of eq. (4.13) is only over those graphs g' that, given a connectivity β on column m of $\mathcal{L}'_{m'}$ lead to a connectivity γ on column $m+1$ of $\mathcal{L}_{m+1}$. Eq. (4.13) may also be written as

$$Z_{m+1}(\gamma) = \sum_{\beta} T(\gamma, \beta) Z_m(\beta), \tag{4.14}$$

where

$$T(\gamma, \beta) \equiv \sum_{g'|\gamma,\beta} \exp(\Delta\mathcal{H}'(\beta, g')) \tag{4.15}$$

is the transfer matrix. From this matrix the dimensionless free energy f and correlation length ξ may be obtained as usual (e.g. Domb[52])). In terms of the dominant eigen values λ_1 and λ_2 one has

$$f = \frac{1}{nm} \log Z_m = \frac{1}{n} \log \lambda_1, \tag{4.16}$$

$$\xi^{-1} = \log(\lambda_1/\lambda_2) \tag{4.17}$$

In the remainder of this section we consider the special case of zero magnetic field: $\mathrm{h} = \mathrm{v} = 0$. We may then obtain the free energy from the transfer matrix of the simple Whitney polynomial. This transfer matrix is also defined by (4.15), however, with connectivities β and γ restricted to non-ghost connectivities: those without sites connected to the ghost site. To illustrate this, we adopt a numbering of connectivities (e.g. as defined in the appendix), where the non-ghost connectivities precede the others. The transfer matrix $\mathbf{T}$ then assumes the following block form

$$\mathbf{T} = \begin{bmatrix} \mathbf{T}^{11} & \mathbf{T}^{12} \\ \mathbf{O} & \mathbf{T}^{22} \end{bmatrix}$$

and $\mathbf{T}^{11}$ is the transfer matrix of the simple Whitney polynomial.

Clearly, all eigenvalues of $\mathbf{T}^{11}$ and $\mathbf{T}^{22}$ are also eigenvalues of $\mathbf{T}$. The free energy is given by (4.16) in terms of the largest eigenvalue of $\mathbf{T}^{11}$. However, it turns out that the second largest eigenvalue of $\mathbf{T}$ is an eigenvalue of $\mathbf{T}^{22}$ and not of $\mathbf{T}^{11}$. As a consequence, the correlation lengths of the simple and extended Whitney polynomials differ, the latter being the larger.

Comparing the Potts model in the original discrete q-state representation[22]) for $q = 2, 3$ and 4 with the *extended* Whitney polynomial, we found, for small system sizes, the corresponding correlation lengths to be identical. Qualitatively, this can be understood as follows (for integral $q \geqslant 2$). Only states of the original Potts model in which $\sigma_i = \sigma_j$ contribute to states of the simple or extended Whitney systems in which sites i and j are connected. One therefore expects that the transfer matrix $\mathbf{T}^{11}$ still contains information concerning the decay of conditional probabilities, such as "what is the probability that $\sigma_i = \sigma_j$ in column m, given that $\sigma_k = \sigma_l$ in column 1." From these probabilities, correlations of the energy–energy type in particular may be obtained. However, decay of spin–spin correlations is not contained in $\mathbf{T}^{11}$: one has projected out conditional probabilities such as: "what is the probability that $\sigma_i = \sigma_j$ where sites i and j belong to columns 1 and m respectively." On the other hand, spin–spin correlations are expected to be contained in the full matrix $\mathbf{T}$: since we have zero field, site j in column m is connected to the ghost site only if: (i) it is connected to site i in column 1; (ii) site i is connected to the ghost site as a consequence of the choice of the boundary conditions in column 1. Finally, we note the following property of the transfer matrix of the original Potts model. Mittag and Stephen[53]) have shown that duality is satisfied only by those eigenvalues which correspond to eigenvectors symmetric under permutation of the q states. These are precisely the eigenvalues which govern the decay of energy–energy correlations. As expected, we did indeed find that the temperature derivative of the correlation length of the simple Whitney polynomial vanishes at criticality, in agreement with duality.

4.3. *Factorization of the transfer matrix*

It is possible to generate the non-symmetric matrix $\mathbf{T}$ directly from the expression (4.15) and diagonalize it by means of a computer. However, there are more efficient ways. One possible way to simplify the calculation is by utilizing symmetry properties of $\mathcal{H}'$ (and hence of $\mathbf{T}$). We can use the translational symmetry in the vertical dimension (if periodic boundaries have been chosen) and the inversion symmetry (which interchanges row labels i and $n + i - 1$). For the calculation of λ_1 it is sufficient to restrict oneself to diagonalisation in the subspace which is completely symmetric with respect to these operations. This follows from the fact that the elements of **T** are non-negative (for u and v non-negative) by application of the Perron–Frobenius theorem. Thus the dimensionality of the eigenvalue problem is reduced by a factor of approximately $2n$. However, a more effective way to reduce computer requirements is the use of sparse matrices. One may use helical boundary conditions and define a sparse matrix corresponding to adding lattice sites one at a time. Alternatively, the matrix **T** introduced above may be written as a product of n sparse matrices in the case of periodic boundary conditions. For the latter boundary conditions finite size estimates of critical exponents converge more rapidly as is seen in section 7.2.

To factorize the transfer matrix, we separate the graph g' into three graphs

$$g' = g_h \cup g_v \cup g_0,$$

where g_h and g_v consist of the horizontal and the vertical bonds in g' respectively; g_0 contains the bonds to the ghost site. Expression (4.15) can now be written as

$$T(\gamma, \beta) = \sum_{(g_h \cup g_v \cup g_0)|\gamma,\beta} \exp(\Delta\mathscr{H}'(\beta, g_h \cup g_v \cup g_0)). \tag{4.18}$$

In order to factorize the exponential function, $\Delta\mathscr{H}'$ is separated into three parts $\Delta\mathscr{H}'_h$, $\Delta\mathscr{H}'_v$ and $\Delta\mathscr{H}'_0$. Imagine that first the new column of n disconnected sites is added to $\mathscr{L}'$, followed by the horizontal bonds g_h. These bonds do not close any loops, so that

$$\Delta\mathscr{H}'_h = n \log(q) + b(g_h) \log(u/q)$$

does not depend on β, the connectivity of the mth column. The present connectivity α of the $(m+1)$th column depends only on β and $g_h : \alpha = \alpha(\beta, g_h)$. Subsequently, the graph g_v is added, leading to an increase of energy

$$\Delta\mathscr{H}'_v = b(g_v) \log(u/q) + l_v(\alpha, g_v) \log(q),$$

where l_v is the number of loop closed by the addition of g_v on $\mathscr{L}_{m+1}$. Let the connectivity of column $m+1$ now be given by δ; it depends only on α and g_v or $\delta = \delta(\alpha, g_v)$. Finally, g_0 is added to the system, leading to a new connectivity $\gamma(\delta, g_0)$ on column $m+1$ and an energy change

$$\Delta\mathscr{H}'_0 = b(g_0) \log(v/q) + l_0(\delta, g_0) \log(q),$$

where l_0 is the number of loops closed by g_0. Since

$$\Delta\mathscr{H}' = \Delta\mathscr{H}'_h + \Delta\mathscr{H}'_v + \Delta\mathscr{H}'_0,$$

eq. (4.18) can be written as

$$T(\gamma, \beta) = \sum_{\alpha} \sum_{\delta} \sum_{g_0|\gamma,\delta} \exp(\Delta\mathcal{H}_0'(\delta, g_0)) \sum_{g_v|\delta,\alpha} \exp(\Delta\mathcal{H}_v'(\alpha, g_v))$$
$$\sum_{g_h|\alpha,\beta} \exp(\mathcal{H}_h'(\beta, g_h))$$

or in matrix notation

$$\mathbf{T} = \mathbf{T}_0 \cdot \mathbf{T}_v \cdot \mathbf{T}_h. \tag{4.19}$$

Along similar lines a further decomposition is realized by splitting each of the graphs g_h, g_v and g_0 into n elementary graphs consisting of one bond variable. Thus we obtain

$$\mathbf{T}_x = \mathbf{T}_{x,n} \cdot \mathbf{T}_{x,n-1} \cdot \ldots \cdot \mathbf{T}_{x,1}, \tag{4.20}$$

where x = 0, v or h. Each of the $3n$ matrices into which $\mathbf{T}$ has now been factorized has a maximum of two elements per column; this fact greatly reduces memory and computer time requirements. Furthermore, for programming convenience $\mathbf{T}_x$ may be written as the nth power of a single matrix. To this purpose, we define a translation operator t on the connectivities by requiring $t\alpha$ to be identical to α apart of a cyclic permutation of the row indices $i \to i+1$.

We further define a permutation matrix $P_{\alpha\beta}$ of which the elements equal 1 if $\alpha = t\beta$ and 0 otherwise. It follows that the α,β element of $\mathbf{P}^i$ is equal to 1 if $\alpha = t^i\beta$ and zero otherwise. The matrix $\mathbf{P}$ enables us to relate the $\mathbf{T}_{x,i}$ to $\mathbf{T}_{x,1}$:

$$\mathbf{T}_{x,i} = \mathbf{P}^{1-i} \cdot \mathbf{T}_{x,1} \cdot \mathbf{P}^{i-1}.$$

Using $\mathbf{P}^n = \mathbb{1}$ expression (4.20) reduces to $\mathbf{T}_x = [\mathbf{P} \cdot \mathbf{T}_{x,1}]^n$.

The matrices $\mathbf{P} \cdot \mathbf{T}_{x,1}$, have, just as $\mathbf{T}_{x,i}$ at most two nonzero elements per column. Thus the information needed to be in memory comprises: which elements of each column are nonzero and what is their value. Multiplication of a vector by the transfer matrix (which, as we will see, is part of the calculation of λ_1) would normally require a number of multiplications proportional to d_n^2; with the present formulation, this number is proportional to nd_n. This exponential reduction is to be compared with the reduction by $4n^2$ employing symmetry as mentioned above. These operations could be performed up to $n = 9$ with the IBM 370/158 computer of the Technische Hogeschool. However, for the special case h = v = 0, all graphs with one or more sites connected to the ghost site carry zero weight. It is therefore possible to perform the diagonalization of the transfer matrix in the corresponding subspace of connectivities. As shown in Appendix I, the number c_n of these "non-ghost" connectivities is considerably smaller than d_n; calculations for zero field could be performed up to $n = 11$.

4.4. Calculation of the dominant eigenvalues of the transfer matrix

The transfer matrix being non-symmetric, the conjugate gradient method employed in ref. 22 is not applicable for the calculation of its eigenvalues in the present case. In this section we describe an alternative method which we devised: a hybrid combination of the Hessenberg method (ref. 54, Chapter 6, §26–29) and direct iteration, sometimes with a shift of origin (ref. 54, Chapter 9, §2–4).

Consider a non-symmetric, real matrix $\mathbf{T}$, assumed to be diagonalisable with eigenvalues $|\lambda_1| \geq |\lambda_2| \geq \ldots$. A few of the dominant eigenvalues are numerically calculated from the sequence of vectors $\boldsymbol{a}_0, \boldsymbol{a}_1, \ldots$ with $\boldsymbol{a}_0$ rather arbitrary and

$$\boldsymbol{a}_{i+1} = c(\mathbf{T} - \mu_i)\boldsymbol{a}_i, \tag{4.21}$$

where c is a normalization constant such that $|c\lambda_1|$ is of order unity; μ_i is determined at each step of the calculation (as explained later) and usually equals zero. We shall say that $\mathbf{T}$ is of effective dimensionality p at stage m if p is the smallest integer such that the $p+1$ vectors $\boldsymbol{a}_{m+i}$, $i = 0, 1, \ldots, p$ are approximately linearly dependent, i.e. $\boldsymbol{a}_m$ can be written as

$$\boldsymbol{a}_m = \sum_{i=1}^{p} \alpha_i \boldsymbol{a}_{m+i} + \boldsymbol{r}$$

with α_i scalars and $\boldsymbol{r}$ a vector of (suitably defined) relative norm smaller than some fixed number ϵ. In the method of direct iteration, one calculates iterates $\boldsymbol{a}_m$ until $p = 1$, at which stage the dominant eigenvalue can be obtained with an error of order ϵ. The value of m for this method is inversely proportional to log $|\lambda_1/\lambda_2|$.

Let us now, for the sake of the argument, assume that $\mathbf{T}$ has a spectrum with closely spaced dominant eigenvalues separated by a gap from the rest of the spectrum:

$$|\lambda_1| \geq |\lambda_2| \geq \cdots \geq |\lambda_p| \gg |\lambda_{p+1}| \geq \cdots. \tag{4.22}$$

Clearly, in this case direct iteration converges very poorly. However, rapidly – after of the order of $1/\log|\lambda_1/\lambda_{p+1}|$ iterations – the effective dimension of $\mathbf{T}$ drops to p. At this point, $\mathbf{T}$ can be represented by an effective matrix $\hat{\mathbf{T}}$ of order p; p being sufficiently small for the application of standard eigenvalue routines. The transfer matrix does not have a spectrum satisfying relation (4.22). Nevertheless, the method described below was found to perform very efficiently: compared with direct iteration, the number of iterations was reduced by a factor three to ten.

Since the matrix $\mathbf{T}$ may be of very large order, the number of vectors $\boldsymbol{a}_i$ to be stored simultaneously must be kept as small as possible. In principle, our method uses p vectors; however, it is not necessary to keep them all in memory. This reduction of storage requirements is accomplished by introducing a projector $\mathbf{P}$ onto a p-dimensional subspace, with $\mathbf{P} \cdot \boldsymbol{x}$ non-vanishing for eigenvectors $\boldsymbol{x}$ corresponding to the part of the spectrum to be calculated. We chose $\mathbf{P}$ such that $\mathbf{P} \cdot \boldsymbol{a}_i$ was given by the mean of the components of $\boldsymbol{a}_i$, together with those $p-1$ components, which on physical grounds, could be expected to be among the largest. The initial choice of p is determined by an arbitrary upper bound; we usually took $p \lesssim 10$. In the process of iteratively calculating the eigenvalues, p is gradually reduced, as described below; eventually $p \approx 3$ in most cases.

Define $y_i \equiv \mathbf{P} \cdot \boldsymbol{a}_i$. The assumption that $\mathbf{T}$ is of effective dimension p implies that $\mathbf{T}$ can be represented by a matrix $\hat{\mathbf{T}}$ such that $y_{i+1} = \hat{\mathbf{T}} \cdot y_i$. From the $p+1$ vectors y_{m+i}, $i = 0, 1, \ldots, p$, a matrix $\mathbf{H}$ equivalent to $\hat{\mathbf{T}}$ can be constructed using the Hessenberg-method. This is done as follows. Choose a particular set of independent vectors $\boldsymbol{x}_1, \ldots, \boldsymbol{x}_p$ (see ref. 54, Chapter 6, §29 for details). Set $\boldsymbol{b}_1 = y_m$ and successively construct $\boldsymbol{b}_2, \boldsymbol{b}_3, \ldots, \boldsymbol{b}_{p+1}$ such that

$$k_{r+1}\boldsymbol{b}_{r+1} = \hat{\mathbf{T}} \cdot \boldsymbol{b}_r - \sum_{i=1}^{r} h_{ir}\boldsymbol{b}_i, \tag{4.23}$$

where k_{r+1} is a scalar chosen so as to normalize $\boldsymbol{b}_{r+1}$, and $h_{1r}, \ldots, h_{rr}$ are determined (note that expression (27.3) in ref. 54, Chapter 6 is incorrect) such that $\boldsymbol{b}_{r+1}$ is orthogonal to $\boldsymbol{x}_1, \ldots, \boldsymbol{x}_r$. The vector $\boldsymbol{b}_{p+1}$, being orthogonal to p independent vectors, vanishes; and eq. (4.23) for $r = 1, \ldots, p$ can be combined into

$$\hat{\mathbf{T}} \cdot \mathbf{B} = \mathbf{B} \cdot \mathbf{H} \tag{4.24}$$

with $\mathbf{B} = [b_{ij}]$ where b_{ij} is component i of $\boldsymbol{b}_j$ and

$$\mathbf{H} = \begin{bmatrix} h_{11} & h_{12} & \ldots & h_{1p} \\ k_2 & h_{22} & \ldots & h_{2p} \\ 0 & k_3 & \ldots & h_{3p} \\ \ldots & \ldots & \ldots & \ldots \\ \ldots & \ldots & \ldots & \ldots \\ 0 & 0 & \ldots k_p & h_{pp} \end{bmatrix}. \tag{4.25}$$

The matrix $\mathbf{H}$ has the upper Hessenberg form and is equivalent to $\hat{\mathbf{T}}$, as it follows from eq. (4.24) if $\mathbf{B}$ is non-singular. At first sight, it seems that in order to calculate $\hat{\mathbf{T}} \cdot \boldsymbol{b}_r$ in eq. (4.23) the matrix $\hat{\mathbf{T}}$ has to be available explicitly, while $\hat{\mathbf{T}}$ is given implicitly only by $y_{i+1} = \hat{\mathbf{T}} \cdot y_i$. However, since $\boldsymbol{b}_r$ can be expressed as a linear combination of the vectors $y_m, \ldots, y_{m+r-1}$, $\hat{\mathbf{T}} \cdot \boldsymbol{b}_r$ follows immediately.

Note that the matrix $\mathbf{H}$, as defined above, depends both on m, the stage of the iteration process, and on p, the effective dimension of $\mathbf{T}$ at that point. We therefore write $\mathbf{H}_m(p)$. It can so be arranged that $\mathbf{H}_m(p-1)$ is the upper left hand diagonal block of order $p-1$ of $\mathbf{H}_m(p)$. The fact that the k_i in the Hessenberg matrix (4.25) are constants chosen so as to normalize the $\boldsymbol{b}_i$, immediately supplies us with a working definition of the effective dimensionality

$$p = \min\{q| \quad |k_2 k_3 \dots k_q| < \epsilon\}. \tag{4.26}$$

Consequently, $\mathbf{H}_m(p)$ is uniquely defined for fixed ϵ, and p is a function of m.

So far we have discussed two ingredients of our algorithm: direct iteration to reduce the effective dimension of $\mathbf{T}$, and the Hessenberg method to find the eigenvalues of the effective matrix $\hat{\mathbf{T}}$. We now turn to the role played by the shift of origin, i.e. $\mu_i \neq 0$ in eq. (4.21), which was introduced to remedy the following problem.

If ϵ was set to the machine accuracy (10^{-16} for IBM double precision arithmetic) we observed that the dominant eigenvalues of $\mathbf{H}_m(p)$ did not converge. Rather, after an initial rapid convergence, numerical instability set in, affecting first the lower part of the spectrum and finally also the dominant part. The minimum value of ϵ which guarantees smooth convergence occurred in the range 10^{-10} to 10^{-12}. Consequently, in the final algorithm, whenever p has to be lowered by one according to the definition of the effective dimension eq. (4.26), an eigenvalue λ of $\mathbf{T}$ still present in $\mathbf{H}_m(p)$ is suppressed in $\mathbf{H}_{m+1}(p-1)$. To reduce the weight of the corresponding right eigenvector in $\boldsymbol{a}_{m+1}$, the latter is calculated using $\mu_m = \lambda$. We note that the best estimate of λ is usually obtained not from $\mathbf{H}_m(p)$ but from some previous Hessenberg matrix, due to the numerical instability mentioned above.

Finally we note that in order to obtain an accurate estimate of the second dominant eigenvalue of $\mathbf{T}$, special precautions have to be taken such that the relative weight of the corresponding right eigenvector in $\boldsymbol{a}_m$ does not become too small as a consequence of the direct iteration. This applies of course only to those cases in which the second dominant eigenvalue cannot be obtained by imposing symmetry properties on the initial vector $\boldsymbol{a}_0$.

5. Analysis of data on finite systems

On the basis of finite-size scaling as discussed in section 2, we generally expect that the free energy, the correlation length and their derivatives of finite systems at their critical coupling can be expanded as

$$P_n = C_0 + \sum_{i=1}^{\infty} C_i n^{t_i}, \tag{5.1}$$

where n is the linear system size and t_1 is the leading critical exponent. The constant C_0, corresponding to an exponent $t_0 = 0$, and possibly originating from a regular contribution in the scaling law of the free energy (section 2), has been taken apart for convenience. The present problem is, given a set of finite-size data points $P_1, P_2, \ldots, P_n$, to estimate t_1, and to do this as accurately as possible. Several ways to attack this problem will be described below.

In many cases, the quantity P_n diverges when $n \to \infty$. In that case, the first term in the sum grows arbitrarily large in comparison with the rest of (5.1) when n increases. Neglecting the other terms we have

$$P_n \approx C_1 n^{t_1}. \tag{5.2}$$

Substitution of two consecutive data points for $n = i$ and $n = i + 1$, respectively into eq. (5.2) gives approximate solutions for C_1 and t_1, which we will denote by $C_1(i)$ and $t_1(i)$. Thus

$$t_1(i) = \log\left(\frac{P_i}{P_{i+1}}\right) \Big/ \log\left(\frac{i}{i+1}\right).$$

This simple procedure referred to as two-point fit. In order to account for a possible regular term in the scaling relation, we may attempt to fit the P_n by

$$P_n \approx C_0 + C_1 n^{t_1}. \tag{5.3}$$

Clearly, in the case $t_1 < 0$ a fit of this type is imperative because a two-point fit would yield the wrong exponent $t_1(n) \to 0$ for $n \to \infty$. But also in the case $t_1 > 0$, we may expect a better (i.e. more rapidly converging) fit if a nonzero C_0 is present. At present, we have three unknowns in (5.3) so that three consecutive data points for $n = i$, $n = i + 1$ and $n = i + 2$ are needed to solve, if possible, for $t_1(i)$, $C_1(i)$ and $C_0(i)$. The solution is performed numerically and referred to as three-point fit.

It is useful to investigate the influence of less singular contributions on the $t_1(n)$. Suppose we have

$$P_n = C_1 n^{t_1} + C_2 n^{t_2} + \ldots .$$

The two-point fit gives

$$t_1(n) = t_1 + \log\left(\frac{1 + (C_2/C_1) n^{t_2 - t_1} + \ldots}{1 + (C_2/C_1)(n+1)^{t_2 - t_1} + \ldots}\right) \Big/ \log\left(\frac{n}{n+1}\right).$$

For large n, this reduces to

$$t_1(n) = t_1 + (t_2 - t_1)(C_2/C_1)n^{t_2-t_1}. \tag{5.4}$$

This demonstrates how two-point fits may be expected to converge with increasing n. Similarly, we investigate the influence of a less singular contribution $C_2 n^{t_2}$ on the three-point fit:

$$P_n = C_0 + C_1 n^{t_1} + C_2 n^{t_2} + \dots. \tag{5.5}$$

Denote by Δ the finite difference operator, i.e.

$$\Delta f_n = f_{n+1} - f_n,$$

for any function f_n. Substituting (5.5) in

$$\Delta P_n = \frac{dP_n}{dn} + \frac{1}{2}\frac{d^2P_n}{dn^2} + \dots, \tag{5.6}$$

we find

$$\Delta P_n = \begin{cases} c_1 t_1 n^{t_1-1} + c_2 t_2 n^{t_2-1} + \dots & \text{if } t_2 > t_1 - 1 \\ c_1 t_1 n^{t_1-1} + \frac{1}{2}c_1 t_1(t_1 - 1)n^{t_1-2} + \dots & \text{if } t_2 < t_1 - 1 \end{cases} \tag{5.7}$$

Applying the two point fit to ΔP_n one again obtains an effective exponent of the form (5.4), where the exponent of the correction term is the larger of $t_2 - t_1$ and -1. The correction term with exponent -1 is a consequence of the approximation (5.6). It can be suppressed by calculating $t_1(n)$ by numerically solving t_1 in

$$\frac{\Delta P_n}{\Delta P_{n-1}} = \frac{\Delta n^{t_1}}{\Delta (n-1)^{t_1}}. \tag{5.8}$$

Thus also three-point fits converge with $n^{t_2-t_1}$ when $n \to \infty$.

The considerations above suggest that even more rapidly converging results can be obtained by extrapolating the $t(n)$ using a three-point fit. These procedures are referred to by iterated two- and three-points fits.

A different approach relies on the observation that the results $t_1(n, q)$ of two- and three-point fits for the q-state Potts model behave remarkably similar for $q < 3$. This suggests a decomposition of the $t_1(n, q)$ into functions of n and q:

$$t_1(n, q) \approx t(q) + x(n)y(q).$$

Since $t(2)$ is known exactly, and $t_1(n, 2)$ are numerically calculated, the $x(n)$ can be obtained, using e.g. the normalisation $y(2) = 1$. Thus we find,

$$t(q) \approx t_1(n, q) + \{t_1(n, q) - t_1(n-1, q)\}/\{1 - x(n-1)/x(n)\}. \tag{5.9}$$

These extrapolations are referred to as two- and three-point similarity fits, depending on the type of the $t_1(n, q)$. Independent of the observation that the $t_1(n, q)$ behave rather similar, similarity fits may be expected to work well if the difference $t_1 - t_2$ is independent of q. For sufficiently high n, when the terms $C_i n^{t_i}$ are negligible for $i > 2$, the similarity fit then gives the correct answer. In this context, it is noteworthy that approximate RG calculations of Burkhardt[14]) and of Nienhuis et al.[55,17]) for the second leading temperature and magnetic exponents indicate that the differences with the leading ones do not depend strongly on q for $q < 4$.

In order to account directly for a next exponent t_2 it is, of course, in principle possible to fit the original finite size data by

$$P_n \approx C_1 n^{t_1} + C_2 n^{t_2}. \tag{5.10}$$

Solutions (if any) can be obtained by substituting four consecutive P_i values in eq. (5.10). These fits proved to be rather fickle in our hands, and the results did not seem to converge faster than the iterated fits. It is, however, possible to take approximate values of the exponent t_2 from e.g. refs. 14 and 17 so that this problem reduces to an ordinary three-point fit after multiplication of the P_n by n^{-t_2}.

Application of these different types of extrapolation may be expected to yield, apart of results for the critical exponents, also an indication of their accuracy (by inspection of the apparent convergence and mutual differences). A further consistency check is possible because we present results based on derivatives of the free energy as well as on those of the correlation length.

6. Numerical results

6.1. *The free energy*

Critical free energies per particle $f(n, q)$ of $n \times \infty$, simple quadratic q-state Potts lattices with periodic boundaries in the finite direction were calculated as described in section 4. Results for $n = 2$–11 and several q are shown in table AII.1. Values for $n = \infty$, which are also shown, were obtained by evaluating exact expressions of Baxter[24]). The data for finite n and $q = 2$ were found to agree accurately with Onsagers exact result as given e.g. by Domb[52]) for $n \times \infty$ Ising lattices.

On basis of the finite-size scaling arguments given in section 2 one expects, at continuous transitions, that $f(n, q)$ converges with increasing n according to

$$f(n, q) \approx f(\infty, q) + an^{-x(q)}, \tag{6.1}$$

so that the three-point fit (section 5) seems appropriate. Indeed, these fits reproduce $f(\infty, q)$ accurately, and yield an exponent x close to 2 for $q<4$. The significance of this result lies in the independent confirmation of the scaling relation for the free energy as given in section 2, and it indicates that the term proportional to n^{-1} is absent for the present class of systems which have periodic boundaries. Fig. 2 shows the exponents x as obtained from the free energies at $n=9$, 10 and 11. They differ only slightly from 2 for $q<4$. This difference may be attributed to a higher order term $bn^{-x'}$ in the free energy. A further analysis of the free energies involves the quantity

$$f'(n, q) \equiv n^2\{f(n, q) - f(\infty, q)\}$$

with $f(\infty, q)$ taken from the exact formulas of Baxter[24]). If the first exponent in the $1/n$ expansion of f is indeed precisely 2, as suggested by the three-point fits, we may search for the next one by three-point fitting

$$f'(n, q) \approx f'(\infty, q) + a'n^{2+x'(q)}.$$

These fits were found to exhibit good apparent convergence to $x' \approx -4$ for $q<4$, indicating that

$$f(n, q) = f(\infty, q) + a_1(q)n^{-2} + a_2(q)n^{-4} + \dots$$

for $q<4$. Results for x' using $n=9$, 10 and 11 are shown in fig. 3.

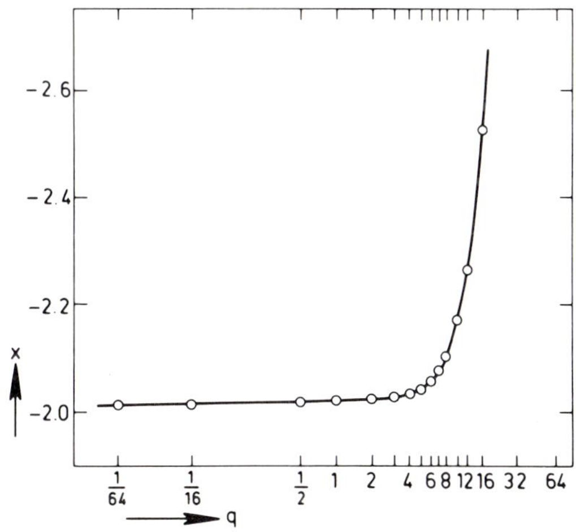

Fig. 2. Effective exponent $x(9, q)$ of the size-dependent part of the free energy (x as defined in eq. 6.1) for q-state Potts models (q on a logarithmic scale). The effective exponent was calculated by three-point fits to the free energies at $n=9$, 10 and 11 as described in section 5. The dependence of the effective exponents $x(n, q)$ on n suggests that $x(\infty, q)$ is very close to -2 for $q<4$. The curve is drawn for visual aid only. The data point at $q=1$ was obtained by averaging between $q=0.95$ and $q=1.05$.

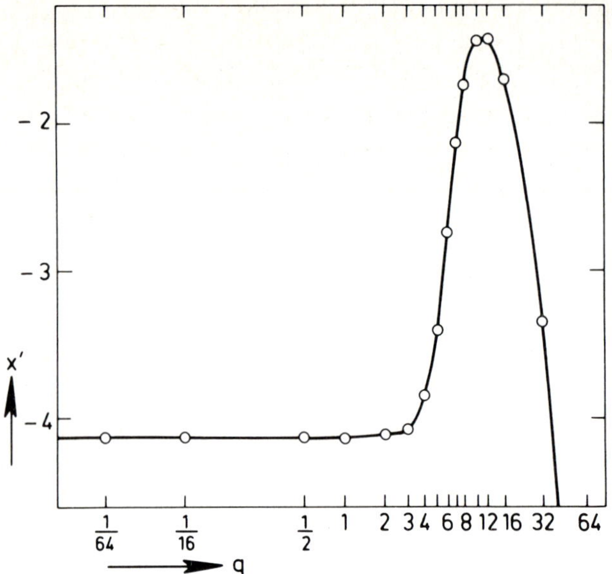

Fig. 3. Effective "correction to scaling exponent" $x'(9, q)$ of the free energy under the assumption that leading exponent is -2. Results for q-state Potts models are shown versus q, with q on a logarithmic scale. The effective correction exponent was obtained by three-point fits to the quantities $f'(q, n)$ as defined in section 6, for $n=9$, 10 and 11. The data point at $q=1$ was averaged between $q=0.95$ and 1.05. The curve is shown for visual aid only. The dependence of $x'(n, q)$ on n suggests that $x'(\infty, q) \approx -4$ for $q<4$. The results shown in this picture and fig. 2 also indicate that the scaling behaviour of the free energy becomes different in the range of $q>4$.

The data in figs. 2 and 3 illustrate, apart from the convergence for $q<4$, also the fact that the scaling behaviour becomes completely different for $q>4$. The data points in fig. 2 increase rapidly with large q; but the results (the effective exponents) of these three-point fits are also found to increase rapidly with the system size (up to $n=11$). The $f(n, q)$ appears to exhibit exponential type convergence with n, in agreement with the behaviour as described in section 3 for systems with a first-order transition described by a zero temperature discontinuity fixed point. It was argued that, for $q>q_c=4$ and sufficiently high n, the free energy may behave like

$$f(n, q) = f(\infty, q) + an^{-d} \exp(-n/\xi), \tag{6.2}$$

and that the correlation length ξ (for q close to q_c) obeys

$$\xi = b \exp(\pi^2/\sqrt{q-q_c}).$$

Here ξ stands for the real physical correlation length of the infinite system; elsewhere we used definition (4.17).

Thus defining

$$\lambda(n, q) = \log(\log(f'(n+1, q)/f'(n, q))),$$

we expect, for sufficiently high n, that

$$\lambda(n, q) = \log(b) + \pi^2/\sqrt{q - q_c}.$$

For q not too far from q_c we may neglect $\log(b)$, so that

$$\lambda^{-2}(n, q) \approx \pi^{-4}(q - q_c). \tag{6.3}$$

This equation is graphically represented in fig. 4 for three values of n, together with data points extrapolated to $n = \infty$, assuming asymptotically exponential behaviour of $f'(n+1, q)/f'(n, q)$ with n. The straight line represents eq. (6.3) with $q_c = 4$. The extrapolated data points agree qualitatively with this line. That is all that might be expected; for $q \gg q_c$ the formula for ξ is no longer valid, and $\log(b)$ is no longer negligible. On the other hand, for $q \approx q_c$ and not too high n, we do not expect to renormalize beyond the neighbourhood of the

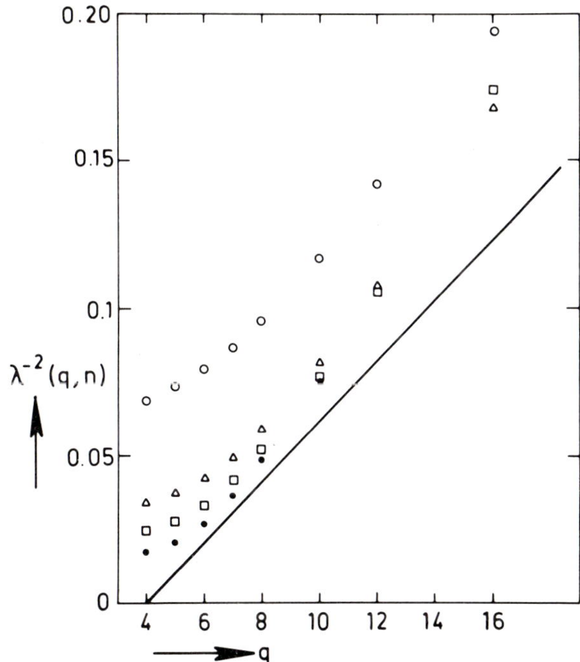

Fig. 4. The quantity $\lambda^{-2}(n, q)$ (defined above eq. 6.3) versus q for $n = 8$, 9 and 10 (shown by ○, △ and □, respectively). For those values of q for which these data points could be fitted by $a + bc^{-n}$, also the extrapolated value a is shown by ●. The straight line represents $\lambda^{-2} = \pi^{-4}(q - q_c)$ with $q_c = 4$; the theory predicts deviations from this behaviour for $q \approx q_c$ as well as for $q \gg q_c$, as discussed in the text

fixed point at q_c as described in section 3, so that eq. (6.2) is no longer valid. In fact we observe both expected deviations in fig. 4; the data presented agree qualitatively with the picture sketched above.

6.2. *The specific heat*

The duality relation[1,24,53]) for an infinitely long Potts strip (N particles) can be expressed in terms of $\hat{u} \equiv (e^k - 1)/\sqrt{q}$, as

$$\hat{u}^{-N} Z_\infty(q, \hat{u}) \leftrightarrow \hat{u}^N Z_\infty(q, \hat{u}^{-1}), \tag{6.4}$$

so that the dimensionless free energy per particle must obey

$$-\log(\hat{u}) + f(q, \hat{u}) = -\log(\hat{u}^{-1}) + f(q, \hat{u}^{-1}).$$

Differentiation with respect to $\hat{u}$ at criticality ($\hat{u} = 1$) yields

$$\left(\frac{\partial f(q, \hat{u})}{\partial \hat{u}}\right)_{\hat{u}=1} = 1,$$

and the critical dimensionless energy per particle follows

$$\left(\frac{\partial f}{\partial K}\right)_{\text{crit}} = \left(\frac{\partial f}{\partial \hat{u}} \frac{\mathrm{d}\hat{u}}{\mathrm{d}K}\right)_{\text{crit}} = 1 + q^{-1/2},$$

which is independent of the strip width n. Thus the critical energy does not depend on n, and no critical exponent can be obtained in this case. This conclusion does not apply to systems that are finite in two directions, because in that case the eq. (6.4) has to be modified.

Thus we have numerically differentiated the free energy in zero field twice to obtain the critical specific heat per particle

$$C(n, q) = K^2 \frac{\partial^2 f}{\partial K^2}.$$

Typical differentiation intervals are $\Delta K/K = 0.01$ to 0.0001; the free energy was calculated at several (up to 7) couplings in such an interval, and these free energy values were fitted by a power series in K. At large intervals, the inaccuracy of this method is dominated by higher-order terms, and at small intervals by numerical rounding errors. Numerical data are presented in Table AII.2 for several values of q. As a first step to analyse these data, we have applied three-point fits $C(n, q) \approx C_0 + C_1 n^{t_1}$ (see section 5); results for the temperature exponent $y_T = t_1/2 + 1$ are shown in fig. 5. Two-point fits were found to give less satisfactory results. This may be attributed to the presence of a regular term in the renormalization transformation and hence a nonzero constant C_0 in eq. (5.1). Such a constant must be present, at least for small q, because then $y_T < 1$ and the critical heat capacity equals C_0. For $q = 2$, C_0 and

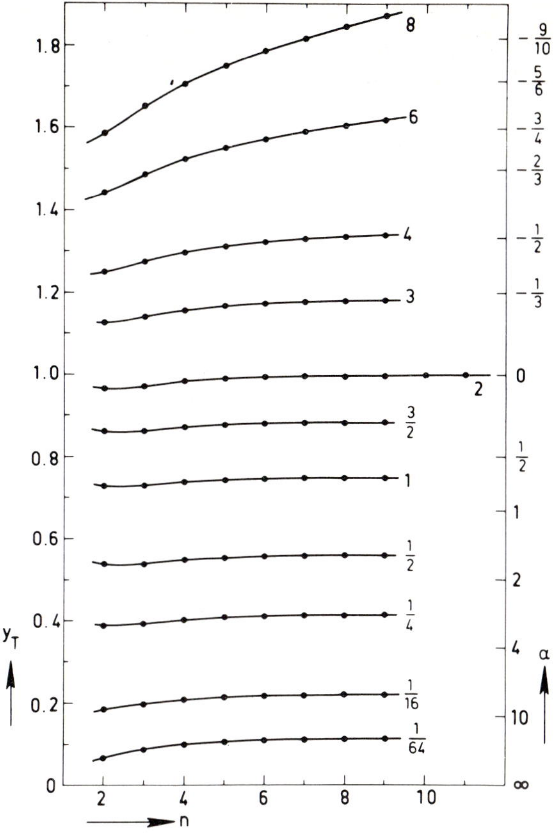

Fig. 5. Effective temperature exponents y_T as obtained by three-point fitting expressions $a + bn^{2y_T-2}$ to the heat capacities of Potts strips of width n. The vertical scale is shown in terms of the heat capacity exponent α (heat capacity $\sim |(T - T_c)/T_c|^{-\alpha}$) at the right hand side. The data points converge well for $q<4$. In contrast, for $q \gg 4$ the effective exponents appear to diverge linearly with increasing n, reflecting exponential divergence of the specific heat.

C_1 diverge to yield a logarithmic divergence. The quantity C_0 has been further analysed by an iterated fit, namely three-point fitting the estimates of C_0 that were obtained by three-point fitting the heat capacities. The results are shown in table I. For $q<2$, the convergence of these fits looks good, so that the data in table I are supposed to represent accurately the critical heat capacity of simple quadratic Potts models. Data for $q = 2$ have been omitted because it is a special case due to the appearance of the logarithmic singularity, and the procedure does not yield convergent results. Some data are shown for $q>2$, but their convergence is relatively poor.

Inspection of the numerical results for y_T from three-point fits (not shown

TABLE I

Constant term in the n-expansion of the critical specific heats of q-state Potts models on $n \times \infty$ lattices. These data were obtained by means of iterated three-point fits to the specific heats as described in the text. Convergence is evident only for $q < 2$. For $y_T < 1$ ($q < 2$) these results apply to the critical heat capacity of the infinite, simple quadratic Potts lattice.

n \ q	1/64	1/16	1/4	1/2
2	−0.962512	−0.916923	−0.754831	—
3	−0.962289	−0.910956	−0.758531	−0.557894
4	−0.962269	−0.910866	−0.757899	−0.555986
5	−0.962266	−0.910856	−0.757839	−0.555832
6	−0.962267	−0.910856	−0.757837	−0.555826
7	−0.962266	−0.910856	−0.757839	−0.555836

n \ q	0.95	1.05	1.5	2.5
2	—	—	—	−8.8315
3	−0.073711	0.080388	1.57317	−2.7608
4	−0.072383	0.078379	1.35628	−3.0512
5	−0.072288	0.078236	1.34329	−3.0431
6	−0.072281	0.078225	1.34185	−2.9972
7	−0.072286	0.078231	1.34214	−2.9517

n \ q	3	3.5	4
2	−4.278	−3.393	−3.070
3	−1.216	−0.359	0.611
4	−1.528	−0.739	0.060
5	−1.525	−0.708	0.188
6	−1.474	−0.595	0.536
7	−1.415	−0.450	1.099

here) suggests that convergence to $y(\infty, q)$ is already realized within a few times 10^{-3} for $q < 4$. The smooth behaviour of $y(n, q)$ with n invites to apply iterated three-point fits. Results are shown in table II for some values of $q < 4$. The case $q = 2$, for which finite-size results were calculated up to $n = 15$ (ref. 22) and for which the limit $y_T = 1$ is known[25]), shows a very shallow minimum near $n = 5.5$ with a depth of 6×10^{-4}. The iterated fits for other values of $q < 4$ suggest rather similar behaviour. Thus an estimate of $y_T(q)$ may be obtained by adding the minor correction $y(\infty, 2) - y(7, 2) \approx 4 \times 10^{-4}$ to $y(7, q)$ (also shown in table II). The similarity of the three-point fits as shown in fig. 5 can also be utilized in an earlier stage of the analysis by

TABLE II

Estimated temperature exponents y_T of q-state, simple quadratic Potts models as obtained from iterated three-point fits to the critical heat capacities of $n\times\infty$ Potts lattices. The lowest entry in each column gives a "best estimate" of y_T as described in the text. Missing entries correspond to extrema in the three-point fitted exponents (so that iterated fits have no solution). For $q\geqslant 4$, the effective exponents appear to diverge with n.

n \ q	1/64	1/16	1/4	1/2	0.95	1.05
2	0.6258	0.1562	0.3877	—	—	—
3	0.1214	0.2316	0.4331	0.5836	0.7574	0.7880
4	0.1151	0.2223	0.4166	0.5620	0.7353	0.7664
5	0.1144	0.2213	0.4154	0.5607	0.7339	0.7650
6	0.1146	0.2213	0.4154	0.5607	0.7339	0.7649
7	0.1145	0.2215	0.4154	0.5608	0.7340	0.7650
	0.1149	0.2219	0.4158	0.5612	0.7344	0.7654

n \ q	1.5	2	2.5	3	3.5	4
2	—	0.9642	1.0464	1.1118	1.1623	1.1996
3	0.9079	1.0217	1.1253	1.2251	1.3265	1.4358
4	0.8880	1.0012	1.1015	1.1952	1.2870	1.3814
5	0.8865	0.9994	1.0995	1.1929	1.2845	1.3795
6	0.8864	0.9994	1.0997	1.1937	1.2870	1.3855
7	0.8865	0.9996	1.1002	1.1951	1.2902	1.3930
8		0.9997				
9		0.9998				
10		0.9999				
11		1.0000				
	0.8869	1	1.1006	1.1955	1.2906	1.3934

application of similarity fits as described in section 5. Results are shown in table III. Apparent convergence and agreement between iterated and similarity fits suggests that the inaccuracy in y_T is not more than a few times 10^{-4} for $q<3$.

We have also tried Padé-approximants in $1/n$ on the $y(n, q)$ as resulting from three-point fits. The results were found consistent with the previous fits, but estimated accuracy (i.e. internal consistency between different approximants) was not so good ($\approx 10^{-3}$). This is not surprising because these Padé-approximants have expansions in integral powers of $1/n$, which is not believed to be the case for $y(n, q)$ in general.

These methods of analysis did not yield convergent results for y_T when applied to heat capacities of Potts model with $q>4$. Instead, effective

TABLE III
Estimated temperature exponents y_T as obtained by three-point similarity fits to finite-size heat capacities for several $q \leq 4$. These results agree well with those in table II. Only a rather qualitative justification is given in the text for this kind of fit, but the results seem to converge remarkably well.

$n \backslash q$	1/64	1/16	1/4	1/2	0.95	1.05
2	0.2036	0.2728	0.4151	0.5380	0.7048	0.7371
3	0.1181	0.2247	0.4176	0.5623	0.7355	0.7666
4	0.1152	0.2228	0.4173	0.5626	0.7359	0.7670
5	0.1149	0.2223	0.4167	0.5621	0.7354	0.7665
6	0.1149	0.2221	0.4163	0.5617	0.7350	0.7660
7	0.1150	0.2220	0.4161	0.5615	0.7347	0.7658
8	0.1149	0.2220	0.4160	0.5614	0.7346	0.7656

$n \backslash q$	1.5	2.5	3	3.5	4
2	0.8695	1.1187	1.2289	1.3323	1.4302
3	0.8882	1.0967	1.1831	1.2618	1.3347
4	0.8883	1.0969	1.1839	1.2638	1.3381
5	0.8879	1.0978	1.1861	1.2677	1.3443
6	0.8875	1.0985	1.1881	1.2714	1.3502
7	0.8873	1.0991	1.1896	1.2743	1.3551
8	0.8871	1.0995	1.1907	1.2767	1.3591

exponents as shown in fig. 5 were found to increase approximately linearly with n, reaching remarkably high values for high q. Such behaviour may be understood in terms of a discontinuity fixed point, leading to an exponentially diverging specific heat as a function of n as described in section 3. Numerical results around criticality for $q = 64$ and several values of n are shown in fig. 6 on a logarithmic scale; they illustrate the development of a δ function type spike at K_c for high n, as expected for a first order phase transition.

6.3. *The susceptibility*

On basis of the symmetry of the Potts states $1, 2, \ldots, q$ the magnetization m of finite systems obeys

$$m = \langle \delta_{\sigma_i 1} \rangle - \frac{1}{q} = \frac{\partial f}{\partial h} - \frac{1}{q} = 0$$

for $h = 0$, in agreement with our numerical results, also for nonintegral q. Thus, for a calculation of the magnetic exponent y_h we have to differentiate

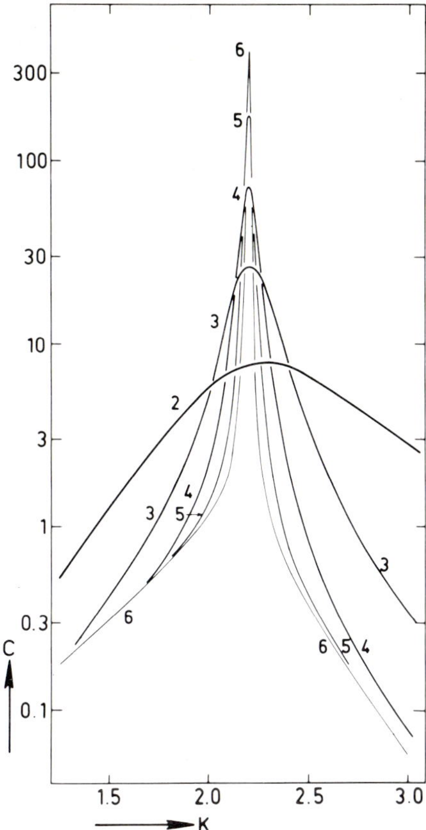

Fig. 6. Heat capacity C of infinitely long 64-state Potts strips versus coupling K, for several values of the strip width n which are shown in the figure. The height and sharpness of the maximum are seen to increase rapidly with n, suggesting a δ-function type contribution to the heat capacity of the infinite system at $K = K_c$.

the free energy once more and analyse the susceptibility per particle

$$\chi(n, q) = \frac{\partial^2 f(n, q)}{\partial h^2}.$$

Numerical differentiation was performed at the critical temperature in a field interval centered about the critical value $h = 0$. As already noted in section 4, the model is well defined only for $h \geq 0$. For $h < 0$, negative statistical weights appear and the model may become unphysical. In fact we observed that for $q < 1$, $h < 0$ and $|h|$ sufficiently large, the largest eigenvalue of $\mathbf{T}$ becomes complex. At this point, an unphysical singularity in f appears. However, for sufficiently small $|h|$ this eigenvalue was always found to be real and sufficiently smooth for the purpose of numerical differentiation. For small q, the proximity of the unphysical singularity imposed very small differentiation intervals, e.g. $\leq 10^{-6}$ for $q = 1/64$ and $n = 9$.

Results for χ are given in table AII.3 for several values of q and n. The susceptibilities may be expected to depend on n according to eq. (5.1) so that

$$\chi(n, q) \approx b_0 + b_1 n^{t_1}.$$

Results of three-point fits in terms of the magnetic exponent $y_h = 1 + t_1/2$ are shown in fig. 7 for several values of q. Note the scale difference between figs. 5 and 7; for $0 < q < 4$ the variation in y_h is only about 1/10 of that in y_T. From these results, we estimate that the convergence to y_h is already within a few times 10^{-3} for $q < 4$. In contrast to the specific heat, the susceptibility of the two-dimensional q-state Potts model diverges at criticality for all $q(\neq 1)$ in the thermodynamic limit. Thus we do not have an a priori reason to expect a non-zero b_0; in fact the three-point fits do not clearly indicate non-zero constants b_0. Hence two-point fits were also applied. Although the y_h results of two-point fits did not appear to converge as rapidly as those of three-point

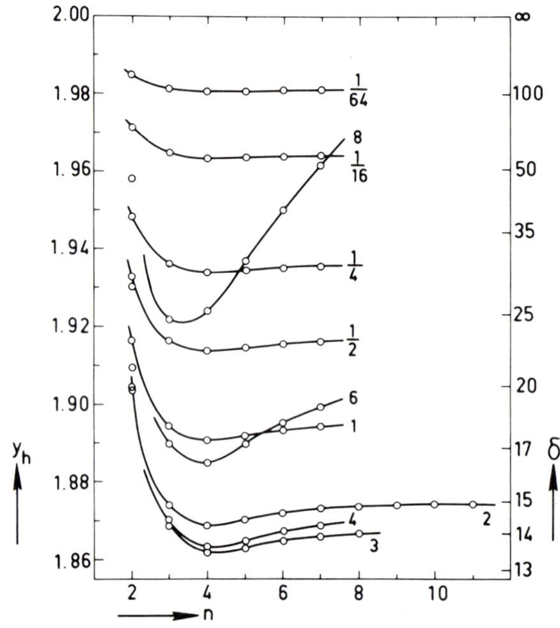

Fig. 7. Effective magnetic exponents y_h as obtained by three-point fitting the expression $a + bn^{2y_h-2}$ to the susceptibilities of Potts strips of width n. The vertical scale is shown in terms of the magnetization exponent δ (magnetizations $\sim h^{1/\delta}$) at the right hand side. The data points suggest rapid convergence for $q<4$. However, for $q \gg 4$ the effective exponents are observed to increase approximately linearly with n, reflecting exponential divergence of the susceptibility. Note that the y_h scale is elongated by a factor 10 in comparison to the y_T scale in fig. 5.

TABLE IV

Estimated magnetic exponent y_h as obtained by iterated two-point fits to the critical susceptibilities of finite, q-state Potts models. The lowest entries in each column show "best estimates" for each q as described in the text. Only results for $q \leqslant 4$ are shown. For $q \gg 4$ the effective exponents appear to diverge with increasing n.

n \ q	1/64	1/16	1/4	1/2	0.95	1.05	1.5
2	1.9882	1.9789	1.9681	1.9666	1.9718	1.9732	1.9789
3	1.9776	1.9579	1.9252	1.9033	1.8799	1.8760	1.8620
4	1.9809	1.9640	1.9354	1.9160	1.8958	1.8926	1.8813
5	1.9814	1.9646	1.9364	1.9175	1.8977	1.8946	1.8836
6	1.9813	1.9647	1.9367	1.9176	1.8979	1.8948	1.8837
	1.9811	1.9645	1.9363	1.9171	1.8972	1.8940	1.8830

n \ q	2	2.5	3	3.5	4
2	1.9835	1.9864	1.9881	1.9890	1.9897
3	1.8509	1.8435	1.8392	1.8378	1.8393
4	1.8731	1.8682	1.8660	1.8662	1.8687
5	1.8757	1.8710	1.8690	1.8693	1.8719
6	1.8758	1.8711	1.8690	1.8693	1.8720
7	1.8756		1.8685		
8	1.8753				
9	1.8752				
10	1.8751				
11	1.8751				
12	1.8750				
	1.875	1.8702	1.8679	1.8684	1.8712

fits, they are better behaved as a function of n. In particular, no extrema appear so that iterated fits may be expected to work well. Results of iterated two-point fits in terms of the magnetic exponent are shown in table IV for some values of q. In the case $q = 2$, we have more data points available[22]) than for other q, and here we clearly observe a very shallow maximum in y_h near $n = 6$ ($y(\infty, 2) - y(6, 2) \approx 8 \times 10^{-4}$). The other available $y(n, q)$ behave rather similar for $q < 4$, but when $q \to 0$ the $y(n, q)$ approach the value 2 and the n-dependence decreases. This suggest to estimate $y(q, \infty)$ by subtracting a very small correction from $y(6, q)$:

$$y(q) \approx y(6, q) - \frac{2 - y(6, q)}{2 - y(6, 2)} [y(6, 2) - y(\infty, 2)].$$

These estimates $y(q)$ are also shown in table IV.

TABLE V

Estimated magnetic exponents y_h as obtained from three-point similarity fits to the critical susceptibilities of finite-size q-state Potts models as described in the text. These results seem to converge well, and agree well with those in table IV. Two-point similarity fits give very similar results.

n \ q	1/64	1/16	1/4	1/2	0.95	1.05
2	1.9816	1.9650	1.9366	1.9171	1.8967	1.8935
3	1.9816	1.9650	1.9366	1.9171	1.8967	1.8935
4	1.9811	1.9644	1.9363	1.9173	1.8976	1.8944
5	1.9814	1.9646	1.9363	1.9173	1.8974	1.8942
6	1.9812	1.9646	1.9365	1.9173	1.8974	1.8942

n \ q	1.5	2.5	3	3.5	4
2	1.8825	1.8709	1.8692	1.8696	1.8716
3	1.8825	1.8709	1.8693	1.8697	1.8716
4	1.8832	1.8700	1.8677	1.8680	1.8708
5	1.8831	1.8702	1.8681	1.8683	1.8710
6	1.8831	1.8701	1.8680	1.8683	1.8710
7			1.8679		

Since also direct fits to the susceptibilities are found to behave rather similar as a function of n, we have also tried two- and three-point similarity fits. Results from three-point similarity fits are given in table V. The two types of similarity fits showed good apparent convergence for $q < 4$, as well as good agreement with the iterated fits (within a few times 10^{-4}).

For $q > 4$, these types of analysis did not yield convergent results. Just as for the heat capacity results for high q, the effective critical exponents increase approximately linearly with n, showing that scaling relations involve exponential functions rather than power laws. The high values of the effective $y_h(n, q)$ for high q correspond to a rapid increase of the critical susceptibility with n. In fig. 8 we have plotted the susceptibility of the $q = 64$ Potts model at the critical coupling versus magnetic field h for several values of the strip with n. These data are suggestive of the development of a δ-function type behaviour for high n, corresponding to a step function singularity in the magnetization at criticality as expected for a first-order transition.

6.4. *The correlation length of the simple Whitney polynomial*

We have also calculated the second largest eigenvalue of the transfer matrix for the simple Whitney polynomial for several values of q and system

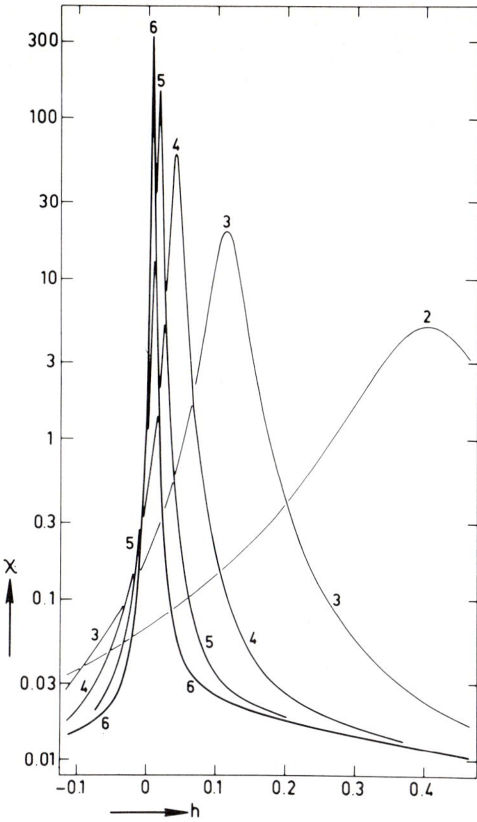

Fig. 8. Susceptibilities χ of infinitely long 64-state Potts strips versus magnetic field h, for several values of the strip width n which are shown in the figure. The height and sharpness of the maximum are seen to increase rapidly with n, suggesting a δ-function type contribution to the susceptibility of the infinite ($n=\infty$) system at $h=0$.

sizes up to $n = 10$, yielding correlation lengths of the Whitney model. Comparison to eigenvalue spectra of transfer matrices of the Potts model with $q = 2$, 3 and 4 (see ref. 22) showed that the second largest eigenvalue for the simple Whitney polynomial is not equal to the second, but to a lower lying eigenvalue of the Potts transfer matrix (see section 4.2). Hence the present results do not apply to the leading correlation length of the Potts model; they will be used to investigate the finite size scaling behaviour of the simple Whitney polynomial.

The dependence of the correlation lengths at criticality on n indicates that they diverge when $n \to \infty$, for all q. For $q < 4$ this divergence is approximately linear with n, while for high q ($q = 64$) it is much stronger and approximately exponential. This does not contradict eq. (3.6) since ξ for the infinite system is

TABLE VI

Estimated temperature exponents y_T as calculated from two-point fits to $d^2\kappa/dK^2$ (table AII.4). These data ($q \leq 4$) agree with earlier results (tables II and III) although the estimated accuracy is less good. For $q = 64$, the y_T estimates increase approximately linearly with n.

n \ q	1/16	0.95	1.05	2	3	4	64
2	2.0255	0.7550	0.7942	1.0398	1.1957	1.3091	2.5042
3	0.8325	0.7547	0.7939	1.0399	1.1988	1.3169	2.7409
4	0.5406	0.7551	0.7919	1.0300	1.1907	1.3133	2.9949
5	0.4249	0.7535	0.7884	1.0223	1.1859	1.3135	3.2480
6	0.3647	0.7512	0.7850	1.0169	1.1836	1.3158	3.5008
7	0.3285	0.7491	0.7820	1.0132	1.1827	1.3190	3.7543
8	0.3047	0.7471	0.7796	1.0105	1.1825	1.3223	4.0088
9	0.2881	0.7455	0.7776	1.0085	1.1827	1.3256	4.2642

derived in a different way. It agrees well with the scaling relations given in sections 2 and 3 for critical and discontinuity fixed points, respectively, and gives support to the applicability of the underlying scaling hypothesis.

Temperature derivatives were obtained by numerical differentiation, similar to the description in section 6.6. The first temperature derivative was found to vanish at criticality as expected on basis of duality and the second derivative of the inverse correlation length

$$\partial^2\kappa/\partial K^2$$

(given in table AII.4) was used to calculate the temperature exponents. Results of two-points fits are shown table VI in terms of y_T. The apparent convergence is less satisfactory than in section 6.2, and especially iterated fits (not shown) were found ill-behaved. Nevertheless the behaviour is sufficiently clear to observe that the two-point fits for $q < 4$ agree with the earlier results for y_T within a margin of a few times 10^{-2}, which is as good as may be expected on basis of the apparent convergence. This margin is much smaller than e.g. the variation of y_T with q, and also much smaller than typical differences between y_T and next leading exponents in the Potts model[14,55]). Thus we conclude that the scaling behaviour of the correlation length of the simple Whitney polynomial is in agreement with that of the heat capacity of the Potts (and Whitney) model.

6.5. *Thermal behaviour of the correlation length of the extended Whitney polynomial*

We have also calculated second largest eigenvalues of extended Whitney polynomials for system sizes up to $n = 8$. Comparison to spectra of transfer matrices for the $q = 2$, 3 and 4 state Potts model (see ref. 22) showed that they are indeed equal to the corresponding second eigenvalues for the Potts model in zero field (which are however $q - 1$ fold degenerate), as might be expected on the basis of the arguments given in section 4.2. Thus the correlation length of the extended Whitney model corresponds to that of the Potts model. Numerical results for the correlation length indicated that its scaling behaviour is in accordance with that described for the simple Whitney polynomial in section 6.4.

Results for the temperature derivative of the inverse correlation length are presented in table AII.5 for several q values. Estimates of the temperature exponent from iterated two-point fits are given in table VII. For $q < 2$, the results seem to converge well and agree within a few times 10^{-4} with the best results from the heat capacities. For $q = 2$, we obtain the exact results $y_T = 1$, since $n^{y_T - 1} \sim \partial\kappa/\partial K = -1$ independent of n. For $q = 3$ again the convergence is somewhat less good, and yields $y_T = 1.206$ which is in agreement with our earlier results. For $q = 4$ we find, even less accurately, $y_T = 1.44$ but still in agreement with the results from section 6.2 within the estimated accuracy of a few times 10^{-2}. In contrast with earlier results for high q, we now find the "discontinuity fixed point" exponent $y_T = 2$ for high q ($q = 64$).

Second derivatives with respect to coupling strength K of the inverse correlation length have also been calculated (not shown in this paper, but available on request), and analysed by two-point fits. Convergence is not as good as for the first derivative, as becomes especially clear when iterated fits are tried: these fits yield even less coherent results. Results of two-point fits in terms of y_T for $n = 8$, $q \leq 4$ differ by a few times 10^{-2} from earlier results in table II, and the change with n is in the right direction (about 10% of the difference with table II per unit in n). The agreement seems significant, indicating that, for $q \leq 4$, also the scaling behaviour of the second derivative

TABLE VII

Estimated temperature exponents y_T as obtained from iterated two-point fits to $d\kappa/dK$ as given in table AII.5. These results agree accurately with those from the heat capacities (tables II and III). For $q = 8$, convergence is not evident in the range of n shown here. Note that the results for $q = 64$ approach the discontinuity fixed point value $y_T = 2$.

n \ q	1/16	1/4	0.95	1.05	3	4	8	64
2	0.2215	0.4144	0.7298	0.7607	1.2204	1.4745	6.400	2.0180
3	0.2216	0.4151	0.7320	0.7629	1.2129	1.4575	25.287	2.0047
4	0.2215	0.4152	0.7328	0.7638	1.2090	1.4483	−11.736	2.0013
5	0.2215	0.4154	0.7333	0.7643	1.2066	1.4423	−4.678	

of the correlation length is in accordance with that of the specific heat for sufficiently high n. For high q ($q = 64$) we find again that the second derivative strongly diverges with increasing n.

6.6. *Magnetic behaviour of the correlation length of the extended Whitney polynomial*

The derivatives of the inverse correlation length of the extended Whitney polynomial has also been calculated with respect to the magnetic coupling strength h, for several values of q, and $n = 2$ through 8. Some results are shown in table AII.6. Results for the magnetic exponent y_h as derived from iterated two-point fits are shown in table VIII. These results exhibit very good apparent convergence ($\approx 10^{-4}$) for $q < 2$. They are in excellent agreement with the results derived from the susceptibility in section 6.3. Data for $q = 2$ are absent because $d\xi/dh$ vanishes for all n. For $q = 3$ and 4 we find somewhat less good apparent convergence ($\approx 10^{-3}$) and especially for the $q = 4$, the agreement with earlier results is less good. We have also calculated second derivatives of the inverse correlation length with respect to h (these are not shown in this paper, but are available on request). Analysis with iterated two-point fits exhibited somewhat less good apparent convergence than in the case of the first derivatives. Differences in terms of y_h with results obtained from first derivatives for the highest available n ranged from 1×10^{-4} for $q = 1/16$ to about 5×10^{-4} near $q = 1$, and to 7×10^{-3} at $q = 4$. At $q = 2$ we find $y_h(2) = 1.8756$, which differs by 6×10^{-4} from the exact value 15/8. The accurate agreement demonstrates that the scaling behaviour of the second derivative is in accordance with that of the first for $q \leqslant 4$.

For high $q(q = 64)$, the analysis produces results similar to those from the

TABLE VIII

Magnetic exponent y_h as obtained from two-point fits to $d\kappa/dh$ as given in table AII.6. These results agree accurately with those from the susceptibilities (given in tables IV and V). The missing entry under $q=8$ corresponds to a rather flat extremum in the two-point fitted exponent to $d\kappa/dh$, so that the iterated fit did not yield a solution. For $q=64$, y_h (just as y_T in table VII) shows a clear trend to the discontinuity fixed point value $y_h=2$.

n \ q	1/16	1/4	0.95	1.05	3	4	8	64
2	1.9633	1.9336	1.8902	1.8864	1.8589	1.8675	1.922	2.0659
3	1.9644	1.9359	1.8962	1.8929	1.8683	1.8752	1.925	2.0024
4	1.9645	1.9363	1.8971	1.8939	1.8686	1.8751	—	2.0005
5	1.9646	1.9363	1.8972	1.8940	1.8678	1.8742	1.926	

temperature derivatives of the correlation length: the first magnetic derivative yields the discontinuity fixed exponent $y_h = 2$, and from the second derivative we find effective exponents rapidly increasing with n.

6.7. *Numerical accuracy*

The number of decimal places of the numerical data in Appendix II was chosen such that firstly the precision is sufficient to perform analyses as given above, and secondly that all digits are justified in view of the non-zero error in floating point arithmetic. We will now discuss this type of numerical error; convergence with increasing n is left out of consideration at present.

Diagonalisation of the asymmetric transfer matrices (as well as subsequent numerical differentiations) involved numerical operations that were performed with the 16 decimal place double precision accuracy of the IBM 370–158 computer. For high values of the linear system size n, the calculations become lengthy so that the error may possibly accumulate to many times 10^{-16}. We can check this for the case $q = 2$, where free energies of finite systems can be independently obtained from the Onsager solution (see e.g. Domb[52]). For zero field (the simple Whitney polynomial) and critical coupling we found differences up to about 10^{-14} for $n \leqslant 11$. From the scatter in data obtained during the last few iterations of the eigenvalue calculations, we may conclude that this is also a typical measure of the accuracy of the free energy for other q-values, with exception of high q, where the inaccuracy increases up to about 10^{-10} for $q = 64$, $n = 11$. Numerical differentiation (twice) in a coupling interval $\Delta K \approx 10^{-2}$ leads to an absolute accuracy in the specific heat $C(n, q)$ of about 10^{-10} for $q \leqslant 4$. For high q, higher derivatives increase rapidly and the differentiation interval has to be chosen relatively small, down to $\approx 10^{-4}$ for $q = 64$, $n = 10$. This leads to an accuracy of about 10^{-2} in the heat capacity; since $C(10, 64) \approx 10^4$ the relative accuracy is still reasonable.

We thus estimate that for most values of $q < 4$, the relative error in C does not exceed 3×10^{-10}. However, for $q = 0.95$ and 1.05, C becomes small, and the relative inaccuracy may reach 10^{-9}. Above $q = 4$, the relative error increases up to about 10^{-6} for $q = 64$. The process of iterated three-point fitting typically adds some five orders of magnitude to this error, as may be calculated algebraically and be observed empirically. Thus we arrive at numerical error estimates in the y_T results from iterated three point fits (table II); 3×10^{-4} for $q = 1/64$, and 1×10^{-4} or better for other $q < 4$. Similarity fits (table III) lead to a somewhat smaller inaccuracy. For $q > 4$, the error is rather irrelevant because of the qualitative character of the analysis.

The accuracy of the second derivatives of the inverse correlation length of the simple Whitney polynomial is comparable to that of the specific heat. In this case the poor convergence dominates completely over the numerical error when calculation of the temperature exponent is attempted.

The numerical error in the largest eigenvalue of the extended Whitney polynomial is larger than that of the simple polynomial, which is probably due to the proximity of the next largest eigenvalue. We estimate the numerical error in the free energies to be about 10^{-12} for $q < 4$, from a comparison with the expression of Domb[52]) for $q = 2$, and from the scatter during the last iteration steps. Furthermore, results for the magnetization m, obtained by numerical differentiation of f to the magnetic field h in an interval Δh, are of the order of $10^{-12}/\Delta h$ or smaller. This agrees well with the value $m = 0$ expected on basis of symmetry, and with the error mentioned above. For $q \gg 4$ we find that the error in f increases rapidly, just as for the simple Whitney polynomial. However, also in this case the semiquantitative analysis does not require high accuracy. For $q < 1$, we had to choose very narrow differentiation intervals, as small as about 10^{-6} for $q = 1/64$, $n = 9$. Thus the absolute error in the susceptibility χ increases to about $10^{-12}/\Delta h^2$, or about unity for $q = 1/64$. However, χ itself increases rapidly for low q, so that the relative error in χ does not increase very much for low q values. Taking into account the widths of the differentiation intervals, and the values of χ, we estimate the relative error in χ to vary from about 3×10^{-7} for $q = 1/64$ to 0.5×10^{-7} for $q = 4$ (not considering cases $q > 4$). The process of iterated two-point fitting now leads to an inaccuracy in y_h which is larger by about two orders of magnitude. This increase is much smaller than that found for the heat capacities. This is related to the fact that we now have used two-point fits, and regular contributions to the heat capacities were very important. Thus we estimate that the numerical error in the y_h values as obtained from iterated two-point fits does not exceed 3×10^{-5} (table IV). The similarity fits (table V) have a comparable or better accuracy.

The calculation of the correlation lengths is liable to inaccuracies originating from two eigenvalues. Further, we do not divide by n, as we did for the free energy. In the case of the simple Whitney polynomial, the numerical error is obviously irrelevant because the errors in the y_T estimates are dominated by poor convergence as a function of n. Thus we will now concentrate on the extended Whitney polynomial. For $q < 4$, we estimate the absolute error in the inverse correlation length κ as a few times 10^{-12}. Differentiation in a typical coupling interval $\Delta K \approx 1 \times 10^{-2}$ leads to an absolute error in $d\kappa/dK$ of a few times 10^{-10}. Because $d\kappa/dK \approx 1$, this is also a good estimate of the relative error. Iterated two-point fitting then yields a numerical error of about 10^{-6} in the exponent y_T.

Differentiation of κ with respect to the magnetic coupling h had to be performed in a relatively small interval, so that the absolute error in $d\kappa/dh$ is larger than that in $d\kappa/dK$, especially for small q. This does, however, hardly affect the relative error because $d\kappa/dh$ becomes large together with the higher derivatives which impose small values for Δh. We estimate the relative error in $d\kappa/dh$ as about 10^{-9} in the range $q \leq 4$. After iterated two-point fitting, this leads to a numerical uncertainty up to about 1×10^{-6} in the y_h results.

On the basis of these error estimates, we have chosen to show the exponents in the tables in four decimal places (in addition to the integer part). This is enough to observe the apparent convergence, and the figures are believed to be correct, except possibly for one unit in the last decimal place in a few cases.

7. Miscellaneous results

7.1. *Square systems*

The transfer matrix constructed in section 4 applies to systems with open ends in the "transfer" direction. This is so because the states of connectivity associated with the values of the transfer matrix indices do not specify how the first and last columns of the lattice are mutually connected by possible paths of bonds. Thus the number of loops that is closed when the first and last

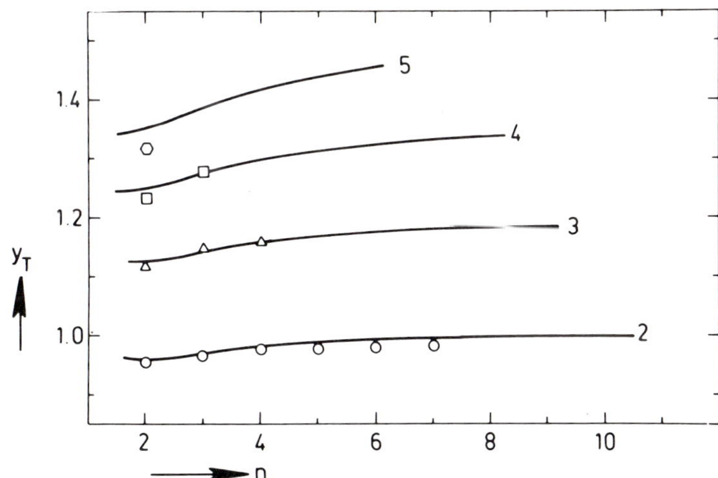

Fig. 9. Comparison between effective temperature exponents as obtained from $n \times \infty$ systems (curves) and $n \times n$ square Potts lattices (data points), all systems having periodic boundaries. The number of states q of the Potts systems is shown for each curve. Different symbols indicate q for the data points; for $q = 2$:○; $q = 3$:△; $q = 4$:□; $q = 5$:⬡. These results were obtained from three-point fits to the heat capacities of systems with linear sizes n, $n+1$, and $n+2$. For low q, the results for $n \times \infty$ systems seem to converge more rapidly than those of $n \times n$ systems.

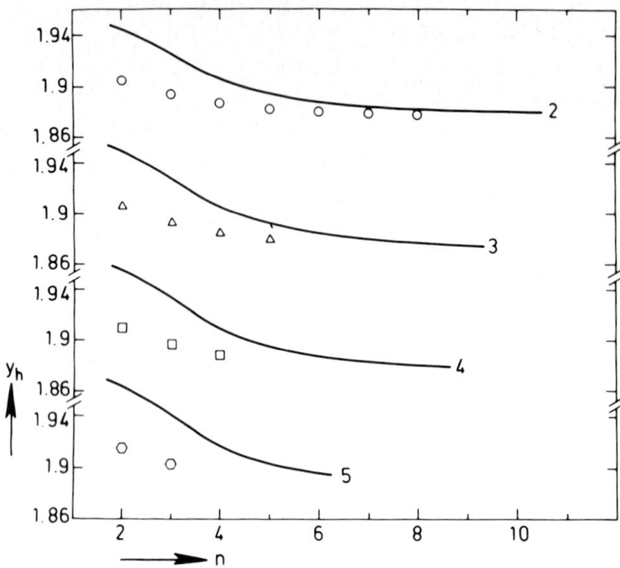

Fig. 10. Comparison between effective magnetic exponents as obtained from $n\times\infty$ Potts models (curves) and $n\times n$ square Potts systems (data points). The number of states q of the Potts system is indicated for each curve. Also the different symbols indicate q; $q=2$:○; $q=3$:△; $q=4$:□; $q=5$:○. These effective exponents were obtained from two-point fits to the critical susceptibilities of Potts systems with linear sizes n and $n+1$. For low q, results for the $n\times n$ systems seem to converge more rapidly than those $n\times\infty$ systems. For clarity, the curves are placed apart by shifting the vertical scale for each q.

columns are coupled together, is not determined. However, the partition function of square $n\times n$ systems with periodic boundaries in both directions can be expressed in the Potts transfer matrix for integral q such as constructed in ref. 22:

$$Z_n = \mathrm{Tr}\,\mathbf{T}^n.$$

Columns of $\mathbf{T}^m$ can be simply obtained by repeated execution of the transfer matrix multiplication algorithm on a vector with only one nonzero component. The computation of this partition function is relatively time-consuming, and only small systems have been analysed. The analysis involved numerical differentiations to coupling K and magnetic field h, yielding specific heats C and susceptibilities χ, for $q=2$, 3 and 4. Results of three-point fits to C and two-point fits to χ, yielding estimates of y_T and y_h, are plotted in figs. 9 and 10, respectively, together with similar results for infinitely long strips taken from section 4. These pictures demonstrate that also square systems yield rapidly converging exponents. The results for the two types of systems are consistent for $q=2, 3$ and 4.

7.2. *Helical systems*

The transfer matrix for Whitney polynomial on a two-dimensional lattice with helical boundary conditions can be constructed by methods similar to those described in section 4 for cylindrical boundary conditions. The renormalization description of finite-size scaling of such systems poses a problem, because helical boundary conditions are not conserved by (square block) renormalization group transformations. However, it is plausible that for sufficiently high n, the thermodynamics of these systems will rapidly become independent on the small displacement of the lattice at the boundary, and we may attempt finite-size scaling. Thus we have also calculated free energies, heat capacities and susceptibilities of infinitely long helical systems. Results for y_T and y_h from three-point fits are shown in figs. 11 and 12 for several values of q, together with results for cylindrical systems. Also helical systems

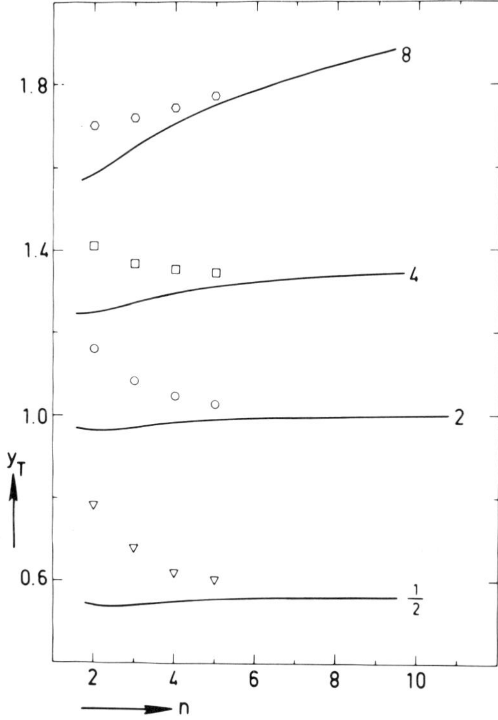

Fig. 11. Comparison between effective temperature exponents as obtained from Potts systems with cylindrical (curves) and helical (data points) boundary conditions. The curves are labelled by the number of states q, and different symbols give q for the data points; ▽: $q=\frac{1}{2}$; ○: $q=2$; □: $q=4$; ⬡: $q=8$. These data were obtained from three-point fits to the heat capacities.

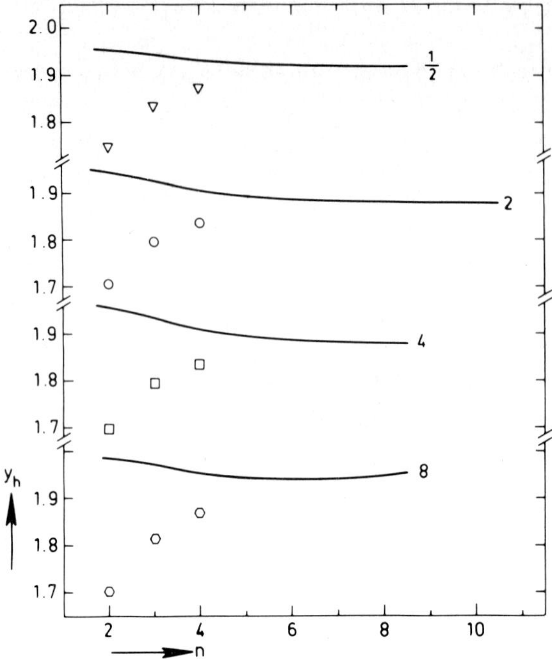

Fig. 12. Comparison between effective magnetic exponents as obtained from Potts systems with cylindrical (curves) and helical (data points) boundary conditions. Each curve is labelled by the number of Potts states q, and also the different symbols for the data point indicate q; $\nabla : q = \frac{1}{2}$; $\bigcirc : q = 2$; $\square : q = 4$; $\bigcirc : q = 8$. These data were obtained from two-point fits to the susceptibilities. For clarity, the curves are placed apart by shifting the vertical scale for each q.

appear to give consistent results for the exponents, although the convergence looks somewhat less good than for cylindrical systems.

7.3. *Anisotropic systems*

By "anisotropic systems" we mean systems with different couplings K_h and K_v between nearest neighbours in the horizontal and vertical directions. On basis of universality arguments we expect that critical exponents do not depend on the ratio K_h/K_v. The generalization $K_h \neq K_v$ is easy to build into the transfer matrix generating algorithm, and we have executed some calculations for the magnetic exponents for two ratios $K_h/K_v \neq 1$.

The duality relation for an anisotropic Whitney polynomial on a square lattice with N sites (N very large) reads[49,56])

$$Z(u_h, u_v) \to u_h^N u_v^N Z(u_v^{-1}, u_h^{-1}).$$

On this basis we expect criticality at $u_h = u_v^{-1}$. These parameters are related to the Potts couplings by

$$u_z = \{\exp(K_z) - 1\}/\sqrt{q}; \quad z = h, v.$$

An interesting limit is $u_h \to \infty$ so that the model becomes continuous in the horizontal (transfer) direction, as explained e.g. in the review article by Kogut[57]), and references therein. When K_v decreases, the correlation length in the vertical direction may be expected to decrease, and the convergence of y_T and y_h with increasing n may become more rapid. We cannot come close to the limit $u_v \to 0$ with our present computer program, because of a rapid increase of the numerical inaccuracy, and slow convergence of the eigenvalue calculation. To investigate this limit, a special program would have to be constructed, by writing the transfer matrix as[57])

$$\mathbf{T} = \mathbb{1} + \lambda \mathbf{t},$$

where λ is an infinitesimal parameter proportional to u_v, and $\mathbf{t}$ is to be diagonalized.

Analysis (not shown here) of actual susceptibility calculations for $q = 2$ and 4, and anisotropy parameter $\alpha \equiv K_v/K_h = 0.5$, 1 and 2, showed indeed that the convergence of y_h was slightly more rapid for $\alpha = 0.5$ (about one step in n). Apart from this, the variation of the y_h estimates with increasing n was very similar for the three values of α. Thus the possibility remains to improve the accuracy of the exponents by means of a program for the continuum limit in future.

8. Discussion

The analysis of the finite-size behaviour of the free energy of the Potts model in section 6.1 is in agreement with the scaling relations which have been presented in sections 2 and 3 for the free energy. This observation supports the technique of derivation of temperature and magnetic critical exponents by finite size scaling. The estimated numerical error in the best y_T results (i.e. those from the iterated three point fits and similarity fits to the heat capacity, and iterated two point fits to the temperature derivative of the correlation length of the extended Whitney polynomial) does not exceed 10^{-4}, and the estimated convergence, and internal consistency between the different results, suggests an inaccuracy of a few times 10^{-4} in y_T for $q < 3$. Fig. 13 shows a comparison between our numerical results and the conjecture of den Nijs[27]). No differences stand out for $q < 3$. In order to make a more profound comparison to this conjecture, it is useful to consider also other expressions that may be fitted to a set of known values $y_i(q)$. Our present purpose is not

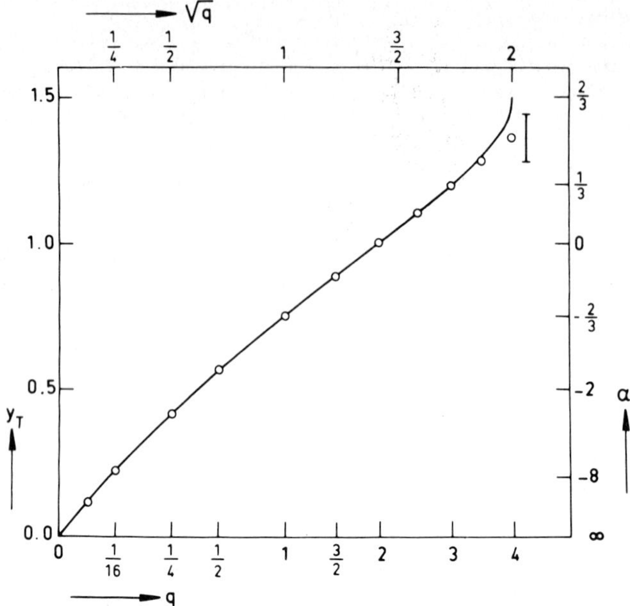

Fig. 13. The conjecture of den Nijs[27]) for the temperature exponent y_T of the Potts model versus q, compared to results from iterated three-point fits to the heat capacities of finite systems (data points). The length of the error bar shown next to the data point at $q = 4$ is two times the difference between the corresponding results in tables II and VII. This difference decreases rapidly with q to very small quantities (a few times 10^{-4}) for $q<2$.

to construct a competing conjecture; for instance, we do not examine what happens to the tricritical branch. Instead, we want to get some idea what numerical accuracy is needed to distinguish between different conceivable conjectures. The conjecture of den Nijs[27]) can be written as

$$\left.\begin{aligned} y_T(q) &= (1.5 - 3p)/(1-p) \\ p &= [\arccos(\sqrt{q}/2)]/\pi \end{aligned}\right\},$$

which is the [1, 1] Padé-approximant in p to three of the four points $y_T(0) = 0$, $y_T(2) = 1$, $y_T(3) = 5/4$, $y_T(4) = 3/2$. This suggests the construction of the [3, 0] approximant to these four y_T data points, since the [1, 2] and [2, 1] approximants are, of course, identical to the [1, 1] approximant, and [0, 3] approximant is unable to produce zeroes in y_T. Thus we have calculated the [3, 0] approximant in p:

$$y'_T(p) = \frac{3}{2} - \frac{4}{5}p - \frac{1}{10}p^2 - \frac{3}{5}p^3.$$

Values of y'_T are shown in table IX for several values of q, together with

TABLE IX

Values of Pade-approximants in $(1/\pi)\arccos(\sqrt{q}/2)$ to known or conjectured thermal and magnetic exponents at $q=0$, 2, 3 and 4. For y_T two approximants are missing: the [1, 2] approximant is identical to the [2, 1] one, and the [0, 3] one cannot fit the four given data points. The [2, 1] approximants for y_T and y_h are identical to the den Nijs[27]) and NPRS[17,28]) conjectures respectively. These are the approximants that give the best argreement with our results for the Potts exponents.

	y_T		y_h			
q	[3, 0]	[2, 1]	[3, 0]	[2, 1]	[1, 2]	[0, 3]
0	0	0	2	2	2	2
1/64	0.108507	0.114871	1.982827	1.981236	1.980897	1.982144
1/16	0.211611	0.221672	1.967044	1.964529	1.964014	1.965965
1/8	0.293326	0.304907	1.954952	1.952057	1.951483	1.953711
1/4	0.403744	0.415712	1.939279	1.936287	1.935720	1.938000
1/2	0.550898	0.561082	1.919789	1.917243	1.916788	1.918705
0.95	0.728268	0.734251	1.898949	1.897454	1.897205	1.898320
1.05	0.760136	0.765277	1.895578	1.894293	1.894082	1.895038
1.5	0.884637	0.886683	1.883733	1.883222	1.883142	1.883521
2	1	1	1.875	1.875	1.875	1.875
2.5	1.102424	1.101783	1.869502	1.869662	1.869684	1.869566
3	1.2	1.2	1.866667	1.866667	1.866667	1.866667
3.5	1.303359	1.305034	1.866647	1.866228	1.866175	1.866487
4	1.5	1.5	1.875	1.875	1.875	1.875

values for the den Nijs conjecture. The differences are rather small, up to about 10^{-2}. However, a comparison between results in tables II, III and VII and the conjecture of den Nijs shows much smaller differences of a few times 10^{-4} for $q<3$. These differences agree well with the estimated accuracy of our results (see above). Hence we may conclude that our results confirm the conjecture of den Nijs. Further, our results indicate that the temperature exponent is very close ($\approx 3\times 10^{-4}$) to 3/4 at $q=1$; they rule out a conjectured exact result by Klein et al.[58]). For $q=3$ our result is less accurate ($\pm 10^{-2}$) but still in good agreement with the expected value $y_T=1.2$. The case $q=4$ stands apart because of the very poor convergence in comparison with $q\leqslant 2$. This effect can be satisfactorily explained[60]) by the renormalization flow picture of Nienhuis et al.[15]) which predicts a marginal eigenvalue, leading to poor convergence and logarithmic anomalies, at a critical q number must be $q=4$ as follows from the exact results of Baxter[24]). We have also tried more sophisticated fits involving logarithmic factors to the heat capacities. This did not produce convincing results; an explanation may involve the anomalously slow renormalization flow near the critical q fixed point. For $n\leqslant 11$, the scaling transformation does not bring the system sufficiently close to the fixed point, so that higher-order terms in the finite-size expansion are important.

For $q \geqslant 4$, our exponentially diverging heat capacity results indicate that the renormalization flow picture implies a zero temperature fixed point, in accordance with the picture of Nienhuis et al.[15]).

Concerning the accuracy of the y_h results for $q < 3$ the same remarks as for y_T apply. Fig. 14 shows a comparison between our results and the conjecture proposed by Nienhuis, Riedel en Schick[17]) and independently by Pearson[28]), or shortly the NPRS conjecture. Differences do not stand out clearly for $q \leqslant 3$ on this scale. A similar testing procedure can be applied for this conjecture. It can be written as the [2, 1] Padé-approximant to the data points $y_h(0) = 2$, $y_h(2) = y_h(4) = 15/8$, $y_h(3) = 28/15$:

$$y_h = (15/8 - 2p + \tfrac{1}{2}p^2)/(1-p)$$

with p as given above.

Values for the NPRS conjecture, together with those for the [3, 0], [1, 2] and [0, 3] Padé-approximants to the four data points, are shown in table IX. A comparison with numerical results for $y_h(q < 3)$ in tables IV, V and VIII shows that the [1, 2] and [2, 1] approximants are much better than the

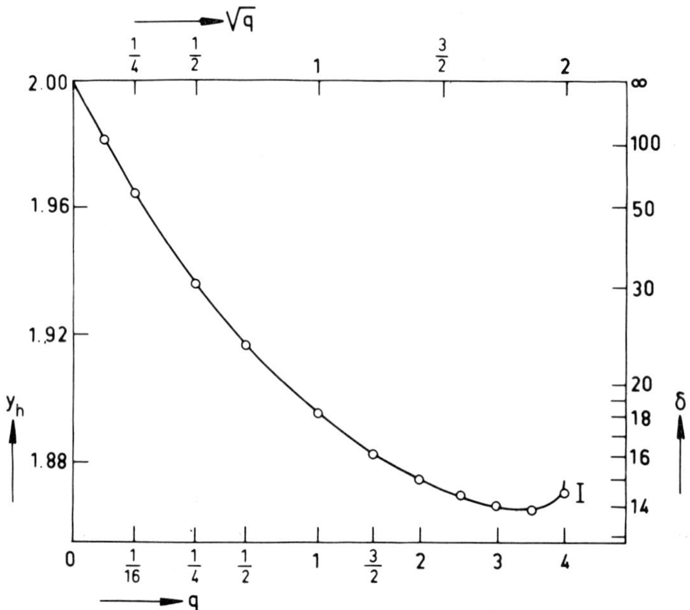

Fig. 14. The NPRS conjecture[17,28]) for the magnetic exponent y_h of the Potts model versus q, compared to results of iterated two-point fits to the susceptibilities of finite systems (data points). The length of the error bar shown next to the data point at $q=4$ is two times the difference between the corresponding results in tables IV and VIII. This difference decreases rapidly with q to very small quantities (a few times 10^{-4}) for $q<2$.

remaining two. Further, the NPRS conjecture agrees better with our results than the [1, 2] approximant does, especially near $q = 1/4$. Thus our results support the NPRS conjecture, and demonstrate that it is correct within a few times 10^{-4} for $q < 3$.

The case $q = 4$ is somewhat peculiar in the sense that the y_h results from the susceptibilities seem to converge well, and seem to converge to 1.871 which is different from the expected value 1.875. However, the former value is also different from the result derived from $d\kappa/dh$, which is 1.874. The most plausible interpretation[60]) is anomalously slow convergence of finite-size results, due again to the marginal eigenvalue at $q = 4$ as discussed above.

For high values of q, the second derivatives of the free energy and the inverse correlation length show qualitatively similar behaviour: exponential divergence with n. In contrast, the first derivatives of the inverse correlation length diverge linearly, in accordance with the discontinuity fixed point exponent[59]) equal to $d = 2$. Apparently, there are continuity conditions in the eigenvalue-coupling diagram which allows exponentially diverging curvatures, but inhibit exponentially diverging slopes.

Acknowledgements

The authors express their gratitude ot Professor J.M.J. van Leeuwen, for his valuable suggestions and his stimulating interest. They have also profited much from discussions with Dr. B. Derrida. Further, they acknowledge discussions with Dr. V.J. Emery, Dr. H.J. Hilhorst, Dr. B. Nienhuis, Professor M. Schick and Professor F.Y. Wu.

These investigations form part of the research program of the "Stichting voor Fundamenteel Onderzoek der Materie (F.O.M.)" which is financially supported by the "Nederlandse Organisatie voor Zuiver Wetenschappelijk Onderzoek (Z.W.O.)". Part of this research was supported by the U.S. National Science Foundation under Grant No. DMR-79-20785.

Appendix I

Numbering the connectivities

The transfer matrix introduced in expression (4.15) has as its indices the n-point connectivities β and γ. These connectivities apply to the n rightmost sites of an $n \times m$ lattice, on which a graph G' specifies which pairs of nearest neighbours are connected by a bond, and also which of the nm sites are connected to the ghost site by a bond (section 4). The connectivity of the mth column describes: (i) which of the n sites of this column are connected to the ghost site (directly or via some path of bonds in the $n \times m$ lattice) and (ii) how the remaining sites of the mth column are divided into groups of sites which are mutually connected via some path of bonds (in the $n \times m$ lattice).

Whenever two sites of the column m are said below to be connected, it will be in this sense. In this appendix we define a one-to-one mapping of connectivities onto consecutive integers 1, 2, 3, ... as employed for numerical calculations involving the transfer matrix. Furthermore, we discuss the number of possible n-point connectivities as a function of n.

Before studying the general case, we consider the simple Whitney polynomial. This follows naturally from a Potts model in the absence of a magnetic field, in which case $v = 0$ in eq. (4.5). Thus all graphs involving bonds to the ghost site carry zero statistical weight, and we may restrict ourselves to the corresponding subset of connectivities. Consequently, the order of the transfer matrix is reduced.

Considering at present the restricted set of connectivities, the state of sites 1 through n of the rightmost column can be described by a (non-unique) sequence of n positive integers $(i_1 i_2 \ldots i_n)$ such that $i_r = i_s$ if and only if sites r and s are connected (via some path of bonds in the $n \times m$ lattice). Consider four sites $r < s < t < u$. If sites r and t are mutually connected, and so are sites s and u, then necessarily all four of them are connected: a path from r to t necessarily intersects a path from s to u. It follows that $(i_1, \ldots, i_n)$ represents a possible state of connectivity, if and only if in all cases that $i_r = i_t$ and $i_s = i_u$, with $r < s < t < u$, also $i_s = i_t$ is satisfied. Integers in a sequence with this property are called well-nested.

Define an auxiliary function ρ on connectivities in terms of sequences of well-nested integers:

$$\rho(i_1 \ldots i_n) = \min\{n+1, j \mid n \geq j > 1 \text{ and } i_1 = i_j\}. \tag{AI.1}$$

The quantity between braces represents the set of numbers consisting of $n+1$ and all numbers j that satisfy the conditions at the right of the symbol $|$. Note that ρ always has value 2 for $n = 1$. The connectivities possess a recursive property by which n-point connectivities are split into disconnected parts: if $(i_1 \ldots i_n)$ is a sequence of well-nested integers with $\rho(i_1 \ldots i_n) = k$, then $(i_2 i_3 \ldots i_{k-1})$ and $(i_k i_{k+1} i_n)$ are again sequences of well-nested integers, which moreover are independent, i.e. the integers occurring in one sequence do not occur in the other. For $k = 2$ or $n + 1$ either one of the sequences $(i_2 \ldots i_{k-1})$ or $(i_k \ldots i_n)$ is empty. This splitting of sequences of well-nested integers makes it possible, with help of the auxiliary function ρ, to define recursively a unique ordering of connectivities. To determine which one of two different connectivities α and β precedes the other one applies the function ρ. If ρ does not provide a decision ($\rho(\alpha)=\rho(\beta)$), then one splits α and β, and ρ is applied to compare the corresponding parts of α and β. The process is continued until a difference is found; if no difference appears before α and β have been completely decomposed, then one concludes that α and β are identical.

More precisely: let α be represented by $(i_1 i_2 \ldots i_n)$ and β by $(j_1 j_2 \ldots j_n)$ and define $k \equiv \rho(\alpha)$. Connectivity α precedes β is one of the following conditions is satisfied:

(i) if $\rho(\alpha) < \rho(\beta)$

(ii) if $\rho(\alpha)=\rho(\beta)$, and $(i_k i_{k+1} \ldots i_n)$ precedes $(j_k j_{k+1} \ldots j_n)$;

(iii) if $\rho(\alpha)=\rho(\beta)$, and $(i_k i_{k+1} \ldots i_n)$ and $(j_k j_{k+1} \ldots j_n)$ represent the same connectivity, and $(i_1 i_2 \ldots i_{k-1})$ precedes $(j_1 j_2 \ldots j_{k-1})$.

For example, in the case $n=2$ there exist only two different connectivities. Their representations are arranged in the order (1 1) and (1 2), according to (i).

It is convenient to make the following definitions. For the total number of possible n-point connectivities we write c_n; we define $c_0 \equiv 1$. The c_n are included in a larger set of numbers $c_{n,l}$, which give the number of n-point connectivities α with $\rho(\alpha) \leqslant l$; clearly $c_{n,n+1}=c_n$, and $c_{nl}=0$ for $l<2$. An immediate consequence of the recursivity is that number of n-point connectivities α with $\rho(\alpha)=k$ is given by $c_{k-2}c_{n-k+1}$. Therefore,

$$c_{nl} = \sum_{k=2}^{l} c_{k-2}c_{n-k+1},$$

and in particular

$$c_n = \sum_{k=2}^{n+1} c_{k-2}c_{n-k+1}. \tag{AI.2}$$

To assign a unique number in the range $1, 2, \ldots, c_n$, corresponding to the ordering defined above, to a connectivity represented by $(i_1 i_2 \ldots i_n)$ we recursively define the function σ:

$$\sigma(i_1 i_2 \ldots i_n) = \begin{cases} 1 \text{ if } n=1 \text{ or if } (i_1 \ldots i_n) \text{ is an empty sequence} \\ c_{n,k-1} + \sigma(i_2 \ldots i_{k-1}) + [\sigma(i_k \ldots i_n) - 1]c_{k-2} \text{ otherwise,} \end{cases} \tag{AI.3}$$

where $k \equiv \rho(i_1 \ldots i_n)$. A function which maps a number in the range $1, 2, \ldots, c_n$ onto a sequence of well-nested integers, and which has σ as its inverse, can also be defined recursively.

From eq. (AI.2) a linear recursion relation for c_n can be obtained as follows.

Introduce the generating function

$$P(x) = \sum_{n=0}^{\infty} c_n x^n.$$

From eq. (AI.2) one then finds

$$P(x) - 1 = x[P(x)]^2.$$

The only acceptable solution of this equation is

$$P(x) = \frac{1 - \sqrt{1-4x}}{2x}.$$

It follows easily from the series expansion of P that

$$c_n = \frac{4n-2}{n+1} c_{n-1}, \quad n \geq 1,$$

so that asymptotically for large n

$$c_n \sim 4^n. \tag{AI.4}$$

Numerical values of c_n are listed in table (AI.1) for n = 1–12. We are now in a position to describe the numbering of the general connectivities which serve as indices of the transfer matrix for the extended Whitney polynomial and the Potts model in a field. Also a general connectivity can be represented as a

TABLE (AI.1)
The total number c_n of restricted n-point connectivities without bonds to the ghost site, and the number d_n of general n-point connectivities.

n	c_n	d_n
1	1	2
2	2	5
3	5	15
4	14	51
5	42	188
6	132	731
7	429	2950
8	1430	12235
9	4862	51822
10	16796	223191
11	58786	
12	208012	

sequence of integers $(i_1i_2 \ldots i_n)$, provided one particular value is reserved for sites that are connected to the ghost site. Thus the restriction $i_j \neq 0$ is dropped: we assign the value $i_j = 0$ whenever site j of column m of $\mathscr{L}_m$ is – directly or indirectly – connected to the ghost site. The "well-nestedness" property now holds only for the nonzero integers. The ordering of the restricted connectivities (the "zero field case") naturally introduces a partial ordering of the general connectivities: if α and β have precisely the same sites connected to the ghost site, they are ordered according to the connectivity of the remaining sites. If different sites are connected to the ghost site, an ordering can simply be obtained from the number and the positions of these sites. Thus a complete ordering for the general connectivities follows. For a more precise formulation, represent a connectivity α by $(i_1i_2 \ldots i_n)$ and a connectivity β by $(j_1j_2 \ldots j_n)$. Denote by s the number of zeroes in α, by t the number of zeroes in β, and by $a = (p_1p_2 \ldots p_{n-s})$ and $b = (q_1q_2 \ldots q_{n-t})$ the sequence of all sites p in α and all sites q in β for which $i_p \neq 0$ and $j_q \neq 0$, respectively. Then α precedes β whenever one of the following conditions is satisfied:

(i) if $s < t$

(ii) if $s = t$, and a precedes b in lexicographic order (i.e. $p_1 = q_1$, $p_2 = q_2, \ldots, p_{k-1} = q_{k-1}$ and $p_k < q_k$ for $1 \leq k \leq n - s$)

(iii) if $s = t$, $(p_1 \ldots p_{n-s}) = (q_1 \ldots q_{n-t})$, and $(i_{p_1}i_{p_2} \ldots i_{p_{n-s}})$ precedes $(j_{q_1}j_{q_2} \ldots j_{q_{n-t}})$ according to the ordering of the restricted set of connectivities.

An auxiliary function ψ which, on basis of condition (ii), assigns a unique integer to a sequence $(i_1i_2 \ldots i_n)$ with precisely s zeroes, is recursively defined as follows:

$$\psi(i_1i_2 \ldots i_n) = \begin{cases} 1 \text{ if } n = 1 \text{ or } s = n \\ \psi(i_2i_3 \ldots i_n) \text{ if } i_1 \neq 0 \\ \binom{n-1}{s} + \psi(i_2i_3 \ldots i_n) \text{ if } i_1 = 0. \end{cases} \tag{AI.5}$$

Further, d_{nl} is defined as the number of n-point connectivities with at most l sites connected to the ghost site:

$$d_{nl} = \sum_{k=0}^{l} \binom{n}{k} c_{n-k},$$

$$d_{nl} = 0 \text{ for } l < 0. \tag{AI.6}$$

The total number of general n-point connectivities $d_n \equiv d_{nn}$ is shown in table (AI.1) for $n = 1, 2, \ldots, 10$. From eqs. (AI.4) and (AI.6) it follows that asymptotically for large n

$$d_n \sim 5^n.$$

Finally a function τ is defined which, according to (AI.5), assigns a unique integer in the range $1, 2, \ldots, d_n$ to a general connectivity represented by a sequence $(i_1 i_2 \ldots i_n)$ with precisely s zeroes, and nonzeroes on positions given by $(p_1 p_2 \ldots p_{n-s})$:

$$\tau(i_1 i_2 \ldots i_n) = d_{n,s-1} + \{\psi(i_1 i_2 \ldots i_n) - 1\} c_{n-s} + \sigma(i_{p_1} i_{p_2} \ldots i_{p_{n-s}}).$$

An inverse function can also be defined, in order to map an integer in the range $1, 2 \ldots d_n$ onto a sequence which represents a general connectivity.

Appendix II

Numerical data

TABLE AII.1

Critical free energies of $n \times \infty$, q-state Potts lattices for several values of n and q. For most values of q, these data were obtained from the simple Whitney polynomial, and could be computed up to $n = 11$. In the special case $q = 2$, data were obtained up to $n = 15$ (see ref. 22) by means of a different algorithm. These data show good apparent convergence with increasing n to the analytical results of Baxter[24]) for all q.

n \ q	1/64	1/16	1/2	0.95
2	−1.120119174989	−0.354283874153	0.915335574730	1.350469158236
3	−0.983897069726	−0.242646201833	0.952916342987	1.353476623045
4	−0.938305675488	−0.205308707222	0.965423202048	1.354472839752
5	−0.917848341758	−0.188585593429	0.970990578598	1.354914681594
6	−0.906921777387	−0.179664094320	0.973949599750	1.355149041841
7	−0.900393234325	−0.174337077013	0.975712782538	1.355288536725
8	−0.896178245847	−0.170899120463	0.976849356333	1.355378400680
9	−0.893297995296	−0.168550388717	0.977625265191	1.355439724384
10	−0.891242331139	−0.166874332373	0.978178682825	1.355483452108
11	−0.889723754951	−0.165636316196	0.978587322901	1.355515734412
∞	−0.882519177979	−0.159764272049	0.980523980355	1.355668662600

n \ q	1.05	2	3	4
2	1.420520199800	1.886426125762	2.193053190795	2.416904586478
3	1.417627080431	1.842546256346	2.121091261780	2.324209117946
4	1.416669704806	1.828157728044	2.097704520030	2.294336863540
5	1.416245392403	1.821819028739	2.087460663806	2.281324350426

TABLE AII.1 (cont.)

n \ q	1.05	2	3	4
6	1.416020416402	1.818468940405	2.082063689903	2.274490915883
7	1.415886534683	1.816478784614	2.078863356335	2.270446977423
8	1.415800296982	1.815198160430	2.076806333203	2.267851302896
9	1.415741452203	1.814324882545	2.075404683689	2.266084394071
10	1.415699494189	1.813702482337	2.074406246134	2.264826740710
11	1.415668519467	1.813243149887	2.073669695186	2.263899533697
12		1.812894445034		
13		1.812623456975		
14		1.812408674418		
15		1.812235551224		
∞	1.415521797633	1.811068985361	2.070187162577	2.259524751387

n \ q	5	6	8	10
2	2.594078501571	2.741090481735	2.977062953660	3.163361458476
3	2.484965133364	2.618413276899	2.832841876752	3.002423836320
4	2.450082121222	2.579492490479	2.787739729634	2.952764206839
5	2.434969521535	2.562721816111	2.768514893224	2.931824411836
6	2.427060432573	2.553976942443	2.758569582214	2.921085806687
7	2.422390810971	2.548827581110	2.752750370495	2.914850159359
8	2.419398669714	2.545534974955	2.749049705287	2.910912817164
9	2.417364626582	2.543300619258	2.746550801879	2.908272420841
10	2.415918434792	2.541714440793	2.744784960172	2.906419364696
11	2.414853227199	2.540547723157	2.743491760863	2.905071647267
∞	2.409849128046	2.535117104257	2.737738040357	2.899522170312

n \ q	12	16	32	64
2	3.317612792242	3.564557172286	4.175880541906	4.807008717583
3	3.143086844571	3.368850632375	3.931470715334	4.518859770517
4	3.089908481002	3.310587328373	3.863986757669	4.447285563554
5	3.067725257548	3.286795265404	3.838531234275	4.423510918714
6	3.056454667110	3.274946903004	3.826893004772	4.414147644918
7	3.049967151003	3.268262924507	3.820922301875	4.410092696525
8	3.045906119167	3.264166394028	3.817630672062	4.408235111945
9	3.043206739886	3.261504168311	3.815727776572	4.407352367265
10	3.041329559268	3.259697198878	3.814590810928	4.406921666466
11	3.039977299401	3.258429071539	3.813894963476	4.406707180785
∞	3.034908520637	3.254520179335	3.812678066251	4.406481187417

TABLE AII.2

Dimensionless specific heats C of $n \times \infty$, q-state Potts lattices at criticality for several values of n and q. For small values of q, these data exhibit apparent convergence to a (q-dependent) constant. In contrast, for high values of q, they show a strongly increasing trend with n.

n \ q	1/64	1/16	1/4	1/2
2	−0.910670257	−0.806558279	−0.564989957	−0.340808193
3	−0.937701070	−0.856430975	−0.639294839	−0.406615592
4	−0.947638006	−0.876343014	−0.673625700	−0.440434303
5	−0.952448688	−0.886545990	−0.693105733	−0.461135256
6	−0.955169784	−0.892578912	−0.705582136	−0.475207861
7	−0.956870060	−0.896489132	−0.714220443	−0.485443408
8	−0.958008787	−0.899190872	−0.720534614	−0.493246290
9	−0.958811542	−0.901147933	−0.725338505	−0.499404559
10	−0.959400263	−0.902618151	−0.729107899	−0.504396457
11	−0.959845799	−0.903755096	−0.732139064	−0.508529803

n \ q	0.95	1.05	1.5	2
2	−0.030269855	0.029640311	0.273373121	0.50858408
3	−0.038627811	0.038294152	0.370731682	0.72109958
4	−0.043485045	0.043432537	0.433472264	0.86831059
5	−0.046736573	0.046927273	0.478834887	0.98075978
6	−0.049108979	0.049509465	0.514029636	1.07195386
7	−0.050938598	0.051521807	0.542594982	1.14876587
8	−0.052404902	0.053149035	0.566511768	1.21516202
9	−0.053613789	0.054501113	0.586998754	1.27365043
10	−0.054632416	0.055648323	0.604858200	1.32592412
11	−0.055505734	0.056638035	0.620645372	1.37318214
12				1.41630572
13				1.45596192
14				1.49266801
15				1.52683333

n \ q	2.5	3	3.5	4
2	0.71794697	0.90816210	1.08339692	1.246469540
3	1.05738995	1.38295688	1.69986733	2.009515580
4	1.30704745	1.75062356	2.19938109	2.653399100
5	1.50679790	2.05692350	2.63078375	3.227919500
6	1.67501736	2.32358135	3.01774032	3.757501860
7	1.82128592	2.56207829	3.37279983	4.255009340
8	1.95124683	2.77925418	3.70344866	4.728040980
9	2.06853898	2.97957117	4.01458418	5.181527090
10	2.17566474	3.16613502	4.30963002	5.618894600
11	2.27442784	3.34121167	4.59109773	6.042648080

TABLE AII.2 (cont.)

n \ q	6	8	16	64
2	1.81079679	2.27956414	3.6696004	7.702726
3	3.19223727	4.30849468	8.3634664	27.03826
4	4.52037899	6.46178070	14.799596	72.97106
5	5.83963562	8.78740866	23.511237	176.3499
6	7.17216182	11.3149105	35.128266	398.4556
7	8.52869218	14.0632952	50.409576	859.3639
8	9.91483564	17.0468885	70.274667	1791.051
9	11.3337577	20.2779493	95.837549	3636.313
10	12.7873700	23.7678915	128.44655	7231.922
11	14.2768936	27.5278727	169.73097	

TABLE AII.3

Critical susceptibilities χ of $n \times \infty$, q-state Potts lattices for several values of n and q. For most values of q these data were obtained from the extended Whitney polynomial, and results could be computed up to $n = 9$. However, for $q = 2$ and 3, a different algorithm was used and data up to somewhat higher n could be derived.

n \ q	1/64	1/16	1/4	1/2
2	-2.663293×10^5	-8.206734×10^3	-2.208870×10^2	-2.770149×10^1
3	-5.931556×10^5	-1.812028×10^4	-4.813786×10^2	-5.993920×10^1
4	-1.046143×10^6	-3.173714×10^4	-8.336935×10^2	-1.030907×10^2
5	-1.623171×10^6	-4.893878×10^4	-1.272435×10^3	-1.563009×10^2
6	-2.323013×10^6	-6.965799×10^4	-1.794685×10^3	-2.191272×10^2
7	-3.144830×10^6	-9.384982×10^4	-2.398558×10^3	-2.912905×10^2
8	-4.087955×10^6	-1.214781×10^5	-3.082545×10^3	-3.725708×10^2
9	-5.151790×10^6	-1.525116×10^5	-3.845358×10^3	-4.627787×10^2

n \ q	0.95	1.05	1.5
2	-6.023821×10^{-1}	4.758567×10^{-1}	2.064346×10^0
3	-1.296608×10^0	1.023707×10^0	4.436474×10^0
4	-2.216408×10^0	1.748474×10^0	7.559750×10^0
5	-3.338981×10^0	2.631576×10^0	1.134417×10^1
6	-4.653132×10^0	3.663974×10^0	1.574743×10^1
7	-6.152063×10^0	4.840215×10^0	2.074467×10^1
8	-7.830462×10^0	6.156047×10^0	2.631663×10^1
9	-9.683760×10^0	7.607817×10^0	3.244679×10^1

TABLE AII.3 (cont.)

n \ q	2	2.5	3	3.5
2	2.117991×10^0	1.900457×10^0	1.669517×10^0	1.469242×10^0
3	4.554241×10^0	4.092690×10^0	3.602888×10^0	3.178483×10^0
4	7.753616×10^0	6.969442×10^0	6.141339×10^0	5.426067×10^0
5	1.161543×10^1	1.043446×10^1	9.196238×10^0	8.131292×10^0
6	1.609363×10^1	1.444457×10^1	1.272848×10^1	1.125909×10^1
7	2.116175×10^1	1.897544×10^1	1.671638×10^1	1.479037×10^1
8	2.679943×10^1	2.400844×10^1	2.114342×10^1	1.871061×10^1
9	3.298921×10^1	2.952756×10^1	2.599538×10^1	2.300732×10^1
10	3.971570×10^1		3.125973×10^1	
11	4.696532×10^1			
12	5.472594×10^1			
13	6.298668×10^1			
14	7.173773×10^1			
15	8.097013×10^1			

n \ q	4	6	8	64
2	1.302714×10^0	8.729978×10^{-1}	6.448399×10^{-1}	6.644713×10^{-2}
3	2.825819×10^0	1.916283×10^0	1.432987×10^0	1.834556×10^{-1}
4	4.833189×10^0	3.310643×10^0	2.505980×10^0	4.373205×10^{-1}
5	7.251600×10^0	5.007076×10^0	3.832712×10^0	9.665300×10^{-1}
6	1.004976×10^1	6.987120×10^0	5.405789×10^0	2.052380×10^0
7	1.321099×10^1	9.242817×10^0	7.226181×10^0	4.249200×10^0
8	1.672267×10^1	1.176875×10^1	9.296861×10^0	8.628000×10^0
9	2.057389×10^1	1.456030×10^1	1.162141×10^1	1.720000×10^1

TABLE AII.4

Second derivative of the inverse correlation length κ of critical, finite size, simple Whitney models, with respect to Potts coupling strength K, for several values of n and q.

n \ q	1/16	0.95	1.05	2
2	−0.792847	3.948989	4.047782	4.618802
3	−2.731687	4.856156	5.138433	7.155418
4	−3.307671	5.622623	6.085301	9.762104
5	−3.368125	6.300672	6.931941	12.36725
6	−3.277155	6.910796	7.700674	14.96181
7	−3.143298	7.467377	8.407904	17.54667
8	−3.002575	7.980972	9.065684	20.12394
9	−2.867566	8.459403	9.682915	22.69545
10	−2.742360	8.908545	10.26627	25.26257

TABLE AII.4 (cont.)

n \ q	3	4	64
2	4.929550	5.130977	6.444482
3	8.666146	9.888715	32.73622
4	12.95531	15.82173	118.8470
5	17.63239	22.74557	361.8633
6	22.64251	30.60055	985.6596
7	27.95441	39.35191	2486.075
8	33.54508	48.97287	5928.705
9	39.39583	59.44077	13549.53
10	45.49097	70.73590	29950.41

TABLE AII.5

Derivatives of the inverse correlation length κ of extended Whitney models at criticality with respect to Potts coupling strength K, for several values of q and n. Results for $q=2$ are equal to -1 (independent of n) and not shown. These data correspond to the spin–spin correlation length of $n\times\infty$, q-state Potts lattices. For most q-values, data were computed up to $n=8$; for $q=64$, only to $n=7$.

n \ q	1/16	1/4	0.95	1.05
2	−0.606367551	−0.704629279	−0.869895882	−0.885763777
3	−0.445728508	−0.563010422	−0.791256688	−0.815313447
4	−0.357275504	−0.478113594	−0.737127528	−0.766183936
5	−0.300692240	−0.420603644	−0.696738407	−0.729176825
6	−0.261091484	−0.378569708	−0.664953865	−0.699838952
7	−0.231668978	−0.346225813	−0.638991709	−0.675731559
8	−0.208857265	−0.320399764	−0.617196919	−0.655391813

n \ q	3	4	8	64
2	−1.08184259	−1.14372631	−1.30053621	−1.71043661
3	−1.14484601	−1.26121480	−1.58023225	−2.48092752
4	−1.19623672	−1.36160028	−1.84528240	−3.27759800
5	−1.23987932	−1.45014438	−2.10031210	−4.08569151
6	−1.27798338	−1.52997057	−2.34823933	−4.89858583
7	−1.31191789	−1.60306648	−2.59095524	−5.71342800
8	−1.34258976	−1.67077815	−2.82974646	

TABLE AII.6

Derivatives of the inverse correlation length κ of finite-size, extended Whitney models at criticality with respect to magnetic field coupling strength h, for several values of q and n. Results for $q=2$ are equal to zero. These data correspond to the spin–spin correlation length of q-state, $n\times\infty$ Potts models. For most q-values, data were computed up to $n=8$; for $q=64$, only to $n=7$.

n \ q	1/16	1/4	0.95	1.05
2	60.959983	13.634261	2.1368675	1.7487141
3	90.340104	20.036205	3.1132225	2.5464743
4	119.33235	26.284567	4.0523983	3.3130591
5	148.04661	32.424184	4.9652705	4.0575641
6	176.54684	38.480869	5.8581792	4.7853043
7	204.87309	44.470657	6.7350837	5.4996041
8	233.05260	50.404376	7.5986591	6.2027116

n \ q	3	4	8	64
2	−0.6445731	−0.9681646	−1.4596318	−1.9201094
3	−0.9371654	−1.4091556	−2.1359458	−2.8667774
4	−1.2168249	−1.8313830	−2.7911381	−3.8153533
5	−1.4871810	−2.2401551	−3.4324494	−4.7660305
6	−1.7503895	−2.6385987	−4.0639170	−5.7179109
7	−2.0078198	−3.0286915	−4.6879903	−6.6703600
8	−2.2604116	−3.4117875	−5.3062914	

References

1) R.B. Potts, Proc. Camb. Phil. Soc. **48** (1952) 106.
2) F.Y. Wu, J. Statist. Phys. **18** (1978) 115.
3) A.N. Berker, S. Ostlund and F.A. Putnam, Phys. Rev. **B17** (1978) 3650.
4) D. Mukamel, M.E. Fisher and E. Domany, Phys. Rev. Lett. **37** (1976) 565.
5) A. Aharony, K.A. Muller and W. Berlinger, Phys. Lett. **38** (1977) 33.
6) A. Aharony and P. Pfeuty, J. Phys. **C12** (1979) L125.
7) A. Aharony, J. Phys. **C11** (1978) L457.
8) J.P. Straley and M.E. Fisher, J. Phys. **A6** (1973) 1310.
9) T. de Neef and I.G. Enting, J. Phys. **A10** (1977) 801.
10) I.G. Enting, J. Phys. **A7** (1964) 1617, 2181.
11) D. Kim and R.I. Joseph, J. Phys. **A8** (1975) 891.
12) I.G. Enting, J. Phys. **A13** (1980) L133.
13) K. Binder, J. Stat. Phys. **24** (1981) 69.
14) Th.W. Burkhardt, Z. Phys. **B39** (1980) 159.
15) B. Nienhuis, A.N. Berker, E.K. Riedel and M. Schick, Phys. Rev. Lett. **43** (1979) 737.
16) B. Nienhuis, E.K. Riedel and M. Schick, J. Phys. **A13** (1980) L31.
17) B. Nienhuis, E.K. Riedel and M. Schick, J. Phys. **A13** (1980) L189.
18) Bambi Hu, J. Phys. **A13** (1980) L321.

19) J. Sólyom and P. Pfeuty, Phys. Rev. **B24** (1981) 218.
20) R.H. Swendsen, Phys. Rev. Lett. **42** (1979) 859.
21) C. Rebbi and R.H. Swendsen, Phys. Rev. **B21** (1980) 4094.
22) M.P. Nightingale and H.W.J. Blöte, Physica **104A** (1980) 352.
23) H.H. Roomany, H.W. Wyld and L.E. Holloway, Phys. Rev. **D21** (1980) 1557.
24) R.J. Baxter, J. Phys. **C6** (1973) L445.
25) L. Onsager, Phys. Rev. **65** (1944) 117.
26) F.Y. Wu, Rev. Mod. Phys. **54** (1982) 235.
27) M.P.M. den Nijs, J. Phys. **A12** (1979) 1857.
28) R.B. Pearson, Phys. Rev. **B22** (1980) 2579.
29) R.J. Baxter, J. Phys. **A13** (1980) L61.
30) R.J. Baxter and F.Y. Wu, Phys. Rev. Lett. **31** (1973) 1294.
31) R.J. Baxter, M.F. Sykes, M.G. Watts, J. Phys. A. **8** (1975) 245.
32) S. Alexander. Phys. Lett. **A54** (1975) 353.
33) E. Domany, M. Schick and J.S. Walker, Phys. Rev. Lett. **38** (1977) 1148.
34) M.P.M. den Nijs, Phys. Rev. **B23** (1981) 6111.
35) J.L. Black and V.J. Emery, Phys. Rev. **B23** (1981) 429.
36) H.W.J. Blöte, M.P. Nightingale and B. Derrida, J. Phys. **A14** (1981) L459.
37) M.E. Fisher, Proc. Enrico Fermi Intern. School of Phys., M.S. Green, ed. (Academic Press, New York, 1971).
38) M.P. Nightingale, Phys. Lett. **59A** (1977) 486.
39) M.P. Nightingale, Proc. Kon. Ned. Ak. Wet. **B82** (1979) 235.
40) F.J. Wegner, Phys. Rev. **B5** (1972) 4529.
41) F.J. Wegner and E.K. Riedel, Phys. Rev. **B7** (1973) 248.
42) M.P. Nightingale and A.H. 't Hooft, Physica **77** (1974) 390.
43) Th. Niemeyer and J.M.J. van Leeuwen in Phase Transitions and Critical Phenomena, C. Domb and M.S. Green, eds. Vol. 6 (Acad. Press, New York, 1976).
44) H. Au-Yang and M.E. Fisher, Phys. Rev. **B11** (1975) 3469.
45) M.N. Barber, M.E. Fisher, Phys. Rev. **A8** (1973) 1124.
46) M. Nauenberg and D.J. Scalapino, Phys. Rev. Lett. **44** (1980) 837.
47) Y. Imry, Phys. Rev. **B5** (1980) 2042.
48) C.J. Hamer and M.N. Barber, J. Phys. **A14** (1981) 2009.
49) H. Whitney, Ann. Math. N.Y. **33** (1932) 688.
50) R.J. Baxter, S.B. Kelland and F.Y. Wu, J. Phys. **A9** (1976) 397.
51) P.W. Kasteleyn and C.M. Fortuin, J. Phys. Soc. Jap. **46** (suppl.) (1969) 11.
52) C. Domb, Adv. in Phys. **9** (1960), §3.5.1.
53) L. Mittag and M.J. Stephen, J. Math. Phys. **12** (1971) 441.
54) J.H. Wilkinson, The Algebraic Eigenvalue Problem (Clarendon, Oxford, 1965).
55) B. Nienhuis, E.K. Riedel and M. Schick, Phys. Rev. **B23** (1981) 6055.
56) H.N.V. Temperley and E.H. Lieb, Pr. Roy. Soc. Lond **A322** (1971) 251.
57) J.B. Kogut, Rev. Mod. Phys. **51** (1979) 659.
58) W. Klein, H.E. Stanley, P.J. Reynolds and A. Coniglio, Phys. Rev. Lett. **41** (1978) 1145.
59) B. Nienhuis and M. Nauenberg, Phys. Rev. Lett. **35** (1975) 477.
60) This explanation has very recently been confirmed: R.H. Swendsen, D. Andelman and A.N. Berker, preprint.

J. Phys. A: Math. Gen. **13** (1980) L169–L174. Printed in Great Britain

LETTER TO THE EDITOR

Finite-size scaling in Hamiltonian field theory

C J Hamer† and Michael N Barber‡§

† Department of Theoretical Physics, Research School of Physical Sciences, Australian National University, Canberra, PO Box 4, ACT 2600, Australia
‡ Department of Physics, University of Washington, Seattle, WA 98195, USA

Received 11 February 1980

Abstract. A new method for investigating the behaviour of lattice Hamiltonian field theories is described. The method uses finite-size scaling to extrapolate finite-lattice results to the infinite chain limit. The technique is illustrated by application to the transverse Ising model and the O(N)-Heisenberg Hamiltonians ($N=2, 3$) in $(1+1)$ dimensions. The accuracy of the method appears comparable to or better than existing approaches.

Hamiltonian field theories on spatial lattices are of interest both in statistical mechanics and field theory (for a review see Kogut 1979). Currently, the only reliable method of investigating the critical behaviour and phase diagrams of such theories is through the analysis of the Rayleigh–Schrödinger perturbation series (see e.g. Hamer *et al* 1978, 1979). Attempts have been made to use renormalisation group techniques (Drell *et al* 1976, Jullien *et al* 1978) but as yet such methods do not offer comparable numerical accuracy even for simple systems.

In this Letter, we propose a new method which uses finite-size scaling (Fisher and Barber 1972) to extract the behaviour of the infinite lattice theory from the way physical quantities of interest vary with lattice size. In particular, we illustrate these ideas by summarising a series of finite-lattice calculations of the mass gap and β-function for the Hamiltonian versions of the two-dimensional Ising and O(N)-Heisenberg ($N=2, 3$) models. Our results indicate that this procedure has an accuracy comparable to or better than that of series methods. Full details and further applications will be reported elsewhere.

We begin with the Ising model. The relevant quantum Hamiltonian (Fradkin and Susskind 1978) is

$$H=(g/2a)\sum_{m=1}^{M}\{1-\sigma_3(m)-x\sigma_1(m)\sigma_1(m+1)\}. \tag{1}$$

Here $\sigma_1(m)$ are Pauli matrices, g is a dimensionless coupling constant (proportional to temperature), a is the lattice spacing, $x=2/g^2$ and the sum is over the M sites of a chain with periodic boundary conditions. In the limit $M\to\infty$, the ground state energy and mass gap of this model are known exactly (Pfeuty 1970). We have investigated the behaviour for finite M both analytically and numerically. Here we focus attention on

§ Permanent address: Department of Applied Mathematics, University of New South Wales, PO Box 1, Kensington, NSW2033, Australia.

0305-4470/80/050169+06\$01.50

the mass gap

$$F(x, M)=(2a/g)(E_1-E_0), \tag{2}$$

where E_0 and E_1 are the energies of the ground state and first excited state respectively.

The mass gap of (1) for finite M can be determined analytically using standard fermion techniques (see e.g. Schultz *et al* 1964). Analysis of this result leads to the following conclusions.

(i) For fixed $x \neq 1$, $F(x, M)$ approaches its known limiting value exponentially fast in M; ie

$$F(x, M)=2|1-x|+\mathrm{O}(\mathrm{e}^{-|1-x|M}), \qquad M\to\infty. \tag{3}$$

(ii) On the other hand, at $x=x_c=1$,

$$F(1, M)=\pi/4M+\mathrm{O}(M^{-3}). \tag{4}$$

(iii) Finally, if we consider a *uniform* limit, $x\to 1$, $M\to\infty$ with $(1-x)M=\mathrm{O}(1)$, then

$$F(x, M)\approx M^{-1}\mathrm{O}(|1-x|M), \tag{5}$$

where the scaling function can be computed exactly, and leads back to (3) and (4) in the appropriate limits.

These results are rather familiar. They are precisely the predictions which follow by extending the results of finite-size scaling from conventional statistical mechanics (Lagrangian field theory) to Hamiltonian field theory.

Let us recall the salient features of this theory (Fisher and Barber 1972). Let $\Psi(g)$ be some quantity which in an infinite system varies near some critical coupling g_c as

$$\Psi(g)=A|\Delta g|^{-\psi}, \qquad \Delta g=g-g_c\to 0. \tag{6}$$

Then in a finite system of *linear* dimension n, finite-size scaling asserts that the behaviour of $\Psi(g; n)$ is described by the ansatz:

$$\Psi(g; n)\approx n^{\psi/\nu}Q_\Psi(n/\xi(g)), \qquad n\to\infty, \Delta g\to 0. \tag{7}$$

Here $\xi(g)$ is the correlation length in the finite system, which diverges at g_c as

$$\xi(g)\simeq\xi_0|\Delta g|^{-\nu}. \tag{8}$$

Thus if g_c is known, (7) immediately implies that

$$\Psi(g_c; n)\simeq\text{const. } n^{\psi/\nu}, \qquad n\to\infty, \tag{9}$$

where for a logarithmic singularity ($\psi=0$) the result is modified to

$$\Psi(g_c; n)\simeq\text{const.} \ln n, \qquad n\to\infty. \tag{10}$$

To apply these results to Hamiltonian field theory we simply make use of the standard relationships between statistical mechanics and field theory (Kogut 1979), in particular, that between the mass gap and the reciprocal of the correlation length. Hence if we identify Ψ in (7) and (9) with $1/\xi=F$, we immediately recover the exact results (4) and (5) provided $\nu=1$, which is true (Pfeuty 1970). Alternatively, ν can be estimated directly for the behaviour of the β-function defined (Hamer *et al* 1979) by

$$\beta(g)/g-(\mathrm{d}/\mathrm{d}g)\ln[(g/2a)F(x)]. \tag{11}$$

1988

Note: The authors have pointed out that in the originally published version eq. (5) contains an error, and that it should read:

$$F(x, M)\approx M^{-1}Q(|1-x|M), \tag{5}$$

In the infinite system this possesses a simple zero at x_c, while in the finite system, finite-size scaling predicts that

$$\beta(g_c, M) \approx M^{-1/\nu}, \qquad M \to \infty, \tag{12}$$

a result which can be confirmed analytically for the Ising Hamiltonian.

To test the predictive capabilities of finite-size scaling in the present context, we show in figure 1 a log–log plot of both the mass gap and the β-function evaluated at $x = x_c = 1$ for several values of M.

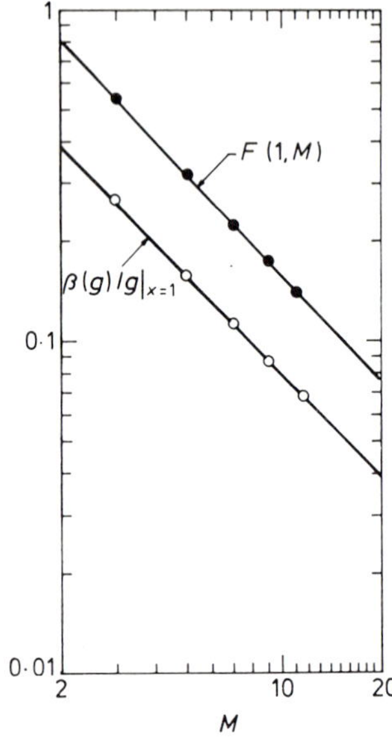

Figure 1. Log–log plots of finite lattice mass gaps (full circles) and β-functions (open circles) evaluated at $x = x_c = 1$ versus M. Straight lines have been drawn through each set of results.

The straight lines indicate that the scaling behaviour is established remarkably quickly. Equation (4) for the mass gap is confirmed to less than one percent, while from the β-function we estimate ν to be unity to a similar precision.

In practice, the critical value x_c of the coupling in the infinite system is often not known. The finite-size scaling form (4) suggests that g_c (or x_c) can be estimated from the sequence of values of x for which successive ratios of $F(x, M)$ and $F(x, M+1)$ exactly scale, i.e. the value of x for which

$$R_M(x) \equiv MF(x, M)/(M-1)F(x, M-1) = 1. \tag{13}$$

In figure 2, we exhibit a plot of $R_M(x)$ against x for various values of M. All curves drop quite rapidly through the value unity near $x = 1$ even for small M. This criterion is actually equivalent to that which follows by extending (Sneddon and Stinchcombe 1979) 'phenomenological renormalisation theory' (Nightingale 1976) to Hamiltonian field theory. It has been used previously to analyse numerical results for the mass gap of (1) (Sneddon and Stinchcombe 1979). As we shall now discuss, the criterion (13) also appears to be a very sensitive probe in more complex systems such as the O(N)-Heisenberg models.

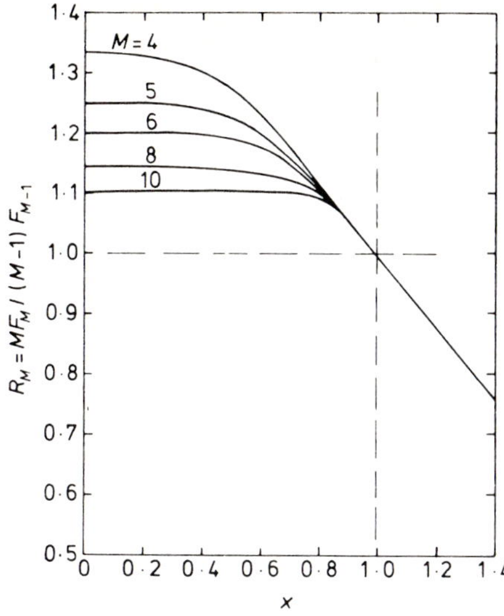

Figure 2. Plot of scaled mass gap ratios $R_M(x)$ versus x for the Ising model.

The lattice Hamiltonian version of these models takes the form (see e.g. Hamer *et al* 1979)

$$H = (g/2a) \sum_{m=1}^{M} \{\boldsymbol{J}^2(m) - x\boldsymbol{n}(m) \cdot \boldsymbol{n}(m+1)\} \tag{14}$$

where the notation is as before, except that $\boldsymbol{J}(m)$ is the angular momentum operator appropriate to $O(N)$ rotational symmetry and $\boldsymbol{n}(m)$ is an N-component spin vector normalised to unity. Strong coupling series for the mass gap and β-function of these models for $N \leq 4$ and $M = \infty$ have been derived and analysed by Hamer *et al* (1978, 1979) and Hamer and Kogut (1979).

Unfortunately, for finite M, it does not seem possible to derive any exact results and one must resort to a numerical procedure. However, unlike the Ising Hamiltonian, (14) has an *infinite* state space even on a finite chain. Thus no numerical procedure can presumably compute exact eigenvalues, and rather we require procedures which are sufficiently rapidly convergent to allow reasonably long chains to be investigated. We have devised two methods with this property.

The first is based on the same approach as used in the strong coupling expansions (Hamer *et al* 1979, Hamer 1979). Decompose (14) as $H = (g/2a)(W_0 - xV)$ where W_0 contains the single-site terms and V the coupling between sites. A set of strong coupling eigenstates of W_0 is now generated by successive applications of the operator V to an unperturbed eigenstate of W_0. Using this basis a finite matrix representation of H is calculated and its eigenvalues determined by standard methods. For a model such as the Ising model (1) with a *finite* state space, this method ultimately terminates yielding the exact eigenenergies of the finite chain. For the $O(N)$-models, sufficiently accurate results follow by perturbing to sufficiently high order ($\sim O(M)$). Most of our results have been obtained by this procedure.

We have also explored to some extent an alternative scheme in which H is reduced to a tri-diagonal form. Eigenenergies and other physical quantities can then be obtained iteratively without any explicit diagonalisation. The method is similar to recursive techniques used extensively in band theory (see e.g. Haydock *et al* 1975) and nuclear physics (see e.g. Whitehead *et al* 1977). In the context of Hamiltonian field theory it appears to have a slight storage advantage over the first method and may allow larger chains to be investigated. This approach has in fact been applied to $Z(2)$ and $Z(3)$-Ising spin systems by Roomany *et al* (1979).

Using these methods we have been able to calculate the energies of the ground state and first excited state for the O(2)-model up to $M=6$. The ratios $R_M(x)$ of successive mass gaps are plotted in figure 3. Their behaviour is remarkable. They drop to within a fraction of one percent of the value 1 at $x\simeq 2$ and then *stay* there. This behaviour is established immediately even for M as low as 3. We regard this as a spectacular demonstration of a region of scale invariance. Such a region is of course expected (Kosterlitz 1974, José *et al* 1977). It is hard to decide the exact value at which the region commences, but we estimate it to be at $x_c = 1\cdot8 \pm 0\cdot1$. This value is in good agreement with the series analysis result of Hamer and Kogut (1979).

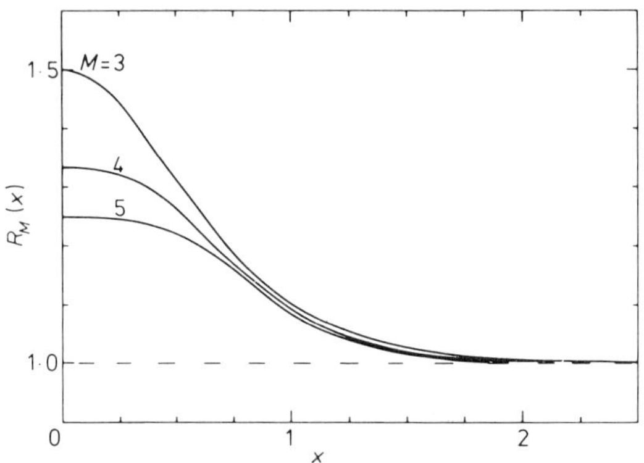

Figure 3. Plot of scaled mass gap ratios $R_M(x)$ versus x for the O(2)-model.

At x_c, one expects (Kosterlitz 1974) that in the infinite system the mass gap varies as

$$F(x) \sim \exp[-a/(x_c - x)^\sigma] \tag{15}$$

with $\sigma = \frac{1}{2}$. It then follows that the β-function has an *algebraic* singularity at x_c. The form (15) is not that usually adopted (see (8)) in the derivation of finite-size scaling results. However, it is easy to adapt the finite-size scaling analysis to incorporate this behaviour. The key prediction for our present purposes is that

$$\beta(g; M)/g|_{x=x_c} \sim (\ln M)^{-(1+\sigma)/\sigma}, \qquad M \to \infty. \tag{16}$$

Unfortunately, the value of β at $x_c \simeq 1\cdot8$ does not obey this relation very well. However, the *minimum* values of the β-function for each M do scale somewhat better and yield an estimate of $\sigma = 0\cdot9 \pm 0\cdot4$. While the accuracy of this value is not very impressive, it should be compared with the value of $\sigma = 0\cdot6 \pm 0\cdot3$ obtained by Hamer and Kogut (1979) via series analysis methods. Clearly results for larger lattices and a more refined method of analysis would be useful in this context. Nevertheless, it remains gratifying that the finite size analysis does give a clear indication (figure 3) of the scale invariant region.

We have finally applied the same method to the O(3)-Hamiltonian. This model has more degrees of freedom than the O(2)-model. Thus we have only been able to compute the mass gap reliably for two- and three-site lattices. The ratio $R_3(x)$ of these results remains distinctly above unity for all x, as one would expect if no transition occurs. While it is obviously unwise to place too much credence on this one calculation, it is significant that the corresponding quantities for the Ising and O(2)-models already exhibit the behaviour confirmed by results from larger chains. We hope to be able to extend the O(3)-results similarly.

In summary, we have described a method of investigating the behaviour of lattice Hamiltonians by scaling finite-lattice quantities to the infinite lattice limit. Extremely accurate results were thus obtained for the Ising model. Undoubtedly, this high accuracy reflects the simplicity of this model. Indeed, the accuracy for the O(2) and O(3) models was much poorer, but the expected behaviour was detected. Thus the method does seem to be a very sensitive qualitative indicator for the presence (or otherwise) of a phase transition, and its nature. We feel that this type of analysis should be useful in the investigation of more complex systems.

One of us (MNB) is grateful to the University of Washington and in particular Professor E K Riedel for their hospitality and to the National Science Foundation for partial support through grant number DMR 77-12676 A02. We would also like to thank Dr R J Baxter and Dr J J Rehr for useful discussions.

References

Drell S D, Weinstein M and Yankielowicz S 1976 *Phys. Rev.* D **14** 487–516
Fisher M E and Barber M N 1972 *Phys. Rev. Lett.* **28** 1516–9
Fradkin E and Susskind L 1978 *Phys. Rev.* D **17** 2637–58
Hamer C J 1979 *Phys. Lett.* **8213** 75–8
Hamer C J and Kogut J 1979 *Phys. Rev.* B **20** 3859–70
Hamer C J, Kogut J and Susskind L 1978 *Phys. Rev. Lett.* **41** 1337–40
—— 1979 *Phys. Rev.* D **19** 3091–105
Haydock R, Heine V and Kelly M J 1975 *J. Phys. C: Solid St. Phys.* **8** 2591–605
José J V, Kadanoff L P, Kirkpatrick S and Nelson D R 1977 *Phys. Rev.* B **16** 1217–41
Jullien R, Pfeuty P, Fields J N and Doniach S 1978 *Phys. Rev.* B **18** 3568–77
Kogut J 1979 *Rev. Mod. Phys.* **51** 659–713
Kosterlitz J M 1974 *J. Phys. C: Solid St. Phys.* **7** 1046–60
Nightingale M P 1976 *Physica* **83** A 561–72
Pfeuty P 1970 *Ann. Phys.* **57** 79–90
Roomany H H, Wyld H W and Holloway L E 1979 *Univeristy of Illinois preprint*
Schultz T D, Mattis D C and Lieb E H 1964 *Rev. Mod. Phys.* **36** 856–71
Sneddon L and Stinchcombe R B 1979 *J. Phys. C: Solid St. Phys.* **12** 3761–70
Whitehead R R, Watt A, Cole B J and Morrison I 1977 *Adv. Nucl. Phys.*

J. Phys. A: Math. Gen. **14** (1981) L163–L168. Printed in Great Britain

LETTER TO THE EDITOR

Directed percolation: a finite-size renormalisation group approach

Wolfgang Kinzel† and Julia M Yeomans‡§

† Institut für Festkörperforschung der Kernforschungsanlage Jülich, 5170 Jülich, Postfach 1913, West Germany
‡ Baker Laboratory, Cornell University, Ithaca, NY 14853, USA

Received 11 February 1981

Abstract. The finite-size renormalisation group technique introduced by Nightingale is applied to the directed percolation problem. The decay of correlations is anisotropic in this model and finite-size scaling is extended to treat such anisotropy. Precise estimates for critical exponents and percolation probabilities are obtained for site, bond and site–bond percolation on the square lattice with bonds directed along the positive axes. Both free boundary conditions, for which the results converge linearly with $1/n$ as $n \to \infty$, and helical boundary conditions, for which, unexpectedly, the results converge linearly with $1/n^3$, are considered.

The percolation problem (see, for example, Stauffer (1979)), which describes lattices with sites or bonds present with probability p and absent with probability $1-p$, is of continuing interest. *Directed* (or oriented) percolation (Smythe and Wierman 1978) is, however, much less studied. The new feature which differentiates directed from ordinary percolation is a restriction on the direction of flow through the 'open' or 'occupied' bonds: for example, the flow of water through a porous medium becomes a directed percolation problem if gravity is important.

In this paper we study the critical properties of directed percolation on the square lattice illustrated in figure 1. Flow along occupied bonds and sites is permitted only along the direction of the arrows (in the forward 'time' direction). Bonds are present

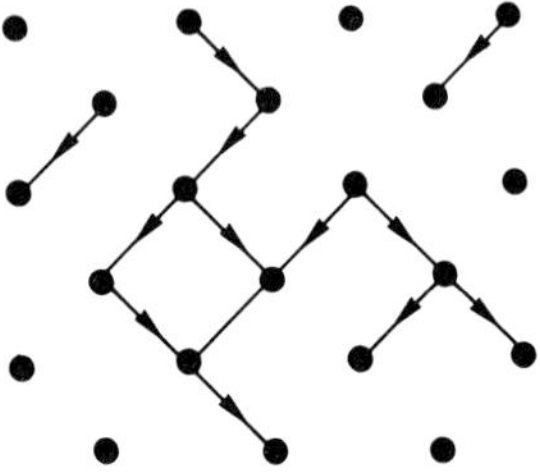

Figure 1. The oriented square lattice. Arrows indicate the allowed directions of flow along the bonds.

§ Address from September 1981: Department of Physics, The University, Highfield, Southampton SO9 5NH, England.

0305-4470/81/050163+06$01.50

with probability p_b and sites with probability p_s, and bond percolation ($p_s = 1$), site percolation ($p_b = 1$) and bond–site percolation (p_s, p_b both variable) are considered.

We utilise the finite-size renormalisation group technique (Nightingale 1976). The directed percolation problem is particularly interesting because the decay of correlations is anisotropic. Specifically, if $\xi_\parallel$ and $\xi_\perp$ are the correlation lengths for decay in the flow direction and perpendicular to it, then we anticipate

$$\xi_\parallel \sim \xi_\perp^\theta \sim (p - p_c)^{-\nu_\parallel} \sim (p - p_c)^{-\nu_\perp \theta} \tag{1}$$

as the critical percolation probability, p_c, is approached. Here, θ is the anisotropy exponent and $\nu_\parallel$ and $\nu_\perp$ the correlation length exponents in the longitudinal and transverse directions respectively. A similar anisotropy of the correlation lengths occurs at a Lifshitz point (Hornreich *et al* 1975), which is currently being studied using similar methods (Yeomans, in preparation).

Both oriented and non-oriented percolation were first defined in the classic paper of Broadbent and Hammersley (1957). Series expansion results of Blease (1977a, b) and Monte Carlo work by Kertész and Vicsek (1980) and Dhar and Barma (1981) yielded estimates for the critical percolation probabilities on several oriented two-dimensional lattices, and indicated that directed and ordinary percolation are in different universality classes. Obukhov (1980) argued that $d = 5$ dimensions provided the upper critical dimension for the directed percolation problem (as opposed to $d = 6$ for ordinary percolation). He derived exponents to leading order in ε for $5 - \varepsilon$ dimensions. Recently Cardy and Sugar (1980) have shown that there is an exact mapping between the directed percolation problem and Reggeon field theory which models the creation, propagation and destruction of a cascade of elementary particles (Grassberger and de la Torre 1979).

Within the finite-size renormalisation group technique (Nightingale 1976, dos Santos and Sneddon 1980), a recursion relation is defined by considering the behaviour of the correlation length under a change in length scale. Thus if ξ_n is the correlation length of a strip of infinite length and of width n sites which is calculated by a transfer matrix technique, one defines the renormalised probability, p', via the relation

$$\xi_{n+1}(p') = [(n+1)/n]\xi_n(p). \tag{2}$$

The fixed point, p_n^*, and critical exponent, ν_n, are then defined as usual by

$$\xi_{n+1}(p_n^*) = [(n+1)/n]\xi_n(p_n^*), \tag{3}$$

$$\frac{1}{\nu_n} = \frac{\ln\left(\left.\frac{d\xi_n}{dp}\right|_{p_n^*} \Big/ \left.\frac{d\xi_{n+1}}{dp}\right|_{p_n^*}\right) - 1}{\ln (n/n+1)}. \tag{4}$$

For $n \to \infty$ one can show that $p_n^* \to p_c$ and $\nu_n \to \nu$ (dos Santos and Sneddon 1980). Note that (3) is equivalent to the assumption of finite-size scaling (Fisher 1972). Precise and apparently accurate values for critical exponents and transition parameters have been obtained using the finite-size renormalisation group for both thermal (Nightingale 1976, Sneddon 1978, dos Santos and Sneddon 1980, Ràcz 1980, Blöte *et al* 1980, Kinzel and Schick 1981) and geometrical (Derrida 1981, Derrida and Vannimenus 1981) models.

However, to treat the directed percolation problem, for which the correlation length is anisotropic, a modification of the approach is necessary. It is the *transverse* correlation length, $\xi_\perp$, which scales linearly with the width of the strip. However the *longitudinal* correlation length, $\xi_{\|}$, is calculated from the transfer matrix. Therefore from (1) and (3) we obtain

$$\frac{\xi_{\|,n+1}(p_n^*)}{\xi_{\|,n}(p_n^*)} = \left(\frac{\xi_{\perp,n+1}(p_n^*)}{\xi_{\perp,n}(p_n^*)}\right)^{\theta_n} = \left(\frac{n+1}{n}\right)^{\theta_n}. \tag{5}$$

This recursion relation contains two unknown parameters p_n^* and θ_n. These may be calculated by a comparison of strips of *three* successive widths, say $n-1$, n and $n+1$.

The transfer matrix technique used to calculate the correlation length in the oriented percolation problem is analogous to that used to treat ordinary percolation by Derrida and Vannimenus (1981). The matrix elements $\langle\psi_N|\mathbf{T}_n|\psi_{N+1}\rangle$ of the transfer matrix $\mathbf{T}_n$ are defined as the probability that row $N+1$ of the lattice is in a configuration $|\psi_{N+1}\rangle$ given that row N is in a configuration $|\psi_N\rangle$ where $|\psi_N\rangle$ describes whether or not the sites in row N are connected to row 1. A simplifying feature of directed percolation (compared with ordinary percolation) is that it is unnecessary to consider paths between sites in rows N and $N+1$ which proceed via earlier rows. The correlation length, ξ_N, is then given by

$$\xi_N = 1/(\ln\lambda_0^{(n)}/\ln\lambda_1^{(n)}) \tag{6}$$

(Camp and Fisher 1972) where $\lambda_0^{(n)}$, $\lambda_1^{(n)}$ are the two largest eigenvalues of $\mathbf{T}_n$. For the percolation problem, as a consequence of the normalisation of the probability, one has $\lambda_0^{(n)} \equiv 1$ and the contribution of the corresponding eigenvector can be factored out of the transfer matrix. The problem is therefore reduced to calculating a single largest eigenvalue, $\lambda_1^{(n)}$.

To calculate $\lambda_1^{(n)}$ a direct iteration technique has been used. Sparse transfer matrices (Domb 1949), which at each step add a single site, prove convenient in speeding the iterations and easing computer storage problems. If the lattice is built up by continued application of a sparse transfer matrix, helical boundary conditions result as illustrated in figure 2. We have studied both helical and free boundary conditions: in the latter case a different sparse matrix must be used to add the edge sites, which have a different connectivity from the sites in the bulk of the strip. For a strip of width n, the sparse transfer matrix is of size $2^n \times 2^n$ and contains 2^{n+1} non-zero elements. The largest strips considered were of width $n = 15$; the calculation took approximately 45 minutes on an IBM 370/168 machine.

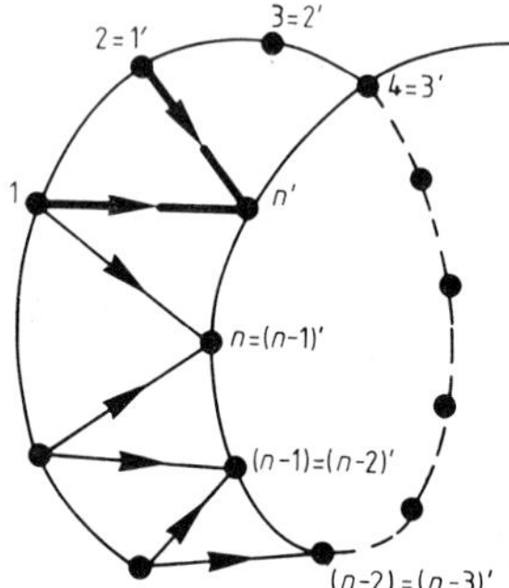

Figure 2. Construction of a strip of width n with helical boundary conditions using a sparse transfer matrix. Each iteration of the matrix adds a site n' together with the two bonds shown by bold lines and sums over the states of site 1.

Table 1 shows results for the site problem on the directed square lattice from calculations on strips of width $n-1$, n and $n+1$ with free boundary conditions. Values are given for p_n^*, the fixed point of the recursion relations, and the corresponding anisotropy and transverse correlation length exponents, θ_n and $\nu_{\perp n}$, obtained for values of n from 4 to 14. Extrapolations calculated from each successive pair of results, assuming that the results tend to a limiting value linearly with $1/n$, are also listed. The extrapolations suggest best values p_c (site) $= 0.7058 \pm 1$, $\theta = 1.581 \pm 1$, $\nu_\perp = 1.094 \pm 1$ where the errors refer to the final significant figure. Similar results were obtained for the bond problem and for the site–bond problem with $p = p_s = p_b$. The extrapolated values of the critical percolation probability and the critical exponents are listed in table 2.

Table 1. p_n^*, θ and $\nu_\perp$ for directed site percolation from calculations on strips of width $n-1$, n, $n+1$ with free boundary conditions. (The extrapolations are calculated from each successive pair of results, assuming that the results tend to a limiting value linearly with $1/n$.)

n	p_n^* (site)	Extrapolated values	θ_n	Extrapolated values	$\nu_{\perp n}$	Extrapolated values
4	0.703 78		1.5389		1.1313	
5	0.703 76	0.703 68	1.5387	1.5379	1.1278	1.1138
6	0.704 17	0.706 22	1.5439	1.5699	1.1230	1.0990
7	0.704 46	0.706 20	1.5483	1.5747	1.1193	1.0971
8	0.704 68	0.706 22	1.5519	1.5771	1.1162	1.0945
9	0.704 83	0.706 03	1.5549	1.5789	1.1138	1.0946
10	0.704 95	0.706 03	1.5574	1.5799	1.1119	1.0948
11	0.705 04	0.705 94	1.5596	1.5816	1.1102	1.0932
12	0.705 10	0.705 76	1.5614	1.5812	1.1089	1.0946
13	0.705 16	0.705 88	1.5629	1.5809	1.1077	1.0933
14	0.705 20	0.705 72	1.5642	1.5811	1.1067	1.0937
Best extrapolated value		0.7058 ± 1		1.581 ± 1		1.094 ± 1

We also studied the bond problem on the directed square lattice with helical boundary conditions. Unexpectedly the results converged linearly with $1/n^3$: intuitively a linear dependence on $1/n^2$, as exhibited by finite systems with periodic boundary conditions (Barber and Fisher 1973), seems more plausible. Extrapolated values are shown in table 2.

The values of p_n^* apparently converge very rapidly for $n \geqslant 9$, suggesting that the extrapolated values give rather accurate estimates for the critical percolation probabilities. An estimate of the accuracy of the extrapolated values of the critical exponents may be obtained by invoking universality. The values of θ obtained in the four different cases are identical (to within the error bars), which is very encouraging. The values of $\nu_\perp$ differ in the last figure, however, suggesting that final values should be $\nu_\perp = 1.093 \pm 5$, $\theta = 1.582 \pm 1$.

Results obtained by previous authors are listed in table 2 for comparison. Our results are in excellent agreement with those quoted by Cardy and Sugar (1980). The results also agree with the series work of Blease (1977a, b) and the Monte Carlo results of Dhar and Barma (1981), but lie outside the error bars of the values of Kertész and

Table 2. Extrapolated results for the critical percolation probability, p_c, the anisotropy exponent, θ, and the transverse and longitudinal correlation length exponents $\nu_\perp$ and $\nu_\parallel$. The results of the present work are compared with those of previous work on directed percolation and branching Markov processes (BMP).

	Boundary	p_c	θ	$\nu_\perp$	$\nu_\parallel$
Site	Free	0.7058 ± 1	1.581 ± 1	1.094 ± 1	1.730 ± 2
Bond–site	Free	0.8228 ± 1	1.582 ± 1	1.095 ± 2	1.732 ± 3
Bond	Free	0.644 ± 1	1.582 ± 1	1.099 ± 1	1.739 ± 2
Bond	Helical	0.6447 ± 1	1.582 ± 1	1.103 ± 1	1.745 ± 2
Previous results:					
Bond, Monte Carlo[a]		0.632 ± 4			1.65 ± 6
Bond, Monte Carlo[b]		0.6445 ± 5			
Bond, series[c]		0.6446 ± 2			1.730 ± 9
BMP, series[d]			1.572 ± 9		1.736 ± 1
BMP, Monte Carlo[e]			1.583 ± 10		1.691 ± 18

[a] Kertész and Vicsek (1980).
[b] Dhar and Barma (1981).
[c] Blease (1977).
[d] Brower *et al* (1978).
[e] Grassberger and de la Torre (1979).

Vicsek (1980), which were also obtained using Monte Carlo simulations. This may be a consequence of the difficulty of applying finite-size scaling arguments to a Monte Carlo simulation on a finite lattice when the decay of correlations in the system is anisotropic.

Finally, figure 3 shows the phase boundary for site–bond percolation on the directed square lattice. The curve plotted is obtained by extrapolation of the results for finite strips.

In conclusion, we have shown that the finite-size renormalisation group can be extended to treat systems with anisotropic correlation lengths. Precise estimates for the correlation length and anisotropy exponents and for the critical percolation probabilities have been obtained for bond, site and site–bond percolation on the oriented square lattice. Results obtained using helical boundary conditions were found to converge linearly with $1/n^3$. This is an unexpected result which deserves further investigation.

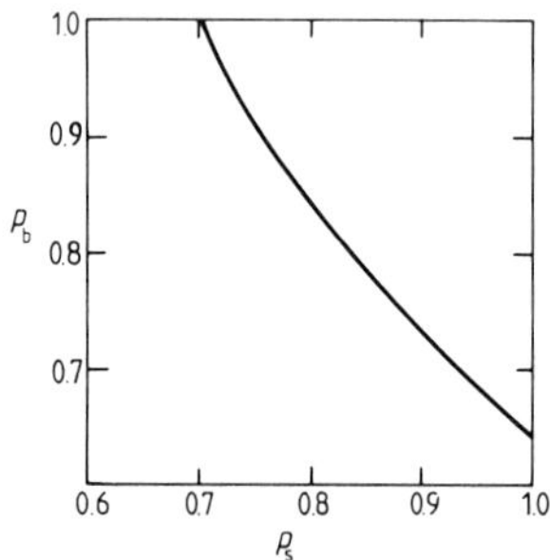

Figure 3. Critical percolation probability for site–bond percolation extrapolated from results obtained for strips of finite width.

It is a pleasure to thank J Kertész, T Vicsek and J L Cardy for interesting discussions, P Reynolds and S Redner for helpful correspondence and M E Fisher for a critical reading of the manuscript and for pointing out the linear dependence on $1/n^3$ of the results obtained using helical boundary conditions.

Part of this work was carried out at the Summer Institute in Theoretical Physics with partial support from the National Science Foundation under Grant No DMR-SC-06328 and the M J Murdoch Charitable Trust. One of us (JMY) acknowledges the support of the National Science Foundation through the Materials Science Centre at Cornell University.

References

Barber M N and Fisher M E 1973 *Ann. Phys.* **77** 1
Blease J 1977a *J. Phys. C: Solid State Phys.* **10** 917
——1977b *J. Phys C: Solid State Phys.* **10** 3461
Blöte H W J, Nightingale M P and Derrida B 1980 *Preprint*
Broadbent S R and Hammersley J M 1957 *Proc. Camb. Phil. Soc.* **53** 629
Brower R, Furman M A and Moshe M 1978 *Phys. Lett.* **76B** 213
Camp W J and Fisher M E 1972 *Phys. Rev.* B **6** 946
Cardy J L and Sugar R L 1980 *J. Phys. A: Math. Gen.* **13** L423
Derrida B 1981 *J. Phys. A: Math. Gen.* **14** L5
Derrida B and Vannimenus J 1981 *J. Phys. Lett.* **41** in press
Dhar D and Barma M 1981 *J. Phys. A: Math. Gen.* **14** L1
Domb C 1949 *Proc. R. Soc.* A **196** 36
Fisher M E 1972 *Proc.* 1970 *Enrico Fermi Summer School Course no* 51, *Varenna, Italy* ed. M S Green (New York: Academic)
Grassberger P and de la Torre A 1979 *Ann. Phys.* **122** 373
Hornreich R M, Luban M and Shtrikman S 1975 *Phys. Rev. Lett.* **35** 1678
Kertész J and Vicsek T 1980 *J. Phys. C: Solid State Phys.* **13** L343
Kinzel W and Schick M 1981 *Phys. Rev.* B in press
Nightingale M P 1976 *Physica* A **83** 561
Obukhov S P 1980 *Physica* **101A** 145
Ràcz Z 1980 *Phys. Rev.* B **21** 4012
dos Santos R and Sneddon L 1980 *Preprint*
Smythe R T and Wierman J C 1978 *Lecture Notes in Mathematics* **671** p 159
Sneddon L 1978 *J. Phys. C: Solid State Phys.* **11** 2823
Stauffer D 1979 *Phys. Rep.* **54** 1

J. Phys. A: Math. Gen. **14** (1981) L5–L9. Printed in Great Britain

LETTER TO THE EDITOR

Phenomenological renormalisation of the self avoiding walk in two dimensions

B Derrida

CEN de Saclay, Boite Postale no 2, 91190 Gif-sur-Yvette, France

Received 15 October 1980

Abstract. Phenomenological renormalisation is used to calculate the exponent ν and the connective constant of the self-avoiding walk problem on a square lattice. A transfer matrix technique is developed for the polymer problem. The results indicate that Flory's value $\nu = 0.75$ is true in two dimensions to extremely high accuracy.

Trying to calculate the critical properties of the two-dimensional self avoiding walk (SAW) by a real space renormalisation is no longer a very new idea (Hilhorst 1976, 1977, Shapiro 1978, Coniglio and Daoud 1979). However since the phenomenological renormalisation (PR) is becoming more and more useful to study two-dimensional systems (Nightingale 1976, 1979, Sneddon 1978, 1979, Nightingale and Blöte 1980, Ràcz 1980, Derrida and Vannimenus 1980a), I found it interesting to use it in this case. The motivations for the present work were provided by the two following facts. First, the PR requires the calculation of the correlation lengths of strips of finite width. The standard technique to do so for spin systems is the transfer matrix method. It was therefore necessary to define a transfer matrix for the SAW. The procedure is similar to one proposed for percolation (Derrida and Vannimenus 1980a). The interest of the definition given here is that it can be generalised to other polymer problems (solution of polymers, branched polymers, vulcanisation). Secondly, the SAW is simple enough to allow calculations for strips of rather large width. The other purpose of this work was to study the convergence of the PR when the width n of the strip increases. Up to now, the convergence law is not well understood except in the Ising case where the critical temperature and the critical exponent ν calculated by the PR with strips of width n and $n-1$ and periodic boundary conditions converge respectively like n^{-3} and n^{-2} to their exact values. (Nightingale 1976, Derrida and Vannimenus 1980b).

The PR method introduced by Nightingale (1976, 1979) is based on finite-size scaling arguments. I merely recall here the principle of the method without repeating its justifications. Suppose that one wants to study the critical properties of a two-dimensional model with coupling constant x. Using the transfer matrix one can calculate the correlation length $\xi_n(x)$ of a strip of width n. The PR consists in writing a renormalisation equation

$$(1/n)\xi_n(x) = (1/m)\xi_m(x') \tag{1}$$

which expresses the changes of the interaction x associated with the change of scale of ratio n/m. The critical point and the exponent ν of the two-dimensional problem can be

0305-4470/81/010005+05\$01.50

calculated from equations (2) and (3):

$$(1/n)\xi_n(x_c) = (1/m)\xi_m(x_c) \tag{2}$$

$$\frac{1}{\nu} = \frac{\ln[\mathrm{d}\xi_n/\mathrm{d}x|_{x_c}(\mathrm{d}\xi_m/\mathrm{d}x|_{x_c})^{-1}]}{\ln(n/m)} - 1. \tag{3}$$

Other exponents can be calculated by similar formulae.

As far as n and m are finite, the method is an approximation which can be improved by choosing n as large as possible and $m = n - 1$ (Nightingale 1976, Derrida and Vannimenus 1980b, dos Santos and Sneddon 1980).

In order to use the PR method for the SAW, one has to define the correlation length as a function of a parameter x in the same way as for spin models. This can be done using the famous $q \to 0$ limit of the classical q-component Heisenberg model (de Gennes 1972, des Cloizeaux 1975, Daoud *et al* 1975)

$$\lim_{q\to 0} \langle \boldsymbol{S}_0 \cdot \boldsymbol{S}_R \rangle = \sum_{p=0}^{\infty} x^p \mathcal{N}_{0R}(p) = G_{0R}(x). \tag{4}$$

Equation (4) relates the correlation function $\langle \boldsymbol{S}_0 \cdot \boldsymbol{S}_R \rangle$ of Heisenberg spins located on sites 0 and R to the number of self-avoiding walks $\mathcal{N}_{0R}(p)$ of length p going from site 0 to site R, x is the nearest-neighbour interaction in the Heisenberg model. When the distance R between the two sites becomes large, the correlation function decreases exponentially in the high-temperature phase. In the $q \to 0$ limit this defines the correlation length $\xi(x)$ for the polymer problem as a function of x which is a chemical potential of monomers

$$G_{0R}(x) \sim \exp(R/\xi(x)). \tag{5}$$

It is now possible to explain how $\xi(x)$ can be calculated for a strip of any width. Suppose that sites 0 and R belong to two columns N_0 and N_R on the strip. If one cuts the strip at column N between N_0 and N_R, the part of the polymer at the left of column N is made of several branches: one site of column N is connected to site 0 of column N_0 whereas some of the other sites of column N are connected by pairs (figure 1). The writing of the transfer matrix needs two steps.

First, one needs the list of all the possible configurations $\mathscr{C}$ at column N. One configuration is defined by the site of column N connected to site 0 and by the pairs of sites connected by the part of the strip at the left of column N. Configurations A and B of figure 1 are examples of such configurations. One can notice that the different branches which reach column N will be connected together by the right part of the strip to form a single polymer. So, some configurations are eliminated (like configuration C of figure 1) where there are crossings between the different branches of the configuration.

For each allowed configuration $\mathscr{C}$, one can define the function $H_N(\mathscr{C})$ by

$$H_N(\mathscr{C}) = \sum_{p=0}^{\infty} x^p \mathcal{N}_N(p, \mathscr{C}) \tag{6}$$

where $\mathcal{N}_N(p, \mathscr{C})$ is the number of ways one can put p monomers at the left of column N in order to realise configuration $\mathscr{C}$ at column N. The transfer matrix T is defined by the

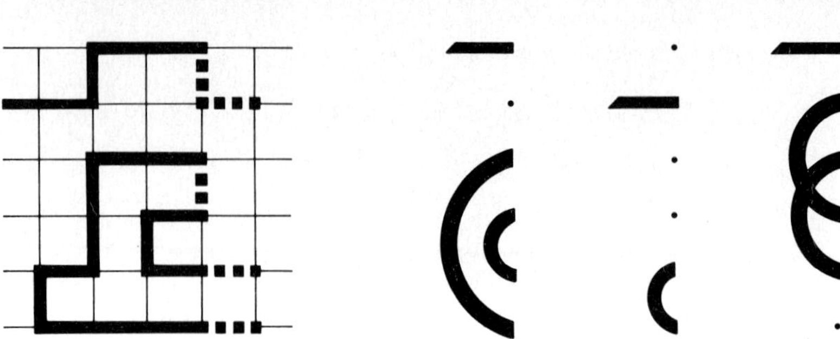

Figure 1. The configurations A and B represent the part of the polymer on the strip at the left of columns N and N + 1 respectively. The dashed links are the monomers one has to add to configuration A at column N to give rise to configuration B at column $N+1$. The number of these monomers is here five. So the matrix element between these configurations A and B is x^5. Configuration C is an example of a forbidden configuration.

set of linear relations which allow the calculation of the $H_{N+1}(\mathscr{C})$ as functions of the $H_N(\mathscr{C})$:

$$H_{N+1}(\mathscr{C}) = \sum_{\mathscr{C}'} T(\mathscr{C}, \mathscr{C}')H_N(\mathscr{C}'). \tag{7}$$

Obviously $T(\mathscr{C}, \mathscr{C}') = x^{t(\mathscr{C}, \mathscr{C}')}$ where $t(\mathscr{C}, \mathscr{C}')$ is the number of monomers one has to add to configuration $\mathscr{C}'$ at column N to give rise to configuration $\mathscr{C}$ at column $N+1$. If there is no way to connect two configurations $\mathscr{C}$ and $\mathscr{C}'$, the matrix element $T(\mathscr{C}, \mathscr{C}')$ is zero.

So the size of the transfer matrix is the number of configurations $\mathscr{C}$ and its elements are either zeros or integral powers of x. Clearly, when the two columns N_0 and N_R are very far from one another, one has

$$G_{0R}(x) \sim [\lambda(x)]^R \tag{8}$$

where $\lambda(x)$ is the largest eigenvalue of the matrix T. So the correlation length $\xi(x)$ can be calculated by

$$\xi(x) = -\frac{1}{\ln(\lambda(x))}. \tag{9}$$

From a practical point of view, the size of the matrix can be reduced using the symmetries of the strip. These reduced sizes S_n for strips of width n with periodic boundary conditions and free boundary conditions are given in tables 1 and 2 respectively. The general expression of these sizes is too complicated to be given here. Let me just mention that for large n, S_n increases like 3^n.

The self avoiding walk was studied here on a square lattice. The results shown in table 1 were obtained by calculating $\xi_n(x)$ for strips with periodic boundary conditions and by using formulae (2) and (3) with $m = n-1$. The values of ν are in much better agreement with Flory's value $\nu = 0.75$ (Domb 1969) than previous real-space renormalisations ($\nu = 0.740$ by Hilhorst 1977 and $\nu = 0.70$ by Shapiro 1978). The discrepancy between those previous works and Flory's value is however, recovered here for narrow strips and is probably due to small size effects.

Table 1. Results of the phenomenological renormalisation for the SAW problem using strips of width n and $n-1$ with *periodic boundary conditions*. S_n is the size of the transfer matrix for a strip of width n once the symmetries have been used. The uncertainty on x_c and ν calculated for all the choices of n and m is less than 10^{-7} for x_c and 10^{-6} for ν. The error bars indicated for the extrapolated values come only from the extrapolation procedure.

n	m	x_c	ν	S_n
2	1	0.347 8104	0.668 473	1
3	2	0.365 3048	0.724 477	2
4	3	0.373 3995	0.739 124	3
5	4	0.376 6329	0.745 005	7
6	5	0.377 9095	0.747 680	13
7	6	0.378 4477	0.748 928	32
8	7	0.378 6984	0.749 527	70
9	8	0.378 8280	0.749 826	179
10	9	0.378 9013	0.749 983	435
11	10	0.378 9459	0.750 067	1142
12	11			2947
Extrapolation		0.379 05 ± 0.000 03	0.7503 ± 0.0002	
Flory's value			0.75	

Table 2 contains the results of the PR for strips with free boundary conditions. The results are worse than in the periodic case and the convergence of ν is not even monotonic. However, for large width the agreement with the periodic case seems to take shape.

The extrapolation of the results of table 1 leads to

$$\begin{aligned} \nu &= 0.7503 \pm 0.0002 \\ x_c &= 0.37905 \pm 0.00003 \end{aligned} \tag{10}$$

and the convergence law is well described by a power law ($n^{-3.5}$) for ν and x_c. The Flory value $\nu = \frac{3}{4}$ and the connective constant $x_c = 0.37900$ given by McKenzie (1976) are out of the error bars of equations (10). However, one cannot be sure that the asymptotic regime in n has been reached. Non-monotonic convergence might occur for larger sizes

Table 2. As table 1 but with *free boundary conditions*. Note that the results for ν do not converge monotonically.

n	m	x_c	ν	
2	1	0.465 5712	0.715 312	1
3	2	0.414 6801	0.745 306	3
4	3	0.398 2330	0.753 958	6
5	4	0.390 8852	0.756 904	16
6	5	0.387 0023	0.757 908	38
7	6	0.384 7187	0.758 139	100
8	7	0.383 2708	0.758 037	256
9	8	0.382 3000	0.757 790	681
10	9	0.381 6200	0.757 483	1805
11	10			4867

as was the case in table 2. In any case, even if Flory's values were not exact, it would be an extremely good approximation to the true exponent.

This work has shown once more that phenomenological renormalisation is a very powerful tool. Small calculations ($n \leq 5$) lead to satisfactory results whereas longer ones give very accurate numbers. It would be of great interest to be able to predict the convergence law of the results for making very accurate extrapolations.

This work was done partly during my visit to the T J Watson Research Center of IBM. I should like to thank J des Cloizeaux, S Kirkpatrick and P Moussa for numerous discussions as well as H J Hilhorst who suggested this calculation to me.

References

des Cloizeaux J 1975 *J. Physique* **36** 281
Coniglio A and Daoud M 1979 *J. Phys. A: Math. Gen.* **12** L259
Daoud M, Cotton J P, Farnoux B, Jannink G, Sarma G, Benoit H, Duplessix R, Picot C and de Gennes P G 1975 *Macromolecules* **8** 804
Derrida B and Vannimenus J 1980a *J. Physique Lett.* **41** in press
—— 1980b *Colloque sur les méthodes de calcul pour l'étude de phénomènes critiques de Carry le Rouet* 1980 (Berlin: Springer)
Domb C 1969 *Adv. Chem. Phys.* **15** 229
de Gennes P G 1972 *Phys. Lett.* **38A** 339
Hilhorst H J 1976 *Phys. Lett.* **A38** 339
—— 1977 *Phys. Rev.* **B16** 1253
McKenzie D S 1976 *Phys. Rep.* **27C** 37
Nightingale M P 1976 *Physica* **83A** 561
—— 1979 *Proc. Koninklijke Nederlandse Akademic van Wetenschappen* **B82** (3) 235
Nightingale M P and Blöte H W J *Physica* to be published
Ràcz Z 1980 *Phys. Rev.* **B21** 4012
dos Santos R R and Sneddon L 1980 *Preprint*
Shapiro B 1978 *J. Phys. C: Solid State Phys.* **11** 2829
Sneddon L 1978 *J. Phys. C: Solid State Phys.* **11** 2823
—— 1979 *J. Phys. C: Solid State Phys.* **12** 3051

J. Physique **43** (1982) 475-483 MARS **1982,** PAGE 475

Classification
Physics Abstracts
05.20 — 05.50

Application of the phenomenological renormalization to percolation and lattice animals in dimension 2

B. Derrida and L. De Seze (*)

Service de Physique Théorique, CEN Saclay, 91191 Gif sur Yvette Cedex, France

(*Reçu le 28 septembre 1981, accepté le 12 novembre 1981*)

Résumé. — Après avoir rappelé la relation entre les lois d'échelles pour les systèmes finis et la renormalisation phénoménologique nous calculons l'exposant ν pour la percolation en dimension 2 et trouvons un bon accord avec la valeur de 4/3 proposée par den Nijs. Nous avons construit les matrices de transfert pour le problème des animaux afin de calculer les longueurs de corrélation. Nous trouvons $\nu = 0{,}640\,8 \pm 0{,}000\,3$ en dimension 2.

Abstract. — We recall the relation between finite-size scaling and the phenomenological renormalization. We calculate the exponent ν in dimension 2 for percolation and find a good agreement with the conjecture 4/3 of den Nijs. For lattice animals, we construct transfer matrices to calculate the correlation lengths and we find

$$\nu = 0.640\,8 \pm 0.000\,3$$

in dimension 2.

1. **Introduction.** — Since Nightingale [1] introduced the phenomenological renormalization (PR) method in the study of the Ising model, many authors used this approach to calculate the critical behaviour of other models like generalized Ising models [2], Ising antiferromagnets in a magnetic field [3, 4], lattice gas models [4], quantum spin systems [5], Lee and Yang singularities [6]. For all these models, the transfer matrix or the hamiltonian can be written obviously. In order to apply the PR method to percolation, it was necessary to show how the idea of transfer matrix could be generalized [7]. Subsequently, other geometrical problems like the self avoiding walk [8] or directed percolation [9] were studied by the PR method. The first motivation of the present work was to continue the work of Derrida and Vannimenus [7] and to find the best prediction of the PR method for the exponent ν of percolation. Once our programs of constructing the transfer matrices for percolation were reliable, it became very easy to study also the problem of lattice animals.

The literature on percolation is considerable : for recent reviews on the field, we refer to the course of Kirkpatrick [10] and to the articles of Stauffer [11] and Essam [12]. However, for a long time, the exponent ν of the two dimensional percolation was not known with enough accuracy to eliminate at least one of the two conjectured values ($\nu = 4/3$ by den Nijs [13] and $\nu = 1.354\,8$ by Klein *et al.* [14]). The situation has been clarified recently by Eschbach *et al.* [15] who found a good agreement with the den Nijs conjecture by increasing the accuracy of Monte Carlo renormalizations which previously seemed to support the predictions of Klein *et al.* [14]. Blöte *et al.* [16] found also a good agreement with $\nu = 4/3$ by interpolating the values of ν for the q state Potts model with q close to 1. It was then interesting to see if the direct application of the PR method to percolation which is much cheaper in computer time than Monte Carlo renormalization [15] could also support the den Nijs conjecture. In the results presented here, we find again an excellent agreement with $\nu = 4/3$. The PR method works also very well in the case of lattice animals and we find accurate estimations for the exponent ν.

The PR method is a consequence of the finite size scaling idea due to Fisher and Barber [17]. The finite-size scaling allows to extract the critical behaviour of infinite systems from the numerical studies done on finite systems [18, 19]. Hamer and Barber [20] have reviewed recently what is known on finite-size scaling. Let us just mention here that it has been used recently to study the XY model [20], the Potts model [16, 21],

(*) Service de Physique du Solide et de Résonance Magnétique, CEN Saclay, 91191 Gif sur Yvette Cedex, France.

the localization problem [22] and the roughening transition [23]. Finite-size scaling is expected to hold only for very large systems. Unfortunately, the systems one can study by computer are rather small. So the corrections to finite-size scaling are not always negligible. When one uses the PR method, due to these corrections, one finds estimations of the critical point of the exponents which depend on the size considered. The size dependence of these predictions is in most cases regular and the extrapolation of the results seems reliable. However, one finds sometimes non monotonic convergences and then the extrapolation becomes much more problematic. We discuss how assumptions on the corrections to finite-size scaling make possible the extrapolation of the results of the phenomenological renormalization.

The paper is organized as follows : in section 2, we examine the different ways of calculating critical points and critical exponents using the finite-size scaling and we recall how the phenomenological renormalization method can be derived from the finite-size scaling.

In section 3 we present the results we have obtained for bond and site percolation using the PR method.

In section 4, we extrapolate the results presented in section 3 and we explain how we can estimate the confidence interval on the extrapolated values of the threshold and of the critical exponent of the 2d percolation.

In section 5, we explain how the transfer matrix method can be used to study the statistics of lattice animals. We compare (in the appendix B) the structure of the transfer matrix for lattice animals with its structure in the case of percolation. Lastly, we present the estimations of the PR method for lattice animals.

2. **Finite-size scaling and phenomenological renormalization.** — To simplify the discussion, we consider an infinite system which is two dimensional and « finite » systems which are infinite strips of finite width n. One can easily generalize what follows to cases where the infinite system has any dimension and where the « finite » systems are finite at least in one direction. Again, for simplicity, we consider that there is only one parameter T (the temperature) in the problem and that the infinite system has a second order phase transition at a critical temperature T_c. For a given physical quantity Q, the typical situation is that Q is singular at T_c in the infinite system

$$[Q_\infty(T) \sim | T - T_c |^{-\omega}]$$

whereas it is regular in the finite systems $[Q_n(T)$ has no singularity]. When n increases, the singularity of $Q_\infty(T)$ starts to develop [for example, if $Q_\infty(T)$ diverges at T_c, $Q_n(T)$ has a maximum which becomes sharper and sharper]. The content of the finite-size scaling hypothesis [17, 20] is to assume the existence of scaling functions F_Q such that :

$$Q_n(T) \simeq Q_\infty(T)\, F_Q(n/\xi_\infty(T)) \qquad (1)$$

where $\xi_\infty(T)$ is the correlation length in the infinite system. This relation is expected to be valid when the size n of the finite system is large and when T approaches T_c. However Brezin [24] has shown recently that relation (1) fails above the upper critical dimensionality. As far as n is finite, $Q_n(T)$ is a regular function of T. This means that F_Q compensates the singularities of $Q_\infty(T)$ and of $\xi_\infty(T)$: if

$$Q_\infty(T) \sim | T - T_c |^{-\omega} \qquad (2)$$

$$\xi_\infty(T) \sim | T - T_c |^{-\nu} \qquad (3)$$

then F_Q must behave like :

$$F_Q(z) \sim z^{\omega/\nu} \quad \text{for} \quad z \to 0\,. \qquad (4)$$

It follows that at T_c, Q_n depends on n as a power law :

$$Q_n(T_c) \sim n^{\omega/\nu} \quad \text{for large } n\,. \qquad (5)$$

If $Q^{(p)}(T)$ is the pth derivative of $Q(T)$, $Q_\infty^{(p)}(T)$ is also singular at T_c

$$\frac{\mathrm{d}^p}{\mathrm{d}T^p} Q_\infty(T) = Q_\infty^{(p)}(T) \sim | T - T_c |^{-\omega-p} \qquad (6)$$

and therefore

$$Q_n^{(p)}(T_c) \sim n^{(\omega+p)/\nu}\,. \qquad (7)$$

Because $Q_n(T)$ is a regular function of T, it has a Taylor expansion around T_c

$$Q_n(T) = \sum_{p=0}^{\infty} Q_n^{(p)}(T_c)\, \frac{(T - T_c)^p}{p\,!}\,. \qquad (8)$$

And using (7) we find that $Q_n(T)$ can be expressed as

$$Q_n(T) \sim n^{\omega/\nu}\, G_Q(n^{1/\nu}(T - T_c)) \qquad (9)$$

where G_Q is a scaling function which is regular around T_c. It is interesting to notice that, usually, the functions F_Q and G_Q depend not only on the physical quantity Q but also on the boundary conditions of the finite system [19]. Recently dos Santos and Stinchcombe [25] have also extended the ideas of finite-size scaling to cases where there are more than one parameter.

At this stage, we see that finite-size scaling gives us many possibilities to calculate the critical properties of a given model. According to what is known on the problem, to the accuracy one can reach in the calculation of $Q_n(T)$, to the available computer time, etc., one chooses one of these possibilities. We recall now what are the usual methods using the finite-size scaling.

A first method [1] (that we shall use in this paper) is the phenomenological renormalization. Let us recall briefly its principle. One starts by calculating the correlation length $\xi_n(T)$ for strips of width n using a transfer matrix. Then one writes the fundamental equation of the PR method

$$\frac{\xi_n(T)}{\xi_m(T')} = \frac{n}{m}. \tag{10}$$

The main hypothesis of the PR *method is to assume that the relation* (10) *between T and T' depends only on the ratio n/m.* Therefore one has

$$\frac{\xi_n(T)}{\xi_m(T')} = \frac{n}{m} = \frac{\xi_\infty(T)}{\xi_\infty(T')}. \tag{11}$$

To justify this relation, one could argue that the strip of width n at temperature T is related by a scaling transformation to a strip of width m at temperature T'. This transformation is a contraction of ratio n/m of the width of the strip and of the correlation length ξ_n which is the characteristic length along the strip.

However it is easy to show that (11) is a direct consequence of the finite-size scaling hypothesis (1). Consider two temperatures T and T' such that

$$\frac{\xi_\infty(T)}{\xi_\infty(T')} = \lambda. \tag{12}$$

For any choice of the scaling factor λ, equation (12) gives us a relation between T and T'. Using expression (1) we can write the ratio $Q_n(T)/Q_m(T')$ as

$$\frac{Q_n(T)}{Q_m(T')} = \frac{Q_\infty(T)}{Q_\infty(T')} \frac{F_Q[n/\xi_\infty(T)]}{F_Q[m/\xi_\infty(T')]}. \tag{13}$$

By choosing $\lambda = n/m$, the function F_Q is eliminated in (13) because its two arguments are identical. Therefore one has in this case

$$\frac{Q_n(T)}{Q_m(T')} = \frac{Q_\infty(T)}{Q_\infty(T')}. \tag{14}$$

By applying this result to the correlation length itself one recovers (11)

$$\frac{\xi_\infty(T)}{\xi_\infty(T')} = \frac{n}{m} = \frac{\xi_n(T)}{\xi_m(T')}. \tag{15}$$

The simplest way to calculate $T_c(n, m)$ using the PR method is to solve the following equation :

$$\frac{\xi_n(T_c)}{n} = \frac{\xi_m(T_c)}{m}. \tag{16}$$

Once $T_c(n, m)$ is known, one can easily calculate the critical exponents from equation (14). For example ν is given by

$$1 + \frac{1}{\nu} = \frac{\log\left[\frac{d\xi_n}{dT}(T_c) \Big/ \frac{d\xi_m}{dT}(T_c)\right]}{\log[n/m]}. \tag{17}$$

In some cases [16, 21], the simplest quantity Q to calculate is not the correlation length ξ. Using relation (5), one can find T_c and the ratio ω/ν by searching the temperature where Q_n behaves as a power law of n. One chooses three widths n, m and p and one calculates $T_c(n, m, p)$ as the solution of

$$\frac{\log[Q_m(T_c)/Q_n(T_c)]}{\log[Q_p(T_c)/Q_n(T_c)]} = \frac{\log[m/n]}{\log[p/n]}. \tag{18}$$

Once $T_c(n, m, p)$ is known, the ratio ω/ν is easily obtained by

$$\frac{\omega}{\nu} = \frac{\log[Q_m(T_c)/Q_n(T_c)]}{\log[m/n]}. \tag{19}$$

Of course, one can use again equation (19) by replacing Q by $Q^{(p)}$ and ω by $\omega + p$. Thus one can calculate ω and ν.

Another method consists in using equation (9). One calculates $Q_n(T)$ for several sizes n. Then one tries to determine what is the best choice of ω, ν and T_c which allows to superpose the curves $n^{-\omega/\nu} Q_n(T)$ as a function of $n^{1/\nu}(T - T_c)$. This approach is more suitable to study disordered systems (random magnets, localization [22] ...). In these problems, the calculated quantities $Q_n(T)$ are obtained by products of random matrices and therefore affected by statistical errors which do not allow to use formulae (17) or (19).

Of course the main difficulty is that the finite-size scaling is valid only for large n and that the numerical calculations can be done only for small n. Therefore the corrections to finite-size scaling cannot be neglected. In section 4 we shall show that reasonable assumptions on the corrections to finite-size scaling allow to give rather accurate predictions for the critical points and the critical exponents.

3. **Application to percolation.** — In this section we present the results we have obtained for the two dimensional percolation. We calculated the correlation lengths of strips as function of the probability p of present bonds or sites in the same way as in reference [7]. We chose two kinds of boundary conditions in the direction perpendicular to the strip : the periodic boundary conditions and the helical boundary conditions [9]. In the case of periodic boundary conditions a strip of width n means that a site can be labelled by two integers (i, j) with $-\infty < i < +\infty$ and $1 \leqslant j \leqslant n$ and that the four neighbours of this site are $(i \pm 1, j)$ and $(i, j \pm 1 + kn)$ [where k is the integer such that $1 \leqslant j \pm 1 + kn \leqslant n$]. In the case of helical boundary conditions, a site can be labelled by a single integer $i(-\infty < i < +\infty)$ and its four nearest neighbours are the sites $i \pm 1$ and $i \pm n$.

The most difficult part of the calculation was the construction of the transfer matrix for large widths. We used a construction similar to the one described

Table I. — *Results of the phenomenological renormalization for site percolation. We have calculated the correlation lengths ξ_n with two kinds of boundary conditions on the strips : periodic boundary conditions and helical boundary conditions. For each choice of n and m, the estimations of p_c and ν were obtained from equations* (16) *and* (17). *The extrapolations were done by assuming a power law convergence* (n^{-x}). *The results are in good agreement with previous works* : [a] = *Ref.* [26], [b] = *Ref.* [13], [c] = *Ref.* [16].

		SITE PERCOLATION			
		Periodic bound. cond.		Helical bound. cond.	
n	m	p_c	ν	p_c	ν
2	1	0.733 89	1.560 4	0.627 12	1.571 2
3	2	0.582 14	1.493 2	0.655 94	1.556 6
4	3	0.590 96	1.471 8	0.613 84	1.432 8
5	4	0.588 70	1.413 9	0.604 39	1.411 1
6	5	0.589 92	1.388 0	0.599 18	1.388 0
7	6	0.590 68	1.371 7	0.596 67	1.374 1
8	7	0.591 27	1.361 6	0.595 29	1.364 5
9	8	0.591 88	1.354 8	0.594 49	1.357 8
10	9			0.593 99	1.353 0
Expected values		0.593 1 ± 0.000 6 [a]	1.333 3 [b] 1.333 ± 0.002 [c]	0.593 1 ± 0.000 6	1.333 3
Extrapolation			1.330 ± 0.005	0.592 7 ± 0.000 2	1.330 ± 0.005
x			2.1 ± 0.4	3.1 ± 0.4	1.8 ± 0.4

Table II. — *The same as table* I *for bond percolation.*

		BOND PERCOLATION			
		Periodic bound. cond.		Helical bound. cond.	
n	m	p_c	ν	p_c	ν
2	1	0.502 60	1.241 0	0.740 19	1.882 5
3	2	0.485 59	1.201 5	0.581 31	1.526 0
4	3	0.491 33	1.237 4	0.530 62	1.386 8
5	4	0.495 63	1.271 0	0.513 67	1.340 6
6	5	0.497 74	1.292 2	0.507 15	1.326 8
7	6	0.498 73	1.304 7	0.504 25	1.323 6
8	7	0.499 21	1.312 1	0.502 77	1.323 8
9	8	0.499 48	1.316 9	0.501 93	1.324 7
Expected values		0.5 exact	1.333 3	0.5	1.333 3
Extrapolation		0.500 0 ± 0.000 2	1.332 ± 0.003	0.500 2 ± 0.000 3	
x convergence law n^{-x}		3.4 ± 0.4	2.3 ± 0.4	3.0 ± 0.3	

in reference [16] for the Potts model. The only difference is that the configurations of connected sites are not exactly the same because we calculate here the correlation length instead of the free energy. Once the transfer matrix is constructed, the largest eigenvalue [7, 27] which gives the correlation length can be easily obtained by multiplying the transfer matrix. The estimations of p_c and ν given in tables I and II were then found by solving equation (16) and from equation (17) for several choices of n and m. As we shall discuss it in section 4, the best choice for n and m is $m = n - 1$ and n as large as possible.

For site percolation (Table I) as well as for bond percolation (Table II) the values found for p_c or ν are close to their expected values. The convergence is often regular and this will allow us to give extrapolated values with small confidence intervals. However in some cases the convergence is not monotonic (p_c for site percolation with periodic boundary conditions or ν for bond percolation with helical boundary conditions). In the next section, we explain the extrapolation method we have used.

4. **Extrapolation.** — The convergence law of the results of the PR method is related to the corrections to the finite-size scaling. As we saw it in section 2, at the critical temperature, the correlation length is linear in $n(\xi_n(T_c) \sim n)$ and its derivatives are power laws. By analogy with the Ising model (see the appendix A), it seems reasonable to assume that the corrections to the finite-size scaling are power laws (we even assume that it is the same power law for the correlation length and its first derivatives)

$$\begin{aligned} \xi_n(T_c) &= A_0\, n(1 + A_1\, n^{-a} + \cdots) \\ \frac{d\xi_n}{dT}(T_c) &= B_0\, n^{1+\frac{1}{\nu}}(1 + B_1\, n^{-a} + \cdots) \qquad (20) \\ \frac{d^2\xi_n}{dT^2}(T_c) &= C_0\, n^{1+\frac{2}{\nu}}(1 + \cdots)\,. \end{aligned}$$

For large n and m, the estimation $T_c(n, m)$ which can be found by solving equation (16) is :

$$T_c(n, m) \simeq T_c - \frac{A_0\, A_1}{B_0}\, \frac{n^{-a} - m^{-a}}{n^{1/\nu} - m^{1/\nu}} + \cdots. \qquad (21)$$

Using this estimation $T_c(n, m)$, one can calculate the estimation $\nu(n, m)$ from equation (17)

$$\nu(n, m) \sim \nu - \nu^2 \left(B_1 - \frac{A_0\, C_0}{B_0^2}\, A_1 \right) \frac{n^{-a} - m^{-a}}{\log(n/m)} + \cdots. \qquad (22)$$

The formulae (21) and (22) give the relation between the corrections to finite-size scaling and the results of the PR method. For a given width n, it is easy to show that the convergence is improved by choosing $m = n - 1$. For this choice one finds that

$$\begin{aligned} T_c(n, n-1) - T_c &\sim a\nu\, \frac{A_0\, A_1}{B_0}\, n^{-a-\frac{1}{\nu}} \\ \nu(n, n-1) - \nu &\sim \nu^2\, a \left(B_1 - \frac{A_0\, C_0}{B_0^2}\, A_1 \right) n^{-a}\,. \end{aligned} \qquad (23)$$

The extrapolated values given in tables I and II were obtained by assuming a power law convergence like in reference [4]. We have plotted

$$\log(T_c(n, n-1) - T)$$

as a function of $\log n$ for several choices of T. The extrapolated value and the confidence intervals were obtained by looking for the values of T for which the curvature was small. In several cases the convergence was monotonic and using this procedure, we could find the extrapolated values, the confidence intervals and the exponent x of the power law convergence. In other cases (p_c for site percolation with periodic boundary conditions or ν for bond percolation with helical boundary conditions) the convergence was not monotonic and we could not find a simple method of extrapolating our results. This non monotonic convergence is probably due to higher order corrections to the finite-size scaling.

The coefficients A_0, B_0, C_0... have probably a physical interpretation. For the XY model in 2 dimension, it has been shown [22, 28] that A_0 was related to the critical exponent η at least at low temperature

$$A_0 = \frac{1}{\pi\eta}\,. \qquad (24)$$

This relation remains probably true for several isotropic systems in dimension 2. For the Ising model $\eta = 1/4$ and $A_0 \doteq 4/\pi$ (appendix A). From references [13] and [29], the conjectured prediction of η for the q state Potts model ($q < 4$) is

$$\eta = \frac{1 - 4\,p^2}{4(1 - p)} \quad \text{with} \quad \cos \pi p = \frac{\sqrt{q}}{2}\,. \qquad (25)$$

For bond percolation on strips with periodic boundary conditions we have calculated the correlation lengths at $p = 1/2$. We found the linear behaviour of $\xi_n(1/2)$ and an estimation of A_0 :

$$A_0 = 1.527\,7 \pm 0.000\,5\,. \qquad (26)$$

This estimation is in good agreement with relation (24) ($1/\pi\eta = 1.527\,9$). It is possible that (24) remains true for a large class of 2d isotropic models [30].

5. **Application to lattice animals.** — The problem of lattice animals [31, 32] lies between percolation and self avoiding walks. One considers s connected sites on a given lattice. One expects that the number

Table III. — *Results of the phenomenological renormalization for lattice animals. They agree well with previous calculations :* [a] = *Ref.* [35], [b] = *Ref.* [36], [c] = *Ref.* [31], [d] = *Ref.* [33], [e] = *Ref.* [37], [f] = *Ref.* [38].

		LATTICE ANIMALS : SITE			
		Periodic bound. cond.		Helical bound. cond.	
n	m	x_c	ν	x_c	ν
2	1	0.381 97	0.706 70	0.324 72	0.741 08
3	2	0.252 65	0.694 53	0.265 82	0.676 79
4	3	0.247 57	0.663 00	0.253 47	0.653 29
5	4	0.245 86	0.648 35	0.249 21	0.644 03
6	5	0.245 79	0.644 12	0.247 59	0.640 67
7	6	0.245 87	0.642 45	0.246 90	0.639 44
8	7	0.245 95	0.641 70	0.246 57	0.639 05
9	8			0.246 41	0.638 98
10	9			0.246 31	0.639 05
Previous works		0.246 0 $\pm$ 0.000 3 [a]	0.57 $\pm$ 0.06 [b] 0.66 [c] 0.61 [d] 0.65 [e] 0.647 $\pm$ 0.02 [f]	0.246 0 $\pm$ 0.000 3	0.57 $\pm$ 0.06 0.66 0.61 0.65 0.647 $\pm$ 0.02
Extrapolation			0.640 8 $\pm$ 0.000 3	0.246 15 $\pm$ 0.000 05	
x			4.2 $\pm$ 0.8	4.3 $\pm$ 0.9	

$\mathcal{N}_s$ of different configurations of these s connected sites behaves for large s like :

$$\mathcal{N}_s \sim s^{-\theta} \mu^s \qquad (27)$$

where μ is a constant which depends on the details of the lattice and θ is a universal exponent which depends only on the dimension d of space. The average size R of these configurations is related to the number of sites s by another universal exponent ν

$$R \sim s^{\nu} . \qquad (28)$$

Recently Parisi and Sourlas [33] have shown that θ and ν in dimension d are not independent $\left(\nu = \dfrac{\theta - 1}{d - 2} \right)$ and are related to the exponent of the Lee and Yang singularity in dimension $d - 2$. However their relation is useless to calculate the exponent ν in dimension 2.

We have calculated this exponent ν by the PR method. To do so it is necessary to define a correlation length for the lattice animals problem. The simplest way is to consider generating functions. If $\mathcal{N}_s(0, \mathbf{R})$ is the number of animals of s sites which occupy both sites 0 and $\mathbf{R}$, one can define the generating function $G_{0\mathbf{R}}(x)$ of these numbers :

$$G_{0\mathbf{R}}(x) = \sum_s x^s \mathcal{N}_s(0, \mathbf{R}) . \qquad (29)$$

The parameter x plays the same role as the probability p in percolation problems or the temperature T in spin problems. Its critical value x_c is equal to μ^{-1}. By analogy with critical phenomena, the function $G_{0\mathbf{R}}(x)$ behaves as a correlation function for $x < x_c$

$$G_{0\mathbf{R}}(x) \sim \exp(-R/\xi(x)) \quad \text{for} \quad R \to \infty .$$

It can be shown that

$$\sum_{\mathbf{R}} G_{0\mathbf{R}}(x) \sim (1 - \mu x)^{\theta - 2} \qquad (30)$$

and

$$\xi(x) \sim (1 - \mu x)^{-\nu} \qquad (31)$$

where θ and ν are defined by equations (27) and (28).

By applying to lattice animals the ideas yet developed for percolation [7], self avoiding walks [8, 34] and the Potts model [16], we have calculated the correlation length $\xi_n(x)$ for lattice animals on strips of width n (appendix B).

Using these correlation lengths $\xi_n(x)$, we have calculated the estimations

$$x_c(n, n - 1) \quad \text{and} \quad \nu(n, n - 1)$$

for several choices of n and for periodic and helical boundary conditions (Table III). The extrapolation was done in the same way as explained in section 4.

Here again we could not extrapolate our results when the convergence was not monotonic (x_c for periodic boundary conditions, ν for helical boundary conditions). Our results are in good agreement with previous works [31, 33, 35, 36, 37]. Even when the convergence is not monotonic, the estimations are very close to the extrapolated values that we give.

6. **Conclusion.** — We have reported here the results of our calculations on 2d percolation and 2d lattice animals using the phenomenological renormalization. The method works very well for both problems and leads to accurate predictions on the critical point and the exponent ν.

We think that the PR is a simple method to study a large class of models in statistical mechanics. Its main advantages are that :

- Only one parameter is renormalized;
- It gives satisfactory results with reasonable calculations;
- The results are usually improved by increasing the width of strips and the convergence seems to be rather rapid (a power law);
- For each width the estimations are obtained without error bars and for this reason it is easier to estimate the confidence intervals on the extrapolated values;
- In principle, the method can be applied to models (like for example frustrated models) where the order parameter is not known *a priori*. The reason is that the correlation length can be calculated from the two largest eigenvalues of the transfer matrix without knowing what correlation function decreases with this correlation length.

However the method presents some difficulties. In particular it is hard to study models in dimension 3 because the size of the transfer matrices becomes too large. Moreover, the sequence of calculated values has sometimes a non-monotonic convergence and does not allow to extract accurate estimations. We must emphasize that these problems of convergence can be completely masked in Monte Carlo simulations by statistical errors.

Acknowledgments. — We are grateful to E. Brezin, J. M. Luck, J. L. Pichard, G. Sarma and D. Stauffer for stimulating discussions.

Appendix A. — In this appendix we give the analytic expression of the large n expansion of the correlation length and of its first derivatives at T_c for strips of Ising spins with periodic boundary conditions. The corrections to finite-size scaling are power laws and it is possible to calculate analytically the convergence law of $K_c(n, m)$ and $\nu(n, m)$.

The correlation lengths for strips of width n with periodic boundary conditions are known exactly (they are recalled in reference [1]). If the nearest neighbour interaction is K :

$$\mathcal{H} = - K \sum_{\langle ij \rangle} \sigma_i \sigma_j \tag{A.1}$$

then the correlation length $\xi_n(K)$ is given by

$$\frac{1}{\xi_n(K)} = \frac{1}{2}[(\gamma_1 + \gamma_3 + \cdots + \gamma_{2n-1}) - (\gamma_0 + \gamma_2 + \cdots + \gamma_{2n-2})] \tag{A.2}$$

where the γ_r are given by

$$\cosh \gamma_r = \frac{\cosh^2 2K}{\sinh 2K} - \cos \frac{r\pi}{n}. \tag{A.3}$$

The critical value K_c is known :

$$K_c = \frac{1}{2} \log (1 + \sqrt{2}).$$

From (A.2) and (A.3), one can find for the Ising model the values of A_0, A_1, B_0, B_1, C_0 and a of equation (20) :

$$\xi_n \sim \frac{4}{\pi} n \left(1 - \frac{\pi^2}{24 n^2} + \cdots\right) \tag{A.4}$$

$$\frac{d\xi_n}{dK} \sim \frac{32}{\pi^2} n^2 \left(1 - \frac{\pi^2}{12 n^2} + \cdots\right) \tag{A.5}$$

$$\frac{d^2\xi_n}{dK^2} = \frac{128}{\pi^3} n^3 (4 - 2 \log 2) + \cdots. \tag{A.6}$$

Then from equations (21) and (22), one finds that :

$$K_c(n, m) = \frac{1}{2} \log (1 + \sqrt{2}) - \frac{\pi^3}{192} \frac{\lambda(\lambda + 1)}{n^3} + \cdots \tag{A.7}$$

$$\nu(n, m) = 1 - \frac{\pi^2 \log 2}{24 n^2} \frac{\lambda^2 - 1}{\log \lambda} + \cdots \tag{A.8}$$

where $\lambda = n/m$. One can notice that one can minimize these corrections by choosing λ as close as possible to 1, i.e. $n = m - 1$.

Appendix B. — In this appendix, we give as an example the transfer matrix for site percolation and lattice animals on a strip of width 4 with periodic boundary conditions.

TRANSFER MATRIX FOR SITE PERCOLATION. — One consider a strip of 4 lines with periodic boundary conditions. We call p the probability that a site is occupied and $q = 1 - p$ the probability that a site is empty. The probability $P_N(p)$ that the first and the Nth column are connected decreases exponentially

$$P_N(p) \sim e^{-N/\xi(p)}. \tag{B.1}$$

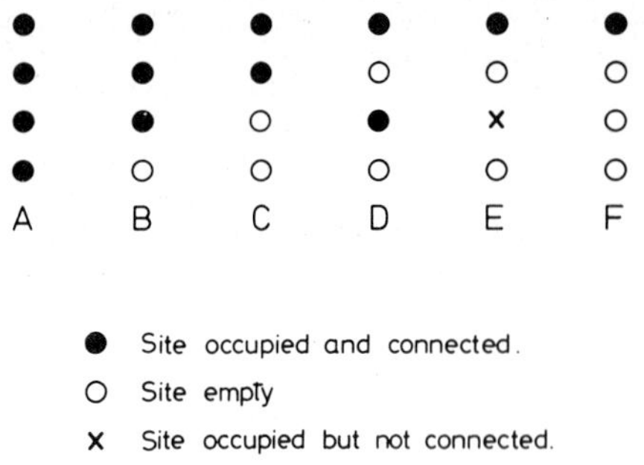

A B C D E F

● Site occupied and connected.

○ Site empty

× Site occupied but not connected.

Fig. 1. — The 6 configurations of a strip of width $n = 4$ with periodic boundary conditions, for site percolation. We call a site connected if it has been connected to column 1 by the portion of the strip included between column 1 and column N.

To calculate the correlation length $\xi(p)$ we follow the same method as in reference [7]. First we have to list all the possible configurations at column N (see Fig. 1). Due to the periodic boundary conditions, it remains only 6 configurations. Then we have to relate the probabilities A_N, B_N, C_N, D_N, E_N, F_N of these configurations at column N to their probabilities at column $N + 1$

$$\begin{aligned} A_{N+1} &= p^4(A_N + B_N + C_N + D_N + E_N + F_N) \\ B_{N+1} &= p^3\, q(4\, A_N + 4\, B_N + 4\, C_N + 4\, D_N + 3\, E_N + 3\, F_N) \\ C_{N+1} &= p^2\, q^2(4\, A_N + 4\, B_N + 3\, C_N + 4\, D_N + 2\, E_N + 2\, F_N) \\ D_{N+1} &= p^2\, q^2(2\, A_N + B_N + D_N) \\ E_{N+1} &= p^2\, q^2(B_N + 2\, C_N + E_N + F_N) \\ F_{N+1} &= pq^3(4\, A_N + 3\, B_N + 2\, C_N + 2\, D_N + E_N + F_N)\,. \end{aligned} \tag{B.2}$$

If $\lambda_1(p)$ is the largest eigenvalue of this matrix, it is clear that each of these probabilities behaves like $[\lambda_1(p)]^N$ for large N. Therefore one has

$$\lambda_1(p) = \exp\left[-\frac{1}{\xi(p)}\right]. \tag{B.3}$$

TRANSFER MATRIX FOR SITE LATTICE ANIMALS. — For percolation, the important quantity was the probability $P_N(p)$ that the column N is connected to column 1. For lattice animals, the important quantity is $G_N(x)$

$$G_N(x) = \sum_s x^s\, \mathcal{N}_s(1, N) \tag{B.4}$$

where $\mathcal{N}_s(1, N)$ is the number of animals of s sites connecting column 1 to column N. As in percolation, we have to list all the possible configurations at column N. We find exactly the same configurations as in percolation (Fig. 1). Each configuration corresponds to a part of sum (B.4). Configuration A corresponds to the sum A_N over all the animals which connect columns 1 and N and occupy the 4 sites of column N; configuration E corresponds to the sum E_N over all the animals which connect columns 1 and N and occupy 2 sites of column N (one of these 2 sites has not been connected to column 1 by the portion of strip included between columns 1 and N). As in percolation, it is easy to write the linear relations between

$$\{ A_N, B_N, \ldots, F_N \}$$

and

$$\{ A_{N+1}, B_{N+1}, \ldots, F_{N+1} \}$$

$$\begin{aligned} A_{N+1} &= x^4(A_N + B_N + C_N + D_N + E_N + F_N) \\ B_{N+1} &= x^3(4\, A_N + 4\, B_N + 4\, C_N + 4\, D_N + 2\, E_N + 3\, F_N) \\ C_{N+1} &= x^2(4\, A_N + 4\, B_N + 3\, C_N + 4\, D_N + 2\, F_N) \\ D_{N+1} &= x^2(2\, A_N + B_N + D_N) \\ E_{N+1} &= x^2(B_N + 2\, C_N + E_N + F_N) \\ F_{N+1} &= x(4\, A_N + 3\, B_N + 2\, C_N + 2\, D_N + F_N)\,. \end{aligned} \tag{B.5}$$

Here again, for large N, $G_N(x)$ behaves like $[\lambda_1(x)]^N$ where $\lambda_1(x)$ is the largest eigenvalue of this matrix. Therefore the correlation length $\xi(x)$ can be calculated by

$$\lambda_1(x) = \exp\left[-\frac{1}{\xi(x)}\right]. \tag{B.6}$$

We see the resemblance between the transfer matrix (B.2) for percolation and the transfer matrix (B.5) for lattice animals. The only difference, at first sight, is that p is replaced by x and q by 1. However the fact that the s sites of an animal are connected imposed constraints. This can be seen by comparing the coefficients of E_N in the two matrices. The site of configuration E which is occupied but not connected plays no role in percolation. On the contrary, for lattice animals, this site must be connected by the portion of the strip which follows the column N. This is why configuration E at column N cannot be followed by configuration F at column $N + 1$ because this site would never be connected.

References

[1] NIGHTINGALE, M. P., *Physica* **83A** (1976) 561.

[2] NIGHTINGALE, M. P., *Proc. Kon. Ned. Ak. Wet.* **B 82** (1979) 235.
SNEDDON, L., *J. Phys.* **C 11** (1978) 2823.

[3] SNEDDON, L., *J. Phys.* **C 12** (1979) 3051.

[4] RACZ, Z., *Phys. Rev.* **B 21** (1980) 4012.
KINZEL, W. and SCHICK, M., preprint (1981).

[5] SNEDDON, L. and STINCHCOMBE, R. B., *J. Phys.* **C 12** (1979) 3761.

[6] UZELAC, K. and JULLIEN, R., *J. Phys.* **A 14** (1981) L-151.

[7] DERRIDA, B. and VANNIMENUS, J., *J. Physique-Lett.* **41** (1980) L-473.
[8] DERRIDA, B., *J. Phys.* **A 14** (1981) L-5.
[9] KINZEL, W. and YEOMANS, J. M., *J. Phys.* **A 14** (1981) L-163.
[10] KIRKPATRICK, S., *Les Houches 1978*, eds. Balian R., Maynard R. and Toulouse G. (North Holland).
[11] STAUFFER, D., *Phys. Rep.* **54** (1979) 1.
[12] ESSAM, J. W., *Rep. Prog. Phys.* **43** (1980) 833.
[13] DEN NIJS, M. P. M., *J. Phys.* **A 12** (1979) 1857.
[14] KLEIN, W., STANLEY, H. E., REYNOLDS, P. J. and CONIGLIO, A., *Phys. Rev. Lett.* **41** (1978) 1145.
[15] ESBACH, P. D., STAUFFER, D. and HERRMANN, H. J., *Phys. Rev.* **B 23** (1981) 422.
[16] BLÖTE, H. W. J., NIGHTINGALE, M. P. and DERRIDA, B., *J. Phys.* **A 14** (1981) L-45.
[17] FISHER, M. E., in *Critical Phenomena*, Proc. Int. School of Physics « Enrico Fermi », Varenna 1970, Course n° 51, ed. M. S. Green (N.Y. Academic) 1971.
FISHER, M. E. and BARBER, N. M., *Phys. Rev. Lett.* **28** (1972) 1516.
SUZUKI, M., *Prog. Theor. Phys.* **58** (1977) 1142.
[18] LANDAU, D. P., *Phys. Rev.* **B 13** (1976) 2997.
[19] HEERMANN, D. W. and STAUFFER, D., *Z. Phys.* **B 40** (1980) 133.
[20] HAMER, C. J. and BARBER, N. B., *J. Phys.* **A 13** (1980) L-169 ; *J. Phys.* **A 14** (1981) 241, 259.
[21] NIGHTINGALE, M. P. and BLÖTE, H. W. J., *Physica* **104A** (1980) 352.
[22] PICHARD, J. L. and SARMA, G., *J. Phys.* **C 14** (1981) L-127 ; *J. Phys.* **C 14** (1981) L-617.
[23] LUCK, J. M., *J. Physique-Lett.* **42** (1981) L-271.
[24] BREZIN, E., preprint (1981).
[25] DOS SANTOS, R. R. and STINCHCOMBE, R. B., preprint (1981).
[26] REYNOLDS, P. J., STANLEY, H. E. and KLEIN, W., *Phys. Rev.* **B 21** (1980) 1223.
[27] For most problems of statistical mechanics, one needs the two largest eigenvalues of the transfer matrix to calculate the correlation length. However, as it was yet said in reference [7], for percolation, the correlation length is related only to the largest eigenvalue of the transfer matrix.
[28] LUCK, J. M., private communication.
[29] NIENHUIS, B., RIEDEL, E. K. and SCHICK, M., *J. Phys.* **A 13** (1980) L-189.
[30] We have tested numerically the relation $A_0 = 1/\pi\eta$ for the q state Potts model in dimension 2. It seems to be true for all values of $q < 4$. It is particularly well verified for small values of q ($q \to 0$) because the linear behaviour of correlation lengths at T_c occurs for small widths.
[31] STAUFFER, D., *Phys. Rev. Lett.* **41** (1978) 1333.
[32] LUBENSKY, T. C. and ISAACSON, J., *Phys. Rev. Lett.* **41** (1978) 829.
[33] PARISI, G. and SOURLAS, N., *Phys. Rev. Lett.* **46** (1981) 871.
[34] KLEIN, D. J., *J. Stat. Phys.* **23** (1980) 561.
[35] GUTTMANN, A. J. and GAUNT, D. S., *J. Phys.* **A 11** (1978) 949.
[36] REDNER, S., *J. Phys.* **A 12** (1979) L-239.
[37] HOLL, K., to be published ;
STAUFFER, D., Intern. Conf. on Disordered Systems to appear in *Springer Lecture Notes in Physics* (1981).
[38] FAMILY, F., private communication.

J. Phys. A: Math. Gen. **15** (1982) L557–L564. Printed in Great Britain

LETTER TO THE EDITOR

A transfer-matrix approach to random resistor networks

B Derrida† and J Vannimenus‡

† SPT, CEN Saclay, 91191 Gif-sur-Yvette, France
‡ GPS, Ecole Normale Supérieure, 24 rue Lhomond, 75231 Paris Cedex 05, France

Received 12 July 1982

Abstract. We introduce a transfer-matrix formulation to compute the conductance of random resistor networks. We apply this method to strips of width up to 40, and use finite size scaling arguments to obtain an estimate for the conductivity critical exponent in two dimensions, $t = 1.28 \pm 0.03$.

1. Introduction

The calculation of the conductivity for a random mixture of insulating and conducting materials is a central problem in the theory of disordered systems, and a number of different approaches to the problem have been proposed (see Kirkpatrick (1979) for a review). In particular, since the work of Stinchcombe and Watson (1976), the singular behaviour of the conductivity near a percolation threshold has repeatedly been studied using renormalisation-group ideas.

The results obtained so far are not so convincing as for the purely geometric connectivity properties. In two dimensions, for instance, very accurate values for the geometric critical exponents have been calculated, in agreement with presumably exact conjectures (Stauffer 1981). On the other hand, one finds in the literature values for the conductivity critical exponent t ranging from 1.0 to 1.43. A discussion of the situation has recently been given by Gefen *et al* (1981).

We propose here a new method of computing the conductance of a resistor network. It is very similar to the transfer-matrix method of statistical mechanics, but involves nonlinear matrix recursion relations. This approach has several features that should make it of general interest in the study of disordered networks:

It gives the conductance of a given network *exactly*, in contrast with commonly used relaxation methods.

It requires much less memory storage than systematic node elimination (Fogelholm 1980), so it is useful for large arrays.

It does not rely on specific properties of the system considered (such as missing resistors), but these may be used to speed up the computations.

As an example, we compute the conductance per unit length of networks consisting of very long strips with resistors placed at random on a square lattice. The results obtained for varying strip widths (up to $N = 40$) are analysed in terms of a finite-size scaling hypothesis. They give an estimate for the two-dimensional conductivity critical exponent $t = 1.28 \pm 0.03$. This value is compared with various recent predictions and other numerical results.

0305-4470/82/100557+08$02.00

2. Transfer-matrix formulation

Let us describe first the matrix method that we have developed to calculate the conductivity of strips of random resistors. We have considered strips of width N for which all the sites of the first horizontal line and all the sites of the $(N+1)$th horizontal line are respectively connected by links of zero resistance (see figure 1(a)). The quantity that we could calculate is the conductivity Σ_N between the 1st and the $(N+1)$th horizontal line per unit horizontal length.

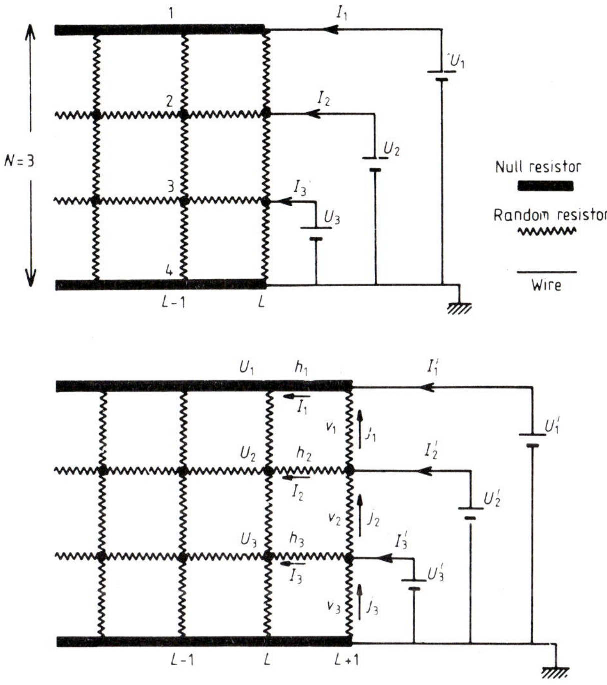

Figure 1. Recursive construction of a strip by adding the horizontal resistors h_i and the vertical resistors v_i.

As usual with transfer matrices, we need to introduce a matrix A_L which characterises the effect of the semi-infinite strip between $-\infty$ and column L. We fix the voltages, U_i, at sites i of column L by connecting these sites to external voltage sources by wires (figure 1). The $(N+1)$th site is always at voltage $U_{N+1}=0$. There will be a current I_i in each wire (see figure 1(a)). The matrix A_L gives by definition the currents I_i as functions of the U_i

$$\begin{pmatrix} I_1 \\ I_2 \\ \vdots \\ I_N \end{pmatrix} = A_L \begin{pmatrix} U_1 \\ U_2 \\ \vdots \\ U_N \end{pmatrix}. \tag{1}$$

We now have to see how the matrix A_L is transformed into a new matrix A_{L+1} when one adds the horizontal resistors h_i and the vertical resistors v_i (see figure 1(*b*)). First, assume that we add only the resistors h_i. We now fix the voltages U'_i at sites i in column $L+1$ by external sources. Then we have

$$U_i = U'_i - h_i I_i. \tag{2}$$

Therefore the currents I_i and the voltages U'_i are related by a matrix B_{L+1}:

$$(I) = B_{L+1}(U'). \tag{3}$$

The equation (2) can be written in a matrix form

$$(U) = (U') - H(I), \tag{4}$$

where the matrix H is diagonal

$$H_{ij} = h_i \delta_{ij}, \tag{5}$$

and the resistor $h_1 = 0$ because all the sites of the first line are connected.

Then one sees that equations (1) and (4) give

$$(U) = (1 + HA_L)^{-1}(U'). \tag{6}$$

Therefore the matrix B_{L+1} is given by:

$$B_{L+1} = A_L(1 + HA_L)^{-1}. \tag{7}$$

Let us now add the vertical resistors v_i. If we impose voltages U'_i at the sites i of column $L+1$, there is a current j_i in the vertical resistor v_i

$$j_i = [U'_{i+1} - U'_i]/v_i. \tag{8}$$

Therefore the current I'_i in the wire connected to the site i of column $L+1$ (to impose the voltage U'_i) is given by

$$\begin{aligned} I'_i &= I_i + j_{i-1} - j_i \\ &= I_i + [1/v_i + 1/v_{i-1}]U'_i - [1/v_i]U'_{i+1} - [1/v_{i-1}]U'_{i-1}. \end{aligned} \tag{9}$$

The formula remains valid for $i = 1$ when one takes $1/v_0 = 0$. The equation (9) has again a matrix form

$$(I') = (I) + V(U'), \tag{10}$$

where V is a tri-diagonal matrix which represents the effect of the vertical bonds at column $L+1$:

$$V_{ij} = [1/v_i + 1/v_{i-1}]\delta_{ij} - [1/v_i]\delta_{j,i+1} - [1/v_{i-1}]\delta_{j,i-1}. \tag{11}$$

It is then clear that the matrix A_{L+1} is given by a recursion relation:

$$A_{L+1} = V + A_L(1 + HA_L)^{-1}. \tag{12}$$

The structure of the matrices H and V is such that the matrix A_L remains always symmetric. So A_L contains $N(N+1)/2$ numbers of interest. It is not surprising that we need only $N(N+1)/2$ numbers to characterise the semi-infinite strip: the semi-infinite strip is equivalent to $N(N+1)/2$ resistors, one for each pair of sites at column L. One can easily show that the conductivity Σ_N per unit length of a strip of width

N is given by:

$$\Sigma_N = \lim_{L\to\infty} A_L(1,1)/L = \lim_{L\to\infty} (A_L^{-1}(1,1))^{-1}/L. \qquad (13)$$

Because the matrices H and V are random and change at every new column L, we did not find any analytic method which allows us to obtain the limit of $A_L(1,1)/L$ for large L. So we simply performed the iterations of (12) numerically for long enough strips in order to approach the limit (13). We have generated randomly, at each column L, a new matrix H and a new matrix V and we have calculated the matrix A_{L+1} using (12). For long lengths, this method would require a lot of matrix inversions. To avoid this difficulty, it is preferable to add the resistors one by one: in this way, the inversion of the matrix becomes simple enough not to require a particular program to invert matrices.

In the present paper, we have chosen to study the case where the resistors are either cut with probability $1-p$ or present with probability p:

$$\begin{aligned} v_i \text{ or } h_i &= 1 && \text{with probability } p \\ &= \infty && \text{with probability } 1-p. \end{aligned} \qquad (14)$$

The fact that some resistors are infinite can be used to speed up the calculations. However, the method described above can be used for any probability distribution of resistors.

3. Results for strips and finite-size scaling analysis

The transfer-matrix formulation presented above is particularly well suited to the case of random strips. It is the finite length L of the strip which can be studied in a given computer time that limits the accuracy of our calculations. The conductance per unit length converges towards its limit value Σ_N for an infinite strip as $L^{-1/2}$. We have found numerically that for a width $N=20$, at the percolation threshold, a length $L=40\,000$ gives a relative accuracy of 4 percent on Σ_N. The computing time in that case was about 1 minute or a CDC−7600.

We hav^ carried out calculations of the conductance for strips of varying widths, at differenı concentrations p of active resistors. The results for concentrations in the vicinity of the percolation threshold ($p=p_c=\frac{1}{2}$) are displayed in figure 2. The data correspond to strips of length $L>5\times10^4$. Longer runs ($L>10^5$) were performed for $p=\frac{1}{2}$, up to 3.6×10^5 for $N=20$ and 2.4×10^5 for $N=30$, except for the largest widths: $L=6\times10^4$ for $N=32$, and 3×10^4 for $N=35$, 37 and 40.

One can see on the figure the change of behaviour of the conductance as a function of strip width, as one crosses the percolation threshold. Below p_c, Σ_N decreases exponentially fast with increasing N, while above p_c it decreases more slowly. One expects for large N a behaviour of the form

$$\Sigma_N \sim (A/N)(p-p_c)^t \qquad (p>p_c), \qquad (15)$$

but it is apparent from figure 2 that for $p=0.52$, values of N much larger than 20 are necessary to observe the asymptotic regime where equation (15) is valid.

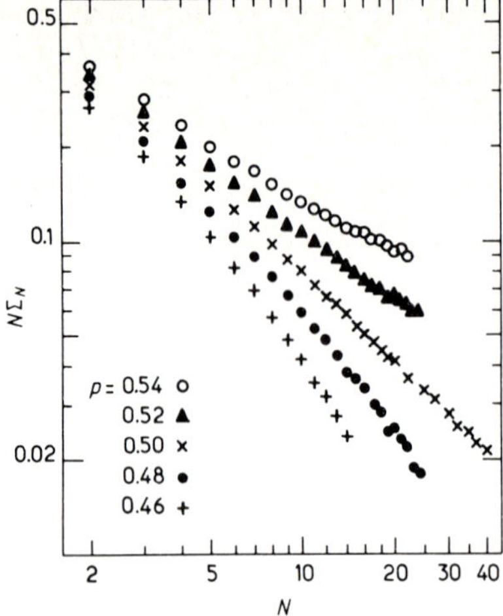

Figure 2. Log–log plot of the conductance per unit of a random resistor strip versus strip width N, in the vicinity of the percolation threshold ($p_c = \frac{1}{2}$ for the square lattice used here).

Precisely at p_c, finite-size scaling arguments (Lobb and Frank 1979 a, b, Mitescu *et al* 1982) suggest that

$$N\Sigma_N \sim CN^{-t/\nu} \qquad (p = p_c), \tag{16}$$

for large N, where ν is the correlation length exponent ($\nu = \frac{4}{3}$ is the accepted value for two-dimensional percolation). Equation (16) implies a linear variation of $\log (N\Sigma_N)$ versus $\log N$, which is indeed well verified on figure 2, for $7 < N < 40$. A least-squares fit of the data in this range yields an estimate for the slope:

$$t/\nu = 0.95 \pm 0.01.$$

The uncertainty comes from the fluctuations between the slopes measured in different sub-ranges of N. We have detected no systematic variation of the slope with N within the precision of the data.

4. Anisotropic media

A class of related problems that may easily be studied by the present approach concerns conduction in anisotropic media. We have considered systems where the probabilities of active resistors are different in the two lattice directions (Blanc *et al* 1980, Nakanishi *et al* 1981). For these systems, the percolation line is known exactly and corresponds to

$$p_V + p_H = 1 \tag{17}$$

if p_V and p_H denote respectively the concentrations of vertical resistors (transverse to the strip) and of horizontal ones (along the strip).

A finite-size scaling form similar to equation (16) is expected to hold for Σ_N on this line, with the same universal value of (t/ν). The data we have obtained for the two cases $p_V = \frac{1}{4}$ and $p_V = \frac{3}{4}$ show that the convergence towards the asymptotic form (equation (16)) is not as rapid as for the isotropic case. This may be seen by plotting for instance $-\log(N\Sigma_N)/\log N$ versus $1/\log N$ (figure 3). The intercepts of the three curves ($p_V = \frac{1}{4}, \frac{1}{2}$ and $\frac{3}{4}$) with the vertical axis are expected to coincide at the common value of t/ν.

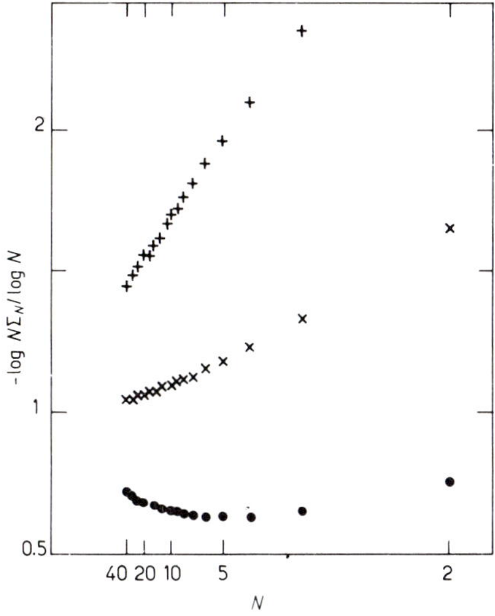

Figure 3. Plot of $-\log(N\Sigma_N)/\log N$ versus $(\log N)^{-1}$, for anisotropic systems on the critical line $p_V + p_H = 1$: $+$, $p_V = \frac{1}{4}$; $\times$, $p_V = \frac{1}{2}$; $\bullet$, $p_V = \frac{3}{4}$.

As is clear from figure 3, the extrapolation of the results is not straightforward for the anisotropic systems: in one case, the curve is non-monotonic, in the other case, the slope is very large. A least-squares fit by a second-degree polynomial in $(1/\log N)$, for various ranges of N, gives as extrapolated values:

$$\begin{aligned} t/\nu &\simeq 1.05 \pm 0.05, \qquad & p_V &= \tfrac{1}{4} \\ &\simeq 0.96 \pm 0.02, & p_V &= \tfrac{1}{2} \\ &\simeq 0.92 \pm 0.06, & p_V &= \tfrac{3}{4}. \end{aligned}$$

This presumably indicates that in the presence of anisotropy the asymptotic regime is not yet reached for the strip widths considered here. Other ways to analyse the data give similar results. The strong dependence of the apparent conductivity exponent on anisotropy has already been noted in the study of finite-size effects (Blanc *et al* 1980) and in renormalisation-group calculations (Lobb *et al* 1981).

5. Discussion of the results

Finite-size scaling has been used in previous calculations of the conductivity (Lobb and Frank 1979b, Sarychev and Vinogradoff 1981, Mitescu *et al* 1982), but on finite networks such as squares. These calculations differ in an important respect from the present one: the conductance per unit length of a random strip is a number, whereas the conductance of finite systems is a random variable with a probability distribution.

In principle, one should renormalise this distribution (Stinchcombe and Watson 1976), though in practice the finite-size scaling analysis is often applied directly to some average. Our approach avoids that problem entirely. In addition, the study of connectivity properties by the transfer-matrix method (Derrida and de Seze 1982) has indicated that strips of moderate width give good results. Squares of much larger linear dimensions are needed to obtain comparable accuracy.

We have analysed carefully the results for the isotropic system at the percolation threshold, to obtain an accurate estimate of (t/ν). A particular motivation comes from the existence in the literature of two simple predictions for this ratio:

$$\text{(i)} \qquad t/\nu = 1, \tag{18}$$

which seems well supported by Monte Carlo simulations (Sarychev and Vinogradoff 1981) and finite-size scaling on squares of size 2 to 14 (Lobb and Frank 1979 b).

$$\text{(ii)} \qquad t/\nu = (\beta+\gamma)/2\nu = 91/96 = 0.9479\ldots, \tag{19}$$

which is a consequence of a recent conjecture for the density of states on the incipient infinite cluster at the percolation threshold (Alexander and Orbach 1982).

The results of various fits that we tried to obtain t/ν fluctuate too much to exclude one of these two predictions. However, the direct measure of the slope of $\log(N\Sigma_N)$ versus $\log N$ (figure 2) yielded $t/\nu = 0.95 \pm 0.01$, and the more elaborate fits in the variable $1/\log N$ (figure 3) gave 0.96 ± 0.02. Therefore, we conclude that a reasonable estimate is

$$t/\nu = 0.96 \pm 0.02,$$

which favours Alexander and Orbach's conjecture.

To confirm this tendency, an increase by at least one order of magnitude in the size of the numerical calculations seems necessary. This might be achieved for instance by combining the transfer matrix approach with an algorithm that first deletes all non-conducting dangling branches (Kirkpatrick 1979).

Acknowledgments

We are indebted to many colleagues for discussions, in particular to L de Seze, S Kirkpatrick, T Lubensky, J P Nadal, R B Pearson and R Rammal. Special thanks are due to E Guyon for communicating and discussing his own work on random conduction.

References

Alexander S and Orbach R 1982 *J. Physique Lett.* **43** to appear
Blanc R, Mitescu C D and Thevenot G 1980 *J. Physique* **41** 387
Derrida B and de Seze L 1982 *J. Physique* **43** 475
Fogelholm R 1980 *J. Phys. C: Solid State Phys.* **13** L571
Gefen Y, Aharony A, Mandelbrot B B and Kirkpatrick S 1981 *Phys. Rev. Lett.* **47** 1771
Kirkpatrick S 1979 *Models of disordered materials* in *Ill-condensed matter, Les Houches 1978* ed R Balian *et al* (Amsterdam: North-Holland) p 321

Lobb C J and Frank D J 1979a *J. Phys. C: Solid State Phys.* **12** L 827
—— 1979b *AIP Conf. Proc.* **58** 308
Lobb C J, Frank D J and Tinkham M 1981 *Phys. Rev.* B **23** 2262
Mitescu C D, Allain M, Guyon E and Clerc J 1982 *J. Phys. A: Math. Gen.* **15** 2523
Nakanishi M, Reynolds P and Redner S 1981 *J. Phys. A: Math. Gen.* **14** 855
Sarychev A K and Vinogradoff A P 1981 *J. Phys. C: Solid State Phys.* **14** L487
Stauffer D 1981 in *Disordered systems and Localization* ed C Castellani *et al* (Berlin: Springer Verlag) p 9
Stinchcombe R B and Watson B P 1976 *J. Phys. C: Solid State Phys.* **9** 3221

J. Phys. A: Math. Gen. **16** (1983) L295–L301. Printed in Great Britain

LETTER TO THE EDITOR

Convergence of finite-size scaling renormalisation techniques

Vladimir Privman and Michael E Fisher

Baker Laboratory, Cornell University, Ithaca, New York 14853, USA

Received 13 April 1983

Abstract. The rate of convergence of the transfer matrix finite-size scaling (or phenomenological renormalisation) method is studied. It is shown both heuristically and numerically that the convergence of estimates for exponents, etc, is governed asymptotically by the leading irrelevant-variable scaling exponent. The more rapid apparent convergence rates observed in many practical calculations for two-dimensional lattices of widths up to ten lattice spacings are attributed to cancellation between various correction terms.

The finite-size rescaling method (or 'phenomenological renormalisation group (RG) technique') introduced by Nightingale (1979) has been used to study the critical behaviour of a variety of the $(d=2)$-dimensional lattice models. (See Nightingale (1982) for an overview and references to the literature.) This method is essentially an application of the finite-size scaling theory of Fisher (1971; Fisher and Barber 1972). Frequently, rather accurate numerical estimates for critical exponents are obtained as judged by comparison with exactly known or reliably conjectured results. The asymptotic rate of convergence of approximants for a critical exponent, say ν, follows from finite-size scaling theory (see below) and should normally be governed by the leading irrelevant bulk RG eigenvalue or scaling exponent, $\lambda_3 \equiv y_3$, which is *negative*. One thus anticipates

$$\nu_{L,L-1} = \nu[1 + CL^{y_3} + E(L)], \tag{1}$$

where $\nu_{L,L-1}$ is obtained from data for lattice strips of width L and $L-1$ lattice spacings (see below), while $E(L)$ denotes terms vanishing more rapidly than $L^{-|y_3|}$ as $L \to \infty$. However, it has been observed by Derrida and DeSeze (1982), Blöte and Nightingale (1982) and Derrida (1981) (see also references quoted in these papers) that the apparent convergence exponent, often called x, found by fitting the finite-size data for $L \lesssim 10$ to the expression

$$\nu_{L,L-1} \simeq \nu(1 + cL^{-x}), \tag{2}$$

has a value $x \gtrsim 2$ for several models for which $|y_3| \lesssim 1$ is expected! Similar effects have been noted in 'phenomenological RG' analyses of Monte Carlo (MC) data, by Binder (1981).

The work reported here attempts to understand the pattern of convergence of finite-size rescaling data and to resolve the seeming contradiction between (1) and (2). First we summarise the derivation of (1) with an emphasis on the variety of sources of corrections, arising from 'bulk' and 'finite-size' effects. Next we use a series analysis technique to estimate y_3 from finite-size data for the $q=3$ Potts model and for percolation problems in $d=2$ dimensions. Finally, we present some evidence that the abnormally large apparent convergence exponent, x, observed for these models, when $L \leqslant 10$, is due to a cancellation between the CL^{y_3} term in (1) and higher-order terms in $E(L)$. We also comment briefly on methods of extrapolating finite size data.

Consider, for illustration, the general bulk ($L=\infty$) RG transformation for the susceptibility, which may be written

$$\chi(g_t, g_h, g_u, \ldots) \approx b^{\gamma/\nu} \chi(g_t b^{y_T}, g_h b^{y_H}, g_u b^{y_3}, \ldots), \tag{3}$$

where b is the RG spatial rescaling factor, while g_t, g_h and g_u are nonlinear scaling fields which approach $t=|T-T_c|/T_c$, $h=H/k_BT$, and a constant u, respectively, in a smooth, analytic manner, when the critical point is approached. (We will comment below on further corrections.) Correspondingly, $y_T, y_H > 0 > y_3 > \ldots$ are the RG fixed point eigenvalues ($\lambda_1, \lambda_2, \lambda_3, \ldots$) or scaling exponents. We will retain explicitly only the leading irrelevant scaling field in (3). For finite-width systems with periodic (or helical) boundary conditions Brézin (1982) has argued that (3) is still approximately valid for finite $D \equiv La \gg a$, where a is the lattice spacing, and that L can be identified as the RG flow parameter, i.e. $b \propto L$ (provided $D/b \equiv La/b \gg a$). A finite-size scaling form is thus obtained as

$$\chi(g_t, g_h, g_u, \ldots; L) = L^{\gamma/\nu} Y_\chi(g_t L^{y_T}, g_h L^{y_H}, g_u L^{y_3}, \ldots) + \Delta\chi(g_t, g_h, g_u, \ldots; L), \tag{4}$$

where one cannot *a priori* exclude the possibility that the bulk nonlinear scaling fields, $g_t, g_h, g_u, \ldots$, are themselves modified by finite-size effects so that they also depend smoothly on, say, L^{-1}: no theoretical study of this point is presently available to our knowledge. Further finite-size corrections depending on the lattice spacing a are contained in $\Delta\chi$ in (4). In general, the only information available on their form is that they become exponentially small when $L \to \infty$ at fixed *non*-critical temperatures, i.e. when $T \neq T_c$. For non-periodic, e.g. free, boundary conditions an expression of the form of (4) can also be advanced: however, the scaling function Y will then contain new arguments with L-dependent, 'surface' scaling fields which result from the generation of surface couplings by the RG transformations (see e.g. Diehl (1982) for a review of this topic and references). In addition, the exponential smallness (at $T \neq T_c$) of other corrections, if true, cannot then be demonstrated so easily. For some further discussion see Fisher (1971, 1973a), and Nakanishi and Fisher (1983).

For quantities such as the free energy or the *finite-size* longitudinal correlation length, $\xi_\parallel$, (calculated, say, from the two largest eigenvalues of the transfer matrix) the relations (3) or (4) hold, with appropriate exponents, only for the singular part, and additional additive background terms must be anticipated; these represent a further source of corrections to the leading scaling behaviour. In what follows we will concentrate on the effect of the leading *bulk* irrelevant scaling field, u. Consider, for example, the estimation of the correlation exponent ν for a ferromagnetic model from the standard expression

$$1 + 1/\nu_{L,L-1} = \ln\left[\frac{\partial\xi(\ldots;L)}{\partial T}\bigg/\frac{\partial\xi(\ldots;L-1)}{\partial T}\right]_c \bigg/ \ln\left(\frac{L}{L-1}\right), \tag{5}$$

where the subscript c denotes evaluation at $T=T_c$ (supposed known) and $H=0$ (Nightingale 1979). On using the scaling form

$$\xi_\parallel \approx L Y_\xi(tL^{1/\nu}, hL^{y_H}, uL^{y_3}), \tag{6}$$

(with, as usual, $\lambda_1 = y_T = 1/\nu$) and the property (Fisher 1971) that Y_ξ is differentiable any number of times at the origin where all the arguments vanish, we obtain after some algebra

$$\nu_{L,L-1} = \nu(1 + CL^{y_3} + E_1 L^{y_3 - 1} + E_2 L^{2y_3} + \ldots), \tag{7}$$

where the higher terms involve powers $n_3 y_3 - n_2$ with $n_2 = 0, 1, 2, \ldots$ and $n_3 = 1, 2, 3, \ldots$. (Note, we are assuming here that u is not a 'dangerous irrelevant variable' as it would be for d exceeding the upper borderline dimensionality, $d_>$ (see Fisher 1973b).) Recently, Aharony and Fisher (1983) studied the effects of the nonlinearities of scaling fields on corrections to scaling: on the basis of their analysis (§ IV) we immediately conclude that if the *exact* T_c is used in (5), only the coefficients of the correction terms in (7) will be affected by the 'bulk' (or L-independent) nonlinearities. Usually, however, T_c itself is also estimated from finite-size data; for example one can solve numerically the relation

$$\xi_\parallel(T_0, H = 0; L)/L = \xi_\parallel(T_0, H = 0; L-1)/(L-1), \tag{8}$$

to obtain an estimate, $T_{0,L}$, for T_c. If we put $t_{L,L-1} \equiv |T_{0,L} - T_c|/T_c$ then (6) gives

$$t_{L,L-1} \propto L^{y_3 - y_T} + \ldots. \tag{9}$$

Allowing for the bulk nonlinearities of the scaling fields, one can show that higher powers of L in (9) have exponents $(n_3 y_3 - n_1 y_T - n_2)$ with $n_1, n_3 = 1, 2, 3, \ldots$ and $n_2 = 0, 1, 2, \ldots$. If we use $T_{0,L}$ in place of T_c in (5), it follows from the above discussion that the general power of L appearing in $\Delta\nu_L = \nu_{L,L-1} - \nu$ is $(n_3 y_3 - n_2 - n_1 y_T)$ with $n_1, n_2 = 0, 1, 2, \ldots$ and $n_3 = 1, 2, \ldots$; this should be compared with (7). Finally, if other bulk irrelevant scaling fields are allowed (with scaling exponents $0 > y_3 > y_4 > y_5 > \ldots$) and if only *bulk* nonlinearities are allowed for, the general power of L in $\Delta\nu_L$ is

$$-n_1 y_T - n_2 + \sum_{j=3} n_j y_j, \tag{10}$$

where $n_1, n_2, \ldots = 0, 1, 2, \ldots$ but at least one of n_j for $j \geq 3$ is non-zero. This result, however, is not really definitive since we have neglected other types of corrections as discussed above in connection with the relation (4). Notice also that we have, as mentioned, assumed $d < d_>$ (=4, 6, etc) in the discussion (see also Brézin 1982) and further supposed that no special relations hold between higher-order scaling exponents which may lead to logarithmic correction factors (see Wegner 1972, and the review by Patashinskii and Pokrovskii 1977).

We proceed next to demonstrate that the L^{y_3} term can actually be detected in numerical sequences which display an apparent convergence rate with $x \simeq 2$ when fitted as in (2). Consider the estimators

$$r_L = \nu + (\nu_{L,L-1} - \nu)/L = \nu + \nu C L^{-(|y_3|+1)} + \ldots, \tag{11}$$

where (1) has been assumed. Asymptotically, the r_L converge in a way similar to that expected for ratios of coefficients constructed for a logarithmic derivative series of a function which has a confluent singularity with exponent $|y_3|$ (see further below). Since, for short series, naive ratio and Padé techniques may be misleading when confluent corrections are involved (see e.g. Baker 1975) we construct a transformed function. Thus consider the sequence

$$a_0 = 1/\nu, \qquad a_j = a_{j-1}/r_{j+1} \qquad (j \geq 1) \tag{12}$$

and the function

$$f(z) = \sum_{n=0}^{\infty} a_n z^n = (\nu - z)^{-1}[1 + A(\nu - z)^{|y_3|} + D(z)], \tag{13}$$

which has a leading divergence with exponent 1 at $z_c = \nu$ and a confluent term with exponent $|y_3|$. Additional terms, denoted by $D(z)$ in (13), will, of course, be present resulting from the transformation and from higher-order corrections in (1). Hopefully, these less singular terms will have a relatively small influence on the analysis: we will, however, comment later on their potentiality for interference. The particular Padé-type method of analysis which we adopt is that proposed recently by Adler, Moshe and Privman (1982a, b) which, in brief, involves transforming the series for $f(z)$ into an expansion in powers of

$$s = 1 - [1 - (z/\nu)]^y, \tag{14}$$

with a variable trial value y. The series for

$$G(y, s) = y(1-s)(\partial/\partial s) \ln f[z(y, s)], \tag{15}$$

is then computed and used to calculate several $[M/N]$ Padé approximants to $G(y, 1)$ and thence a family of curves

$$\Gamma(y) = G^{[M/N]}(y, s = 1) \simeq G(y, 1), \tag{16}$$

in the (y, Γ) plane. A region of 'convergence' or 'confluence' of the $\Gamma(y)$ curves is expected close to the point $(|y_3|, 1)$ in the (y, Γ) plane: see Adler *et al* (1982a, b) for details.

Consider first the $q = 3$ Potts model and let us study a sequence of estimates for β/ν obtained by Hamer (1982) from the expression

$$(\beta/\nu)_{L,L-1} = (1-L)[M_L(T_c)/M_{L-1}(T_c) - 1], \tag{17}$$

where $M_L(T)$ is a quantity with the critical behaviour of magnetisation which vanishes with exponent β when $L = \infty$. The exact value of β/ν is known (Den Nijs 1979, Nienhuis, Riedel and Schick 1980, Nienhuis 1982), and this exponent replaces ν in (11)–(14). (It is easy to verify that $(\beta/\nu)_{L,L-1}$ of (17) converges to β/ν with the same general pattern of corrections as for $\nu_{L,L-1}$.) In figure 1 are plotted $\Gamma(y)$ curves obtained from several near-diagonal Padé approximants. Also indicated, on the y axis, is the recent estimate, $|y_3| = 0.68 \pm 0.13$, obtained from analysis of several low-temperature series expansions (Adler and Privman 1982); this range is consistent with earlier $|y_3|$ estimates (see Adler and Privman (1982) for references). The point with coordinates $(y = 0.68, \Gamma = 1)$ is marked on the plot by a cross. The $\Gamma(y)$ curves display a clear confluence close to this point: this provides a definite indication of the presence of the L^{y_3} term in the $(\beta/\nu)_{L,L-1}$ sequence. The confluence region in figure 1, if taken at face value, suggests a rather narrow $|y_3|$ range; however, the series analysed is relatively short and large systematic errors may arise in this method from background terms (see Adler *et al* 1982b). Thus it is inadvisable to propose confidence limits narrower than indicated by the original series analysis.

Consider, as a second example, the $\nu_{L,L-1}$ sequences of (5) for percolation which were calculated by Derrida and DeSeze (1982). Here only results for their longest $\nu_{L,L-1}$ sequence will be discussed: this was obtained for site percolation with helical boundary conditions. (Other $\nu_{L,L-1}$ sequences give similar results; some of the corresponding critical point, $(p_c)_{L,L-1}$, data are discussed briefly below.) The exact value of

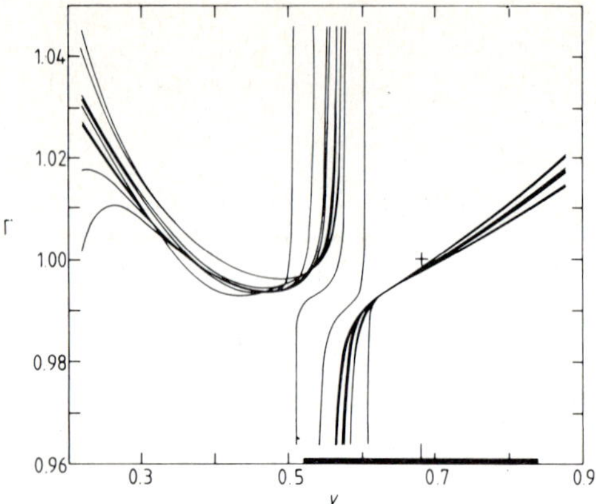

Figure 1. Plots of $\Gamma(y)$ derived from the sequence of estimates for β/ν for the $q=3$ Potts model, obtained with [2/6], [3/5], [4/4], [5/3], [6/2], [2/5], [3/4], [4/3] and [5/2] Padé approximants. The cross indicates the series estimate $|y_3|=0.68$, $\Gamma=1$; the corresponding range is shown on the axis.

ν is known (Den Nijs 1979, Nienhuis *et al* 1980, Nienhuis 1982). Several series and MC studies of the leading corrections to scaling for percolation have been reported and the results cluster in two non-overlapping ranges of y (see Adler *et al* (1982b) for a review) as obtained from the relations $y=\theta/\nu$ and $y=\Omega\beta\delta/\nu$, where θ and Ω are temperature-like and field-like bulk confluent exponents. The above relations are meaningful only for corrections due to irrelevant scaling fields. Since there is a theoretical argument (Aharony and Fisher 1983, Margolina, Djordjević, Stauffer and Stanley 1983) which suggests that the higher-value range is observed when corrections due to the nonlinearity of g_t are substantial, we concentrate on the lower range of estimates. Thus $|y_3|=0.94\pm0.11$ (Adler *et al* 1982a) is typical of both the series estimates and the MC results of Stauffer (1981). The corresponding curves $\Gamma(y)$ are plotted in figure 2 where the point ($y=0.94$, $\Gamma=1$) is marked by a cross and the quoted $|y_3|$ range is shown on the y axis. One observes a confluence region consistent with the central series estimate. Again, one should probably not be tempted to narrow the uncertainties quoted in the series analysis: indeed, systematic errors may be especially large in this case owing to a linear term, $D_1(\nu-z)$, in $D(z)$ of (13), whose presence is anticipated since it represents merely an additive constant background in $f(z)$. One may try to study the resulting interference by dividing out different terms (see Adler and Privman 1982) but the available series are too short for such analysis to be convincing. Independent evidence for the presence of both terms comes from a study of $(p_c)_{L,L-1}$ for bond percolation (where the exact value is $p_c=\frac{1}{2}$) in which two 'loose' confluences were found: one at $y\simeq1$ and another in the range of y consistent with the lower half of the range $|y_3|+1/\nu=1.69\pm0.11$ derived from series.

The above studies demonstrate that the expected leading correction term, varying as L^{y_3}, is indeed present in the finite-size sequences. There is additional convincing evidence that for $L\leqslant10$ the asymptotic form for $L\gg1$ has not been achieved in existing finite-size data. Thus several finite-size sequences vary non-monotonically (see e.g. Derrida and DeSeze 1982). Even for some monotonic sequences there are problems: thus if one determines an exponent $x\equiv x_\nu$ from an exponent sequence

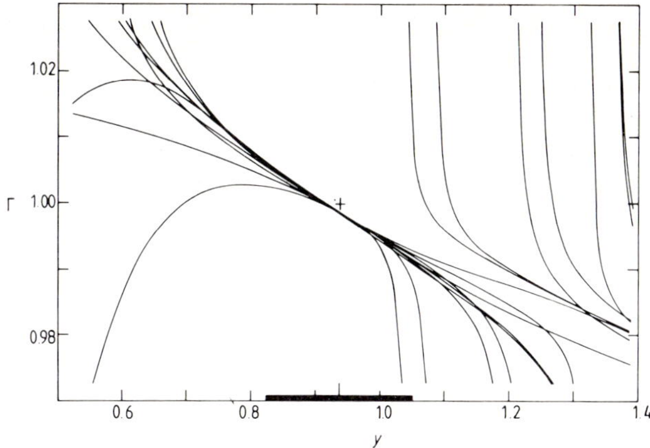

Figure 2. Plots of $\Gamma(y)$ obtained from estimates for ν for site percolation on a square lattice with helical boundary conditions (using the same Padé approximants as in figure 1). The cross indicates the series estimate $|y_3| = 0.94$, $\Gamma = 1$ with the associated range being shown on the axis.

$\nu_{L,L-1}$ and another exponent $x \equiv x_{T_c}$ from estimates for T_c as in (8), one should find $x_{T_c} - x_\nu \simeq 1/\nu$. (Compare (7) and (9).) In practice, however, this difference takes other values (see e.g. Derrida 1981). In other cases x depends strongly on the boundary conditions (see e.g. Kinzel and Yeomans 1981). Unfortunately, the pattern of corrections to scaling in the $d = 2$ Ising spin-$\frac{1}{2}$ model, where $\nu_{L,L-1}$ is known exactly (Onsager 1944, Derrida and DeSeze 1982), is effectively 'pathological' since *all* non-analytic corrections seem to vanish identically (Aharony and Fisher 1980).

Since the sequences with $L \leqslant 10$ appear too short to fit reliably to more than one power-law term, an indirect method of estimating the relative amplitude of the contribution of L^{y_3} term has been attempted. To this end, consider

$$\tilde{\nu}_L = [L^{|y_3|}\nu_{L,L-1} - (L-1)^{|y_3|}\nu_{L-1,L-2}]/[L^{|y_3|} - (L-1)^{|y_3|}] \qquad (18)$$

(where ν now denotes the exponent, etc appropriate for the particular sequence considered). The term in L^{y_3} cancels in $\tilde{\nu}_L$ which should thus approach ν as $L \to \infty$ at a rate determined by the higher powers in (7) but with modified coefficients. (In practice, one must calculate using several $|y_3|$ values based on the range of series estimates.) One finds in all the examples that $|\tilde{\nu}_L - \nu|$ usually exceeds $|\tilde{\nu}_{L,L-1} - \nu|$ while the ratio $|\tilde{\nu}_L - \nu|/|\nu_{L,L-1} - \nu|$ ranges from below 1 to 10. This suggests that there is generally some cancellation between the CL^{y_3} term in $\nu_{L,L-1}$ and one or more higher-order terms from $E(L)$ (see (1)) which leads to as much as an order of magnitude reduction in the deviation $\Delta\nu_L$ at $L \simeq 10$ over what the leading term alone would give. Such cancellation also leads fairly naturally to the observed inequality $x > |y_3|$. Notice, however, that there is probably no general underlying mechanism of cancellation, beyond fortuitous numerical coincidence, since there is at least one counter-example in which one does find x close to the expected $|y_3|$ value (Uzelac and Jullien 1981).

A related topic is the actual technique of extrapolation of the finite-size sequences to the $L \to \infty$ limit. An effective method was developed by Hamer and Barber (1981). In each iteration of their technique, a power-law correction is, ideally, cancelled. However, no effective method is known for the systematic study of exponents and the coefficients of correction terms. (The method we have used above is useful only for estimating y_3 in models with exactly known leading exponents or critical points.)

Recently, Barber (1983) proposed allowing for additional scaling fields by extending the parameter space (see also Yeomans and Fisher 1983). Although his methods are somewhat limited, in that they require prior knowledge of aspects of the phase diagram, they do seem to be a step in the right direction. Likewise, in series analysis it has proven very useful (Chen, Fisher and Nickel 1982) to allow for more variables when studying non-analytic, confluent correction terms. Basically, the idea in both cases is to fit to a multisingular scaling form like (3).

Helpful discussions with M Barma, E Brézin, D A Huse, M P Nightingale and R Pandit are appreciated. One of us (VP) is grateful for the financial support of the Rothschild Fellowship Foundation. Further support by the National Science Foundation†, in part through the Materials Science Center at Cornell University, is gratefully acknowledged.

References

Adler J, Moshe M and Privman V 1982a *Phys. Rev.* B **26** 1411
—— 1982b in *Percolation Structures and Processes* ed G Deutscher, R Zallen and J Adler (Bristol: Adam Hilger) p 397
Adler J and Privman V 1982 *J. Phys. A: Math. Gen.* **15** L417
Aharony A and Fisher M E 1980 *Phys. Rev. Lett.* **45** 679
—— 1983 *Phys. Rev.* B **27** 4394
Baker G A 1975 *Essentials of Padé Approximants* (New York: Academic)
Barber M N 1983 *University of New South Wales preprint*
Binder K 1981 *Z. Phys.* B **43** 119
Blöte H W J and Nightingale M P 1982 *Physica* **112A** 405
Brézin E 1982 *J. Physique* **43** 15
Chen J-H, Fisher M E and Nickel B G 1982 *Phys. Rev. Lett.* **48** 630
Den Nijs M P M 1979 *J. Phys. A: Math. Gen.* **12** 1857
Derrida B 1981 *J. Phys. A: Math. Gen.* **14** L5
Derrida B and DeSeze L 1982 *J. Physique* **43** 475
Diehl H W 1982 *J. Appl. Phys.* **53** 7914
Fisher M E 1971 in *Critical Phenomena, Proc. Enrico Fermi School* vol 51, ed M S Green (New York: Academic) p 1
—— 1973a *J. Vac. Sci. Technol.* **10** 665
—— 1973b in *Renormalization Group in Critical Phenomena and Quantum Field Theory: Proc. Conf.* ed J D Gunton and M S Green (Temple University Press) p 65
Fisher M E and Barber M N 1972 *Phys. Rev. Lett.* **28** 1516
Hamer C J 1982 *J. Phys. A: Math. Gen.* **15** L675
Hamer C J and Barber M N 1981 *J. Phys. A: Math. Gen.* **14** 2009
Kinzel W and Yeomans J M 1981 *J. Phys. A: Math. Gen.* **14** L163
Margolina A, Djordjević Z V, Stauffer D and Stanley H E 1983 *Boston University preprint*
Nakanishi H and Fisher M E 1983 *J. Chem. Phys.* **78** 3279
Nienhuis B 1982 *J. Phys. A: Math. Gen.* **15** 199
Nienhuis B, Riedel E K and Schick M 1980 *J. Phys. A: Math. Gen.* **13** L189
Nightingale M P 1979 *Proc. Kon. Ned. Akad. Wet.* B **82** 235
—— 1982 *J. Appl. Phys.* **53** 7927
Onsager L 1944 *Phys. Rev.* **65** 117
Patashinskii A Z and Pokrovskii V L 1977 *Usp. Fiz. Nauk* **121** 55 (1977 *Sov. Phys. Usp.* **20** 31)
Stauffer D 1981 *Phys. Lett.* A **83** 404
Uzelac K and Jullien R 1981 *J. Phys. A: Math. Gen.* **14** L151
Wegner F J 1972 *Phys. Rev.* B **5** 4529
Yeomans J M and Fisher M E 1983 to be published

† Through grant no DMR 81-17011.

A NEW METHOD TO COMPUTE THE SPECTRUM OF LOW-LYING STATES IN MASSLESS ASYMPTOTICALLY FREE FIELD THEORIES ☆

M. LÜSCHER
Institut für Theoretische Physik, Universität Bern, Sidlerstrasse 5, CH-3012 Bern, Switzerland

Received 12 July 1982

A systematic universal expansion for the masses of the low-lying stable particles in pure Yang–Mills gauge theories and non-linear σ-models is proposed.

1. The masses m_i of the stable particles in (four-dimensional) pure Yang–Mills gauge theories and (two-dimensional) non-linear σ-models are proportional to the Λ-parameter:

$$m_i = C_i \Lambda_{\overline{\mathrm{MS}}} . \tag{1}$$

If the system is confined to a compact space [‡1], the numbers C_i become functions of the linear size L of the space. In this letter a systematic expansion of the C_i's is obtained, the expansion parameter being

$$z = m_0 L, \quad m_0\text{: mass gap} . \tag{2}$$

The method is here explained for the simplest case, the O(n) non-linear σ-model. Results and details of the computations in the Yang–Mills case will be given elsewhere. I shall first (section 2) define the expansion and then show how to use it to obtain C_i at $L = \infty$ (section 3).

2. The O(n) non-linear σ-model describes n-component spin fields $s(x)$, $s(x)^2 = 1$, on a two-dimensional space–time manifold M. The action is

$$S = \frac{n}{2f_0} \int_{\mathrm{M}} \mathrm{d}^2x \, |h|^{1/2} h^{\mu\nu} \partial_\mu s \cdot \partial_\nu s , \tag{3}$$

where $h_{\mu\nu}$ is the metric on M and f_0 the (bare) coupling constant. Ultimately, we are interested only in the case, where M is the ordinary flat Minkowski space, but for the purpose of generating the expansion alluded to above, we shall take

$$\mathrm{M} = \mathbf{R} \times \mathrm{S}^1 . \tag{4}$$

Here, $\mathbf{R}$ refers to time and S^1 to space. The volume (circumference) of S^1 is denoted by L.

Because space is compact, the momentum operator has a discrete spectrum of eigenvalues

$$P = (2\pi/L)\nu, \quad \nu \in \mathbf{Z} . \tag{5}$$

The hamiltonian also has a discrete spectrum and its eigenstates group into irreducible multiplets of the internal symmetry group O(n).

A first observation now is that the eigenvalues and eigenvectors of the hamiltonian can be computed in ordinary perturbation theory. For example, one may regularise with a lattice cutoff and then set up a straightforward hamiltonian formalism. Alternatively, one can compute suitable euclidean correlation functions using dimensional regularisation [1] and expand for large time separations (to avoid superficial infrared divergencies one should add *compact* extra dimensions to regularise [2]). It then turns out that the ground state is not degenerate and that the lowest excited states make up an O(n)-vector, i.e. these states occur as intermediate states in the spin field two-point function. They have vanishing momentum and their energy up to one-loop order is

☆ Work supported in part by Schweizerischer Nationalfonds.
‡1 *Not* space–time, i.e. time remains an unrestricted real variable.

0 031-9163/82/0000–0000/$02.75 © 1982 North-Holland

$$m_0 = L^{-1}\{f(n-1)/2n + f^2[(n-1)(n-2)/2n^2] \times (4\pi)^{-1}[\ln \mu^2 L^2 - \ln 4\pi - \Gamma'(1)] + ...\}. \quad (6)$$

Here, f denotes the renormalized coupling constant in the dimensional regularisation scheme with minimal subtraction (MS) [3] and μ is the normalization mass.

Eq. (6) has three remarkable features.

(a) m_0 vanishes for $f \to 0$. This behaviour is due to the fact that the classical ground state $s(x)$ = constant is degenerate.

(b) In the abelian case $n = 2$, the second term vanishes and m_0 is exactly proportional to L^{-1} as it ought to be.

(c) m_0 is a physical quantity and therefore renormalization group invariant:

$$[\mu\partial/\partial\mu + \beta(f)\partial/\partial f]m_0 = 0, \quad (7)$$

$$\beta(f) = -\sum_{\nu=0}^{\infty} b_\nu f^{\nu+2}, \quad (8)$$

$$b_0 = (n-2)/2\pi n, \quad b_1 = (n-2)/(2\pi n)^2. \quad (9)$$

The coefficients b_0 and b_1 are taken from ref. [4]. b_2 is also known [5], but not needed here.

We now eliminate the coupling constant f in favour of the dimensionless, renormalization group invariant parameter $z = m_0 L$. To this end, note first that

$$\Lambda_{\overline{\mathrm{MS}}} = \Lambda_{\mathrm{MS}} \exp\{\tfrac{1}{2}[\ln 4\pi + \Gamma'(1)]\}, \quad (10)$$

$$\Lambda_{\mathrm{MS}} = \mu(b_0 f)^{-b_1/b_0^2} \exp(-1/b_0 f) \times \exp\left(-\int_0^f dx\,[1/\beta(x) + 1/b_0 x^2 - b_1/b_0^2 x]\right). \quad (11)$$

Next, from eq. (6) we have

$$f = z\,2n/(n-1) - z^2[2n/(n-1)]^2[(n-2)/n] \times (4\pi)^{-1}[\ln \mu^2 L^2 - \ln 4\pi - \Gamma'(1)] + \dots. \quad (12)$$

When inserted into eq. (11), the following formula is obtained:

$$m_0/\Lambda_{\overline{\mathrm{MS}}} = K(n)(z e^{\pi/z})^{(n-1)/(n-2)} \times \left(1 + \sum_{\nu=1}^{\infty} a_\nu z^\nu\right), \quad (13)$$

$$K(n) = (4\pi)^{-1} \exp[-\Gamma'(1)] \times [\pi^{-1}(n-2)/(n-1)]^{1/(n-2)}. \quad (14)$$

The coefficients a_ν can be computed by extending the series (6). They are pure numbers independent of μ and L, because m_0, $\Lambda_{\overline{\mathrm{MS}}}$ and z are renormalization group invariant. In other words, the expansion (13) is *universal*.

3. The problem is to compute $C_0 = m_0/\Lambda_{\overline{\mathrm{MS}}}$ in the infinite volume limit, i.e. for $z \to \infty$ (provided m_0 does not vanish in this limit). The small z expansion (13) therefore does not seem to be very useful. That this impression is false, can nicely be demonstrated in the large n limit, which is exactly solvable.

At $n = \infty$, the mass gap m_0 is determined from the large-n saddle point condition

$$\ln(m_0/\Lambda_{\overline{\mathrm{MS}}}) = F(z), \quad z = m_0 L, \quad (15)$$

$$F(z) = \int_0^\infty \frac{dt}{t} \exp(-tz^2) \sum_{k=1}^{\infty} \exp(-k^2/4t). \quad (16)$$

Eq. (15) is an implicit equation for m_0, which has a unique solution $m_0(L)$. $m_0(L)$ is monotonically decreasing and

$$\lim_{L\to\infty} m_0 = \Lambda_{\overline{\mathrm{MS}}}. \quad (17)$$

For small z, on the other hand, we have

$$F(z) = \pi/z + \ln z - \ln 4\pi - \Gamma'(1) + \sum_{\nu=1}^{\infty} (-1)^\nu \binom{2\nu}{\nu} \zeta_R(2\nu+1)(z/4\pi)^{2\nu} \quad (18)$$

and thus obtain the large-n limit of eq. (13):

$$m_0/\Lambda_{\overline{\mathrm{MS}}} = (4\pi)^{-1} \exp[-\Gamma'(1)]\, z \exp(\pi/z) \times \left(1 + \sum_{\nu=1}^{\infty} a_{2\nu} z^{2\nu}\right), \quad (19)$$

$$a_2 = -2(4\pi)^{-2}\zeta_R(3),$$

$$a_4 = 2(4\pi)^{-4}[3\zeta_R(5) + \zeta_R(3)^2], \quad \text{etc.} \quad (20)$$

{ζ_R denotes Riemann's zeta function (ref. [6], § 9.522)}.

The expansions (18) and (19) are absolutely convergent for $|z| < 2\pi$. We can therefore accurately calculate C_0 for (say) $z \leqslant 4$ by evaluating the series (19) keeping only the first few terms. This is shown in fig. 1. Besides the rapid convergence, the most remarkable fact this plot reveals is that $C_0(z)$ is practically

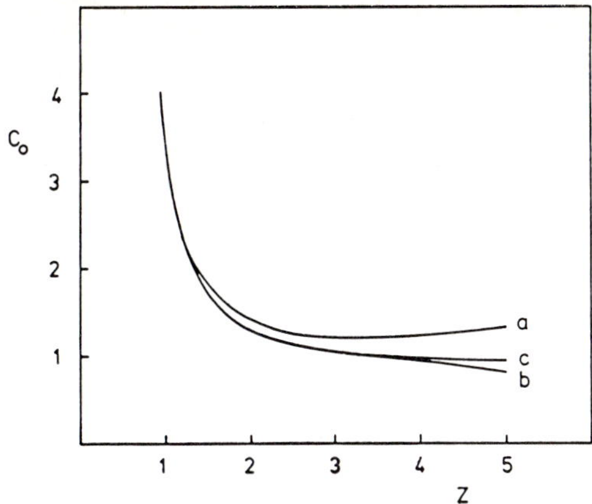

Fig. 1. Plot of $C_0(z) = m_0/\Lambda_{\overline{\mathrm{MS}}}$ versus z in the large-n limit [eq. (19)]. Curve "a" is obtained from the leading term alone, "b" includes the contribution of a_2 and "c" also includes a_4.

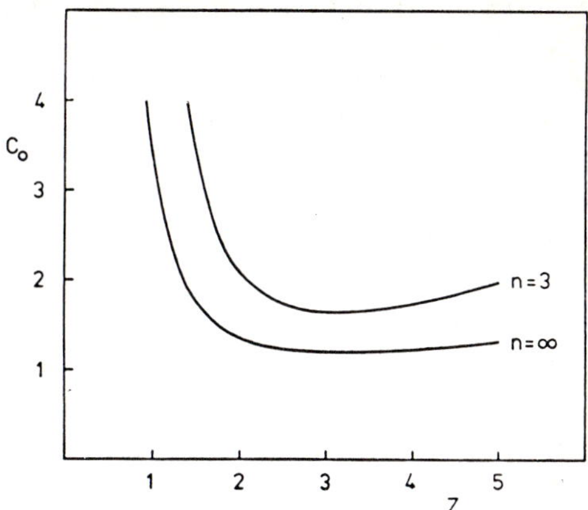

Fig. 2. Plot of $C_0(z) = m_0/\Lambda_{\overline{\mathrm{MS}}}$ versus z for $n = 3$ and $n = \infty$, keeping only the leading term in eq. (18) [cp. eq. (22)]. The curves for intermediate values of n are sandwiched between the $n = 3$ and $n = \infty$ curves.

equal to its limiting value $C_0(\infty) = 1$ already at $z = 3$. More precisely, the deviation $C_0(z) - 1$ vanishes exponentially as $z \to \infty$:

$$m_0/\Lambda_{\overline{\mathrm{MS}}} = 1 + \mathrm{O}(\mathrm{e}^{-z}) \,. \tag{21}$$

At $z = 3$ it is about 8% and then shrinks to about 2% at $z = 4$. This behaviour is not really surprising in view of the general experience that finite size corrections become negligible as soon as L is larger than three to four times the correlation length.

The moral, which can be drawn from the large n limit is:

(i) It is not necessary to extrapolate $C_0(z)$ all the way up to $z = \infty$, because the limiting value $C_0(\infty)$ is approached quickly. More precisely, $C_0(z)$ at (say) $z = 3$ can be expected to be a good approximation to $C_0(\infty)$.

(ii) Only a few terms in the small z expansion (13) are needed to get accurate values for $C_0(z)$ in the interval $0 < z \leqslant 4$.

We now apply these rules in the case $n = 3$. The leading order formula

$$m_0/\Lambda_{\overline{\mathrm{MS}}} = (8\pi^2)^{-1}\exp[-\Gamma'(1)]\,z^2\exp(2\pi/z) \qquad (n = 3) \,, \tag{22}$$

is plotted in fig. 2. As expected, $C_0(z)$ is rapidly decreasing for $0 < z < 2$ and then quickly becomes flat giving a value

$$m_0 \simeq 1.7\Lambda_{\overline{\mathrm{MS}}} \qquad (n = 3) \tag{23}$$

in the region $z = 3 \dots 4$ [the result (23) may very well be off the true value by 20% as would be the case in the large-n limit, when only the lowest term in the z-expansion is used to estimate $C_0(z)$ at $z = 3 \dots 4$].

It is interesting to compare eq. (23) with the number obtained from a Monte Carlo simulation of the lattice O(3) σ-model [7,8]:

$$m_0 \simeq 100\Lambda_{\mathrm{L}} \qquad (n = 3) \,. \tag{24}$$

Here, Λ_{L} denotes the lattice Λ-parameter

$$\Lambda_{\mathrm{L}} = a^{-1}(b_0 f_0)^{-b_1/b_0^2}\exp(-1/b_0 f_0) \times [1 + \mathrm{O}(f_0)] \tag{25}$$

and a is the lattice spacing. Noting

$$\Lambda_{\mathrm{L}}/\Lambda_{\overline{\mathrm{MS}}} = 32^{-1/2}\exp[-\pi/2(n-2)] \,, \tag{26}$$

one gets from eq. (24)

$$m_0 \simeq 3.7\Lambda_{\overline{\mathrm{MS}}} \qquad (n = 3) \,, \tag{27}$$

i.e. a value, which is about twice as large as our result (23). There is no reason to be discouraged, however, because eq. (23) is only the lowest approximation and because the Monte Carlo result may very well be wrong by a factor of 2 [‡2].

4. The method proposed in this letter is applicable to a number of interesting physical quantities, in particular, to mass ratios and form factors at zero momentum. The main limitation on the accuracy that can be attained seems to be one's ability to do perturbation theory. For example, to get the next to leading term in eq. (13), requires the computation of the three-loop β function and the finite parts of a set of two-loop diagrams.

‡2 The Monte Carlo method involves an extrapolation of numerically computed data at $\beta = n/f_0 \simeq 1.5$ up to $\beta = \infty$. Because the corrections to the exact result at $\beta = \infty$ are *powers* of β^{-1}, they are not easily seen in the small interval of β, which is accessible to the computer. An error of a factor of 2 may therefore occur even if the data seem to scale perfectly.

References

[1] G. 't Hooft and M. Veltman, Nucl. Phys. B44 (1972) 189.
[2] M. Lüscher, Ann. Phys. 142, to be published.
[3] G. 't Hooft, Nucl. Phys. B61 (1973) 455.
[4] E. Brézin and J. Zinn-Justin, Phys. Rev. B14 (1976) 3110.
[5] E. Brézin and S. Hikami, J. Phys. A11 (1978) 1141.
[6] I.S. Gradshteyn and I.M. Ryzhik, Table of integrals, series and products (Academic Press, New York, 1965).
[7] S.H. Shenker and J. Tobochnik, Phys. Rev. B22 (1980) 4462.
[8] G. Fox, R. Gupta, O. Martin and S. Otto, Nucl. Phys. B205 [FS5] (1982) 188.

Nuclear Physics B210[FS6] (1982) 133–141

FINITE SIZE SCALING AND LOW MASS GLUEBALLS*

Richard C. BROWER

Physics Department, University of California, Santa Cruz, CA 95064, USA

Michael CREUTZ

Brookhaven National Lab., Upton, Long Island, NY 11973, USA

Michael NAUENBERG[1]

Institute for Theoretical Physics, University of California, Santa Barbara, CA 92106, USA

Received 15 March 1982

We propose finite lattice effects as a probe of the glueball mass spectrum, and give an analysis of the recent SU(2) Monte Carlo data of Brower, Nauenberg and Schalk in terms of a gas of free glueballs. For L^4 lattices with $L=4$, 5, 6 fits are made to $\xi(m=1/a\xi)$ which indicate a rather large effective number of degrees of freedom (i.e. statistical degeneracy where a spin J counts as $2J+1$) from 5 to 15 states. As the degeneracy is increased, the central glueball mass increases from $m=(1.3\pm0.2)\sqrt{\kappa}$ at degeneracy 5 to about $m=(1.9\pm0.2)\sqrt{\kappa}$ at degeneracy 15, relative to the SU(2) string tension $\sqrt{\kappa}$.

Recent Monte Carlo simulations for lattice quantum chromodynamics have come close to experimental confrontation for predicted mass ratios in the continuum limit. For SU(2) and SU(3) quantum chromodynamics without quarks, calculations of the ratio of the square root of the string tension ($\sqrt{\kappa}$) relative to the minimal subtracted $\Lambda_{\overline{\text{ms}}}$ parameter of logarithmic scaling violations [1, 2] have given

$$\begin{aligned} \Lambda_{\overline{\text{ms}}}/\sqrt{\kappa} &= (1.3\pm0.2)(0.199) \text{ for SU(2)}\,, \\ \Lambda_{\overline{\text{ms}}}/\sqrt{\kappa} &= (0.5\pm0.15)(0.289) \text{ for SU(3)}\,. \end{aligned} \tag{1}$$

In SU(3), with $\sqrt{\kappa}=420$ MeV, this is a suitable Λ parameter ($\Lambda_{\overline{\text{ms}}}\simeq 60$ MeV); however, the experimental uncertainty for $\Lambda_{\overline{\text{ms}}}$ does not encourage much more streneous computations of this ratio.

Within quarkless QCD, a more definitive test should be the prediction of the lowest mass glueball and its degeneracy (i.e. statistical weight $2J+1$ for spin J). After all, the glueball is a salient non-perturbative effect of QCD, absent in earlier quark models. Even a 10% bound on its mass could in principle prove fatal to the standard QCD theory.

* Supported by a grant from the National Science Foundation.

[1] Permanent address: University of California, Santa Cruz, CA 95064.

However, even for quarkless SU(2) Monte Carlo, the computation of the glueball mass has appeared extremely difficult. By measuring the plaquette–plaquette correlation function at separation L, and assuming

$$\langle \mathrm{Tr}\,(U_{P_1})\,\mathrm{Tr}\,(U_{P_2})\rangle - \langle \mathrm{Tr}\,(U_P)\rangle^2 = C_L\,\mathrm{e}^{-L/\xi_{GB}} \qquad (2)$$

Bhanot and Rebbi [2] estimated the SU(2) glueball mass $m \equiv 1/a\xi$ relative to their string tension* (κ) to be

$$m/\sqrt{\kappa} = 3 \pm 1\,, \qquad (3)$$

and Berg [3] by a similar method obtained

$$m/\sqrt{\kappa} = 3.7 \pm 1.2\,. \qquad (4)$$

A major difficulty is the need to have good statistical accuracy so that the small exponential *difference* in the connected part can be measured.

Here we attempt a new approach to measuring the glueball mass or inverse correlation length. We look at the finite lattice behavior of the one-point function,

$$E_P(L, \beta) = \tfrac{1}{2}\langle \mathrm{Tr}(U_P)\rangle_L \qquad (5)$$

in a periodic box of volume L^4, and bare coupling $\beta = 4/g_0^2$. Subsequently, we will show that the leading exponential corrections are due to glueball propagation to a plaquette's periodic image:

$$E_P(L, \beta) - E_P(\infty, \beta) \sim C_L\,\mathrm{e}^{-L/\xi}\,. \qquad (6)$$

Moreover, Brower, Nauenberg and Schalk [4] have demonstrated that this difference obeys the scaling behavior dictated by finite lattice scaling theory and asymptotic freedom for $L \geqslant 4$ and $\beta = 4/g_0^2 \geqslant 2.05$,

$$E_P(L, \beta) - E_P(\infty, \beta) = \frac{1}{L^4}\,\varepsilon\left(\frac{L}{\xi}\right), \qquad (7)$$

where the correlation length obeys the scaling law

$$\xi = C\left(\frac{6\pi^2}{11}\beta\right)^{51/121} \mathrm{e}^{-(3\pi^2/11)\beta}\,. \qquad (8)$$

We now turn to a model of these finite scaling effects based on a free gas of glueballs.

A GLUEBALL GAS

Formulated in a periodic box of finite temporal extent, a path integral represents the partition function for a finite temperature quantum field theory. The period in

* Bhanot and Rebbi's estimate for $\sqrt{\kappa}$ for SU(2) is 1.2 times bigger, so that their glueball mass is actually $3.6\sqrt{\kappa}$ in terms of the string tension of ref. [1], which we use throughout.

time corresponds to the inverse physical temperature. For a cool system, a dilute gas of the lightest particles in the theory will dominate the finite size effects. In this way, we can obtain information on the light particle spectrum.

These ideas can be formalized with the transfer matrix expression for the path integral

$$Z=\int \mathrm{d}U\, \mathrm{e}^{-S(U)}=\mathrm{Tr}\,(\hat{T}^{L_0})\,. \tag{9}$$

Here L_0 is the number of discrete time intervals in our lattice and the trace is over the quantum mechanical Hilbert space. The hamiltonian $\hat{H}$ is defined by

$$\hat{T}=\mathrm{e}^{-a\hat{H}}, \tag{10}$$

where a in the lattice spacing. In a continuous time limit, this establishes the connection between a path integral and a canonical formalism.

Working on a hyper-rectangular lattice of dimension $L_\mu=(L_0, \boldsymbol{L})$, we assume that the lowest eigenstate of the hamiltonian is an isolated vacuum state with energy per unit volume

$$\mathscr{E}(a, \boldsymbol{L})=\frac{1}{V_3}\langle 0|\hat{H}|0\rangle\,, \tag{11}$$

where V_3 is the spatial volume

$$V_3=\prod_{\mu=1}^{3} L_\mu\,. \tag{12}$$

Using dimensional transmutation, we place all coupling constant dependence in the lattice spacing a. We adopt a renormalization scheme whereby the lowest glueball mass is held fixed.

We now assume that the first excited states of the system are single glueballs of momentum $\boldsymbol{q}$ and energy $E(\boldsymbol{q})$. In a finite box the momentum assumes only discrete values, however for the leading behavior we can replace sums over $\boldsymbol{q}$ with continuous integrals. Putting the vacuum and one glueball state into the trace in eq. (9), we find

$$Z=\exp\left(-V_4\mathscr{E}(a, \boldsymbol{L})\right)\left\{1+rV_3\int\frac{\mathrm{d}^3q}{(2\pi)^3}\,\mathrm{e}^{-L_0aE(\boldsymbol{q})}+\text{higher states}\right\}. \tag{13}$$

Here we define the four-dimensional volume of our lattice

$$V_4=\prod_{\mu=1}^{4} L_\mu\,. \tag{14}$$

The factor r represents the degeneracy of the first state; for a spin J state we have

$$r=2J+1\,. \tag{15}$$

The glueball energy above the vacuum, $E(\boldsymbol{q})$, should become relativistic for small lattice spacing; consequently, we assume the form

$$E(\boldsymbol{q}) = m + \frac{\boldsymbol{q}^2}{2m} + \mathrm{O}(\boldsymbol{q}^4) . \tag{16}$$

At strong coupling the rest and kinetic masses need not be equal; we assume that the coupling is small enough that such deviations are negligible. Inserting the spectrum (16) into eq. (13), we obtain

$$Z = \exp\left(-V_4 \mathscr{E}(a, \boldsymbol{L})\right)\left\{1 + rV_3\left(\frac{ma}{2\pi L_0}\right)^{3/2} \mathrm{e}^{-maL_0}\left(1 + \mathrm{O}\left(\frac{1}{maL_0}\right)\right)\right\} . \tag{17}$$

We now must take account of the finite size dependence of the vacuum energy density $\mathscr{E}$. Such corrections should also be exponential in the glueball mass

$$\mathscr{E}(a, \boldsymbol{L}) = \mathscr{E}(a, \infty) + \mathrm{O}(\mathrm{e}^{-maL_i}) . \tag{18}$$

In eq. (17) we found the finite size correction exponential in L_0. A transfer matrix in the ith direction will give the contribution exponential in L_i. We now consider all components of L_μ to be equal, i.e. we work on an L^4 lattice. Each direction contributes equally to the finite size effect, and we conclude

$$Z(a, L^4) = Z(a, \infty)\left\{1 + 4r\left(\frac{maL}{2\pi}\right)^{3/2} \mathrm{e}^{-maL}\left(1 + \mathrm{O}\left(\frac{1}{maL}\right)\right)\right\} . \tag{19}$$

We now use the connection between Z and the internal energy per plaquette

$$E_{\mathrm{P}} = \frac{1}{6L^4} \frac{\mathrm{d}}{\mathrm{d}\beta} \log Z , \tag{20}$$

and use the renormalization group equation to convert the derivative with respect to β into a derivative with respect to a:

$$\frac{\mathrm{d}}{\mathrm{d}\beta} = \frac{g_0^3}{8\gamma(g_0)} a \frac{\mathrm{d}}{\mathrm{d}a} , \tag{21}$$

where $\gamma(g_0)$ in the Gell-Mann–Low function

$$a \frac{\mathrm{d}}{\mathrm{d}a} g_0(a) = \gamma(g_0) . \tag{22}$$

Combining these equations gives the result

$$E_{\mathrm{P}}(L) - E_{\mathrm{P}}(\infty) = \frac{2^{1/2} r (ma)^{5/2}}{48\pi^{3/2} L^{3/2}} \frac{g_0^3}{\gamma(g_0)} \mathrm{e}^{-maL}\left(1 + \mathrm{O}\left(\frac{1}{maL}\right)\right) . \tag{23}$$

To further study the corrections to this formula, we model the glueball gas using the partition function for a free scalar field of mass $m = 1/a\xi$:

$$Z = \int \mathrm{d}\phi_n \exp\left(-\tfrac{1}{2}\sum_n \{(\Delta_\mu \phi_n)^2 + (ma)^2 \phi_n^2\}\right). \tag{24}$$

Again from scaling (22), we compute

$$E_P = \frac{1}{6L^4}\frac{\partial}{\partial\beta}\log Z = \frac{g_0^3 (ma)^2}{16\gamma(g_0)}\langle\phi_n^2\rangle_L , \tag{25}$$

where the finite lattice propagator is

$$\langle\phi_n^2\rangle_L = \frac{1}{L^4}\sum_{q_\mu}\frac{1}{(ma)^2 + 4\sum_\mu \sin^2(aq_\mu/2)} . \tag{26}$$

The sum extends over $aq_\mu = 2\pi l_\mu/L$ for $l_\mu = 1, 2, \cdots, L$. This may be recognized as simply the infinite lattice propagator for the glueball from $x = 0$ to all the periodic images at $x_n = aL(n_1, n_2, n_3, n_4)$ by using the identity,

$$\frac{1}{(aL)^4}\sum_{q_\mu} = \int_{-\pi/a}^{\pi/a}\frac{\mathrm{d}^4 q}{(2\pi)^4}\sum_{n_\mu} \mathrm{e}^{iq\cdot x_n} . \tag{27}$$

The leading scaling contribution to $\varepsilon(L/\xi) = L^4(E_P(L) - E_P(\infty))$ for large $aLm = L/\xi$ is easily evaluated from the nearest image,

$$\int \frac{\mathrm{d}^4 q}{(2\pi)^4}\frac{\mathrm{e}^{iaLq_0}}{m^2 + q_0^2 + \boldsymbol{q}} \simeq \int \frac{\mathrm{d}^3 q}{(2\pi)^3}\frac{\mathrm{e}^{-aLE(\boldsymbol{q})}}{2E(\boldsymbol{q})} , \tag{28}$$

and the approximation $aE(\boldsymbol{q}) \simeq 1/\xi + \frac{1}{2}\xi a^2 \boldsymbol{q}$. The result for $r = 1$ and $\gamma(g_0^2)/g_0^3 \simeq 11/24\pi^2$ is

$$\varepsilon_0(L/\xi) \simeq \tfrac{1}{2}\sqrt{\pi}\tfrac{1}{11}(L/\xi)^{5/2}\,\mathrm{e}^{-L/\xi} , \tag{29}$$

in accord with our transfer matrix result (23). However, as we will see shortly, this approximation is inadequate for the values of L/ξ we have in our Monte Carlo data. Most importantly, it misses the next nearest images at $\sqrt{2}L$ and $\sqrt{3}L$. But even for rather large values of L/ξ the polynomial factor $(L/\xi)^{5/2}$ is an important effect. A similar factor of $(L/\zeta)^{3/2}$ has been neglected in earlier work [2, 4].

Fortunately it is not difficult to include exactly all scaling contributions of our glueball gas. After several manipulations, we obtain

$$\varepsilon(x) = \tfrac{1}{11}r \cdot \tfrac{1}{16}x^3 \int_0^\infty \mathrm{d}\lambda\, \mathrm{e}^{-x/2\lambda}\left\{\left(\sum_{n=-\infty}^{\infty} \mathrm{e}^{-n^2 x\lambda/2}\right)^4 - 1\right\} , \tag{30}$$

where $x = L/\xi = amL$ and r is the statistical weight. The contribution to the first three nearest images at L, $\sqrt{2}L$ and $\sqrt{3}L$ is

$$\varepsilon_3(x) = \tfrac{1}{11}r \cdot \tfrac{1}{16}x^3 \int_0^\infty \mathrm{d}\lambda\, \mathrm{e}^{-x/2\lambda}[\mathrm{e}^{-\lambda x/2} + 3\,\mathrm{e}^{-\lambda x} + 4\,\mathrm{e}^{-3\lambda x/2}] . \tag{31}$$

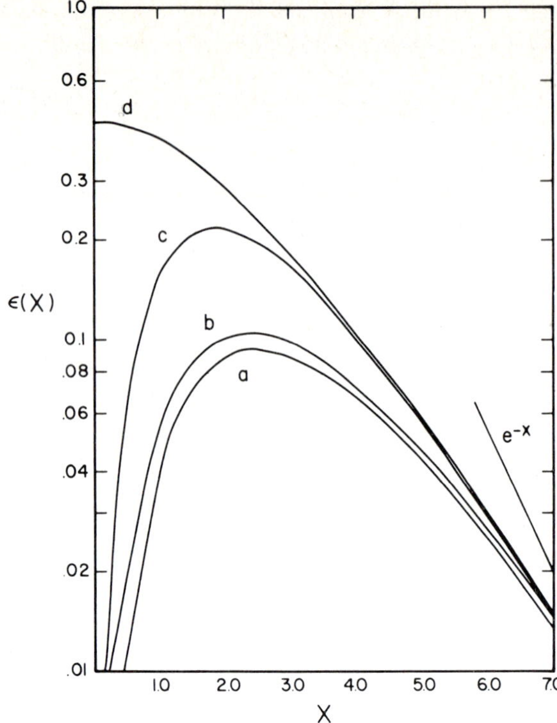

Fig. 1. Contributions to the scaling function $\varepsilon(x)$ for $x = L/\xi$ from a single glueball of mass $m = 1/a\xi$. Curve (a) is the leading term for $x \to \infty$, (b) is the nearest image at distance L, (c) is the first three images at L, $\sqrt{2}L$ and $\sqrt{3}L$ and (d) is the full scaling contribution.

In fig. 1, we plot for $r = 1$ the successive approximations: (a) the asymptotic equation (29) for the nearest image $\varepsilon_0(x)$; (b) the full relativistic contribution of the nearest image $\varepsilon_1(x)$; (c) the first three terms $\varepsilon_3(x)$, and (d) the exact scaling contribution.

We can draw some interesting conclusions. The maximum for $x \simeq 2\frac{1}{2}$ seen in the leading asymptotic term is only present if the first few terms dominate. Clearly as L/ξ gets small, contributions of the high mass spectrum (excited glueballs, threshold, etc.) come in, and our model should not be used. (So the physics of the maximum in the scaling data is an open question, discussed briefly in the conclusion.) On the other hand even for $x \geqslant 3$, the images at $\sqrt{2}L$ and $\sqrt{3}L$ are significant. We shall use the exact formula for definiteness, but fortunately the images at $2L$ and beyond are negligible for $x \geqslant 3$ where we use it. At $2L$ the corrections are of the same order as neglected effects due to glueball scattering and where our free gas should not be trusted.

Our model has two parameters, the degeneracy r and the mass m measured in units of the Λ parameter. For convenience we plot the scaling data of ref. [4] on a log–log plot so that the two parameters represent translating the curve upward for increasing degeneracy, and leftward for increasing mass. The parameter x_0 is defined with an arbitrary normalization

$$x_0 = L/\xi_0\,, \qquad \xi_0^{-1} = 100\left(\frac{6\pi^2}{11}\beta\right)^{51/121} e^{-(3\pi^2/11)\beta}\,. \tag{32}$$

When the first two terms of the renormalization group function dominate the coupling dependence of the correlation length, x_0 is proportional to x. (The value $x = x_0$ corresponds to $m = 1.3\sqrt{\kappa}$ in terms of the string tension [1].)

There is a great deal of ambiguity if both r and m are allowed to vary simultaneously. However, a small degeneracy, such as a single $J = 0$ glueball is inconsistent with the magnitude of ε. If we fix the degeneracy at $r = 5$ (e.g. one $J = 2$ state) the glueball mass is quite tightly constrained to $m \simeq (1.3 \pm 0.2)\sqrt{\kappa}$ as shown in fig. 2, but solutions with larger degeneracies and correspondingly larger masses are allowed. For example, fig. 2 also shows that a degeneracy of $r = 15$ with $m = (1.9 \pm 2)\sqrt{\kappa}$ is certaintly consistent with the data.

These results are not implausible. As you allow for higher degeneracy, the central mass of the glueballs increases. We could state the result as the assertion that the

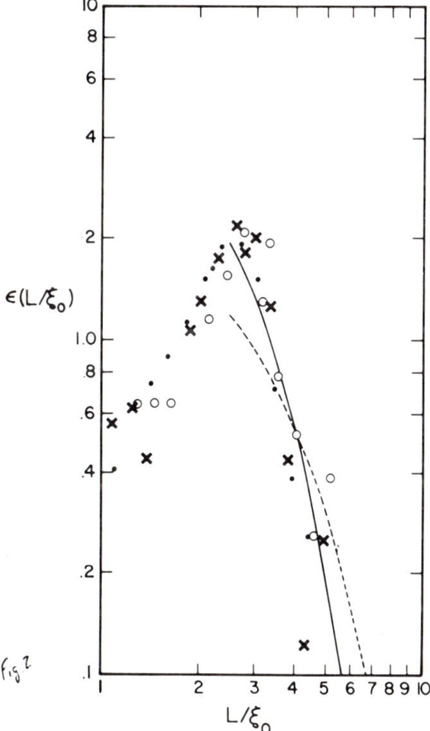

Fig. 2. The finite lattice scaling data for $\varepsilon(x_0)$ for $L = 4$ (·), 5 (×) and 6 (○) with the mass scale $m_0 = 1/a\xi_0$ set equal to $1.3\sqrt{\kappa}$. The dotted line is a fit to $m_{GB} = 1.3\sqrt{\kappa}$ and degeneracy $r = 5$, and the solid line is $m_{GB} = 1.9\sqrt{\kappa}$ and $r = 15$. The scatter in the data points reflects their statistical uncertainty.

spectrum grows to 15 states as you move to a mass on the order of $1\frac{1}{2}$ times the lowest state. Bag models have this property. For example, Donoghue, Johnson and Li find [6] a glueball spectrum with states, m_0, $1.3m_0$, $1.5m_0$, $1.8m_0$ with degeneracies 6, 6, 11, 21, respectively, in agreement with our trend. Indeed, a better fit to our data would be the superposition of a few low mass states plus a large number of higher mass states. This tends to give a steeper envelope. However, we don't feel fits with more parameters are useful. Nonetheless, two features do stand out.

First, rather high degeneracies are favored. Since most glueball models begin with nearly degenerate $J=0$ and $J=2$ states, degeneracies of six or more are sensible. Roughly speaking, this reflects the spin states of adding two or more vector gluons [$(2J+1)\times(2J+1)=9$ for two vectors] to form the color singlet glueball.

Second, we get a much lower mass than earlier Monte Carlo estimates [2, 3]. The earlier estimates of $m \sim (3\text{–}4)\sqrt{\kappa}$ used the two-point function with poorer statistics. In their analysis, power corrections in L/ξ, were omitted. On the other hand, our estimate is consistent with the strong coupling expansion of Münster* $m=(1.8\pm0.8)\sqrt{\kappa}$ for SU(2). Ref. [4] also suggested $m \leqslant 1.2\sqrt{\kappa}$ for the lowest glueball state which is nearly satisfied. Their argument is based on the plausible idea that at the onset of precocious scaling ($\beta > 2.05$) the largest correlation length should exceed one lattice spacing ($\xi > 1$).

If we notice that $m/\sqrt{\kappa}$ increases by as much as a factor of 2 going from SU(2) to SU(3) in the strong coupling estimate [6] our lowest mass becomes $m \sim 1000 \pm 500$ MeV adjusted to SU(3), where 50% error reflects the uncertainty in $\sqrt{\kappa}$. While the mass looks a little small compared to Regge expectations for a $J=2$ state at about $\simeq 1400$ MeV, there is no conflict. Perhaps, the lowest glueball is a pseudoscalar and by mixing with the η is usually sensitive to adding light quarks.

Finally, we should emphasize that the peak and turn over in the scaling data on the left (weak coupling) side in fig. 2 is totally beyond the scope of our glueball model. On a lattice with a finite time axis ($L_0 \ll L$) one expects a deconfinement transition into a gluonic phase at high temperature (or weak coupling) [7]. Very likely, our peak in $\varepsilon(x)$ represents an analogous phenomenon. From the peak, we can define a critical "radius" R_c (or inverse effective "temperature" $T_c = 1/R_c$):

$$R_c = aL|_{\text{peak}} = \frac{1}{(0.5\pm0.1)\sqrt{\kappa}}. \tag{33}$$

Curiously, the finite temperature studies of deconfinement have also given the same number, $T_c \approx (0.5\pm0.1)\sqrt{\kappa}$. A more precise understanding of our scaling function near the peak could be of great help in alleviating the degeneracy-mass ambiguity [8].

* Münster [6] estimates by a Padé of strong coupling expansions that $m=(1.8\pm0.8)\sqrt{\kappa}$ for SU(2) and $m=3\sqrt{\kappa}$ for SU(3).

Another improvement would be to introduce an asymmetric lattice of volume L_0L^3, which is shorter in the time direction ($L_0 < L$). This would suppress the contribution of the spatial images. Since images at or beyond $2L$ give small contributions, asymmetric lattices with $L_0 \leq \frac{1}{2}L$ should be adequate. Moreover, these lattices allow direct contact with the finite temperature studies, so the deconfinement transition can be more easily identified and controlled. The asymmetry also allows some separation of different spin-parity components, which may be very useful in view of the high degeneracy of the glueball spectrum. In this respect, the two-point correlation function also has a distinct advantage, so a combined approach using both methods may be best. Finally, our determination of the glueball mass to string tension ratio $m/\sqrt{\kappa}$ cannot be pushed farther without a parallel improvement in the errors for $\sqrt{\kappa}$.

In conclusion, our finite lattice image technique provides a useful tool for deciphering the glueball spectrum, but better statistics and the accommodation of SU(3) and quark effects are necessary to confront the experiments. Still our results already favor low masses (from $1.2\sqrt{\kappa}$ to $2.0\sqrt{\kappa}$) and high degeneracies (from 5 to 15) as a prediction of SU(2) gauge theory.

One of us (RCB) would like to thank the Harvard theory group for hospitality during the duration of this research.

References

[1] M. Creutz, PRL, 45 (1980) 313;
Phys. Rev. D21 (1980) 2308
[2] G. Bhanot and C. Rebbi, Nucl. Phys. B180 [FS2] (1981) 469
[3] B. Berg, Phys. Lett. 97B (1980) 401
[4] R.C. Brower, M. Nauenberg and T. Schalk, Santa Cruz preprint UCSC TH 81 140, SCIPP/81/1
[5] J. Donoghue, K. Johnson and B. Li, Phys. Lett. B, to be published
[6] B. Münster, Nucl. Phys. B190 [FS3] (1981) 439; (E, A: B200 [FS4] (1982) 536; E: to be published)
[7] L.D. McLerran and B. Svetitsky, Phys. Lett. 98B (1981) 195;
J. Kuti, J. Polonyi and K. Szlachanyi, Phys. Lett. 98B (1980) 199;
J. Engels, F. Karsch, I. Montvay and H. Satz, Phys. Lett. 101B (1981) 89
[8] J. Engels, F. Karsch, I. Montvay and H. Satz, Phys. Lett. 102B (1981) 332

Nuclear Physics B235[FS11] (1984) 123–134

COMPLEX ZEROES OF THE $d = 3$ ISING MODEL: FINITE-SIZE SCALING AND CRITICAL AMPLITUDES

Enzo MARINARI

Service de Physique Théorique, Centre d'Etudes Nucléaires de Saclay, 91 191 Gif-sur-Yvette, Cedex, France

Received 15 November 1983

By means of an MC simulation of lattices of size 4^3 to 8^3, we study the distribution of the complex zeroes closest to the real β axis of the $d = 3$ Ising model. We observe they do scale and we measure in this way ν and A_+/A_-. We obtain the prediction $A_+/A_- = 0.45 \pm 0.07$.

1. Introduction

The problem of zeroes in the partition function of a statistical system on a finite lattice is receiving a lot of attention [3, 2]. Both exact studies of small systems [2], and the use of methods like the Bethe approximation [3], are suggesting that a lot of physical information can be extracted from the distribution of zeroes; we will specialize here to the $d = 3$ (simple cubic) Ising model. In this case, for a system of size N^3, if $N \geqslant 5$ an exact resolution of the model seems impossible [2]. We will apply an MC integration method trying to follow the line of zeroes that, in the $N \to \infty$ limit, are pinching the real β axis at the critical inverse temperature β_c.

In ref. [1] the SU(2) lattice gauge theory was analysed, and the location of the zero closest to the real coupling constant axis was found for a 4^4 system. We show here that this method is very efficient in order to study the scaling behaviour of *different* zeroes, close enough to the real β axis, for increasing lattice size. The main observation is that, if the complex zeroes are generating a real singularity in the infinite-volume limit, the oscillating factor one has to face when moving in the complex plane will not behave as N^d, but (for example the $d = 3$ Ising model) as $N^{(d-1/\nu)} \sim N^{1.4}$. This claim is supported by results for the first two zeroes on $N = 5$–8 lattices; the agreement between these results and the scaling behaviour hypothesized in ref. [3] is remarkable. We find, for the critical amplitudes ratio, $A_+/A_- = 0.45 \pm 0.07$. The total CPU time used for obtaining these results is ~ 5 hours of CDC 7600.

In sect. 2 we describe the MC procedure [1] we used in order to compute the partition function for complex β. In sect. 3 we give our numerical results, and in sect. 4 we work out their physical implications.

2. Monte Carlo integration for complex β

Our goal is to compute, by Monte Carlo integration, the zeroes of the partition function $Z(\beta)$, for β complex, having a small $\operatorname{Im}\beta$. These are the zeroes that are pinching the real β axis in the infinite-volume limit, giving rise to the critical behaviour of the system. We will apply to the $d=3$ Ising model the method used in ref. [1] for the SU(2) lattice gauge theory. Our action will be

$$S=-\sum_{ij}{}'(\sigma_i\sigma_j-1), \tag{2.1}$$

where $\sigma_i=\pm 1$, and the sum runs over first neighbours. The partition function is defined by

$$Z(\beta)=\sum_{C}\mathrm{e}^{-\beta S(C)}, \tag{2.2}$$

where the sum runs over all possible configurations the system can assume. An exact computation of $Z(\beta)$ was performed in ref. [2] for a 4^3 system; in this case the number of configurations one has to consider is $\sim 10^{19}$, and only by exploiting all the symmetries of the system it was possible to solve the 4^3 system in a realistic computer time. But already a 5^3 system is absolutely impossible to solve in an exact way (one should consider $\sim 10^{37}$ configurations) and the number of states of an 8^3 system is $\sim 10^{154}$ *. That is why, to go to larger lattices, we have to use a random sampling.

The partition function (2.2) can be expressed as a sum over the energies W the system can reach, where the factor $\mathrm{e}^{-\beta W}$ is weighted by the density of states with energy W, N_W:

$$Z(\beta)=\sum_{W}N_W\mathrm{e}^{-\beta W}. \tag{2.3}$$

The energy W can assume values ranging from 0 (for an ordered configuration) up to $2dN^3$. We explore now, by means of a probabilistic algorithm, the phase space of the system at a given real β (we assume that we start from a configuration that has already been led to thermal equilibrium), and record the energy states assumed by the system (properly normalising the distribution function obtained in this way). We are computing in this way the normalised energy distribution function

$$F_W(\beta)=\frac{f_W(\beta)}{Z(\beta)}=\frac{N_W\mathrm{e}^{-\beta W}}{Z(\beta)}, \tag{2.4}$$

* Also using the transfer matrix formalism one would count an unrealistic number of configurations ($\sim 10^{20}$).

such that

$$\sum_W F_W(\beta) = 1 . \tag{2.5}$$

Let us now compare the two expressions

$$F_W(\beta) Z(\beta) e^{\beta W} = F_W(\beta') Z(\beta') e^{\beta' W},$$

that is

$$F_W(\beta) = F_W(\beta') e^{-(\beta-\beta')} \frac{Z(\beta')}{Z(\beta)}, \tag{2.6}$$

and integrate this relation over W, by applying the normalization condition (2.5):

$$\frac{Z(\beta)}{Z(\beta')} = \sum_W F_W(\beta') e^{-(\beta-\beta')W} . \tag{2.7}$$

This is the relation that will be useful in the following. We will consider now the case $\beta' \in R$, and β will be made complex. The sense of this analytic continuation is clear in (2.7); $F_W(\beta')$ will be the output of MC runs, done at some fixed β'. Moreover, if we define $\beta = \eta + i\xi$; $\eta, \xi \in R$, we see that in the RHS of (2.7) we have an oscillating factor $[\cos(\xi W) + i \sin(\xi W)]$ and a damping contribution $e^{-(\eta-\beta')W}$. The mean value of W is a number of order $\frac{1}{2}$ times the volume of the system; this makes apparent the difficulty of going far from the real axis (increasing ξ implies faster oscillations), and of going to larger lattices. But also suggests that, in order to take care of the damping factor, we compute (since our goal is determining the zeroes of $Z(\beta)$ in a situation in which we know there are no zeroes for $\beta \in R$), instead of (2.7), the quantity

$$\frac{Z(\beta)}{Z(\mathrm{Re}\beta)} = \frac{Z(\beta)}{Z(\beta')} \left\{ \frac{Z(\mathrm{Re}\beta)}{Z(\beta')} \right\}^{-1}$$

$$= \frac{\sum_W F_W(\beta') e^{-(\eta-\beta')W} [\cos(\xi W) - i \sin(\xi W)]}{\sum_W F_W(\beta') e^{-(\eta-\beta')W}} . \tag{2.8}$$

We will eventually determine the zeroes by looking at the minima of

$$\left| \frac{Z(\beta)}{Z(\mathrm{Re}\beta)} \right|^2 = \frac{\left(\sum_W F_W(\beta') e^{-(\eta-\beta')} \cos(\xi W) \right)^2 + \left(\sum_W F_W(\beta') e^{-(\eta-\beta')W} \sin(\xi W) \right)^2}{\left(\sum_W F_W(\beta') e^{-(\eta-\beta')W} \right)^2} . \tag{2.9}$$

3. Numerical analysis

We studied the $d=3$ Ising model on lattices of size 4^3, in order to compare our results with the exact ones of ref. [2], and of sizes going from 5^3 to 8^3. We used a heat-bath algorithm to flip the spins of our system (for a description of the method see ref. [5]). We chose randomly the spin to be updated; we recorded the energy of the system after any $5N^3$ updates of spins. For any lattice size we performed $10^5 N^3$ updates (not recording the energies) to bring the system to thermal equilibrium; eventually we performed $5 \cdot 10^5 N^3$ updates at two β' values (0.230 and 0.235) for the 4^3 lattice, and $10^6 N^3$ updates at $\beta' = 0.230$ for the sizes $N = 5$–8. In table 1 we give our average energies for the different lattice sizes.

We start by discussing our tests on a 4^3 lattice: see figs. 1 and 2. In this case the exact result is known [2], and the zeroes of Z have been computed with high precision [4]. The first check concerns the capability of our procedure to reproduce complex zeroes when the exact energy distribution function $F_W(\beta)$ is used. In ref. [2] the coefficients C_n of the polynomial

$$Z(u) = \sum_{n=0}^{96} C_n u^n, \tag{3.1}$$

where

$$u = e^{-4\beta},$$

are given; we can now easily compute $F_W^{\text{th}}(\beta)$, inserting it in eq. (2.8), and using our numerical procedure to determine the zeroes of Z.

Secondly we performed $5 \cdot 10^5 N^3$ updates at $\beta' = 0.230$ and 0.235 (after $10^5 N^3$ equilibrating flips) and we computed the corresponding zeroes. We want to emphasize also that if we had wanted to use the improved β's available (so as to minimize the damping factor of (2.8)), the numerical integration we performed was

TABLE 1
$\langle S \rangle$ (see (2.1)) at $\beta = 0.23$ for the different lattices sizes

N	$\langle S \rangle$
5	0.5281(9)
6	0.5360(7)
7	0.5388(5)
8	0.5411(7)

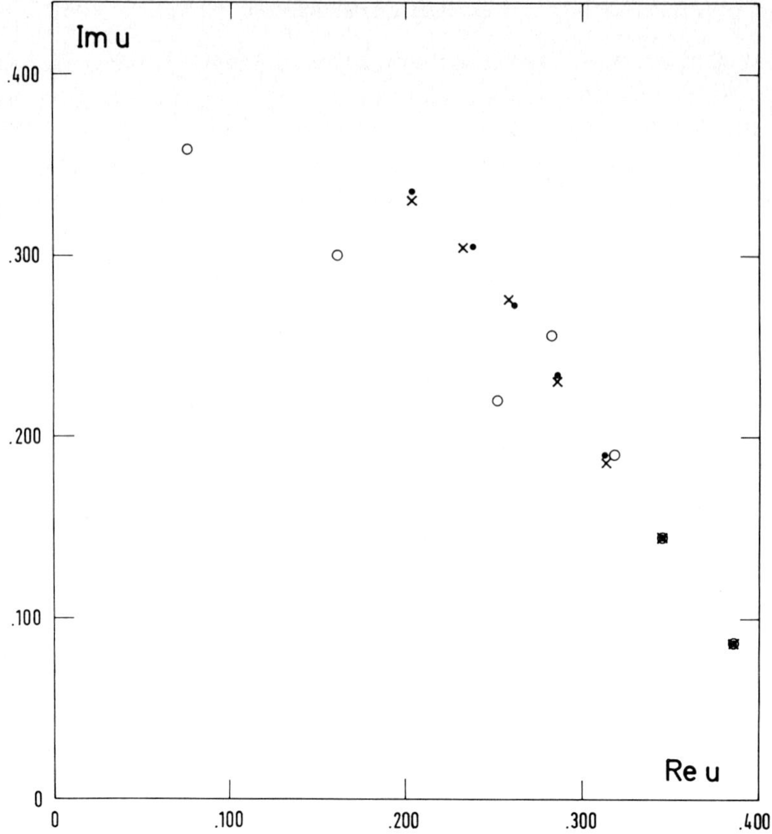

Fig. 1. Im $u^{(i)}_{N,0}$ versus Re $u^{(i)}_{N,0}$ for the first seven zeroes on the 4^3 lattice: comparison between the exact result [4] (dots), the numerical integration of the exact energy distribution (crosses), and MC results for 10^5 energy values (open circles). $\beta' = 0.230$.

supposed to converge at the correct result independently of β'. That is why we did our check at two different β's. We looked at the first seven zeroes $u^{(i)}_{N=4,0}$, $i = 1,\ldots,7$; for all of them the exact result was very close to the one obtained from $F^{\text{th}}_{W}(\beta')((|u^{(i)\,\text{exact}}_{4,0}| - |u^{(i)F^{\text{th}}_{W}}_{4,0}|)/|u^{(i)\,\text{exact}}_{4,0}| < 1\%$, for $i = 1,\ldots,7$), both for $\beta' = 0.230$ and for $\beta' = 0.235$.

As far as the MC determination of the zeroes is concerned, it is apparent from figs. 1, 2 that the first two zeroes ($u^{(1)}_{4,0}$, $u^{(2)}_{4,0}$) do coincide, for both β's, inside our error, with the exact ones. An error of the order of one percent is observed for the third zero, while precision is lost for the farther ones.

Now when studying larger lattices, we have reasons to be worried: the oscillating factor we are trying to integrate, $\sim \cos(\xi W)$, can be written as $\cos(\xi_v VE)$, where V is the volume of the system, E is the integration variable, going from zero to one,

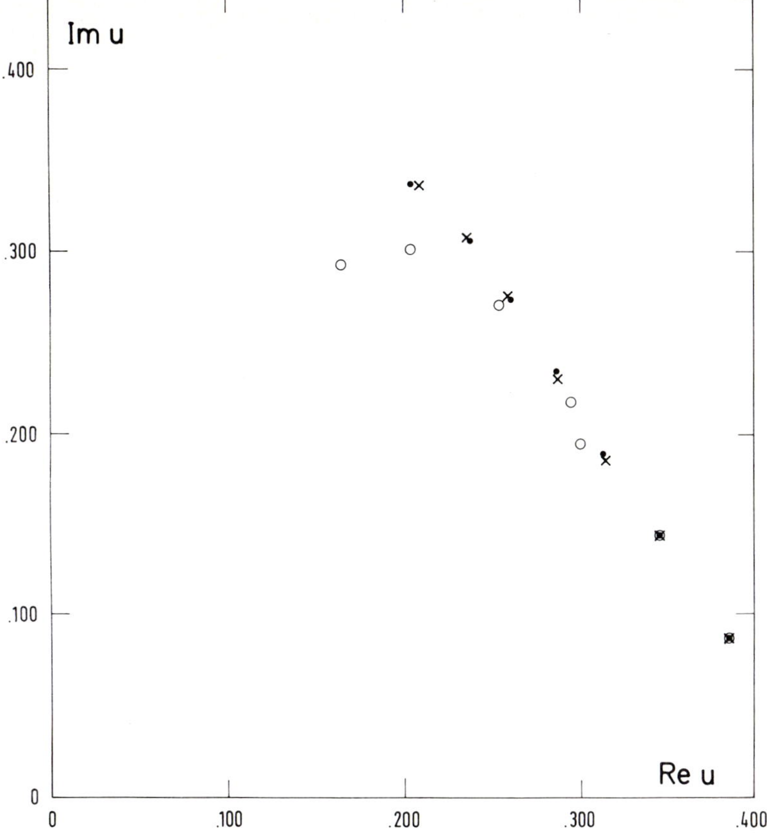

Fig. 2. Same as in fig. 1, but $\beta' = 0.235$.

and ξ_v is the distance of the interesting zero from the real u axis in the considered volume. The magnitude of the oscillations will dramatically increase with the volume of the system. The situation is in our case nicer than that just because of the presence, in the infinite-volume limit, of a zero on the real U axis; since ξ_N is supposed to behave as [3] $N^{-1/\nu}$, the amplitude of the oscillations will not increase as N^3, but as $\sim N^{1.4}$. That means, for example, that when going from $N = 4$ to $N = 8$, it will not increase by a factor 8, but just by a factor of ~ 2.5.

In tables 2 and 4 we give our results for the first and the second zero, for $N = 5$–8. We defined our error in the following way: we compute the zero in two clusters of energy values, each one given from $4 \cdot 10^5 N^3$ spin updates, and separate them by $10^5 N^3$ updates. We assume our statistical precision is one in which these two results coincide and also coincide with the one we get from considering all the $10^6 N^3$ spin updates. We consider our statistics not large enough to obtain reliable information on zeroes farther than the second one.

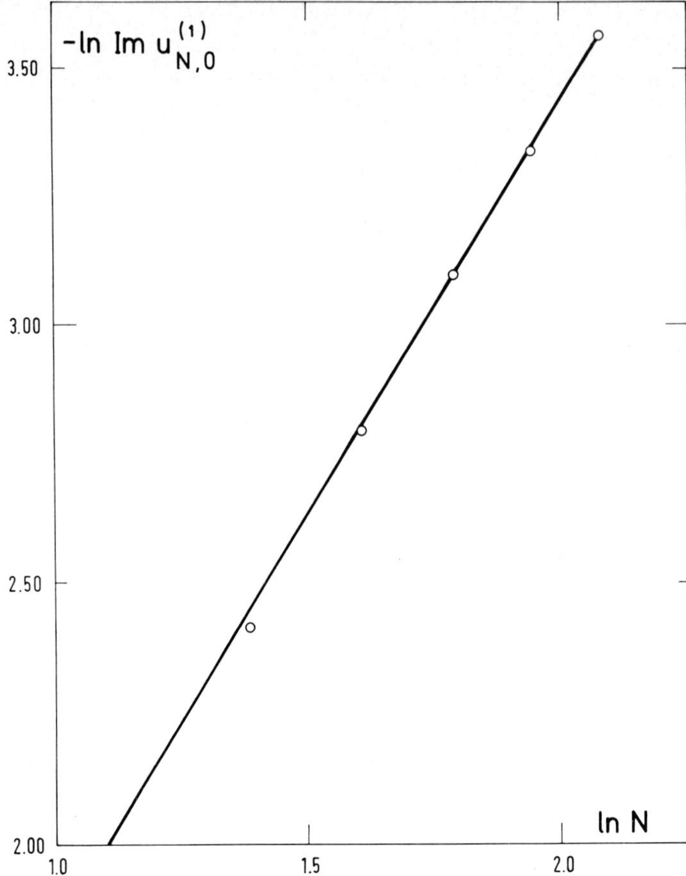

Fig. 3. $-\ln \mathrm{Im}\, u^{(1)}_{N,(0)}$ versus ln N, for $N = 4$–8.

4. Interpretation of results

The physical conclusions we will draw here will follow the analysis performed in ref. [3]. The scaling picture [6] is applied to the location of the zeroes of $Z(\beta)$ in the complex β plane.

Our first result follows from the relation

$$\tau \sim CN^{-1/\nu}, \tag{4.1}$$

where τ is the distance of the closest zero from the physical singularity at $N = \infty$, $\tau = u_c(N) - u_c(\infty)$.

In fig. 3 we plot $-\ln(\mathrm{Im}\, u^{(1)}_{N,0})$ versus ln N: the linearity of the results is quite impressive. In table 3 we give the values of $\nu(N, N')$ for $N = 4, \ldots, 7$, $N' > N$, when assuming $\tau = \mathrm{Im}\, u^{(1)}$. In table 4 we use $\tau = |u^{(1)}_{N,0} - u_c|$ with $u_c = 0.4120$ [11]: in the

TABLE 2
First zero: $u \equiv e^{-4\beta}$

N	$\mathrm{Re}\,\beta^{(1)}_{N,0}$	$\mathrm{Im}\,\beta^{(1)}_{N,0}$	$\mathrm{Re}\,u^{(1)}_{N,0}$	$\mathrm{Im}\,u^{(1)}_{N,0}$
4	0.2327	0.0561	0.3844	0.0877
5	0.2307	0.0384	0.3927	0.0608
6	0.2289	0.0283	0.3977	0.0452
7	0.2277	0.0220	0.4006	0.0353
8	0.2270	0.0175	0.4023	0.0282

Error is 1 on the last digit in columns one and two and 2 on the last digit in columns 3 and 4.

TABLE 3
$v(N, N') = (\ln N' - \ln N)/(\ln \mathrm{Im}\, u^{(1)}_{N,0} - \ln \mathrm{Im}\, u^{(1)}_{N',0})$

N \ N'	4	5	6	7	8
4		0.610(9)	0.612(6)	0.615(4)	0.611(5)
5			0.614(4)	0.619(9)	0.612(9)
6				0.624(20)	0.611(14)
7					0.596(31)
8					

TABLE 4
As in table 3, but $\ln \mathrm{Im}\, u^{(1)}_{N,0} \to \ln(\mathrm{Im}\, u^{(1)^2}_{N,0} + (\mathrm{Re}\, u^{(1)}_{N,0} - u_c)^2)^{1/2}$

N \ N'	4	5	6	7	8
4		0.610(6)	0.612(4)	0.617(3)	0.616(3)
5			0.614(10)	0.621(6)	0.618(5)
6				0.628(16)	0.621(9)
7					0.612(22)
8					

scaling limit the two definitions are equivalent. We obtain that ν from table 4 is closer to the true value ($\nu = 0.631 \pm 0.001$, from (9)) than the one from table 3; this is for all values of the lattice size. More information is contained in table 4 than in table 3, which can help explain this phenomenon.

A systematic smallness of ν computed from our small lattices ($\sim 1\%$) is apparent; given our error bars we cannot try to study the way in which $\nu(N, N') \to \nu(\infty)$.

TABLE 5
The second zero

N	Re $u^{(2)}_{N,0}$	Im $u^{(2)}_{N,0}$
4	0.3444	0.1433
5	0.366(3)	0.101(2)
6	0.379	0.076(3)
7	0.386	0.054
8	0.390	0.045

Error is 1 on the last digit, if not indicated in parenthesis.

A last remark on this point is in order: it is clear that, for increasing N, all the zeroes close enough to the real β axis should scale according to (4.1). In fig. 4 we plot $-\ln \operatorname{Im} u^{(i)}_{N,0}$ for $i = 1, 2$; also the second zero has the expected behaviour.

The following result is based on the relation

$$\tan[(2-\alpha)\varphi] = \frac{\cos \pi\alpha - A_-/A_+}{\sin \pi\alpha}, \tag{4.2}$$

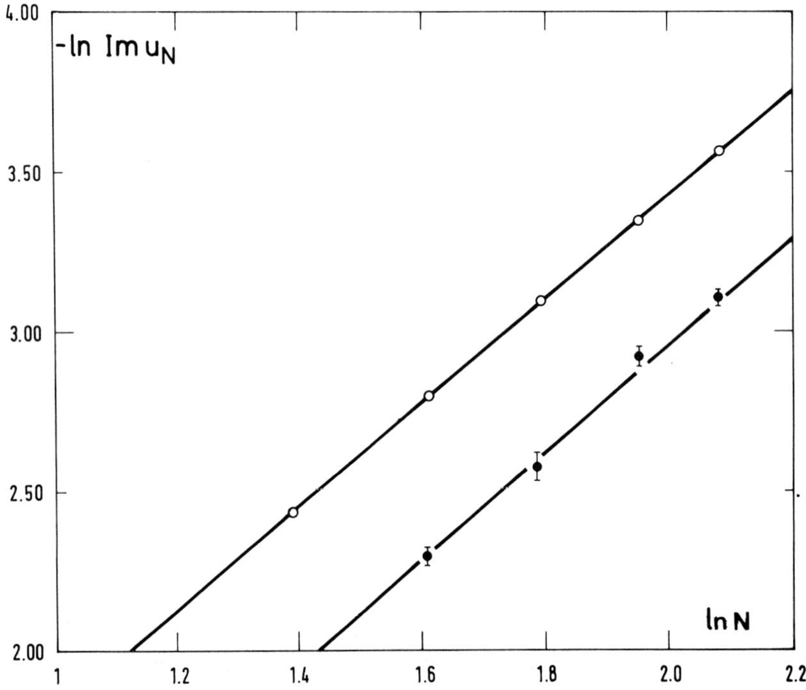

Fig. 4. As in fig. 3, but dots for $u^{(1)}_{N,(0)}$ and crosses for $u^{(2)}_{N,0}$, $N = 5$–8.

TABLE 6

$\varphi_N \equiv \tan^{-1}\{(\mathrm{Im}\, u^{(2)}_{N,0} - \mathrm{Im}\, u^{(1)}_{N,0})/(\mathrm{Re}\, u^{(2)}_{N,0} - \mathrm{Re}\, u^{(1)}_{N,0})\}$

N	φ
5	$56.5^{+4.3}_{-4.5}$
6	$58.7^{+3.9}_{-4.3}$
7	$52.0^{+3.8}_{-3.7}$
8	$53.8^{+4.2}_{-4.3}$

obtained in ref. [3]. Here α is the specific heat exponent (we will use as input the value $\alpha = 2 - d\nu = 0.107 \pm 0.003$), A_-/A_+ the critical amplitude ratio for the specific heat, and φ is the angle at which the complex zeroes deviate from the real u axis (in the complex u plane) in the infinite-volume limit (at zero magnetic field). We give in table 6 the φ_N we get for our lattice sizes, from the first and second zeroes. The errors are quite large (obviously enough the precision of the order of one percent in the second zero is reflected by a large indetermination of the slope, see fig. 5). From our data we feel quite comfortable with the assumption that the N dependence of the slope is by far considerably smaller than our statistical indetermination; so we compute

$$\bar{\varphi}_{5,8} = (55.3 \pm 1.5)^\circ\,, \tag{4.3}$$

and average φ on the lattices of sizes from 5 to 8. We get in this way, from (4), the estimate

$$\frac{A_+}{A_-} = 0.45(7)\,, \tag{4.4}$$

where the error has a statistical meaning, and the systematic error coming from the finiteness of the lattice is not taken into account (but seems to be much smaller than the statistical one). For a comparison with experimental data, see for example ref. [8] (in particular table 3 and fig. 2 therein); the experimental values range from 0.36 and 0.63, and the comparison with our value is fair enough. On the theoretical ground a computation [7] at the order ε^2 gives for A_+/A_- a value of 0.48 (lowering the value [10] at the order ε, 0.55); also in this case the comparison with our value is more than reasonable.

We wish to thank C. Itzykson and J.B. Zuber for their enthusiastic, generous and continuous help. Interesting discussions with D. Beysens, H. Flyberg, R. Pearson and J. Zinn-Justin are gratefully acknowledged. We thank C. Itzykson and G. Parisi for a critical reading of the manuscript.

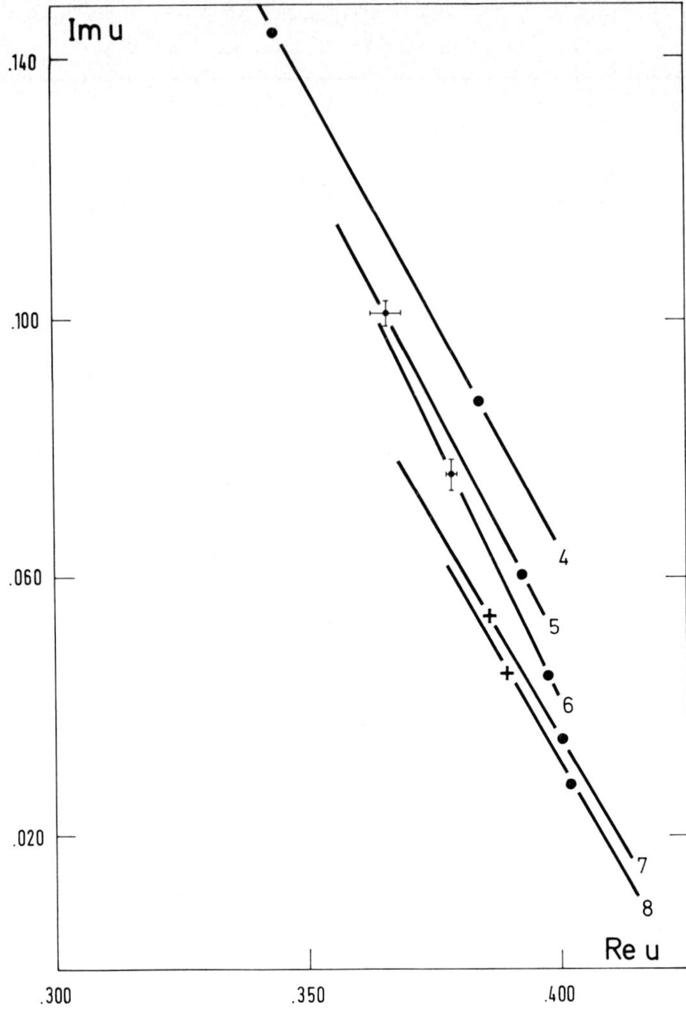

Fig. 5. Determination of φ_N.

References

[1] M. Falcioni, E. Marinari, M.L. Paciello, G. Parisi and B. Taglienti, Phys. Lett. 108B (1982) 331
[2] R.B. Pearson, Phys. Rev. B26 (1982) 6285
[3] C. Itzykson, R.B. Pearson and J.B. Zuber, Nucl. Phys. B220[FS8] (1983) 415;
J.B. Zuber, Talk at Workshop on nonperturbative field theory and QCD, Trieste, December 1982, Saclay preprint, SPhT/83/004 (January 1983)
[4] R.B. Pearson, private communication
[5] E. Marinari, G. Paladin, G. Parisi and A. Vulpiani, 1/f noise, disorder and dimensionality, Saclay preprint, SPhT/83/96 (July 1983), J. de Phys., to be published
[6] M.E. Fisher and M.N. Barber, Phys. Rev. Lett. 28 (1972) 1516
[7] C. Bervillier, Phys. Rev. B14 (1976) 4964

[8] D. Beysens, A. Bourgou and P. Calmettes, Phys. Rev. A26 (1982) 3589
[9] J.C. Le Guillou and J. Zinn-Justin, Phys. Rev. B21 (1980) 3976;
B.G. Nickel, in Phase transitions, status of the experimental and theoretical situation, Cargèse 1980, eds. M. Lévy, J.C. Le Guillou and J. Zinn-Justin (Plenum, New York, 1981);
J. Zinn-Justin, J. de Phys. 42 (1981) 783
[10] P.C. Hohenberg, A. Aharony, B.I. Halperin and E.D. Siggia, Phys. Rev. B13 (1976) 2896
[11] C. Domb, in Phase transitions and critical phenomena, vol. 3, eds. C. Domb and M.S. Green (Academic Press, 1974);
J. Zinn-Justin, private communication

3. Conformal Invariance and Finite-Size Scaling

Introduction

Although conformal invariance was first applied to the study of critical behavior in 1970 [1], it is only quite recently that its usefulness in understanding finite-size scaling amplitudes has been realized. This has been concurrent with considerable analytical progress in two dimensions [2–4]. The basic idea is simple: that correlation functions at a critical point should transform simply not only under scale transformations, as discussed in chapter 1, but also under the more general conformal transformations, which correspond locally to scale transformations, with a rescaling factor b which depends on position. Once this is understood, the basic strategy follows that of applications of conformal invariance to two-dimensional potential problems: use a conformal mapping to relate correlation functions in one geometry to those in a simpler geometry.

One of the most useful geometries to consider is that of a cylinder, that is, an infinitely long strip of width L with periodic boundary conditions. This is because it is readily suited for the application of transfer matrix methods. The first result of conformal invariance is that the correlation lengths ξ_i in the strip are related to the scaling dimensions x_i of operators by

$$\xi_i^{-1} \sim 2\pi x_i/L. \tag{3.1}$$

Note that the fact that $\xi_i \propto L$ is simply a result of finite-size scaling. What is remarkable is that the amplitude is determined. Relation (3.1) was in fact first conjectured from exact and numerical studies on various models. However, the predictions of conformal invariance go far beyond this. Recall that ξ_i^{-1} is given in terms of the ratio of leading eigenvalues of the transfer matrix, and therefore (3.1) gives some information about the leading eigenvalues. In fact, conformal invariance predicts the whole spectrum of the transfer matrix (at least of those states which become degenerate as $L \to \infty$). In sections 3.1 and 3.2 we give a self-contained account of the ideas of conformal invariance which lead to this result. This account should complement those in the selected papers, which arrive at the same conclusions by slightly different routes.

3.1. Conformal invariance in two dimensions

To begin with, consider the plane $\mathbb{R}^2$. (Very few results of conformal invariance extend to dimensionality $d \geqslant 2$; this is because there are then only a finite number of generators.) In two dimensions it is useful to consider coordinates $(z, \bar{z})$ with a metric $\mathrm{d}s^2 = \mathrm{d}z\,\mathrm{d}\bar{z}$. Note that in this metric we must distinguish upper and lower indices on tensors. In two dimensions it is simple to discuss scaling operators which are not rotational scalars, by introducing the "complex" scaling dimensions $(h, \bar{h})$ (actually real numbers) so that the two-point function is

$$\langle \phi(z, \bar{z})\phi(0, 0)\rangle = z^{-2h}\bar{z}^{-2\bar{h}}, \tag{3.2}$$

so that the real scaling dimension is $x = h + \bar{h}$, while $h - \bar{h} = s$ is the spin. Equivalently, we may say that under a RG transformation corresponding to the combined dilatation and rotation $z \to z' = \lambda z$, $\phi(z) \to \lambda^h \bar{\lambda}^{\bar{h}} \phi(z')$.

Now consider a generalized RG transformation, where the new lattice is some arbitrary deformation of the old one, not simply a rescaling. (Such a procedure, although hard to carry out in practice, is possible in principle.) A given point with coordinates x^μ measured on the old lattice will have new coordinates x'^μ on the new one, where $x'^\mu = x^\mu + \alpha^\mu(x)$, and we take $\alpha^\mu(x)$ to be an infinitesimal differentiable function of x^μ. Under such a generalized RG transformation, even if the system on the old lattice were governed by the fixed point Hamiltonian $\mathscr{H}^*$, that on the new lattice will not be. The change may be expanded in terms of derivatives of α^μ, and the most relevant operator will be of the form

$$\delta\mathscr{H} = -\frac{1}{2\pi} \int \frac{\partial \alpha^\mu}{\partial x_\nu} T_{\mu\nu}(x)\, \mathrm{d}^2 x. \tag{3.3}$$

This introduces the *stress tensor* $T_{\mu\nu}(x)$. (The factor $1/2\pi$ is conventional.) The main assumption here (which can be proved perturbatively in quantum field theory) is that such a *local* quantity $T_{\mu\nu}(x)$ exists. We may think of the interactions on the lattice introducing a quasi-elasticity: under a strain $\partial\alpha^\mu/\partial x_\nu$, the change in the energy is the integral of strain $\times$ stress. The strain tensor can be decomposed into three parts: an antisymmetric part corresponding to a local rotation, a dilatation piece proportional to $g^{\mu\nu}$, and a traceless symmetric part representing the effects of shear. Since $\delta\mathscr{H}$ is supposed to vanish for the first two types of transformation, the stress tensor must be traceless and symmetric.

Why introduce such general transformations? The reason is that in order to make the dilatation and rotation $z' = \lambda z$ inside some region and not to disturb the boundary conditions at large distances, we must consider more general transformations along the boundary of the region. For example, suppose we have an operator $\phi(0)$ at $z = 0$. Surround this by two regions $|z| < R_1$ and $R_1 < |z| < R_2$. Make a transformation $x^\mu \to x'^\mu = x^\mu + \alpha^\mu(x)$ which reduces to $z' = z + (\lambda - 1)z$ inside $|z| < R_1$ and for which $\alpha^\mu = 0$ for $|z| > R_2$. The change in the Hamiltonian is of the form (3.3) and is limited to the annulus $R_1 < |z| < R_2$. On integrating (3.3) by parts, the surface term from the outer boundary goes away, since $\alpha^\mu = 0$ there, and the bulk term involves $\int \alpha^\mu \partial^\nu T_{\mu\nu}\, \mathrm{d}^2 x$. Since α^μ is almost completely arbitrary in the annulus, we must have $\partial^\nu T_{\mu\nu} = 0$. The remaining term is $\int \alpha^\mu T_{\mu\nu}\, \mathrm{d}S^\nu$ over the inner boundary.

In the coordinate system $(z, \bar{z})$ the condition that $T_{\mu\nu}$ is symmetric, traceless and conserved implies that it has just two independent components: $T_{zz} = T(z)$, which depends on z but not $\bar{z}$, and $T_{\bar{z}\bar{z}} = \bar{T}(\bar{z})$, depending on $\bar{z}$ only. In that case the surface term can be written as a contour integral,

$$\delta\mathscr{H} = \frac{1}{2\pi \mathrm{i}} \int_{\mathrm{C}} \alpha(z)\, T(z)\, \mathrm{d}z - \frac{1}{2\pi \mathrm{i}} \int_{\mathrm{C}} \overline{\alpha(z)}\, \bar{T}(\bar{z})\, \mathrm{d}\bar{z}, \tag{3.4}$$

where C is the circle $|z| = R_1$, and in our case, $\alpha(z) = (\lambda - 1)z$. If we now consider how the correlation function $\langle \phi(0) \cdots \rangle$ of $\phi(0)$ with other operators outside $|z| = R_1$ transforms:

$$\langle \phi(0) \cdots \rangle_{\mathscr{H}^*} = \lambda^h \bar{\lambda}^{\bar{h}} \langle \phi(0) \cdots \rangle_{\mathscr{H}^* + \delta\mathscr{H}}, \tag{3.5}$$

we see, comparing terms $\mathrm{O}(\lambda - 1)$ and $\mathrm{O}(\bar{\lambda} - 1)$, that

$$\frac{1}{2\pi \mathrm{i}} \int_{\mathrm{C}} z T(z)\, \phi(0)\, \mathrm{d}z = h\phi(0) \tag{3.6}$$

with a similar equation involving $\bar{T}$ and $\bar{h}$. The meaning of (3.6) is that arbitrary correlation functions of either side with operators outside C are equal.

The above equation has immediate consequences for the operator product expansion (OPE) of $T(z)$ with $\phi(0)$. Because $T(z)$ is analytic, this must include only integer powers of z, of the form

$$T(z)\,\phi(0) = \sum_{n=-\infty}^{\infty} z^{-2+n}\phi^{(n)}(0). \tag{3.7}$$

This defines the operators $\phi^{(n)}(0)$. Comparison with (3.6) now shows that $\phi^{(0)}(0) = h\phi(0)$. Similarly, using translational invariance, one can show that $\phi^{(1)}(0) = \partial_z\phi(0)$. The other $\phi^{(n)}$ are, however, not in general just derivatives. Note that $\phi^{(n)}$ has scaling dimension $h+n$. Since scaling dimensions cannot be arbitrarily large and negative (in most cases they cannot even be negative), $\phi^{(n)}$ must vanish for sufficiently large negative n. Operators for which $\phi^{(n)} = 0$ for *all* $n < 0$ are called *primary*. They play a special role in what follows. Given a general operator, we can construct a primary operator by making the OPE with T. In fact, all the scaling operators can be obtained by taking the primary operators ϕ and making repeated OPEs with T (and $\bar{T}$). We thus get operators $\phi^{(n_1,n_2,\ldots,\bar{n}_1,\bar{n}_2,\ldots)}$ whose scaling dimensions differ from those of ϕ by positive integers. These operators constitute the conformal tower corresponding to ϕ.

For a primary operator, the only singular terms in the OPE are

$$T(z)\,\phi(0) = \frac{h}{z^2}\phi(0) + \frac{1}{z}\partial_z\phi(0) + \cdots. \tag{3.8}$$

We may then run through the previous arguments with a *general* analytical transformation $z \to z + \alpha(z)$ in $|z| < R_1$. The change in $\mathscr{H}$ is given by (3.4). However, (3.8) shows that only $\alpha'(0)$ and $\alpha(0)$ enter the result, by Cauchy's theorem. We conclude that a primary operator transforms under $z \to z + \alpha(z)$ according to

$$\phi(0) \to \phi(0) + \left[h\alpha'(0) + \bar{h}\bar{\alpha}'(0)\right]\phi(0) + \phi(\alpha(0), \bar{\alpha}(0)). \tag{3.9}$$

This result, true for infinitesimal $\alpha(z)$, is easy to generalize to a finite transformation $z' = f(z)$:

$$\phi(z, \bar{z}) \to f'(z)^h\,\overline{f'(z)}^{\bar{h}}\,\phi(z', \bar{z}'). \tag{3.10}$$

Of course such an analytic transformation is conformal, in that it corresponds locally to a dilatation plus a rotation. Thus primary operators (and their correlation functions) transform in a simple way under conformal transformations.

It is important to realize that T itself is not a primary operator. In fact the short-distance expansion of T with itself must have the form

$$T(z)T(0) = \frac{c/2}{z^4} + \frac{2}{z^2}T(0) + \frac{1}{z}\partial_z T(0) + \cdots. \tag{3.11}$$

The coefficient of the z^{-2} term results from the fact that T has scaling dimensions (2, 0). There is no z^{-3} term by $z \to -z$ symmetry. However the z^{-4} term must be present because $\langle T(z)T(0)\rangle$ is non-zero. Unlike other operators which we are free to normalize as in (3.2), the normalization of T is fixed by its definition (3.3). Thus c is a universal number characteristic of the theory: its role turns out to be very important. The additional term in (3.11) means that T does not transform according to (3.9). Instead

$$T(z) \to T(z) + 2\alpha'(0)T(0) + \alpha(0)\,\partial_z T(0) + \frac{c}{12}\alpha'''(0). \tag{3.12}$$

The integrated form analogous to (3.10) is more complicated:

$$T(z')\,\mathrm{d}z'^2 = T(z)\,\mathrm{d}z^2 + \frac{c}{12}\{z', z\}\,\mathrm{d}z^2, \tag{3.13}$$

where $\{z', z\} = \{f(z), z\}$ is the so-called Schwarzian derivative $[f'''f' - \frac{3}{2}f''^2]/f'^2$. We see that c measures the amount by which T fails to transform as a tensor under a general conformal transformation.

3.2. Cylindrical geometry

The finite conformal transformation which transforms the plane (actually with the origin removed) into the cylinder is $z' = (L/2\pi)\ln z$. Applying the transformation law (3.1) we have

$$\langle\phi(z', \bar{z}')\phi(0, 0)\rangle_{\text{cyl}} = \left(\frac{2\pi}{L}\right)^{2(h+\bar{h})}\left(\mathrm{e}^{2\pi z'/L}\right)^{h}\left(\mathrm{e}^{2\pi z'/L}\right)^{\bar{h}}$$
$$\times\langle\phi\left(\mathrm{e}^{2\pi z'/L}, \mathrm{e}^{2\pi\bar{z}'/L}\right)\phi(1, 1)\rangle_{\text{plane}}. \tag{3.14}$$

Writing $z' = u + \mathrm{i}v$ and taking the limit $u \to \infty$ this becomes (on using 3.2)

$$\sim\left(\frac{2\pi}{L}\right)^{2x}\mathrm{e}^{-2\pi xu/L}\,\mathrm{e}^{-2\pi svi/L}. \tag{3.15}$$

In the transfer matrix formalism, we treat u as an "imaginary time" coordinate, and $0 < v < L$ as a periodic space coordinate. The partition sum over "times" $u_1 < u < u_2$ is equal to an appropriate matrix element of the transfer matrix $\hat{t} \equiv \mathrm{e}^{-\hat{H}}$ raised to the power $u_2 - u_1$. In this language, local scaling "operators" like $\phi(u, v)$ become true operators $\hat{\phi}(v)$ acting on the same vector space as does the transfer matrix. If we denote the eigenvalues and eigenstates of $\hat{H}$ by E_n and $|n\rangle$, then the above correlation function can be expressed as

$$\sum_n\langle 0|\hat{\phi}(v)|n\rangle\,\mathrm{e}^{-(E_n-E_0)u}\langle n|\hat{\phi}(0)|0\rangle. \tag{3.16}$$

That is, correlation lengths in the cylinder are given by differences of eigenvalues of the logarithm of the transfer matrix. Comparing with (3.15) we see that to each primary operator corresponds one eigenstate of $\hat{H}$ with eigenvalue $E_0 + 2\pi x/L$. This is the result (3.1). As well as $\hat{H}$, there is a momentum operator $\hat{k}$ which generates translations around the cylinder. This commutes with $\hat{H}$. From the v-dependence in (3.15) we see that the eigenstate corresponding to ϕ has momentum $2\pi s/L$.

What about the ground state energy E_0? This gives the free energy per unit length of a very long cylinder. From the general theory of finite-size scaling (see 1.10) we expect that

$$E_0 \sim AL + CL^{-1}, \tag{3.17}$$

where A is the non-universal bulk term, but C should be universal. If we use (3.13) we find that

$$\langle T\rangle_{\text{cyl.}} = \langle \bar{T}\rangle_{\text{cyl.}} = \frac{c}{24}\left(\frac{2\pi}{L}\right)^2, \tag{3.18}$$

where we assume that $\langle T\rangle = \langle \bar{T}\rangle = 0$ in the plane. Now E_0 is given by

$$E_0 = -\frac{1}{2\pi}\int_0^L \langle T_{uu}\rangle \, \mathrm{d}v = -\frac{L}{2\pi}\left[\langle T\rangle + \langle \bar{T}\rangle\right]_{\text{cyl.}} \tag{3.19}$$

so that $C = -\pi c/6$. To understand this, recall the definition of $T_{\mu\nu}$ (3.3). If we make the non-conformal transformation $(u \to (1+\lambda)u,\ v \to v)$ then the Hamiltonian changes by an amount $-(\lambda/2\pi)\int_0^L T_{uu}\,\mathrm{d}v$ per unit length. This must be balanced by an explicit change in the free energy per unit length of $-\lambda E_0$. From this follows (3.19). In field theory terms, this is nothing but the statement that the Hamiltonian operator H is the space integral of the time–time component $\hat{T}_{uu}$ of the stress tensor (energy–momentum tensor).

We have seen that the L^{-1} terms in the ground state energy E_0 and the leading gaps $E_n - E_0$ are related to the value of c and to the values of the scaling dimensions x of the primary operators respectively. What about the higher states? In fact there is a one-to-one correspondence between the scaling operators of the theory and those eigenstates of $\hat{H}$ whose gaps scale like L^{-1}. To see this, recall that, in the plane, given a scaling operator ϕ (not necessarily primary) we construct the operator $\phi^{(n)}$ from the term $\mathrm{O}(z^{n-2})$ in the OPE of $T(z)$ with ϕ. We can write, using Cauchy's theorem

$$\phi^{(n)} = \frac{1}{2\pi \mathrm{i}}\int_C (z - z_1)^{1-n}\ T(z)\ \phi(z_1)\ \mathrm{d}z. \tag{3.20}$$

This equation can be transformed to the cylinder, using (3.13); taking $|z_1| \ll |z|$ we find that

$$\phi^{(n)}(u_1, v_1)_{\text{cyl.}} = \frac{1}{L}\int_0^L \mathrm{d}v\ \mathrm{e}^{2\pi \mathrm{i} n v/L} T(u, v)_{\text{cyl.}}\ \phi(u_1, v_1)_{\text{cyl.}} \tag{3.21}$$

for $n \neq 0$, and for $u_1 \ll u$. In the transfer matrix formalism, we can associate with each operator $\hat{\phi}$ an eigenstate of $\hat{H}$:

$$|\phi\rangle = \lim_{u\to\infty}\left(\frac{\mathrm{e}^{-u\hat{H}}\hat{\phi}\,|0\rangle}{\|\mathrm{e}^{-u\hat{H}}\hat{\phi}\ |0\rangle\|}\right). \tag{3.22}$$

From (3.21) then, if we define the operators

$$\hat{L}_n = \frac{1}{L}\int_0^L \mathrm{e}^{-2\pi \mathrm{i} n v/l}\hat{T}(v)_{\text{cyl.}}\ \mathrm{d}v \tag{3.23}$$

we can write

$$|\phi^{(n)}\rangle \propto L_{-n}|\phi\rangle \quad (n \geqslant 1). \tag{3.24}$$

Thus in general

$$|\phi^{(n_1, n_2, \ldots;\, \bar{n}_1, \bar{n}_2, \ldots)}\rangle \propto \hat{L}_{-n_1}\hat{L}_{-n_2}\cdots \hat{\bar{L}}_{-\bar{n}_1}\hat{\bar{L}}_{-\bar{n}_2}\cdots |\phi\rangle. \tag{3.25}$$

The state on the rhs has energy $E_0 + (2\pi/L)(x + \Sigma n_i + \Sigma \bar{n}_i)$ and momentum $(2\pi/L)(s + \Sigma n_i - \Sigma \bar{n}_i)$. If two such states are not linearly independent, one can show that the corresponding operators are not independent. Thus the spectrum of $\hat{H}$ exhibits the whole spectrum of scaling dimensions, with the correct degeneracies.

Although this is not directly relevant to the subject of finite-size scaling, it would be remiss not to mention two theoretical advances related to the above ideas [2–4]. First, from (3.11) one can show that the operators $\hat{L}_n$ satisfy the Virasoro algebra:

$$\left[\hat{L}_n, \hat{L}_m\right] = (n-m)\hat{L}_{n+m} + \frac{c}{12}n(n^2-1)\delta_{n,-m}. \tag{3.26}$$

Moreover the states $|\phi\rangle$ (for ϕ primary) satisfy $\hat{L}_n|\phi\rangle = 0$ for $n \geqslant 1$. Thus the whole space spanned by $|\phi\rangle$ and its conformal tower gives a highest-weight representation of this algebra. Such representations have been studied extensively; in particular the requirements of unitarity for $c < 1$ are very severe [3]. They lead to a quantization of the allowed values of c, and a finite list of possible primary scaling dimensions $(h, \bar{h})$.

The second development is based on the notion that once we know the complete spectrum of $\hat{H}$ we can calculate the partition function of a finite cylinder, a torus. The trivial observation that we can form a torus by imposing periodic boundary conditions on a finite cylinder in two different ways gives rise to the idea of modular invariance. For the models with $c < 1$ this leads to severe constraints on which scaling dimensions in the finite list can appear [5]. These constraints have recently been solved in general, enabling a classification [6] of all unitary models with $c < 1$.

To return to our story, the theoretical ideas outlined above have led to a growing industry in numerical studies of two-dimensional models. As in the case of ordinary finite-size scaling, calculations are easier to perform in the quantum Hamiltonian limit. Since this corresponds to a limit of extremely anisotropic couplings, rotational invariance is lost, and therefore so is conformal invariance. For models where anisotropy is not relevant, however, we expect this to be restored by an appropriate rescaling of space v versus time u. Thus the basic relations are replaced by

$$E_0 \sim AL \sim \frac{\pi c \gamma}{6L} \tag{3.27}$$

and

$$E_n - E_0 \sim \frac{2\pi x \gamma}{L} + \frac{2\pi\gamma}{L}(n + \bar{n}), \tag{3.28}$$

where γ depends on the model, and is non-universal. It can be interpreted as the "speed of light". There are then two approaches: either evaluate only ratios of universal quantities, like x/c, or, using the second term in (3.27), note that γ can be evaluated by looking at excited states (corresponding to non-primary operators). Of course, this involves more numerical work.

The literature of numerical studies falls into two groups:

– Those papers which examine exactly solved models (usually those with $c < 1$) and confirm the above predictions. Amongst these we have chosen one (paper 3.6) which uses Bethe ansatz methods to push the analytical approach as far as possible, before resorting to number crunching *. As a result calculations are possible for very large L. They test with great

* In fact, it is probably possible to do the whole calculation analytically. See F. Woynarovich, Phys. Rev. Lett. 59 (1987) 259, 1264.

precision the predictions of conformal invariance.

– The second, smaller group consists of papers which study new, unsolved models. As an example we have chosen one (paper 3.5) * which studies a model which turns out to have $c = 1$. Here relatively brute force methods must be used. Nevertheless we see that exponents can be calculated quite accurately. It is of great interest to study models with $c > 1$ ** as only partial classifications so far exist.

We have also included some more theoretical papers (3.7, 3.8) which show how different boundary conditions may influence the spectrum of the transfer matrix, and how corrections to scaling enter in a calculable way. Finally, a (hopefully) provocative paper (3.9) suggesting that the same thing can be done for dimensions $d > 2$. Recently [7], this has been shown to work, at least approximately, for $d = 3$.

References

[1] A.M. Polyakov, Pis'ma v Zh. Eksp. & Teor. Fiz. 12 (1970) 538 [JETP Lett. 12 (1970) 381].
[2] A.A. Belavin, A.M. Polyakov and A.B. Zamolodchikov, Nucl. Phys. B 241 (1984) 333.
[3] D. Friedan, Z. Qiu and S. Shenker, Phys. Rev. Lett. 52 (1984) 1575.
[4] J.L. Cardy, in: Phase Transitions and Critical Phenomena, Vol. 11, eds. C. Domb and J.L. Lebowitz (Academic Press, London, 1987).
[5] J.L. Cardy, Nucl. Phys. B 270[FS16] (1986) 186.
[6] A. Cappelli, C. Itzykson and J.-B. Zuber, Nucl. Phys. B 280[FS18] (1987) 445.
[7] F.C. Alcaraz and H. Herrmann, J. Phys. A Lett., to be published.

* This model has also been studied, with similar results, by F.C. Alcaraz and M.N. Barber, J. Phys. A 20 (1987) 179.

** Such models have been studied by the Bonn group: see, e.g., G. von Gehlen and V. Rittenberg, J. Phys. A 20 (1987) 1309.

J. Phys. A: Math. Gen. **16** (1983) L657–L664. Printed in Great Britain

LETTER TO THE EDITOR

The relation between amplitudes and critical exponents in finite-size scaling

Peter Nightingale†§ and Henk Blöte‡

† Department of Physics, FM-15, University of Washington, Seattle, Washington 98195, USA

‡ Laboratorium voor Technische Natuurkunde, Technische Hogeschool Delft, PO Box 5046, 2600GA Delft, The Netherlands

Received 25 August 1983

Abstract. The relation between critical exponents and the amplitude of the correlation length divergence at the critical point of two-dimensional systems as a function of finite size is investigated and generalised in two ways. Correlations more general than those of order–order type are included. A form appropriate for anisotropic systems is proposed. We present (a) exact results for the Ising and Gaussian models and (b) numerical results for the symmetric eight-vertex (Baxter), continuous q-state Potts, and continuous N-component cubic models.

In this letter we consider two-dimensional lattices, infinite in one direction and of finite size n in the other. Denote by κ_n the inverse correlation length, in the sense defined below, of the system in the infinite direction. Suppose that all parameters such as temperature and symmetry-breaking field are set to the critical values of the truly infinite system. If we furthermore assume the case of a continuous transition, then κ_n behaves for large n as

$$\kappa_n \simeq A/n. \tag{1}$$

The divergence of the correlation length as shown by this equation is a fundamental feature of a phase transition. In numerical studies in critical phenomena, accurate estimates of critical properties can be derived from the behaviour of the correlation length as a function of system size with the help of finite-size scaling or phenomenological renormalisation (Nightingale 1982 and references therein).

In a study of the two-dimensional XY model Luck (1982) derived the remarkable relation‖

$$A = 2\pi x, \tag{2}$$

where x, the anomalous dimension, is the exponent which describes how the spin–spin correlation function g of the infinite system at criticality decays as a function of distance r:

$$g(r) \sim r^{-2x}. \tag{3}$$

§ Address as from September 1, 1983: Department of Physics, University of Rhode Island, Kingston, Rhode Island 02881, USA.

‖ This relation was first demonstrated by J L Pichard and G Farma (1981 *J. Phys. C: Solid State Phys.* **14** L617) in the case of Anderson localisation.

1988

Note: The authors have pointed out that in the originally published version the second footnote contains an error:

G. Farma should have been written G. Sarma.

Using (2), one need only calculate A to find x; no derivatives of κ_n are required as in the more usual finite-size scaling procedure. Equation (2) has been numerically verified for the spin–spin correlation function of the q-state Potts model in two dimensions (Derrida and de Seze 1982). Results were published only for two special cases: percolation and the Ising model, i.e. $q=1$ and $q=2$.

Luck's result, equation (2), can be reproduced easily (Thouless 1982) with the spin-wave approximation to the two-dimensional XY model, i.e. the Gaussian model As it stands (2) applies to isotropic models only. To investigate the effect of anisotropy we explicitly consider the anisotropic Gaussian model. The reduced Hamiltonian (i.e. with a factor $-1/k_BT$ included) for this model is

$$\mathcal{H}=-K_1\sum^{(1)}[\phi(\boldsymbol{r})-\phi(\boldsymbol{r}')]^2-K_2\sum^{(2)}[\phi(\boldsymbol{r})-\phi(\boldsymbol{r}')]^2 \qquad (4)$$

where the first sum is over all nearest-neighbour bonds in one direction and the second sum is the same in the other direction of a square lattice. The continuous variables ϕ assume values from $-\infty$ to ∞. The analogue of the spin–spin correlation function in the XY model for a strip of width n on a cylinder is given by

$$\langle\cos\phi(\boldsymbol{0})\cos\phi(\boldsymbol{r})\rangle\sim r^{-2x}\,\mathrm{e}^{-A_1 r/n}. \qquad (5)$$

Here we assume that the strip is infinite in the direction of K_1 and also that $\boldsymbol{r}$ points along this direction. The exponent and amplitude in (5) are

$$x=1/4\pi(K_1K_2)^{1/2} \qquad \text{and} \qquad A_1=1/2K_1. \qquad (6)$$

Equations (5) and (6) follow from (20) of Cardy and Nightingale (1983), the generalisation of which to anisotropic interactions is straightforward. Since (5) and (6) also hold with K_1 and K_2 interchanged, one finds

$$A\equiv(A_1A_2)^{1/2}=2\pi x, \qquad (7)$$

where A_2 is the amplitude of the inverse correlation length when K_2 is in the infinite direction. This is the generalisation to anisotropic lattices of equation (2).

For the case that the strip makes an arbitrary angle θ with the direction of K_1, we may make use of known properties of the Gaussian model in continuous space. After a transformation to a coordinate system with the y axis parallel to the infinite direction, and a Fourier transformation, a cross term proportional to the wavenumbers k_x and k_y appears. Since $k_x=0$ is the dominant term, the cross term vanishes and it follows immediately that

$$A_\theta=2\pi x/(\sqrt{A_2/A_1}\cos^2\theta+\sqrt{A_1/A_2}\sin^2\theta). \qquad (8)$$

Equations (7) and (8) were derived for the Gaussian model. However, their validity extends to all critical models in the Gaussian universality class, which includes almost all known two-dimensional models (see e.g. Kadanoff and Brown 1979, Knops 1980 and den Nijs 1981 and references therein). This is shown by the following argument.

Finite-size scaling can be derived from renormalisation group theory, assuming that $1/n$ is an additional scaling field (Suzuki 1977, Blöte and Nightingale 1982). To be precise, the assumption is that, if corrections to scaling are ignored, the renormalisation group equations of the finite system are the same as those of the system in the thermodynamic limit; the scaling of $1/n$ is an immediate consequence of the length rescaling under renormalisation. Equation (1) follows if the only non-zero, relevant scaling field is $1/n$. Also, the universality of the amplitude A is obtained from this assumption; i.e. the presence of non-zero irrelevant scaling fields shows up only in corrections to scaling of the $1/n$ behaviour of the inverse correlation length. Finally, the argument is completed by observing that models within the same universality class differ only by the values of the irrelevant scaling fields.

Note that A in (5) is a universal amplitude, while in general only *ratios* of amplitudes are universal. This is a consequence of the fact that physical parameters couple to scaling fields in a system-dependent way, and this is where $1/n$ is an exception. Generally, unknown constants are introduced into amplitudes, and these cancel only in suitably chosen amplitude ratios.

The discussion above holds for correlations of operators which can be identified with spin-wave operators in the Gaussian model. It also applies to the case of the dual vortex operators, but not necessarily to the mixed spin-wave–vortex case. We shall treat various different operators below; indices will be used to identify the associated amplitudes and anomalous dimensions.

From the assumption that, with respect to critical behaviour, lattice and continuum models differ only in an irrelevant way (in the renormalisation group sense) it follows that equation (8) for the general strip orientation is valid also for lattice models. Furthermore, we expect its validity to extend to general lattices such as honeycomb or triangular. The angle θ is then defined with respect to one of the principal directions. These generically depend in an unknown way on the anisotropic interactions.

Spin–spin correlations in the Ising model canot be related in any obvious way to correlations in the Gaussian model that are of simple spin-wave or vortex type. Therefore our derivation does not apply directly to this case. Yet (7) does hold for this model, as we now show. Again the two interaction constants in the square lattice are denoted by K_1 and K_2. A straightforward calculation, starting from the exact solution of Onsager (see e.g. Domb 1960), gives

$$A_{1\mathrm{m}} = \tfrac{1}{4}\pi(\sinh 2K_2/\sinh 2K_1)^{1/2} \tag{9}$$

for the amplitude of the inverse correlation length in the K_1 direction. At criticality one has $(\exp 2K_1 - 1)(\exp 2K_2 - 1) = 2$. Interchanging subindices 1 and 2, we find that $A_{1\mathrm{m}}$ and $A_{2\mathrm{m}}$ indeed satisfy (7), since x_{m}, the anomalous dimension of the order parameter, equals $\frac{1}{8}$.

Equation (9) was obtained for the correlation length in the Ising model associated with spin–spin correlations. The largest and second-largest eigenvalues of the transfer matrix are the ones that appear in the expression for this length. They are found in the sectors which are respectively even and odd under spin inversion. In the case of energy–energy correlations the pertinent eigenvalues are both in the even sector. Again from the exact solution one finds

$$A_{1\mathrm{T}} = 2\pi(\sinh 2K_2/\sinh 2K_1)^{1/2}, \tag{10}$$

in agreement with (7) and $x_{\mathrm{T}} = 1$ for the anomalous dimension of the energy. The independence of anisotropy of the ratio $A_{1\mathrm{m}}/A_{1\mathrm{T}}$ is likely to be a universal feature.

Next we present the results of numerical calculations. First we consider the symmetric eight-vertex, or Baxter, model (Baxter 1972). Formulated in terms of Ising spins $s_i = \pm 1$, the reduced Hamiltonian of this model reads

$$\mathscr{H} = K_2 \sum_{(i,j)} s_i s_j + K_4 \sum_{(i,j,k,l)} s_i s_j s_k s_l, \tag{11}$$

the first sum being over next-nearest-neighbour bonds on a square lattice, the second over elementary plaquettes. There are three relevant exponents: the thermal anomalous dimension x_T; and x_m and x_p, which pertain to magnetisation and polarisation. They are given by Baxter (1972), Barber and Baxter (1973), Baxter and Kelland (1974) and Baxter (1980):

$$x_T = (2/\pi)\cos^{-1}\tanh 2K_4, \qquad x_p = \tfrac{1}{4}x_T, \qquad x_m = \tfrac{1}{8}. \qquad (12a, b, c)$$

We found the associated amplitudes A_T, A_m and A_p of the inverse correlation lengths from the ratios of the largest eigenvalue of the transfer matrix and the largest subdominant eigenvalues belonging to eigenvectors which have the following symmetry properties. In the magnetic case the desired eigenvector is odd under spin inversion. Similarly, A_p is obtained by imposing oddness under polarisation inversion, i.e. by flipping a sublattice of next-nearest-neighbour spins. The eigenvector associated with the thermal amplitude A_T is even under both of these transformations.

Figures 1(*a*), (*b*) and (*c*) are plots of $n\kappa_n - A$ against x_T for $n = 4, 6, \ldots, 16$, where A is obtained from the relations (7) and (12). The poor convergence at both extremes of x is to be expected in view of previous calculations (Nightingale 1977). In table 1 extrapolated estimates of the amplitudes are compared with the conjectured exact values. Assuming power law convergence according to $A = n\kappa_n + cn^b$, estimates of A can be obtained from three consecutive values of κ_n (Blöte and Nightingale 1982). Such three-point fits were made to the values obtained for $n = 12$, 14 and 16. The results clearly indicate that in the Baxter model equation (7) is satisfied for the thermal, magnetic and polarisation amplitudes.

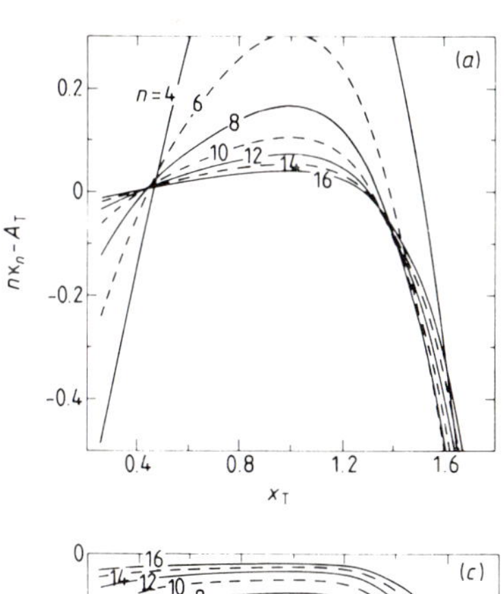

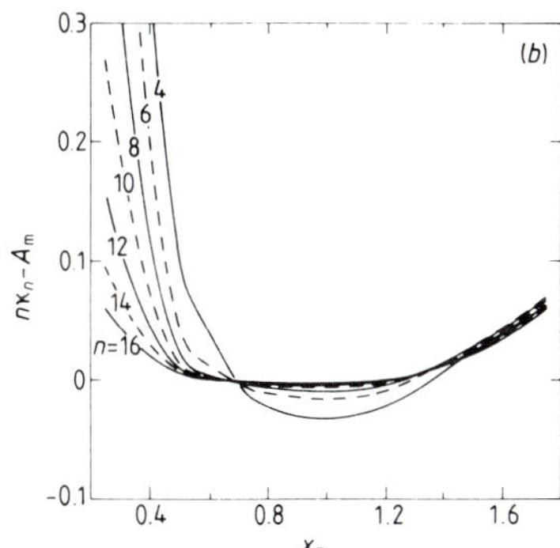

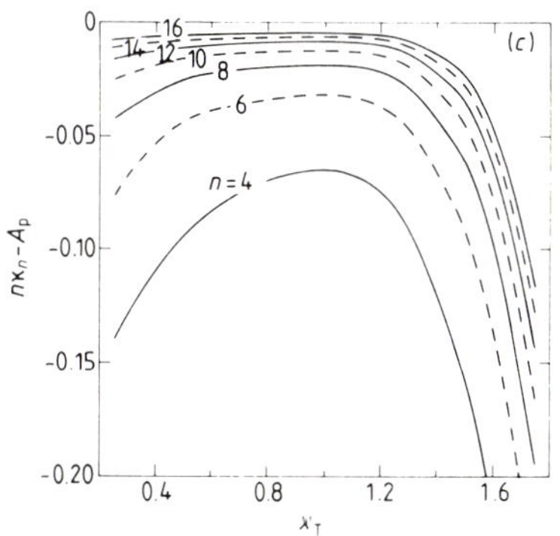

Figure 1. The difference between the calculated and conjectured exact amplitudes A_T (*a*), A_m (*b*), and A_p (*c*) as a function of x_T in the Baxter model for $n = 4, 6, \ldots, 16$.

Table 1. Three-point ($n = 12$, 14, 16) extrapolated estimates (where possible) of the temperature, magnetic and polarisation amplitudes are compared with the conjectured exact values, obtained on the basis of Baxter's exact results, for various values of the four-spin interaction of the Baxter model.

K_4	A_T	Exact	A_m	Exact	A_p	Exact
−0.8					2.7249	2.7432
−0.4					2.2972	2.2970
−0.2	7.8409	7.8422	0.7867	0.7854	1.9607	1.9605
−0.1	7.0788	7.0779	0.7855	0.7854	1.7696	1.7695
0.0	6.2839	6.2832	0.7855	0.7854	1.5710	1.5708
0.1	5.4892	5.4885	0.7855	0.7854	1.3723	1.3721
0.2	4.7248	4.7242	0.7854	0.7854	1.1812	1.1811
0.4	3.3781	3.3784	0.7856	0.7854	0.8447	0.8446
0.8	1.5919	1.5937	0.7857	0.7854	0.3975	0.3984

The next model for which we present numerical results is the continuous q-state Potts model, again on the square lattice. We used the formulation of the model as a Whitney polynomial (Kasteleyn and Fortuin 1969, Baxter 1973). The correlation lengths were computed by employing a transfer matrix as introduced in previous finite-size calculations for the Potts model (Blöte *et al* 1981, Blöte and Nightingale 1982). For the thermal amplitude A_T we made use of the transfer matrix of the *simple* Whitney polynomial (i.e. no ghost site included). The case of the magnetic amplitude A_m was treated with the *extended* polynomial (i.e. including a ghost site).

The thermal and magnetic exponents of the Potts model are given by the generally accepted conjectures (den Nijs 1979, Nienhuis *et al* 1980, Pearson 1980, den Nijs 1981, Black and Emery 1981, Nienhuis 1982a, den Nijs 1983)

$$x_T = 3/x - 1, \qquad x_m = (1-y^2)/4x \tag{13a, b}$$

with $x = 2 - y$ and $\cos(\pi y/2) = \frac{1}{2}\sqrt{q}$.

In figures 2(a, b) we show the differences $n\kappa_n - A$ of the estimated and exact values of the amplitudes against x_T. This was done for $n = 2, 3, \ldots, 10$ and $n = 2, 3, \ldots, 8$

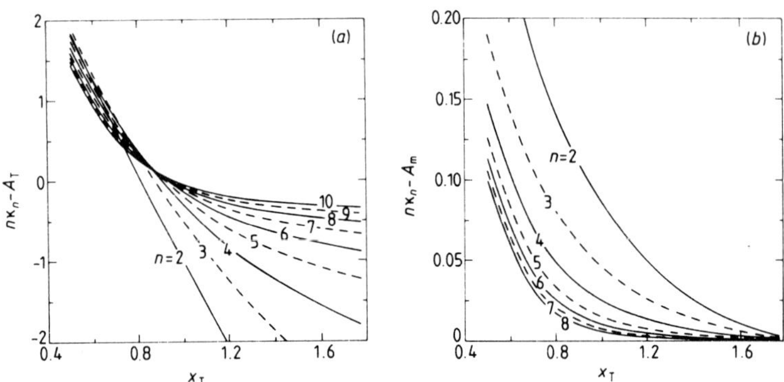

Figure 2. The difference between the calculated and conjectured exact amplitudes A_T (a), and A_m (b) as a function of x_T in the Potts model for $n = 2, 3, \ldots, 10$ (thermal) and $n = 2, 3, \ldots, 8$ (magnetic).

in the thermal and magnetic cases respectively. Results for A_T of three-point extrapolations are shown in table 2. For the magnetic case, we also checked relation (7) for anisotropic lattices. Table 3 contains the extrapolated results. As in the Baxter model, the numerical results agree with equation (7).

Table 2. Three-point ($n = 8, 9, 10$) extrapolated estimates of the temperature amplitude A_T are compared with the conjectured exact results for various values of the number of states q of the Potts model.

q	A_T	Exact
0.0625	11.193	11.174
0.95	7.973	7.953
1.05	7.775	7.758
2.00	6.284	6.283
4.00	3.240	3.142

Table 3. Three-point ($n = 6, 7, 8$) extrapolated estimates of the magnetic amplitude A_m. For different values of the anisotropy $(e^{K_2}-1)/(e^{K_1}-1)$, the geometric mean A_m of A_{1m} and A_{2m} is compared with the conjectured exact results for various values of the number of states q of the Potts model.

Anisotropy	$\sqrt{2}$		2		4		1	
q	A_{1m}	A_m	A_{1m}	A_m	A_{1m}	A_m	A_m	Exact
0.0625	0.263 19	0.222 90	0.310 78	0.222 89	0.433 21	0.222 75	0.222 90	0.222 87
0.25			0.551 55	0.400 39			0.400 41	0.400 32
0.95			0.871 27	0.644 43			0.644 49	0.644 32
1.05			0.896 57	0.664 29			0.664 35	0.664 18
2.00	0.907 52	0.785 94	1.048 17	0.785 89	1.398 79	0.784 88	0.785 94	0.785 40
3.00			1.115 92	0.842 82			0.842 77	0.837 76
4.00			1.132 21	0.857 24			0.856 97	0.785 40

Finally we also applied analogous numerical techniques to the N-component cubic model on an $n \times \infty$ strip. The reduced Hamiltonian is a sum over all pairs of nearest-neighbour sites of a square lattice

$$\mathcal{H} = K \sum_{(i,j)} \boldsymbol{\sigma}_i \cdot \boldsymbol{\sigma}_j, \tag{14}$$

where the $\boldsymbol{\sigma}$ are discrete vectors of unit length and $2N$ possible orientations: parallel or antiparallel to N Cartesian axes. We (Blöte and Nightingale, to be published) constructed a transfer matrix for this model which treats N as a continuous parameter. A thermal correlation length (analogous to the one of the simple Whitney model mentioned above) was calculated for linear system sizes up to $n = 8$, and for $N = \frac{1}{64}, \frac{1}{32}, \ldots, \frac{1}{2}$ ($N = 1$ and $N = 2$ reduce to Ising models). Critical couplings were obtained by scaling of the correlation length (see e.g. Nightingale 1982). The results showed good apparent convergence; estimates are shown in table 4. These data were used to compute the amplitude A_T. Again, the apparent convergence is good, and the results together with conjectured exact values are listed in table 4. The exact numbers were

Table 4. Finite-size results for the critical coupling strength and for the amplitude of the N-component cubic model. Differences from the conjectured exact amplitude are of the same order as numerical uncertainties due to extrapolation.

N	K_c	A_T	Exact
$\frac{1}{64}$	0.005 935	4.241	4.217
$\frac{1}{32}$	0.011 89	4.269	4.245
$\frac{1}{16}$	0.023 91	4.325	4.301
$\frac{1}{8}$	0.048 29	4.439	4.414
$\frac{1}{4}$	0.098 46	4.672	4.647
$\frac{1}{2}$	0.204 6	5.167	5.138

obtained from (7) and a conjecture of Cardy and Hamber (1980, see also Nienhuis 1982b), according to which the thermal anomalous dimension reads $(-2<N<2)$

$$x_T = 4\pi/[2\pi - \cos^{-1}(-N/2)] - 2. \qquad (15)$$

The agreement between estimated and conjectured results is close. Assuming, in view of the evidence presented above for other models, that (7) holds for the cubic model, this implies a verification of the exponent conjecture. We also mention that independent calculation of x_T from temperature derivatives of free energy and correlation length agrees with the above results to within a few parts in one hundred, consistent with apparent uncertainties in the extrapolation procedures.

In conclusion, all our results, both analytic and numerical, support equation (7), which establishes a relation between finite-size amplitudes of inverse correlation lengths associated with correlations of various types of operators and the corresponding anomalous dimensions.

Instructive conversations with Alma Johnson, Marcel den Nijs, John Rehr, Michael Schick, David Thouless, Bernard Derrida and Henk Hilhorst are kindly acknowledged.

This work was supported in part by the US National Science Foundation under grant no DMR-79-20785, by the Dutch 'Stichting voor Fundamenteel Onderzoek der Materie', and by the Einstein Center for Theoretical Physics, Weizmann Institute of Science, Israel.

References

Barber M N and Baxter R J 1973 *J. Phys. C: Solid State Phys.* **6** 2913
Baxter R J 1972 *Ann. Phys.* **70** 193
—— 1973 *J. Phys. C: Solid State Phys.* **6** L445
—— 1980 in *Fundamental Problems in Stat. Mech. V* ed E G D Cohen (Amsterdam: North-Holland)
Baxter R J and Kelland S B 1974 *J. Phys. C: Solid State Phys.* **7** L403
Black J L and Emery V J 1981 *Phys. Rev.* B **23** 429
Blöte H W J and Nightingale M P 1982 *Physica* **112A** 405
Blöte H W J, Nightingale M P and Derrida B 1981 *J. Phys. A: Math. Gen.* **14** L45
Cardy J L and Hamber H W 1980 *Phys. Rev. Lett.* **45** 499
Cardy J L and Nightingale M P 1983 *Phys. Rev.* B **27** 4256
Derrida B and de Seze J 1982 *J. Physique* **43** 475
Domb C 1960 *Adv. Phys.* **9** 191
Kadanoff L P and Brown A C 1979 *Ann. Phys.* **121** 318

Kasteleyn P W and Fortuin C M 1969 *J. Phys. Soc. Japan* **46** (suppl) 11
Knops H J F 1980 *Ann. Phys.* **128** 448
Luck J M 1982 *J. Phys. A: Math. Gen.* **15** L169
Nienhuis B 1982a *J. Phys. A: Math. Gen.* **15** 199
—— 1982b *Phys. Rev. Lett.* **49** 1062
Nienhuis B, Riedel E K and Schick M 1980 *J. Phys. A: Math. Gen.* **13** L189
Nightingale M P 1977 *Phys. Lett.* **59A** 486
—— 1982 *J. Appl. Phys.* **53** 7927
den Nijs M P M 1979 *J. Phys. A: Math. Gen.* **12** 1857
—— 1981 *Phys. Rev.* B **23** 6111
—— 1983 *Phys. Rev.* B **27** 1674
Pearson R B 1980 *Phys. Rev.* B **22** 2579
Suzuki M 1977 *Prog. Theor. Phys.* **58** 1142
Thouless D J 1982 *Private communication*

J. Phys. A: Math. Gen. **17** (1984) L385–L387. Printed in Great Britain

LETTER TO THE EDITOR

Conformal invariance and universality in finite-size scaling

John L Cardy

Department of Physics, University of California, Santa Barbara, California 93106, USA

Received 5 March 1984

Abstract. The universal relation between critical exponents and the amplitude of the correlation length divergence as a function of finite size at the critical point of two-dimensional systems is shown to be a consequence of conformal invariance. Both periodic and free boundary conditions are considered.

In studies of the finite-size scaling behaviour of two-dimensional systems on infinitely long strips of finite width, a remarkable universality has been observed. The theory of finite-size scaling (Barber 1983) predicts that the inverse correlation length $\varkappa_n$ (measured in lattice spacings), for a strip of width n lattice spacings, should behave, at the critical point of the infinite system, as

$$\varkappa_n \sim A/n \tag{1}$$

where the amplitude A is universal in the usual sense (Privman and Fisher 1984). However, exact and numerical calculations (Luck 1982, Derrida and de Seze 1982, Nightingale and Blöte 1983, Privman and Fisher 1984) also suggest that the ratio A/x, where x is the scaling dimension of the operator concerned, is equal to 2π for several different operators in a wide variety of isotropic two-dimensional models.

In this letter, we point out that this result is a simple consequence of conformal covariance of the correlation functions, which is believed to hold in the continuum limit at the critical point (Polyakov 1970, 1974). In two dimensions, conformal invariance implies that the correlation functions of a local scalar operator $\varphi(z)$ satisfy

$$\langle\varphi(z_1)\varphi(z_2)\rangle = |w'(z_1)|^x|w'(z_2)|^x\langle\varphi(w(z_1))\varphi(w(z_2))\rangle \tag{2}$$

where $z \to w(z)$ is an arbitrary conformal transformation, and x is the scaling dimension of φ. Now consider the particular transformation

$$w = \ln z \tag{3}$$

which maps the whole z-plane into the strip $|\mathrm{Im}\, w| \leq \pi$, with periodic boundary conditions. Using the result that in the infinite plane

$$\langle\varphi(z_1)\varphi(z_2)\rangle \sim |z_1 - z_2|^{-2x} \tag{4}$$

and writing $z_j = \exp(y_j + \mathrm{i}\theta_j)$, $(|\theta_j| \leq \pi)$, it follows from (2) that

$$\langle\varphi(y_1+\mathrm{i}\theta_1)\varphi(y_2+\mathrm{i}\theta_2)\rangle_S \sim \exp\{-x\ln[\exp(y_1-y_2)+\exp(y_2-y_1)-2\cos(\theta_1-\theta_2)]\} \tag{5}$$

where the correlation function on the left-hand side is evaluated in the strip geometry.

0305-4470/84/070385+03\$02.25

The inverse correlation length $\varkappa$ is defined by

$$\int_{-\pi}^{\pi} d\theta_1 \langle \varphi(y_1+i\theta_1)\varphi(y_2+i\theta_2)\rangle_S \sim \exp(-\varkappa|y_1-y_2|) \tag{6}$$

as $|y_1-y_2|\to\infty$. Comparing with (5) we see that $\varkappa = x$.

However, this is measured in physical units in which the width of the strip is 2π. If we now introduce a lattice, the lattice spacing will be $2\pi/n$. Hence $\varkappa_n$ as measured in inverse lattice spacings equals $2\pi x/n$, which is the desired result.

The same methods may be applied to the case of free boundary conditions. The same transformation (3) maps the half plane $\mathrm{Re}\, z \geq 0$ onto the strip $|\mathrm{Im}\, w| \leq \frac{1}{2}\pi$, with the same boundary conditions. We then find

$$\langle \varphi(y_1+i\theta_1)\varphi(y_2+i\theta_2)\rangle_S \sim \exp[x(y_1+y_2)]\langle \varphi[\exp(y_1+i\theta_1)]\varphi[\exp(y_2+i\theta_2)]\rangle_{S'} \tag{7}$$

where S′ refers to the surface geometry. The theory of surface critical phenomena (Binder 1983) now implies that the right-hand side has the scaling form† $F(\exp(y_1-y_2), \theta_1, \theta_2)$, where, for $e^{y_1} \gg e^{y_2}$,

$$F(\exp(y_1-y_2), \theta_1, \theta_2) \sim a(\theta_1, \theta_2)\exp[-x_s(y_1-y_2)] \tag{8}$$

where x_s is the scaling dimension of the corresponding surface operator (equal to $\frac{1}{2}\eta_\parallel$ for the order parameter). If we now define the inverse correlation length in the strip by integrating (7) over θ_1 and θ_2 we find $\varkappa = x_s$. Since the width in physical units is now π, the inverse correlation length measured in inverse lattice spacings is

$$\varkappa_n \sim \pi x_s/n \tag{9}$$

or, equivalently, $\varkappa_n \sim \pi\eta_\parallel/2n$ for the order parameter correlation length.

The above calculations may be easily verified for the Gaussian model, where conformal invariance is explicit. In that case, $\eta_\parallel = 2\eta$, so the amplitude A is the same for the two different boundary conditions. For the Ising model, however, $\eta_\parallel = 1$, so we predict $\varkappa_n \sim \pi/2n$ for free boundaries.

Finally, we mention that recently Friedan *et al* (1983) have used conformal invariance and unitarity to constrain severely possible exponents in two dimensions. The requirement of unitarity appears to exclude several interesting models (percolation, continuous q-state Potts, continuous N-component cubic models), for which the universality of A/x has been numerically verified (Derrida and de Seze 1982, Nightingale and Blöte 1983). This suggests that such models are conformally invariant but not unitary.

This work was supported by the National Science Foundation under Grant No PHY-80-18938.

References

Barber M N 1983 *Phase Transitions and Critical Phenomena* vol 8 ed C Domb and J Lebowitz (London: Academic)

Binder K 1983 *Phase Transitions and Critical Phenomena* vol 8 ed C Domb and J Lebowitz (London: Academic)

† Conformal invariance may also be used to constrain the form of F.

Derrida B and de Seze J 1982 *J. Physique* **43** 475
Friedan D, Qiu Z and Shenker S 1983 *Talk presented at Workshop on Vertex Operators, Berkeley* to be published
Luck J M 1982 *J. Phys. A: Math. Gen.* **15** L169
Nightingale M P and Blöte H W J 1983 *J. Phys. A: Math. Gen.* **16** L657
Polyakov A M 1970 *Pisma ZhETP* **12** 538 (1970 *JETP Lett.* **12** 381)
—— 1974 *ZhETP* **66** 23 (1974 *JETP Lett.* **39** 10)
Privman V and Fisher M E 1984 *Phys. Rev.* B in print

Conformal Invariance, the Central Charge, and Universal Finite-Size Amplitudes at Criticality

H. W. J. Blöte
Laboratorium voor Technische Natuurkunde, 2600 GA Delft, The Netherlands

John L. Cardy
Department of Physics, University of California, Santa Barbara, California 93106

and

M. P. Nightingale
Department of Physics, University of Rhode Island, Kingston, Rhode Island 02881
(Received 21 November 1985)

We show that for conformally invariant two-dimensional systems, the amplitude of the finite-size corrections to the free energy of an infinitely long strip of width L at criticality is linearly related to the conformal anomaly number c, for various boundary conditions. The result is confirmed by renormalization-group arguments and numerical calculations. It is also related to the magnitude of the Casimir effect in an interacting one-dimensional field theory, and to the low-temperature specific heat in quantum chains.

PACS numbers: 64.60.Fr, 05.70.Jk, 68.55.–a, 75.40.–s

The principle of conformal invariance at a critical point has been shown to be remarkably powerful, especially in two dimensions.[1,2] Universality classes appear to be characterized by a single dimensionless number c, the conformal anomaly or the value of the central charge of the Virasoro algebra.[3] It was shown by Friedan, Qiu, and Shenker[2] that unitarity constrains those values of c less than unity to be quantized. For such theories, the critical exponents are given by the Kac formula,[4] and the correlation functions are determined.[1,5] For various models, c has been determined indirectly by use of exact information on exponents and correlation functions obtained by other means.[1,2,5] In this Letter we give a simple means of determining c.

The free energy (measured in units of k_BT) per unit length of an infinitely long strip of width L at criticality has the finite-size scaling form $F=fL+f^\times+\Delta/L+\ldots$, where f is the bulk free energy per unit area, and $\frac{1}{2}f^\times$ is the surface free energy, which vanishes in the case of periodic boundary conditions. It has been argued, from the assumption that L^{-1} is a scaling field which does not require the introduction of a metric factor,[6,7] that Δ is universal. We find that

$$\Delta=\begin{cases}-\pi c/6, & \text{periodic boundary conditions,} \quad (1)\\ -\pi c/24, & \text{free or fixed boundary conditions,} \quad (2)\end{cases}$$

where, in the last case, the order parameter is fixed to the *same* value on either side of the strip.

These results have several other interesting physical interpretations. Since F corresponds to the ground-state energy of a $(1+1)$-dimensional quantum field theory in a finite volume, Eq. (2) also gives the magnitude of the Casimir effect[8] in such a theory. The partition function of a classical system of finite width with periodic boundary conditions may also be interpreted as the Feynman path integral for an infinitely long quantum chain at finite temperature $T\propto L^{-1}$. In that case Eq. (1) gives the leading $T\to 0$ correction to the free energy, from which may be deduced the specific heat C. In fact the conformal result applies only if the two-dimensional classical system is rotationally invariant at large distances. This is equivalent to the requirement that the dispersion relation for gapless excitations of the quantum chain is of the form $\omega\sim vk$ with $v=1$. The case $v\neq 1$ can be accommodated by a suitable rescaling of time versus length for the quantum chain. The result is $C\sim \pi c k_B^2 T/3\hbar v$. This is confirmed by exact results for the spin-$\frac{1}{2}$ XXZ chain[9] ($c=1$) and for the anisotropic spin-$\frac{1}{2}$ XY model in a critical transverse field[10] ($c=\frac{1}{2}$). In three dimensions, the analog of Δ is the interaction energy (in units of k_BT) per unit area of two plates immersed in a critical system.[11] Universality in this case was verified by Monte Carlo techniques.[12] The same constant also plays a role in determining the thickness of gravity-thinned, critical wetting layers.[13] Two-dimensional analogs of these systems, which would allow an experimental determination of c, are conceivable.

A system at a critical point is governed by a reduced fixed-point Hamiltonian[14] $\mathscr{H}^*$. Under a coarse graining in which lengths are rescaled uniformly, the form of the Hamiltonian is invariant. For short-range interactions, the Hamiltonian remains at the fixed point also under conformal transformations, which correspond to a *nonuniform* rescaling and rotation. Transformations with a shear component, however, modify the Hamiltonian. The response of $\mathscr{H}$ to such an infinitesimal transformation of the form $x^\mu \to x^\mu + \alpha^\mu$ is

$$\delta \mathscr{H} = -\frac{1}{2\pi}\int \frac{\partial \alpha^\mu}{\partial x_\nu} T_{\mu\nu}\, d^2x. \tag{3}$$

This defines[15] the stress tensor $T_{\mu\nu}$. In complex coordinates $(z,\bar{z})$ the only nonzero components are $T_{zz} = T(z)$ and $T_{\bar{z}\bar{z}} = \bar{T}(\bar{z})$. The conformal anomaly number c may then be defined by[1,2]

$$\langle T(z) T(z')\rangle_c = (c/2)(z-z')^{-4}. \tag{4}$$

Even if T is subtracted so that $\langle T\rangle = 0$ in the infinite plane, it is nonzero in the strip. As we now show, its value is related to Δ. Consider the nonconformal transformation $u' = u(1-\lambda)$, $v' = v(1+\lambda)$, where $\lambda << 1$, and (u, v) measure distances along and across the strip, respectively. According to Eq. (3)

$$\delta\langle \mathscr{H}\rangle = -(\lambda/2\pi)\int_{-\infty}^{\infty} du \int_0^L dv \langle -T_{uu} + T_{vv}\rangle$$
$$= (\lambda L/\pi)\int_{-\infty}^{\infty} (\langle T\rangle + \langle \bar{T}\rangle)\, du \tag{5}$$

for the translationally invariant case of periodic boundary conditions. Invariance of the partition function implies that this is compensated by a change in F, which is $-2\lambda\Delta/L$. Hence we find $\Delta = (L^2/\pi)\langle T\rangle$, since $\langle T\rangle = \langle \bar{T}\rangle$ by symmetry. Now the response of $\langle T\rangle$ to $\delta\mathscr{H}$ is

$$\delta\langle T(0,0)\rangle$$
$$= -(\lambda/\pi)\int_{-\infty}^{\infty} du \int_0^L dv \langle T(0,0) T(u,v)\rangle_c, \tag{6}$$

using $\langle T\bar{T}\rangle_c = 0$. The connected correlation function $\langle TT\rangle_c$ in the strip may be found[16] by conformal transformation of the infinite-plane result Eq. (4) using the transformation $w = u + iv = (L/2\pi)\ln z$. The result is

$$\langle T(0) T(w)\rangle_c = (c/2)(\pi/L)^4[\sinh(\pi w/L)]^{-4}.$$

The integral is divergent as $w \to 0$, but the final result is independent of the particular method of regularization. The integrals over u and v in Eq. (6) are then elementary, and one obtains $\delta\langle T\rangle = \lambda\pi^2 c/3L^2$. This is to be compared with $\delta\langle T\rangle = \delta(\pi\Delta/L^2) = -2\pi\Delta\lambda/L^2$, and the result in Eq. (1) follows.

In the case of free or fixed boundary conditions, the correlation function in the strip is found[16] with use of the transformation $w = (L/\pi)\ln z$ from the upper half plane. In the latter geometry, the $\langle TT\rangle$ correlation function is as in Eq. (4), while[17] $\langle T(z_1)\bar{T}(\bar{z}_2)\rangle = (c/2)(z_1 - \bar{z}_2)^{-4}$. However, this term does not contribute to Δ. Thus the only difference between the two cases is that L is replaced by $2L$. This accounts for the factor of 4 difference between Eqs. (1) and (2).

The results in Eqs. (1) and (2) agree with exact results for the Gaussian model[18] $(c=1)$ and the Ising model[19] $(c=\frac{1}{2})$. Equation (1) has also been verified[20] for all the theories in the unitary classification of Friedan, Qiu, and Shenker.[2] In fact, it is possible to calculate the free energy in an arbitrarily shaped parallelogram with periodic boundary conditions,[20] of which the infinitely long strip is only a special case.

The result in Eq. (1) can be verified in a modified Gaussian model with reduced Hamiltonian

$$\mathscr{H} = \frac{1}{2}K\sum_{k=1}^{m-1}\sum_{l=1}[(\phi_{k,l} - \phi_{k+1,l})^2 + (\phi_{k,l} - \phi_{k,l+1})^2] + i\alpha\sum_{k=1}^{m-1}(\phi_{k,l} - \phi_{k+1,1}), \tag{7}$$

where the $\phi_{k,l} \in R$ are located at the sites of a simple square lattice on a cylinder, i.e., $\phi_{k,1} = \phi_{k,L+1}$ for $k = 1, \ldots, m$, subject to the constraints $\phi_{1,1} = \phi_{1,s}$ and $\phi_{m,1} = \phi_{m,s}$ for $s = 2, \ldots, L$. The second term in Eq. (7) represents a defect line. A duality transformation changes $K \to K^{-1}$ in the above Hamiltonian, while the last term becomes $(\alpha/K)\sum_{k=1}^{m-1}(\phi_{k,1} - \phi_{k,L})$. This term may be eliminated by a shift $\phi_{k,l} \to \phi_{k,l} + \alpha l/L$, which adds a constant $-\alpha^2(L-1)/2KL$ to the free energy per unit length F. This modifies Δ to

$$\Delta = -\frac{\pi}{6} + \frac{\alpha^2}{2K}. \tag{8}$$

The defect line is equivalent to charges $\pm\alpha$ at $k=1$ and $k=m$, respectively. As $m \to \infty$, this is equivalent to a charge -2α at infinity, as considered by Dotsenko and Fateev.[5] They found $c = 1 - 24\alpha^2$, when $K = 1/8\pi$, in agreement with Eqs. (1) and (8).

From (8) we derive the value of Δ for the q-state Potts model $(0 \leq q \leq 4)$ as follows. The critical Potts model can be represented as an F model.[21] With the usual labeling of the vertices (see Fig. 2 of Ref. 22), the vertex weights are $(\omega_1, \ldots, \omega_6) = (1,1,1,1,z^\pi + z^{-\pi}, z^\pi + z^{-\pi})$, where $q^{1/2} = 2\cosh\theta$ and $z = e^{\theta/2}$. As noted in Ref. 21, cylindrical boundary conditions lead to a seam of vertices with modified weights $\omega_3' = e^{2\theta}$ and $\omega_4' = e^{-2\theta}$. In the body-centered solid-on-solid representation of the F model,[23] the weight of the modified vertices is $e^{2\theta(n_2 - n_4)}$, where n_2 and n_4 are the column heights at next-nearest-neighbor sites straddling the seam. This corresponds to $\alpha = 2i\theta$ in Eq. (7). The presence of "external" sites[21] leads to the restriction of constant column height at $k = 1, m$, as introduced above. Under renormalization, this body-

TABLE I. Numerical results for Δ as a function of the four-spin interaction K_4 of the Baxter model, obtained from data for $L=4,6,\ldots,16$. The exact result in this case is $-\pi/6 \approx -0.523\,599$.

K_4	Δ	K_4	Δ
−1.0	−0.525	0.1	−0.523 604
−0.8	−0.524	0.2	−0.523 604
−0.6	−0.524	0.3	−0.523 602
−0.4	−0.5239	0.4	−0.523 590
−0.3	−0.523 69	0.6	−0.524
−0.2	−0.523 608	0.8	−0.525
−0.1	−0.523 604	1.0	−0.57
0.0	−0.523 604		

centered solid-on-solid model flows to the Gaussian model with[22,24,25] $K=\pi(2-y)$, where $q^{1/2}=2\times\cos(\pi y/2)$ and $0\leqslant y\leqslant 2$. Topological objects, such as charges, remain unrenormalized. From Eq. (8) we therefore find

$$\Delta=-\frac{\pi}{6}+\frac{\pi y^2}{2(2-y)}. \quad (9)$$

The q dependence of c that we then infer from Eq. (2) agrees with that derived by Kadanoff, and quoted in Ref. 2, and with that conjectured by Dotsenko and Fateev.[5]

The same argument can be used for the O(n) model on a hexagonal lattice introduced by Nienhuis,[26] which can be mapped onto a six-vertex model on a Kagomé lattice, which in turn may be represented by a solid-on-solid model. Again a defect line has to be introduced to obtain the correct weights for cylindrical boundary conditions, and the argument proceeds in the same way as above. Renormalization maps the O(n) model onto a Gaussian model with interaction[24] $K=\pi(2-y)$, where $n=2\cos(\pi y/2)$, and $-2\leqslant y\leqslant 0$. The value of Δ agrees with Eq. (1) if the n dependence of c conjectured by Dotsenko and Fateev[5] is used.

The F model and the critical Baxter model[27] both renormalize onto a Gaussian model with no defect line, and so we expect $\Delta=-\pi/6$ universally for these cases, in agreement with the idea[2] that models with continuously varying exponents have $c=1$.

Finally, we present numerical results supporting the expressions derived for Δ for periodic boundary conditions. The results were obtained from the free energy per site of infinitely long strips of increasing width L, by standard extrapolation techniques.[28] The models in the universality class of the O(n) model that we studied are the continuous n-component cubic model defined by Blöte and Nightingale,[29] in the two cases considered there: $L'=0$ and $A=0$ ($e^{-L'}=\cosh K$), where L' and K are the coefficients of the quadratic and quartic terms in the Hamiltonian.

For the Baxter model in the Ising spin representation[30] we varied the four-spin and the two equal next-nearest-neighbor interactions K_4 and K_2, along the critical line. As shown in Tables I–III, the results agree very well with the theory in all cases, particularly for those values of the parameters where also in previ-

TABLE II. Numerical results for Δ for the q-state Potts model compared with exact results derived in the text; free-energy data of Blöte and Nightingale (Ref. 28) for $L=2,\ldots,11$.

q	Δ	Exact
$\frac{1}{64}$	0.869 148	0.869 154
$\frac{1}{16}$	0.708 251	0.708 256
$\frac{1}{2}$	0.233 420	0.233 438
0.95	0.018 4268	0.018 4267
1.05	−0.017 6778	−0.017 779
2	−0.261 796	−0.261 799
3	−0.418 92	−0.418 88
4	−0.525 30	−0.523 599

TABLE III. Numerical results for Δ as a function of n for two special cases ($L'=0$ and $A=0$) of the n-component cubic model (Ref. 29) extrapolated from data for $L=2,\ldots,8$, compared with exact results derived in the text.

n	$\Delta(A=0)$	$\Delta(L'=0)$	Exact
−1	0.312		0.3142
$-\frac{1}{2}$	0.146		0.1461
$-\frac{1}{4}$	0.0711		0.0711
$-\frac{1}{8}$	0.0352		0.0351
$-\frac{1}{16}$	0.0175		0.017 46
$-\frac{1}{32}$	0.0087		0.008 70
$-\frac{1}{64}$	0.0044		0.004 35
$\frac{1}{64}$	−0.0043	−0.004 34	−0.004 33
$\frac{1}{32}$	−0.0087	−0.008 67	−0.008 66
$\frac{1}{16}$	−0.0173	−0.001 73	−0.001 727
$\frac{1}{8}$	−0.0344	−0.0344	−0.0344
$\frac{1}{4}$	−0.0682	−0.0682	−0.0681
$\frac{1}{2}$	−0.1343	−0.134	−0.1340
1	−0.262	−0.262	−0.2618
2	0.523	−0.525	−0.5236

ous calculations[6,29,31] the asymptotic behavior was observed to set in for those system sizes considered here.

It is a pleasure to acknowledge useful conversations with Jill Bonner, Hubert Knops, Gerhard Müller, Marcel den Nijs, Michael Peskin, and Richard Scalettar. One of us (M.P.N.) thanks the Institut voor Theoretische Fysica of the Katholieke Universiteit Leuven, where part of this work was done, for its hospitality. This research was supported by the U.S. National Science Foundation under Grants No. PHY83-13324 at the University of California at Santa Barbara and No. DMR-8406186 at the University of Rhode Island, by NATO under Grant No. 198/84, and by the Dutch Stichting voor Fundamenteel Onderzoek der Materie.

Note added.—After this paper was submitted for publication, we learned that Affleck[32] has also obtained the result in Eq. (1).

[1]A. A. Belavin, A. M. Polyakov, and A. B. Zamolodchikov, J. Stat. Phys. **34**, 763 (1984), and Nucl. Phys. **B241**, 333 (1984).

[2]D. Friedan, Z. Qiu, and S. Shenker, Phys. Rev. Lett. **52**, 1575 (1984).

[3]M. A. Virasoro, Phys. Rev. D **1**, 2933 (1970).

[4]V. G. Kac, in *Group Theoretical Methods in Physics*, edited by W. Beiglbock and A. Bohm, Lecture Notes in Physics Vol. 94 (Springer-Verlag, New York, 1979), p. 441.

[5]Vl. S. Dotsenko and V. A. Fateev, Nucl. Phys. **B240**, 312 (1984), and **B251**, 691 (1985).

[6]M. P. Nightingale and H. W. J. Blöte, J. Phys. A **16**, L657 (1983).

[7]V. Privman and M. E. Fisher, Phys. Rev. B **30**, 322 (1984).

[8]H. B. G. Casimir, Proc. K. Ned. Akad. Wet. **51**, 793 (1948).

[9]M. Takahashi, Prog. Theor. Phys. **50**, 1519 (1973), and **51**, 1348 (1974). For reviews see, e.g., J. C. Bonner, in *Magnetostructural Correlations in Exchange-Coupled Systems*, edited by R. D. Willett *et al.* (Reidel, Dordrecht, The Netherlands, 1985).

[10]S. Katsura, Phys. Rev. **127**, 1508 (1962).

[11]M. E. Fisher and P.-G. de Gennes, C.R. Acad. Sci. **287**, 207 (1978).

[12]K. K. Mon and M. P. Nightingale, Phys. Rev. B **31**, 6137 (1985), and unpublished work.

[13]M. P. Nightingale and J. O. Indekeu, Phys. Rev. Lett. **54**, 1824 (1985); J. O. Indekeu, M. P. Nightingale, and W. V. Wang, to be published. Here it is exactly verified for the anisotropic Ising model on a triangular lattice that spatial rescaling restores the universality of Δ.

[14]"Hamiltonian" here means "action" in field-theory language.

[15]The factor of $1/2\pi$ is consistent with the normalization used in Refs. 1 and 2.

[16]J. L. Cardy, J. Phys. A **17**, L385 (1984).

[17]J. L. Cardy, Nucl. Phys. **B240**, 514 (1984).

[18]For the continuum and various lattice versions of the Gaussian model, this calculation can be done exactly. Details will be published elsewhere. The result for free boundary conditions is equivalent to that of L. Brink and H. B. Nielsen, Phys. Lett. **43B**, 319 (1973).

[19]A. E. Ferdinand and M. E. Fisher, Phys. Rev. **185**, 832 (1969); H. Au-Yang and M. E. Fisher, Phys. Rev. B **11**, 3469 (1975).

[20]J. L. Cardy, to be published.

[21]R. J. Baxter, S. B. Kelland, and F. Y. Wu, J. Phys. A **9**, 397 (1976).

[22]M. P. M. den Nijs, Phys. Rev. B **27**, 1674 (1983).

[23]H. van Beijeren, Phys. Rev. Lett. **38**, 993 (1977); H. J. F. Knops, Ann. Phys. (N.Y.) **128**, 448 (1981).

[24]B. Nienhuis, J. Stat. Phys. **34**, 731 (1984).

[25]The functions $\theta(y)$ and $K(y)$ are multivalued. The correct branch of $\theta(y)$ is chosen by comparison with the Ising case $q = 2$.

[26]B. Nienhuis, Phys. Rev. Lett. **49**, 1062 (1982).

[27]R. J. Baxter, Ann. Phys. (N.Y.) **70**, 193 (1972).

[28]H. W. J. Blöte and M. P. Nightingale, Physica (Amsterdam) **112A**, 405 (1982).

[29]H. W. J. Blöte and M. P. Nightingale, Physica (Amsterdam) **129A**, 1 (1984).

[30]L. P. Kadanoff and F. Wegner, Phys. Rev. B **4**, 3989 (1971); F. Y. Wu, Phys. Rev. B **4**, 2312 (1971).

[31]M. P. Nightingale, J. Appl. Phys. **53**, 7927 (1982).

[32]I. Affleck, following Letter [Phys. Rev. Lett. **56**, 746 (1986)].

Universal Term in the Free Energy at a Critical Point and the Conformal Anomaly

Ian Affleck
Department of Physics, Princeton University, Princeton, New Jersey 08544
(Received 6 December 1985)

We show that the leading finite-size correction to $\ln Z$ for a two-dimensional system at a conformally invariant critical point on a strip of length L, width β ($\beta << L$), is $(\pi/6)c(L/\beta)$, where c is the conformal anomaly. Equivalently, the leading low-temperature correction to the free energy of a one-dimensional quantum system is $-(\pi/6)cL(kT)^2/\hbar v$, where v is the effective "velocity of light." The latter formula is used to check recently derived critical theories of spin-s quantum chains against Bethe-*Ansatz* solutions.

PACS numbers: 64.60.Fr, 05.30.Ch, 05.70.Jk, 75.40.-s

Conformal invariance powerfully constrains the critical behavior of two-dimensional classical (and one-dimensional quantum) systems.[1,2] Critical theories are parametrized by the conformal anomaly c, which is the central charge in the Virasoro algebra obeyed by the energy-momentum tensor:

$$-i[T(x_-),T(x_-')]=\delta(x_- - x_-')T' - 2\delta'(x_- - x_-')T + (c/24\pi)\delta'''(x_- - x_-') \qquad [T=(T_{00}-T_{01})/2]. \quad (1)$$

For $c<1$ a discrete set of values are allowed by unitarity[2] (reflection positivity): $c=1-6/m(m+1)$, $m=3,4,5,\ldots$. These are realized by the Ising ($c=\frac{1}{2}$), tricritical Ising ($c=\frac{7}{10}$), three-state Potts ($c=\frac{4}{5}$), tricritical three-state Potts ($c=\frac{6}{7}$), and other models. A complete classification has not been given for $c \geq 1$ except when a continuous symmetry G is assumed.[3,4] For $G=U(1)$ we get the Gaussian model ($c=1$) which describes a wide variety of critical phenomena ($q=4$ Potts model, X-Y model, Coulomb gas, $s=\frac{1}{2}$ antiferromagnet,. . . .) For $G=SU(n)$ the possible values of c are $(n^2-1)k/(n+k)$, $k=1,2,3,\ldots$. These describe Wess-Zumino σ models[5] and antiferromagnetic chains (and perhaps also two-dimensional statistical models). In all the above cases, all scaling dimensions are known exactly.

Conformal invariance can also be used to study finite-size effects in two-dimensional statistical systems[6] or finite-temperature effects in one-dimensional quantum systems. These are related because a $(1+1)$-dimensional quantum field theory at temperature T is given by a Euclidean-space functional integral on a strip of width $\beta=1/T$ (in the imaginary time direction). Correlation functions behave as

$$\langle\phi(x,0)\phi(y,0)\rangle \sim e^{-|x-x|/l_\phi} \qquad (2)$$

with $l_\phi = 2\pi d_\phi/\beta$, where d_ϕ is the scaling dimension of of ϕ and periodic boundary conditions are imposed on strip of width β.[6] This has been generalized to other boundary conditions and the interfacial tension (difference in free energy per unit length for periodic and antiperiodic boundary conditions) has been related to the scaling dimension of a disorder operator, in some cases.

In this work, we will derive a simple, general formula for the leading finite-width correction to $\ln Z$:

$$(\ln Z)/L = \text{const}\times\beta + \pi c/6\beta + O(1/\beta^2). \qquad (3)$$

Here β is the width of a strip with periodic boundary conditions and c is the conformal anomaly. (The length L is taken to infinity.) Equivalently, for a one-dimensional quantum system we obtain the same formula but now $1/\beta = T$ and the correction is scaled by the effective "velocity of light" v (which occurs in the low-energy excitation spectrum):

$$F/L = \epsilon_0 - \pi c T^2/6v + O(T^3). \qquad (4)$$

(We set $\hbar$ and Boltzmann's constant equal to 1.) Note that for the statistical problems (assumed to be defined on a square lattice with rationally invariant couplings) the speed of light is $v=1$.

The proof of this result rests on the definition of c as the response of a theory to curving of the two-dimensional space. If Z is the partition function on a space with metric $q_{\mu\nu}$ then[7]

$$-g^{\mu\nu}\frac{\delta \ln Z}{\delta g^{\mu\nu}} = g^{\mu\nu}\langle T_{\mu\nu}\rangle = \frac{c}{48\pi}[R(x)+\mu^2]. \qquad (5)$$

Here $T_{\mu\nu}$ is the energy-momentum tensor and the first equation follows from the canonical definition of $T_{\mu\nu}$. $R(x)$ is the curvature scalar and the second equation follows because R is the only invariant function of the metric with the right dimension (L^{-2}). μ^2 is a constant (of dimension L^{-2}) and c is an arbitrary constant but it can be shown[8] to be the same one which appears in the Virasoro algebra by variation of $\ln Z$ a second time with respect to the metric:

$$\langle T(x)T(x')\rangle = c/2(x-x')^4 + \text{less singular terms}. \qquad (6)$$

But this leading singularity in the operator product expansion can also be determined from the Virasoro algebra and thus the constant c must be the same. Equation (5) can now be integrated to find[7]

$$-\ln Z=(c/48\pi)\int d^2x\,(\tfrac{1}{2}\,\partial_a \ln\rho\,\partial_a \ln\rho+\mu^2\rho)+\alpha. \tag{7}$$

Here we have chosen the metric in conformal gauge, $g_{\mu\nu}=\rho(x)\delta_{\mu\nu}$, and α is a dimensionless constant. We assume that the manifold has no boundaries so that there are no "surface terms" to worry about.

Let us now fix the space Γ_0 to be a strip of length L in the x direction, $0\leq x\leq L$ $(L\to\infty)$, and width β. We assume periodic boundary conditions in the y direction. Boundary conditions in the x direction are immaterial because $L\to\infty$. In what follows it is actually convenient to impose Dirichlet boundary conditions in the x direction. Equation (7) would then be corrected by (L independent) boundary terms. If we consider an arbitrary manifold Γ that can be obtained from Γ_0 by a conformal transformation $w=f(z)$ then $\ln Z_\Gamma$ is given by $\ln Z$ on Γ_0 with a metric $\rho=|\partial f/\partial Z|^2$. Thus[9]

$$\ln Z_{\Gamma_0}-\ln Z_\Gamma =(c/48\pi)\int d^2x\,[\tfrac{1}{2}(\partial_a \ln\rho)^2+\mu^2(\rho-1)].$$

Consider now $W=e^{-2\pi z/\beta}$. This maps Γ_0 onto Γ, the annulus of outer radius 1, inner radius $e^{-2\pi L/\beta}\to 0$, with Dirichlet boundary conditions. (Note that the boundaries of Γ_0 at $y=\pm\beta/2$ are mapped onto the same line in Γ.) We expect $\ln Z_\Gamma$ to remain finite as $L\to\infty$ (Γ simply becomes the unit disk with vanishing conditions at the origin and its boundary). So we conclude that

$$\lim_{L\to\infty}\frac{\ln Z_{\Gamma_0}}{L}=\text{const}\times\beta+\frac{c\pi}{6\beta}.$$

The first term, proportional to the area, is nonuniversal (depends on μ^2), but the second *is* universal, depending only on the conformal anomaly, c.

We can immediately check this formula by applying it to the Gaussian model with $c=1$. The simplest way of doing this is to use the equivalence of the classical partition function on a strip of width β with the one-dimensional quantum partition function at temperature $T=1/\beta$. Thus

$$\frac{\ln Z_G}{L}=-\frac{F}{TL}=-\frac{\epsilon_0}{T}-\int_{-\infty}^{\infty}\frac{dp}{2\pi}\ln[1-e^{-|p|/T}]$$
$$=-\frac{\epsilon_0}{T}+\frac{\pi}{6}T. \tag{8}$$

The ground-state energy per unit length is ultraviolet-cutoff dependent but the leading T-dependent part of F is universal. [It determines the heat capacity per unit length, $C/L=(\pi/3)T$.] Another simple check is the Ising model at the critical point ($c=\frac{1}{2}$). The partition function on the strip is that of a one-dimensional system of free fermions[10] with dispersion relation

$$\epsilon(p)=2|\sinh^{-1}\sin p/2|=|p|+O(p^3). \tag{9}$$

In the sector with an even fermion number the allowed values of p are $(n+\frac{1}{2})2\pi/L$ $(n=0,\pm1,\pm2)$ and in the sector with an odd fermion number they are $n2\pi/L$ $(n=0,+-1,+-2\ldots)$. However in the limit $L\to\infty$ sums over discrete moments are replaced by integrals so that this feature can be ignored, giving

$$\frac{\ln Z}{L}=-\frac{\epsilon_0}{T}+\int\frac{dp}{2\pi}\ln(1+e^{-\epsilon(p)/T}). \tag{10}$$

Extracting the term linear in T gives

$$\frac{\ln Z}{L}=-\frac{\epsilon_0}{T}+\int_{-\infty}^{\infty}\frac{dp}{2\pi}\ln[1+e^{-|p|/T|}]+O(T^3); \tag{11}$$

note that the second term is $\ln Z$ for a relativistic *Majorana* fermion (no antiparticle). Evaluating the integral gives

$$\frac{\ln Z}{L}=-\frac{\epsilon_0}{T}+\frac{\pi T}{12}+O(T^3) \tag{12}$$

in agreement with Eq. (3). The equivalent term for the three-state Potts model, tricritical Ising model, etc., can be read off. We see that the conformal anomaly c can be directly measured experimentally!

As a nontrivial application of this result let us consider antiferromagnetic quantum spin chains,

$$H=\sum_{n=1}^{N}P(\mathbf{S}_n\cdot\mathbf{S}_{n+1}),\quad \mathbf{S}_n^2=s(s+1), \tag{13}$$

where P is some polynomial of degree $\leq 2s$. It was argued elsewhere[11] that for choices of P such that H is antiferromagnetic and has gapless excitations, it is described at low energies by the SU(2), $k=2s$ Wess-Zumino σ model [equivalently by the SU(2) Kac-Moody algebra with central charge $k=2s$]. For this model[13] $c=3s/(1+s)$. The spin chain has a relativistic low-energy behavior with some effective (nonuniversal) "speed of light" or "Fermi velocity" v. This enters the universal term in the free energy in a manner determined by dimensional analysis. Thus, the low-temperature heat capacity should behave as

$$\frac{C}{L}=\frac{\pi sT}{(1+s)v}+O(T^3). \tag{14}$$

Note that the $s=\frac{1}{2}$ chain is equivalent, at low energies, to a free boson as argued long ago.[12] However, the higher-spin chains have a specific heat which is, in general, a fractional multiple of that for a free boson, demonstrating that the higher-s models contain interacting bosons. For each value of s there is one choice of polynomial P which renders H integrable[13]:

$$P(\mathbf{S}\cdot\mathbf{S}')=\sum_{l=1}^{2s}2sa_lP_l, \tag{15}$$

where P_l is a projector onto total spin l and $a_l=\sum_{k=1}^{l}1/k$. Comparison of the Bethe-*Ansatz* solution with the proposed critical theory is difficult because one approach gives only the spectrum and the other only the Green's functions. However, the velocity of light and low-temperature heat capacity are known from the Bethe *Ansatz*, allowing a check on the critical theory.[14] These are[15]

$$V=\pi/2 \quad [\text{for all } s],$$

$$\frac{C}{L}=\tfrac{2}{3}T \quad (s=\tfrac{1}{2}),$$

$$=T \quad (s=1),$$

$$=\tfrac{2}{3}T-\frac{2T}{\pi^2}\sum_{n=1}^{2s-1}\int_0^{a_n}\left[\frac{1}{x}\ln(1-x)+\frac{1}{1-x}\ln x\right] \quad (s\geq\tfrac{3}{2});$$

$$a_n=\sin^2[\pi/2(s+1)]/\sin^2[\pi(n+1)/2(s+1)].$$

We see that the exact values for $s=\frac{1}{2}, 1$ agree with our prediction. The indefinite integral cannot be evaluated exactly and so we have calculated it numerically for $s=\frac{3}{2}$ (five significant digits), 2, and $\frac{5}{2}$ (three significant digits), finding agreement with the prediction. (Thus the sum of definite integrals apparently can be done exactly.)

I would like to thank F. D. M. Haldane, E. Lieb, and E. Witten for helpful discussions. This work was supported in part by the Alfred P. Sloan Foundation and by the National Science Foundation under Grant No. PHY80-19754.

[1]A. A. Belavin, A. M. Polyakov, and A. B. Zamolodchikov, J. Stat. Phys. **34**, 763 (1984), and Nucl. Phys. **B241**, 333 (1984).

[2]D. Friedan, Z. Qiu, and S. Shenker, Phys. Rev. Lett. **52**, 1575 (1984), and in *Vertex Operators in Mathematics and Physics*, edited by J. Lepowsky, S. Mandelstam, and I. M. Singer (Springer-Verlag, New York, 1984), p. 419.

[3]V. Knizhnik and A. Zamalodchikov, Nucl. Phys. **B247**, 83 (1984).

[4]I. Affleck, Phys. Rev. Lett. **55**, 1355 (1985).

[5]E. Witten, Commun. Math. Phys. **92**, 455 (1984).

[6]J. L. Cardy, J. Phys. A **17**, L385 (1984), and L957 (1984), and Nucl. Phys. **B240**, [FS12] 514 (1984).

[7]See, for example, A. Polyakov, Phys. Lett. **103B**, 207 (1981).

[8]D. Friedan, in *Recent Advances in Field Theory and Statistical Mechanics*, edited by J.-B. Zuber and R. Stora, Les Houches Summer School Proceedings Session 39 (North-Holland, Amsterdam, 1984), p. 839.

[9]This approach was used to calculate the free energy of the Gaussian model on a strip with Dirichlet or Neumann boundary conditions by M. Lüscher, K. Symanzik, and P. Weisz, Nucl. Phys. **B173**, 365 (1980).

[10]T. D. Schultz, D. C. Mattis, and E. Lieb, Rev. Mod. Phys. **36**, 856 (1964).

[11]I. Affleck, to be published.

[12]A. Luther and I. Peschel, Phys. Rev. B **12**, 3968 (1975).

[13]P. Kulish and E. Sklyanin, in *Integrable Quantum Field Theories*, edited by J. Hietarinta and C. Montonen, Lecture Notes in Physics Vol. 151 (Springer-Verlag, New York, 1982), p. 61; P. Kulish and N. Yu. Reshetikhin, Lett. Math. Phys. **5**, 393 (1981); L. Takhtajan, Phys. Lett. **87A**, 479 (1982); J. Babudjian, Phys. Lett. **90A**, 479 (1982).

[14]For $s=\frac{1}{2}$ this consistency check is well known. See, for example, F. D. M. Haldane, to be published.

[15]J. Babudjian, Nucl. Phys. **B215**, 317 (1982).

1988

Note Added By Author

Since publication of this paper it has been realized that while the integrable higher-spin chains are in the $k=2s$ universality class, generic higher-spin Hamiltonians, including the realistic Heisenberg Hamiltonian, are in the $k=1$ universality class, for all half-integer s (they are massive for integer s).

J. Phys. A: Math. Gen. 19 (1986) L779–L784. Printed in Great Britain

LETTER TO THE EDITOR

Conformal invariance and the phase transition of a spin chain with three-spin interaction

M Kolb†‡ and K A Penson‡

† Laboratoire de physique de la matière condensée, Ecole Polytechnique, 91128 Palaiseau Cedex, France
‡ Institut für Theorie der kondensierten Materie, Freie Universität Berlin, Arnimallee 14, 1000 Berlin 33, West Germany

Received 11 June 1986

Abstract. The ground-state properties of a quantum spin chain with three-spin couplings are investigated using finite-size calculations. The results are compared with the predictions of conformal invariance for finite systems at the critical point. The critical exponents η_E and η_S of energy density and spin correlation functions are calculated and estimated to be $\eta_E \simeq 1.30$ and $\eta_S \simeq 0.25$. The results suggest that this model is conformally invariant with a central charge very close to or equal to one.

Recently, models of systems with multibody interactions have received considerable attention. It is known by now that n-body interactions lead to critical behaviour with n-dependent universality classes. In two dimensions (2D) cases in point are the exactly solvable Baxter ($n = 2, 4$) (Baxter 1972) and the Baxter–Wu ($n = 3$) (Baxter and Wu 1973) models.

To study phase transitions systematically as a function of the multiplicity n of interactions a spin model with anisotropic n-spin couplings has been introduced (Penson *et al* 1982, Turban 1982). Here we consider the 1D quantum version (Suzuki 1976) of this model with the Hamiltonian

$$H = -J_n \sum_{l=1}^{L} s_{l+1}^x s_{l+2}^x \ldots s_{l+n}^x - h \sum_{l=1}^{L} s_l^z \tag{1}$$

where s^x and s^z are components of spin-$\frac{1}{2}$ operators. Self-duality determines the value of the critical ratio h/J_n exactly, $(h/J_n)_c = 2^{-(n-1)}$. The energy and the transverse field h are measured in units of J_n and we furthermore set $J_n = 1$.

Since the introduction of this model there have been several extensions, including the use of Potts variables, coupling through other spin components and different and competing multiplicities (Turban and Debierre 1982, Penson 1984, Turban 1985, Kolb and Penson 1985). For $n = 3$ there have been attempts to determine the character of the transition. It is generally agreed (Penson *et al* 1982a, b, Turban 1982, Debierre and Turban 1983, Maritan *et al* 1984) that there is a second-order phase transition but there is no consensus as yet concerning the precise values of the critical exponents. For example, the estimates of the exponent ν vary in the range 0.72–0.77 depending on the method used (Igloi *et al* 1983, 1986). The purpose of this letter is to study the finite-size behaviour of this model in the light of the recently developed apparatus based on conformal invariance (Luck 1982a, b, c, Cardy 1986). The question of conformal invariance is particularly intriguing for the present model as the ordered ground state of its 2D classical version is anisotropic in space.

0305-4470/86/130779+06$02.50

The developments relating conformal invariance and finite systems started with a remarkably simple formula linking finite-size amplitudes to critical exponents. The correlation function exponent η of an infinite 2D classical isotropic system at criticality and the amplitude A of the inverse correlation length, $\xi^{-1} = A/L$, of the same system in a finite strip of width L are related through (Pichard and Sarma 1981, Luck 1982a, b, c, Derrida and de Sèze 1982, Nightingale and Blöte 1983, Cardy 1984a, b)

$$A = \pi\eta. \tag{2}$$

It has been shown that for quantum models (Penson and Kolb 1984) somewhat weaker relations hold. For different operators X and Y the ratios of their respective amplitudes A_X and A_Y are universal and equal to the ratios of their exponents η_X and η_Y

$$A_X/A_Y = \eta_X/\eta_Y. \tag{3}$$

For quantum systems the inverse correlation length is equal to the corresponding gap in the energy spectrum, $\xi_\alpha^{-1} = E_{1\alpha} - E_0$, $\alpha = X, Y$. E_0 is the ground-state energy. In actual calculations one uses for X and Y the spin and energy density operators, respectively (Penson and Kolb 1984, Alcaraz *et al* 1985, Burkhardt and Guim 1985, Guimaraes and Drugowich de Felicio 1986, von Gehlen *et al* 1986, von Gehlen and Rittenberg 1986). For quantum systems equation (3) replaces equation (2) because the Hamiltonian can be multiplied by an arbitrary overall factor without changing its critical properties. This ambiguity is removed if the quantum system is conformally invariant. The spectrum of single-particle excitations then has the form $\varepsilon(k) \equiv E(k) - E_0 = v_s k$ with a sound velocity $v_s = 1$ (Blöte *et al* 1986). In this case equation (2) holds for quantum systems as well. Alternatively, when $v_s \neq 1$ equation (2) takes the form

$$A = \pi\eta v_s. \tag{4}$$

Conformal invariance actually tells us much more than this (Cardy 1986). The structure of higher-lying energy levels is given by the formula (Cardy 1986, Rittenberg 1986)

$$\begin{aligned} &\varepsilon_\alpha(k) = (2\pi/L)(x_\alpha + r + x'_\alpha + r')v_s \qquad r, r' = 0, 1, 2, \ldots \\ &k = (2\pi/L)(x_\alpha + r - x'_\alpha - r') \end{aligned} \tag{5}$$

for periodic boundary conditions where, in our case, $x_\alpha = x'_\alpha = \eta_\alpha/4$. α labels the different symmetries of the excitations—here we distinguish between spin- (η_S)- and energy- (η_E)-type excitations. Equation (4) follows from equation (5) setting $r = R' = 0$. The x_α are anomalous dimensions of irreducible representations of an operator algebra. For unitary theories the characteristic central charge or conformal anomaly c, as well as the x_α, is quantised ($c < 1$) (Belavin *et al* 1984, Friedan *et al* 1984, Huse 1984). Additional consequences of the conformal invariance have recently been analysed (Itzykson *et al* 1986, Saleur 1986).

Now we want to apply these concepts to the Hamiltonian system given by equation (1). The transition separates a disordered phase from an ordered phase with a fourfold degenerate ground state. The strategy is to calculate critical exponents by conventional finite-size methods and compare the results with the predictions of the amplitude exponent relations, equation (4). Equation (5), furthermore, provides a number of consistency tests.

We have diagonalised the Hamiltonian matrix of equation (1) numerically using standard methods for periodically bounded chains up to length $L = 15$. The Lanczos tridiagonalisation scheme has been used and the symmetries of the Hamiltonian

1988

Note: The authors have pointed out that the originally published version contains an error three lines below eq. (5), and that this line should read:

.... Equation (4) follows from equation (5) setting $r = r' = 0$.

were explored. The Hamiltonian commutes with the translation operator T (which shifts the chain by one lattice spacing) and with the parity operators $P_f = \Pi_{m=1}^{L/3} s_{3m+f-2}^x s_{3m+f-1}^x$, $f = 1, 2, 3$, which satisfy $P_f P_{f+1} = P_{f+2}$ (modulo 3). These parities do not commute with the translation operator, but the projection operator $P = (1 + P_1 + P_2 + P_3)/4$ does. Accordingly, the states can be classified by their wavevector and into either singlet or triplet states. The triplet states have wavevectors k and $k \pm 2\pi/3$. This can be seen directly from applying the operator $P^{\pm} = \Sigma_{f=1}^{3} e^{\pm 2\pi i f} P_f$ to a state with a given wavevector k, $|k\rangle = \Sigma_{l=1}^{L} e^{2\pi i k l} T^l |0\rangle$. $P^{\pm}|k\rangle$ yields—if it does not vanish—a vector of wavevector $k \pm 2\pi/3$. For finite systems the classification in a singlet subspace and a triplet subspace only holds for $L = 3, 6, 9, \ldots$. Therefore we restrict our study to these values.

We first use the phenomenological renormalisation group (PRG) (Barber 1983) to determine the energy density correlation function exponent $\eta_E = 4 - 2/\nu$ from ν (the correlation length exponent) and then the spin correlation function exponent η_S from η_E and the amplitude ratios, equation (3). The excitations of energy type are singlets (as is the ground state with wavevector $k = 0$) and the excitations of spin type are triplets. The PRG is performed between sizes L and $L - 3$ for $L = 6, 9, 12, 15$. The amplitude ratios for η_S are estimated from size L at the fixed point of the PRG. In figure 1 the calculated values for η_E and η_S are plotted as a function of $1/L$. They yield—if extrapolated to $L \to \infty$—the estimates $\eta_E \simeq 1.30$ and $\eta_S \simeq 0.25$.

In order to calculate the exponents directly using conformal invariance, we first have to calculate the sound velocity v_s. This is done from equation (5) (von Gehlen

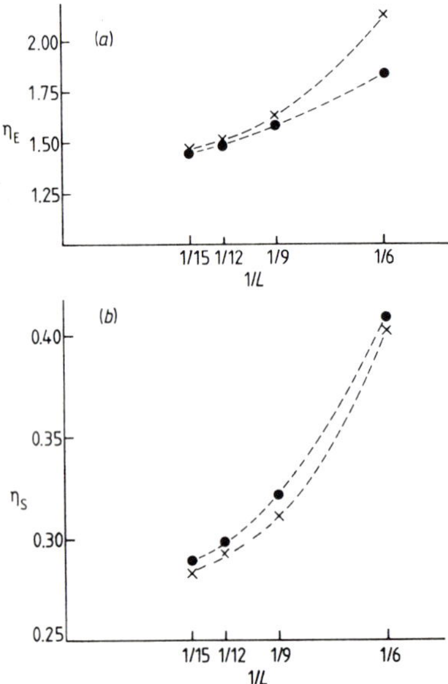

Figure 1. Critical exponents η_E and η_S against $1/L$ calculated from the PRG (sizes L and $L-3$) and using the amplitude exponent relations. In (a) the exponent $\eta_E = 4 - 2/\nu$ is calculated (i) from ν obtained by the PRG (●) and (ii) using $\varepsilon_E(0) = (2\pi/L)(\eta_E/2)v_s$ (×) where v_s is obtained from $\varepsilon_S(2\pi/L) - \varepsilon_S(0) = (2\pi/L)v_s$. In ($b$) the exponent η_S is calculated (i) from η_E and the amplitude ratio $A_S/A_E = \eta_S/\eta_E$ (●) and (ii) is obtained from $\varepsilon_S(0) = (2\pi/L)(\eta_S/2)v_S$ with the same V_s as in (a) (×).

and Rittenberg 1986, Cardy 1986):

$$\varepsilon_\alpha(k=2\pi/L)-\varepsilon_\alpha(k=0)=(2\pi/L)v_s \tag{6}$$

both for energy (singlet) and for spin (triplet) excitations. In this and all subsequent analyses of the energy spectrum, the ratio (h/J_3) was set to the exact critical value of the infinite system, $(h/J_3)_c=0.25$. The results for v_s are presented in figure 2. Both the series from the singlet and the triplet levels appear to converge towards a value of $v_s \simeq 0.42$. This is our first indication that the system is conformally invariant.

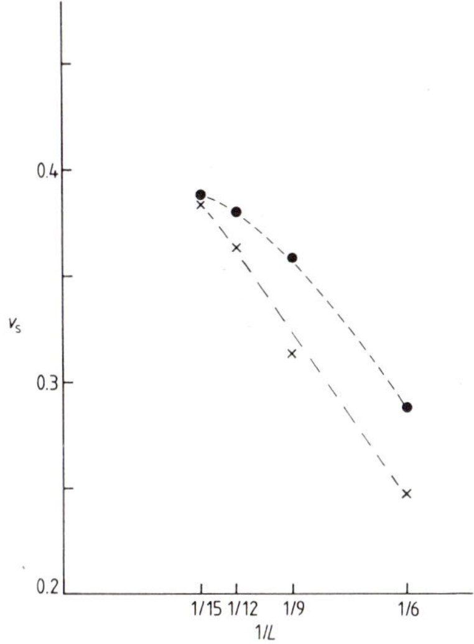

Figure 2. The sound velocity v_s of a conformally invariant system can be determined from $\varepsilon_\alpha(2\pi/L)-\varepsilon_\alpha(0)=(2\pi/L)v_s$ where α stands either for triplet (spin) (●) or singlet (energy) (×) excitations. The finite-size estimates at the critical point $h/J_3=0.25$ are shown for both types of excitation. They converge towards a limiting value $v_s \simeq 0.42$.

The central charge c, which for $c<1$ determines the quantised set of x_α can be determined from the finite-size correction of the ground-state energy (Blöte *et al* 1986, Affleck 1986):

$$E_0=[Le_0-\pi c/6L]v_s \tag{7}$$

where e_0 is the energy density of the infinite system. In order to eliminate e_0 the difference $E_0^L/L-E_0^{L'}/L'$ with $L'=L-3$ is calculated. For v_s the L-dependent values shown in figure 2 are used, both from the singlet and the triplet spectrum. The resulting series for the central charge c are shown in figure 3. Both curves suggest that c is very close to 1.

Higher-lying energy levels of the three-spin system can be compared with the predictions of equations (5) as well. Figure 4 shows all the low-lying excitations up to $\varepsilon_L \simeq 6/L$. We have selected all singlet and triplet levels with wavevectors $k=0$ and $k=2\pi/L$. The extrapolation of the finite-L estimates are consistent with the theoretical values of equation (5). Levels above $L\varepsilon_L \simeq 6$ are difficult to analyse because the convergence is much more slow and because the Lanczos method becomes less reliable.

We conclude that the spectrum of the quantum Ising model with three-spin interaction is consistent with predictions of conformal invariance, equation (5). Using this

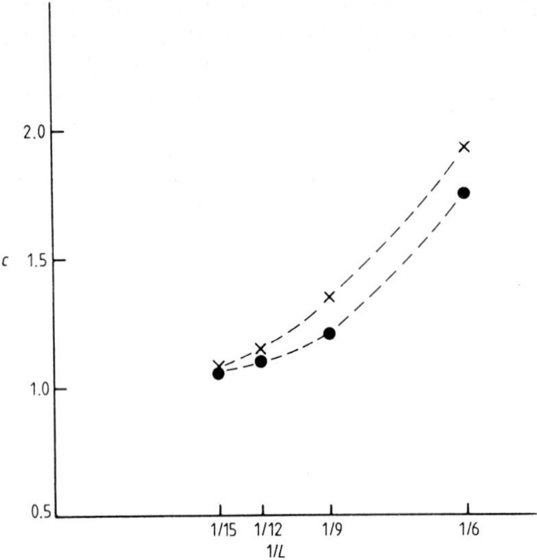

Figure 3. The central charge c calculated from the finite-size corrections to the ground-state energy. Equation (7) is used for two sizes, L and $L-3$, and v_s is taken from figure 2 (●, spin), (×, energy). The $L \to \infty$ estimate for c is very close to $c = 1$.

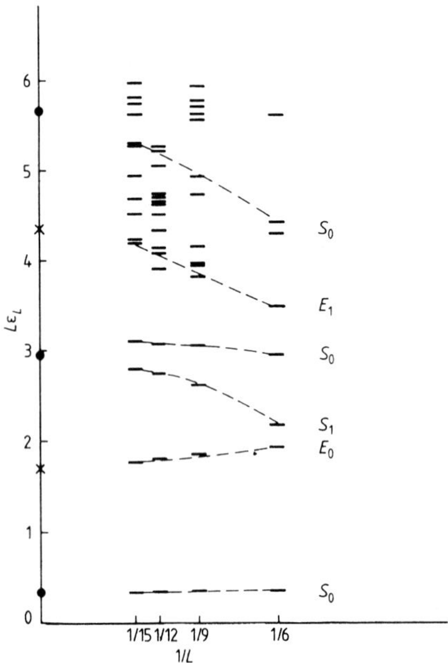

Figure 4. Spectrum of all low-lying excitations $L\varepsilon_L$ ($L\varepsilon_L < 6$) at the critical point. The broken lines connect the singlet (energy) excitations with wavevector $k = 0$, $2\pi/L(E_0, E_1)$ and the triplet (spin) excitations with $k = 0$, $2\pi/L(S_0, S_1)$. The levels predicted from equation (5) for r, $r' = 0$, 1 are indicated for singlet (×) and triplet (●) excitations. The value $v_s = 0.42$ is used, as estimated from extrapolating the finite L values of figure 2 to $L = \infty$.

fact the exponents η_E and η_S are calculated. The value of η_E is consistent with the direct estimate from the PRG. The central charge c is estimated to be very close to 1 (multispin models which are known to have $c = 1$ are the Baxter and the Baxter-Wu models). While the convergence of the data is sufficient to draw qualitative conclusions, it is not accurate enough to clearly identify the universality class for this model.

We have benefited from interesting discussions with G von Gehlen, V Rittenberg, H Saleur and T T Truong. The Deutsche Forschungsgemeinschaft has supported this project.

Note added. After completion of this work, we received a preprint by F C Alcaraz and M N Barber on the three-spin Ising model. Their results differ from ours numerically because they calculate v_s from excitations in the $k = 0$ sector, but their conclusions are very similar to ours.

References

Affleck I 1986 *Phys. Rev. Lett.* **56** 746
Alcaraz F C, Drugowich de Felicio J R, Köberle R and Stick F 1985 *Phys. Rev.* B **32** 7469
Barber M N 1983 *Phase Transitions and Critical Phenomena* vol 8, ed C Domb and J L Lebowitz (New York: Academic) p 146
Baxter R J 1972 *Ann. Phys., NY* **70** 193
Baxter R J and Wu F Y 1973 *Phys. Rev. Lett.* **21** 1294
Belavin A A, Polyakov A M and Zamolodchikov A B 1984 *Nucl. Phys.* B **241** 333
Blöte H W J, Cardy J L and Nightingale M P 1986 *Phys. Rev. Lett.* **56** 742
Burkhardt T W and Guim T 1985 *J. Phys. A: Math. Gen.* **18** L33
Cardy J L 1984a *J. Phys. A: Math. Gen.* **17** L385, L961
—— 1984b *Nucl. Phys.* B **240** 514
—— 1986 *Phase Transitions and Critical Phenomena* vol 11, ed C Domb and J L Lebowitz (New York: Academic) in press
Debierre M and Turban L 1983 *J. Phys. A: Math. Gen.* **16** 3571
Derrida B and de Sèze L 1982 *J. Physique* **43** 475
Friedan D, Qui Z and Shenker S H 1984 *Phys. Rev. Lett.* **52** 1575
Guimaraes L G and Drugowich de Felicio J R 1986 *J. Phys. A: Math. Gen.* **19** L341
Huse D A 1984 *Phys. Rev.* B **30** 3908
Igloi F, Kapor D V, Solyom J and Skrinjar M 1983 *J. Phys. A: Math. Gen.* **16** 4067
Igloi F, Kapor D V, Skrinjar M and Solyom J 1986 *J. Phys. A: Math. Gen.* **19** 1189
Itzykson C, Saleur H and Zuber J B 1986 *Preprint*
Kolb M and Penson K A 1985 *Phys. Rev.* B **31** 3147
Luck J M 1982a *J. Phys. A: Math. Gen.* **15** L169
—— 1982b *Nucl. Phys.* B **210** 111
—— 1982 *J. Physique* **42** L275
Maritan A, Stella A and Vanderzande C 1984 *Phys. Rev.* B **29** 519
Nightingale M P and Blöte H W J 1983 *J. Phys. A: Math. Gen.* **16** L657
Penson K A 1984 *Phys. Rev.* B **29** 2404
Penson K A, Jullien R and Pfeuty P 1982a *J. Physique* **43** 582
—— 1982b *Phys. Rev.* B **26** 6334
Penson K A and Kolb M 1984 *Phys. Rev.* B **29** 2854
Pichard L and Sarma G 1981 *J. Phys. C: Solid State Phys.* **14** L127, L617
Rittenberg V 1986 *Proc. Int. Symp. on Conformal Groups and Structures, Clausthal* unpublished
Saleur H 1986 *Preprint*
Suzuki M 1976 *Prog. Theor. Phys.* **56** 1454
Turban L 1982 *J. Phys. C: Solid State Phys.* **15** L65
—— 1985 *J. Phys. A: Math. Gen.* **18** 2313
Turban L and Debierre J M 1982 *J. Phys. C: Solid State Phys.* **15** L129
von Gehlen G and Rittenberg V 1986 *J. Phys. A: Math. Gen.* **19** 2439
von Gehlen G, Rittenberg V and Ruegg H 1986 *J. Phys. A: Math. Gen.* **19** 107

Conformal Invariance and the Spectrum of the *XXZ* Chain

Francisco C. Alcaraz,[a] Michael N. Barber,[b] and Murray T. Batchelor
Department of Mathematics, The Faculties, Australian National University, Canberra, Australian Capital Territory 2601, Australia

(Received 21 November 1986)

Numerical solutions of the Bethe-*Ansatz* equations for the eigenenergies of the *XXZ* Hamiltonian on very large chains are used to identify, via conformal invariance, the scaling dimensions of various two-dimensional models. With periodic boundary conditions, eight-vertex and Gaussian model operators are found. The scaling dimensions of the Ashkin-Teller and Potts models are obtained by the exact relating of eigenstates of their quantum Hamiltonians to those of the *XXZ* chain with modified boundary conditions. The irrelevant operators governing the dominant finite-size corrections are also identified.

PACS numbers: 64.60.−i, 05.50.+q, 75.10.Hk, 75.10.Jm

Two ideas unify the theory of critical phenomena in two dimensions. These are conformal invariance[1,2] and the notion that there exist general models to which specific models of physical interest can be related by appropriate transformations.[3-5] Examples of such "central theories" are the Coulomb (lattice) gas[3] and the generalized Gaussian model.[4] While analysis[3-6] of these generalized models can yield detailed information on the critical behavior of the related physical models, including critical exponents, the possible universality classes are not predicted by this formalism.[7]

The universality classes of two-dimensional [or equivalently (1+1)-dimensional] theories are constrained, however, by conformal invariance.[2,8] Within conformal theory, these classes are characterized by a single dimensionless number c, the central charge or conformal anomaly of the associated Virasoro algebra, the irreducible representations of which determine the operator algebra describing the critical behavior. If c is less than unity, unitarity restricts[8] c to the values $c=1-6/m(m+1)$, $m=3,4,5,\ldots$. For such theories, which include the Ising ($c=\frac{1}{2}$) and three-state Potts ($c=\frac{4}{5}$) models amongst others, the operator algebra is finite; the anomalous dimensions being given by the Kac formula.[9] The limiting value $c=1$ is of particular interest. This class includes the four-state Potts model and models[10] such as the eight-vertex and Askin-Teller models exhibiting continuously varying exponents.

In this Letter, we identify the possible operators of $c=1$ theories from a study of the spectrum of the one-dimensional quantum *XXZ* model:

$$H_{XXZ}=-\frac{\gamma}{2\pi\sin\gamma}\sum_{i=1}^{L}(\sigma_i^x\sigma_{i+1}^x+\sigma_i^y\sigma_{i+1}^y+\Delta\sigma_i^z\sigma_{i+1}^z). \tag{1}$$

Here $\sigma^x,\sigma^y,\sigma^z$ are Pauli matrices and $\Delta=-\cos\gamma$, $\gamma\in[0,\pi]$. In the bulk limit, $L\to\infty$, this Hamiltonian is massless with critical exponents varying continuously with Δ.[10] The prefactor in (1) is included to ensure that the resulting equations of motion are conformally invariant.[11] Its precise value can be inferred[12] from the known[13] energy-momentum dispersion relation. The value of the conformal anomaly follows[14] from the behavior of the ground-state energy $E_0(\Delta,L)$ for periodic boundary conditions as $L\to\infty$. Analytical[12] and numerical results[15] yield

$$E_0(\Delta,L)/L=e_\infty-\pi/6L^2+o(L^{-2}), \tag{2}$$

confirming[16] $c=1$.

The major advantage of the *XXZ* model over other possible Hamiltonians with $c=1$ is that its spectrum on a finite lattice can be calculated by the Bethe *Ansatz*. The Bethe-*Ansatz* solution of the infinite *XXZ* chain is well known.[10,17] Surprisingly, the method has received relatively little attention as a *numerical* procedure for computation of the spectra of finite chains.[15,18] We have found[19,20] that numerical solution of Bethe-*Ansatz* equations is feasible for quite large lattices up to $L\sim512$ not only for the ground state[15,18] but also for various excited states. In addition, the method can be extended[20] to yield eigenvalues of (1) subject to the generalized boundary condition

$$\sigma_{L+1}^x\pm i\sigma_{L+1}^y=e^{\pm i\varphi}(\sigma_1^x\pm i\sigma_1^y),\quad \sigma_{L+1}^z=\sigma_1^z, \tag{3}$$

where φ is an arbitrary angle. For all φ, H can be block-diagonalized into disjoint sectors labeled by $n=\sum\sigma^z/2$.

Our subsequent analysis and identification of critical operators rely on the predictions[2,21] of conformal invariance concerning the spectrum of a critical quantum Hamiltonian in a finite strip. The key results can be summarized as follows: To each primary operator $\mathcal{O}$, with scaling dimension x and spin s, in the operator algebra there exists a set of states in the quantum Hamiltonian. For a chain with periodic boundary conditions, the energy and momentum of these states are given by

$$E_{n,n'}(L)=E_0(L)+2\pi(x+n+n')L^{-1}+o(L^{-1}),\quad n,n'=0,1,\ldots, \tag{4a}$$

$$P_{n,n'}=2\pi(s+n-n')/L,\quad n,n'=0,1,\ldots, \tag{4b}$$

respectively. From these relations and our eigenvalue data we are able to estimate with very high precision various anomalous dimensions for any coupling. Typical estimates of several anomalous dimensions are shown for $\gamma=\pi/6$ ($\Delta=-\sqrt{3/2}$) in Table I. Similar accuracy is possible for other values of γ except near $\gamma=0$, where it is necessary to extend the calculations to larger lattices because of the slower convergence with L.

Let us discuss initially our findings for even L and periodic boundary conditions ($\phi=0$). The lowest-energy state in each sector yields, through (4), a set of operators $\mathcal{O}_{n,0}$ with scaling dimensions $x_{n,0}=n^2x_p$ where $x_p=(\pi-\gamma)/2\pi$. The operators $\mathcal{O}_{1,0}$ and $\mathcal{O}_{2,0}$ correspond to the polarization and energy operators of the eight-vertex model.[5] This identification of $\mathcal{O}_{1,0}$ confirms that made from an analytical treatment[12] of the Bethe-*Ansatz* equations. We have also obtained the lower levels in the conformal blocks associated with these operators, in accord with (4). Stringlike solutions of the Bethe-*Ansatz* equations yield excited states corresponding to further operators $\mathcal{O}_{n,m}$ with dimensions

$$x_{n,m}=n^2x_p+m^2/4x_p,\quad n,m=0,1,2,\ldots. \tag{5}$$

These operators are the analogs of the Gaussian-model operators[4,5] composed of a spin-wave excitation of index n and a "vortex" excitation of vorticity m. Hence, we can identify the operator $\mathcal{O}_{0,1}$ as the crossover operator of the eight-vertex model or equivalently the energy operator of the Ashkin-Teller model. The operator $\mathcal{O}_{0,2}$ is irrelevant for $\gamma>0$ but becomes marginal at $\gamma=0$. As we shall see, this operator is the main determinant of finite-size corrections for both the XXZ and Ashkin-Teller models as well as for the logarithmic corrections that appear in the four-state Potts model. Finally, the spectrum contains a state which gives $x=2$ for all γ. The associated operator is thus marginal corresponding to the four-spin coupling of the eight-vertex model. Its presence in $c=1$ theories results in the appearance of a line of critical points, the operator itself governing motion along the critical line.[4-6]

The identification of $\mathcal{O}_{0,1}$ as the Ashkin-Teller energy operator can be confirmed directly since it is possible to derive[20] the eigenvalues of the critical quantum Ashkin-Teller model[22] with four-spin coupling λ on an M-site chain exactly from those of a $2M$-site XXZ chain with $\Delta=-\lambda$ subject to (3) for particular values of ϕ. High-precision estimates of the anomalous dimensions of the Ashkin-Teller model then follow from (4). In addition to identifying $x_{0,1}$, we find[20] that the difference in ground-state energy for $\varphi=\pi$ and $\varphi=0$ gives the mass-gap amplitude corresponding to the Ashkin-Teller polarization operator with dimension $x_{0,1/2}$, the indices in (5) being extended to fractional values. Similarly, we are able to locate all of the parafermions present in the Ashkin-Teller model. Our numerical results suggest that the dimension of the spin-$\frac{1}{4}$ parafermion is $x_{1,1/4}$, which clarifies previous results.[23]

To obtain the order and disorder operators of the Ashkin-Teller model it is necessary to apply the boundary condition, $\sigma_{L+1}^x=\sigma_1^x$, $\sigma_{L+1}^y=-\sigma_1^y$, $\sigma_{L+1}^z=\sigma_1^z$, to the XXZ chain. The "magnetic" [$Z(2)$-charged] sector of the Ashkin-Teller model is then located in the ground-state sector of the XXZ model. While the Bethe-*Ansatz* is no longer applicable, the required eigenvalues can be computed easily by the Lanczos method for chains up to $L=20$. Extrapolation of the resulting gaps gives, for all γ, the values $\frac{1}{8}$ and $\frac{5}{8}$ for the dimensions of the magnetic order and spin-$\frac{1}{2}$ parafermion operators of the Ashkin-Teller model, respectively.

The generalized boundary conditions (3) are also of interest because they connect the XXZ model to the q-state Potts model. Specifically, the eigenvalues in the ground-state sector of the quantum Hamiltonian[24] of the critical q-state Potts model on a lattice of M sites can be related[20] exactly to eigenvalues of a $2M$-site XXZ chain with $\Delta=-\cos\gamma=-\frac{1}{2}\sqrt{q}$ and $\varphi=2\gamma$. For $\varphi\neq0$, (2) is no longer valid. The coefficient of the L^{-2} term becomes $-\pi c(\varphi)/6$, where our results strongly suggest that $c(\varphi)=1-12x_{0,\varphi/2\pi}$. Setting $\varphi=2\gamma$ reproduces the value of the conformal anomaly of the Potts model,[14,25] namely, $c(q)=1-6\gamma^2/\pi(\pi-\gamma)$. New higher-energy states also appear in the XXZ spectrum corresponding to (new) Potts operators. In particular, the Potts energy operator (dimension x_ε) is associated with a string state in the ground-state sector of the XXZ model. Similarly, the eigenstates of the Potts Hamiltonian associated with the

TABLE I. Finite-lattice estimates of anomalous dimensions of the XXZ chain for $\gamma=\pi/6$, $x_p=\frac{5}{12}$, $x_{n,m}=n^2x_p+m^2/4x_p$.

L \ (n,m)	$(0,\frac{1}{2})$	$(1,0)$	$(1,\frac{1}{4})$	$(0,1)$	Marginal
8	0.155 19	0.397 93	0.436 06	0.4791	1.5699
16	0.152 17	0.405 94	0.444 60	0.5074	1.7522
32	0.151 04	0.410 63	0.448 97	0.5289	1.8538
64	0.150 55	0.413 27	0.451 28	0.5453	1.9126
128	0.150 30	0.414 74	0.452 54	0.5579	1.9477
256	0.150 17	0.415 57	0.453 24	0.5677	1.9690
$x_{n,m}$	0.15	0.416	0.454 16	0.6	2.0

order and spin-$1/q$ parafermion operators (dimensions x_σ and x_{pf}) can be obtained from the XXZ Hamiltonian by application of (3) with $\phi=\pi$ and $\phi=2\pi/q$, respectively. As a result, we are able to compute eigenenergies of the Potts Hamiltonian on very large lattices for any q. In contrast, conventional finite lattice calculations[24] are restricted to $M \le 10$. Table II lists the resulting estimates of x_ε, x_σ, and x_{pf} for $q=4$. Cardy[26] has recently shown that these estimates should converge as $1/\ln M$. Allowing for such a convergence rate yields the "extrapolated" values quoted in Table II. These are in excellent agreement with the expected results. More generally, for arbitrary $q=4\cos^2\gamma$, our numerical estimates[20] are in full agreement with the expressions

$$x_\varepsilon=(\pi+2\gamma)/2(\pi-\gamma),$$

$$x_\sigma=(\pi^2-4\gamma^2)/8\pi(\pi-\gamma),$$

$$x_{pf}=(\pi-\gamma)/2\pi+(\pi^2-q^2\gamma^2)(\pi-\gamma),$$

thereby confirming predictions from Coulomb-gas calculations[27] and the identifications made by Dotsenko.[25]

We have also investigated numerically the corrections to (4a). These corrections arise[21] since a lattice Hamiltonian such as (1) deviates from the conformally invariant Hamiltonian H^* of the continuum theory by terms involving irrelevant operators, i.e.,

$$H=H^*+\sum_{j=1}^{n} a_j O_j+\dots, \qquad (6)$$

where a_j are coupling constants. Among these operators are those associated with the conformal block of the identity operator O_I, the leading operator of which has dimension $x_I=4$. For the periodic case ($\varphi=0$), our numerical results indicate that the dominant correction terms in (4a) can be accounted for by considering in addition to O_I the operator $O_{0,2}$, with scaling dimension $\bar{x}\equiv x_{0,2}=2\pi/(\pi-\gamma)$. It is, however, necessary to go beyond the first-order perturbation calculation of the corrections performed by Cardy.[21] More generally, we obtain[20] for the eigenenergy, $E_{n,m}$, associated with dimension $x_{n,m}$, the expansion

$$\frac{E_{n,m}}{L}\approx e_\infty+2\pi L^{-2}\left(x_{n,m}-\frac{1}{12}+\sum_{\substack{k=0\\(k,l)\neq(0,0)}}^{\infty}\sum_{l=0}^{\infty}\alpha_{k,l}L^{-k(\bar{x}-2)-2l}\right), \qquad (7)$$

where e_∞ is the ground-state energy per particle of the infinite lattice and the periodic ground state corresponds to $(n,m)=(0,0)$ and $x_{0,0}=0$. The coefficients $\alpha_{k,l}$ depend[20,21] on the couplings a_i and the operator-product expansion coefficients $c_{i,j,k}$. In particular, $\alpha_{1,0}$ depends linearly on $c_{n,m;n,m;0,2}$. In the Gaussian model,[4] this coefficient vanishes unless $(n,m)=(0,1)$. It appears that this selection rule remains true for the XXZ chain. Specifically, we find that the corrections to the amplitude corresponding to the energy operator of the Ashkin-Teller model, the operator $O_{0,1}$ in the XXZ model, are of the form $L^{-2}(b_0L^{-2\gamma/(\pi-\gamma)}+b_1L^{-2})$ while for all the other operators leading corrections are $L^{-2}(b_0\times L^{-4\gamma/(\pi-\gamma)}+b_1L^{-2})$. In this case the dominant correction term switches at $\gamma=\pi/3$ which accounts for the behavior found by Hamer.[12] At $\gamma=\pi/3$, the leading correction is $O((\ln L)/L^4)$. It is also interesting to observe that as $\gamma\to 0$, $x_{0,2}\to 2$ and more powers in (7) corresponding to higher values of k become important. This gives a simple visualization of the reason for the poor convergence rate observed in finite lattice calculations around the four-state Potts point ($\gamma=0$). Strictly at $\gamma=0$ the number of equally important corrections tends to infinity and the original couping constants in (6) renormalize giving rise to the logarithmic corrections.[26]

We have analyzed the correction terms for the generalized boundary conditions. In particular, for the Potts model with $q>2$, the leading corrections for the energy and order-parameter amplitudes are of order $L^{-(2-x')}$,

TABLE II. Estimates of anomalous dimensions of the four-state Potts model.

M	x_ε	x_σ	x_{pf}
4	0.771 229	0.143 407	0.459 772
8	0.722 621	0.139 056	0.474 226
16	0.684 992	0.136 751	0.483 573
32	0.657 247	0.135 233	0.490 105
64	0.636 473	0.134 106	0.494 965
128	0.620 490	0.133 217	0.498 747
256	0.607 858	0.132 493	0.501 786
512	0.597 638		
Extrapolated	0.501 ± 0.002	0.126 ± 0.002	0.529 ± 0.003
Exact	0.5	0.125	0.531 25

where $x' = 2(\pi+\gamma)/(\pi-\gamma)$. This value is precisely that of the second thermal exponent of the Potts model.[25,28]

It is a pleasure to acknowledge profitable discussions with Professor R. J. Baxter and Dr. C. J. Hamer. This work was supported in part by the Australian Research Grants Scheme, the Commonwealth Department of Education, and by Fundação de Amparo à Pesquisa do Estado de São Paulo, Brazil.

---

[a]Permanent address: Departamento de Física, Universidade de São Carlos, CP 616, 13560 São Carlos, S.P., Brasil.

[b]Address from January 1, 1987: Institute for Theoretical Physics, University of California, Santa Barbara, Santa Barbara, CA 93106.

[1]A. M. Polyakov, Pis'ma Zh. Eksp. Teor. Fiz. **12**, 538 (1970) [JETP Lett. **12**, 381 (1970)]; A. A. Belavin, A. M. Polyakov, and A. B. Zamolodchikov, Nucl. Phys. **B241**, 333 (1984).

[2]For a recent review see J. L. Cardy, in "Phase Transitions and Critical Phenomena, edited by C. Domb and J. L. Lebowitz (Academic, New York, to be published).

[3]See, e.g., B. Nienhuis, J. Stat. Phys. **34**, 731 (1984), and references cited therein.

[4]L. P. Kadanoff and A. C. Brown, Ann. Phys. (N.Y.) **121**, 318 (1979).

[5]M. P. M. den Nijs, Phys. Rev. B **23**, 6111 (1981).

[6]H. J. F. Knops, Ann. Phys. (N.Y.) **128**, 448 (1980).

[7]As emphasized by Nienhuis (Ref. 3), these calculations rely on crucial but unproven renormalization-group and universality assumptions; only if these are valid are the results exact.

[8]D. Friedan, Z. Qui, and S. Shenker, Phys. Rev. Lett. **52**, 1575 (1984).

[9]V. G. Kac, in *Group Theoretical Methods in Physics*, edited by W. Beiglbock and A. Bohm, Lecture Notes in Physics Vol. 94 (Springer-Verlag, New York, 1979).

[10]See, e.g., R. J. Baxter, *Exactly Solved Models in Statistical Mechanics* (Academic, New York, 1982).

[11]G. v. Gehlen, V. Rittenberg, and H. Ruegg, J. Phys. A **19**, 107 (1986).

[12]C. J. Hamer, J. Phys. A **18**, L1133 (1985), and **19**, 3335 (1986).

[13]J. D. Johnson, S. Krinsky, and B. M. McCoy, Phys. Rev. A **8**, 2526 (1973).

[14]H. W. J. Blöte, J. L. Cardy, and M. P. Nightingale, Phys. Rev. Lett. **56**, 742 (1986); I. Affleck, Phys. Rev. Lett. **56**, 746 (1986).

[15]L. V. Avdeev and B.-D. Dörfel, J. Phys. A **19**, L13 (1986).

[16]In general, the coefficient of the L^{-2} term is $-\pi c/6$ (see Ref. 14).

[17]C. N. Yang and C. P. Yang, Phys. Rev. **150**, 321, 327 (1966).

[18]The little work that has been reported is limited to the isotropic case ($\Delta=1$) and concerns mainly the computation of ground-state correlation functions [see, e.g., W. Grieger, Phys. Rev. B **30**, 344 (1984); J. Borysowicz, T. A. Kaplan, and P. Horsch, Phys. Rev. B **31**, 1590 (1985)].

[19]M. T. Batchelor, M. N. Barber, and P. A. Pearce, to be published.

[20]F. C. Alcaraz, M. N. Barber, and M. T. Batchelor, to be published.

[21]J. L. Cardy J. Phys. A **17**, L385, L957 (1984), and Nucl. Phys. **B240 [FS12]**, 514 (1984).

[22]M. Kohmoto, M. den Nijs, and L. P. Kadanoff, Phys. Rev. B **24**, 5229 (1981).

[23]G. v. Gehlen and V. Rittenberg, J. Phys. A (to be published).

[24]See, e.g., C. J. Hamer, J. Phys. A **14**, 2981 (1981). Hamer related the quantum-Hamiltonian Potts model to the *XXZ* chain via standard transformations (see Ref. 10), but did not keep track of the effect of boundary conditions.

[25]Vl. S. Dotsenko, Nucl. Phys. **B235 [FS11]**, 54 (1984).

[26]J. Cardy, J. Phys. A (to be published); F. Woynarovich and P. Eckle, to be published. The amplitude of the correction term is universal. Our results confirm Cardy's predictions.

[27]M. den Nijs, Phys. Rev. B **27**, 1674 (1983); B. Nienhuis and H. J. F. Knops, Phys. Rev. B **32**, 1872 (1985).

[28]B. Nienhuis, J. Phys. A **15**, 199 (1982).

J. Phys. A: Math. Gen. 17 (1984) L961–L964. Printed in Great Britian

LETTER TO THE EDITOR

Finite-size scaling in strips: Antiperiodic boundary conditions

John L Cardy

Department of Physics, University of California, Santa Barbara, California 93106, USA

Received 3 October 1984

Abstract. The exact values of the universal amplitude A relating the correlation length $\xi \sim L/A$ to the strip width L at the critical point of Ising and 3-state Potts models are obtained for the case of antiperiodic, or twisted, boundary conditions. Predictions are also made for the finite-size scaling behaviour of the interfacial tension.

For an infinitely long two-dimensional strip of finite width L, the theory of finite-size scaling (Barber 1983) predicts that the inverse correlation length ξ^{-1} at the critical point of the infinite system should tend to zero like A/L, where the amplitude A is universal, but may depend on the boundary conditions. In a previous letter (Cardy 1984a), we allowed that A is equal to $\pi\eta$ and $\frac{1}{2}\pi\eta_{\parallel}$ for periodic and free boundary conditions respectively, for any isotropic model. The result for the periodic case has been confirmed by exact and numerical calculations in many models (Luck 1982, Derrida and de Seze 1982, Nightingale and Blöte 1983, Privman and Fisher 1984).

In this letter we calculate A for the case of antiperiodic or twisted boundary conditions for two examples where such a boundary condition is appropriate, namely the Ising (Z_2) and 3-state Potts (Z_3) models. First, consider an Ising model defined on the strip $0 \leqslant x \leqslant L$, $-\infty \leqslant y \leqslant +\infty$ with antiperiodic boundary conditions $\sigma(x+L, y) = -\sigma(x, y)$, where $\sigma(x, y)$ is the local value of the spin. This problem may be transformed into one with periodic boundary conditions by inserting an antiferromagnetic seam, that is by changing the sign of the exchange interaction on those bonds which intersect an arbitrarily chosen path which begins at $y = -\infty$ and ends at $y = +\infty$. This is equivalent to inserting two disorder operators μ (dual to σ) at $y = \pm\infty$ (Kadanoff and Ceva 1971). Explicitly,

$$\langle\sigma(x_1, y_1)\sigma(x_2, y_2)\rangle_A = \lim_{\substack{y_1' \to -\infty \\ y_2' \to +\infty}} \frac{\langle\mu(x_1', y_1')\sigma(x_1, y_1)\sigma(x_2, y_2)\mu(x_2', y_2')\rangle}{\langle\mu(x_1', y_1')\mu(x_2', y_2')\rangle} \tag{1}$$

where the correlation functions on the right-hand side are evaluated in the strip with periodic boundary conditions. For $y_2 - y_1 \to \infty$ this quantity should decay like $\exp[-(y_2 - y_1)/\xi]$.

The correlation functions in (1) may also be written in an operator notation, introducing the transfer matrix for the strip $\hat{T} = e^{-\hat{H}a}$. $\hat{H}$ is the infinitesimal y-translation operator in the continuum limit, as the lattice spacing $a \to 0$. Introducing the eigenvalues E_n and corresponding eigenstates $|n\rangle$ of $\hat{H}$, (1) can be rewritten for $|y_2 - y_1| \gg L$ as

$$\frac{\langle 0|\mu|1\rangle \exp[-E_1(y_1 - y_1')]\langle 1|\sigma|2\rangle \exp[-E_2(y_2 - y_1)]\langle 2|\sigma|1\rangle \exp[-E_1(y_2' - y_2)]\langle 1|\mu|0\rangle}{\langle 0|\mu|1\rangle \exp[-E_1(y_2' - y_1')]\langle 1|\mu|0\rangle} \tag{2}$$

where E_1 and E_2 are the lowest eigenvalues of $\hat{H}$ such that the matrix elements $\langle 0|\mu|1\rangle$ and $\langle 1|\sigma|2\rangle$ are non-zero. In writing (2) we have normalised $\hat{H}$ so that $E_0=0$, and suppressed the x dependence. From this expression we see that $\xi^{-1}=E_2-E_1$. Now E_1 gives the correlation length of the $\langle\mu\mu\rangle$ correlation function in the strip, which is the same as that of $\langle\sigma\sigma\rangle$, by duality. Thus $E_1=\pi\eta/L=\pi/4L$ (Cardy 1984a). To evaluate E_2 we may consider the same correlation function as appears in the numerator of (1), but in the limit $y_2-y_1\to\infty$ with $y_1'-y_1$ and $y_2'-y_2$ fixed, and assume that the same intermediate state $|2\rangle$ will dominate. In this limit, we can use the short-distance expansion (Kadanoff and Ceva 1971) $\sigma\mu\sim\psi$ (where ψ is the Ising fermion) to argue that the dependence on y_2-y_1 is the same as that of the correlation function

$$\langle\psi(x_1,y_1)\psi(x_2,y_2)\rangle. \tag{3}$$

In the bulk, it is known (Dotsenko 1984) that this correlation function behaves like $(z_1-z_2)^{-1}$, where $z_j=x_j+\mathrm{i}y_j$. Note that this changes sign when z_1 and z_2 are interchanged, consistent with the fermionic interpretation of ψ. The conformal invariance methods (Cardy 1984a) then imply that in the strip this correlation function decays with an inverse correlation length $E_2=\pi/L$. Thus in this case $\xi^{-1}=3\pi/4L$.

An analogous argument holds for the Z_3 model, for which σ takes the values 1, $\exp(2\pi\mathrm{i}/3)$, $\exp(-2\pi\mathrm{i}/3)$, and we impose twisted boundary conditions $\sigma(x+L,y)=\exp(2\pi\mathrm{i}/3)\sigma(x,y)$. In that case the short-distance product of an order and a disorder operator gives a parafermion ψ (Fradkin and Kadanoff 1980) whose spin is $\frac{1}{3}$. This means that the bulk correlation function $\langle\psi(z_1)\psi(z_2)\rangle$ behaves like $(z_1-z_2)^{-2\Delta}(\bar{z}_1-\bar{z}_2)^{-2\bar{\Delta}}$ where $\Delta-\bar{\Delta}=\frac{1}{3}$. In the Z_3 model, the values of Δ and $\bar{\Delta}$ allowed by conformal invariance are (Dotsenko 1984, Friedan *et al* 1984) 0, $\frac{2}{5}$, $\frac{7}{5}$, 3, $\frac{1}{15}$, $\frac{2}{3}$. We conclude that $\Delta=\frac{2}{5}$, $\bar{\Delta}=\frac{1}{15}$, so that ψ has scaling dimension $x=\Delta+\bar{\Delta}=\frac{7}{15}$. For this case, then, $E_2=14\pi/15L$ and $E_1=4\pi/15L$, so that $\xi^{-1}=2\pi/3L$.

The complete set of predictions for A/π for the three kinds of boundary conditions are shown in table 1. We now compare these results with existing numerical data. Burkhardt and Guim (1984a, b) have studied isotropic Ising models in strips with periodic and free boundary conditions, and the quantum Ising chain with all three types of boundary conditions. Our results agree with theirs. Note that for the quantum model the amplitude A itself is not universal (see Penson and Kolb 1984 and Nightingale and Blöte 1984); but ratios of amplitudes with different boundary conditions should be.

Gehlen *et al* (1984) have considered anisotropic quantum Ising models and quantum Z_3 models. For Ising models they found that A was independent of the boundary conditions in clear contradiction to both our results and those of Burkhardt and Guim. For the Z_3 model they found amplitudes approximately in the ratios $\frac{2}{5}$:1:1 for periodic,free and twisted boundary conditions respectively, in agreement with our exact results.

Table 1. Values of A/π for Ising (Z_2) and 3-state Potts (Z_3) models. The values of $\eta_{\parallel}$ used in the second row come from Cardy (1984b) and McCoy and Wu (1967) for the Z_2 case.

	Z_2	Z_3
Periodic	$\frac{1}{4}$	$\frac{4}{15}$
Free	$\frac{1}{2}$	$\frac{2}{3}$
Twisted	$\frac{3}{4}$	$\frac{2}{3}$

Duality arguments may also be used to predict the beahviour of the interfacial tension Σ at the critical point in a finite system. This quantity may be defined in terms of the difference between the free energies per unit length of strips with periodic and antiperiodic boundary conditions. For the infinite system, Σ is supposed to vanish like $(T-T_c)^\nu$ in two dimensions (Widom 1972). Standard finite-size scaling arguments then imply that $\Sigma \sim B/L$ for a strip of width L at T_c, with B a universal amplitude. Σ may also be calculated in terms of correlation functions of disorder variables

$$\Sigma = -\lim_{y_2-y_1\to\infty} (y_2-y_1)^{-1} \ln\langle\mu(x_1, y_1)\mu(x_2, y_2)\rangle. \tag{3}$$

However, this correlation function is dual to $\langle\sigma\sigma\rangle$, whose behaviour we know (Cardy 1984a). We conclude that $B=\pi\eta$ for this definition of Σ.

An alternative definition of Σ is given by the difference in free energies for the two cases of fixed boundary conditions: (a) when the spins on either side are fixed in the same state; (b) when they are fixed in different states. In that case (3) still holds, but $\langle\mu\mu\rangle$ is now dual to the correlation function $\langle\sigma\sigma\rangle$ evaluated with *free* boundary conditions. In that case, then, $B=\frac{1}{2}\pi\eta_{\|}$. This definition of Σ is natural for the q-state Potts model. For the percolation limit $(q\to 1)$ it may be shown that the probability that the two sides of a strip of width L and length L' are connected by at least one path is

$$P = 1-\exp(-L'\Sigma) \tag{4}$$

in the limit $L'\to\infty$, with the above definition of Σ. With the exact value $\eta_{\|}=\frac{2}{3}$ (Cardy 1984b) we therefore predict that

$$P \sim 1-\exp(-\pi L'/3L) \tag{5}$$

in the limits $L\to\infty$, $L'/L\to\infty$, at the percolation threshold p_c.

This work was supported by the National Science Foundation under Grant No PHY83-13324. The author is extremely grateful for the hospitality offered by the Service de Physique Théorique, CEN-Saclay, where this work was initiated, to T Burkhardt for communicating his resutls prior to publication, and to B Derrida for useful discussions.

References

Barber M N 1983 *Phase Transitions and Critical Phenomena* vol 8 ed C Domb and J Lebowitz (London: Academic)
Burkhardt T and Guim 1984a to be published
—— 1984b *Math. Gen.*
Cardy J L 1984a *J. Phys. A: Math. Gen.* **17** L385
—— 1984b *Nucl. Phys.* B [*FS*] to appear
Derrida B and de Seze J 1982 *J. Physique* **43** 475
Dotsenko Vl S 1984 *Nucl. Phys.* B **235** [*FS11*] 54
Fradkin E and Kadanoff L P 1980 *Nucl. Phys.* B **170** [*FS1*] 1
Friedan D, Qiu Z and Shenker S 1984 *Phys. Rev. Lett.* **52** 1575
Gehlen G V, Hoeger C and Rittenberg V 1984 *J. Phys. A: Math. Gen.* **17** L469
Kadanoff L P and Ceva H 1971 *Phys. Rev.* D **3** 3918
Luck J M 1982 *J. Phys. A: Math. Gen.* L169
McCoy B and Wu T T 1967 *Phys. Rev.* **162** 436

Nightingale M P and Blöte H W 1983 *J. Phys. A: Math. Gen.* **16** L657
Penson K and Kolb M 1984 *Phys. Rev.* B **29** 2854
Privman V and Fisher M E 1984 *Phys. Rev.* B **30** 322
Widom B 1972 *Phase Transitions and Critical Phenomena* vol 2, ed C Domb and M S Green (London: Academic)

J. Phys. A: Math. Gen. **19** (1986) L1093–L1098. Printed in Great Britain

LETTER TO THE EDITOR

Logarithmic corrections to finite-size scaling in strips

John L Cardy

Department of Physics, University of California, Santa Barbara, CA 93106, USA

Received 23 June 1986

Abstract. The corrections to the finite-size scaling behaviour of the eigenvalues of the transfer matrix of a critical theory defined on an infinitely long strip of finite width, which occur when the Hamiltonian contains a marginal operator, are computed using conformal invariance. They show a calculable universal logarthmic character. For the four-state Potts model they agree with numerical data.

One of the most immediate applications of conformal invariance in two-dimensional critical behaviour results from the observation (Cardy 1984) that, if we denote the transfer matrix of a strip of width L, with periodic boundary conditions, by $\exp(-\hat{H})$, then the eigenvalues E_n of $\hat{H}$ are related to the scaling dimensions x_n of the scaling operators of the theory by

$$E_n - E_0 \sim 2\pi x_n / L \tag{1}$$

in the limit $L \to \infty$. In addition (Blöte *et al* 1986, Affleck 1986), the ground state energy E_0, which is the free energy per unit length, is related to the conformal anomaly number c, which plays an important role in the formal classification scheme (Friedan *et al* 1984), by

$$E_0 = AL - \pi c/6L + o(L^{-1}) \tag{2}$$

where A is the bulk free energy per unit area, a non-universal constant.

By now these relations have been verified for a large number of models and they form a powerful means of investigating new ones. The corrections to (1) due to the presence of irrelevant operators have already been considered (Cardy 1986a, b). As expected, they are of order $L^{-1-|y|}$, where y is the renormalisation group eigenvalue of the irrelevant operator. Similar corrections are to be expected in (2). However, in the presence of a marginally irrelevant operator, one typically expects logarithmic terms. In this letter, we calculate the leading logarithmic corrections to the results in (1) and (2) and show that they have a universal form. Our main results are contained in (11) and (22).

To be specific, let us consider the effect of adding a term $(-g)\Sigma_r \phi(r)$ to the fixed-point Hamiltonian. The non-linear scaling field g is supposed to have scaling dimension $x = 2 - y$, where y is its renormalisation group eigenvalue. Under a change of length scale it satisfies the renormalisation group equation

$$dg/dl = (2-x)g - \pi b g^2 + O(g^3) \tag{3}$$

0305-4470/86/171093+06$02.50

where $x=2$ and $b>0$ if g is marginally irrelevant. In this case the solution for large l is

$$g(l)=\frac{g}{1+\pi bgl}. \tag{4}$$

In addition, the other scaling variables u_n are supposed to satisfy the renormalisation group equations

$$\mathrm{d}u_n/\mathrm{d}l=(2-x_n)u_n-2\pi b_n g u_n+\mathrm{O}(g^2u_n). \tag{5}$$

The numbers b and b_n are universal if ϕ is normalised so that its two-point function is $|r|^{-2x}$. In fact, they are related to operator product expansion coefficients or, equivalently, they give the normalisation of the three-point functions

$$\langle\phi(r_1)\phi(r_2)\phi(r_3)\rangle=-b/|r_{12}|^x|r_{23}|^x|r_{31}|^x \tag{6}$$

and

$$\langle\phi(r_1)\phi_n(r_2)\phi_n(r_3)\rangle=-b_n/|r_{12}|^x|r_{13}|^x|r_{23}|^{2x_n-x}. \tag{7}$$

The easiest way to see this is to perform a Kosterlitz-type renormalisation group calculation (Kosterlitz 1974) on the expansion of the free energy in the infinite system. An example of such a calculation for the random Potts model is given in Ludwig (1986).

We discuss the form of the corrections to (1) as predicted by the renormalisation group. Since the left-hand side is an inverse correlation length, under a rescaling it transforms according to

$$\xi^{-1}(g,L^{-1})=\mathrm{e}^{-l}\xi^{-1}(g(l),L^{-1}\mathrm{e}^{l}). \tag{8}$$

Choosing $\mathrm{e}^l=L$ and using (4), we obtain the scaling prediction

$$E_n-E_0\sim L^{-1}\Phi_n\left(\frac{g}{1+\pi bg\ln L}\right) \tag{9}$$

where Φ_n is a universal function. In Cardy (1986a, b) we showed that the O(g) correction to (1), for arbitrary x, is given by

$$E_n-E_0\sim 2\pi x_n/L+gb_nL(2\pi/L)^x+\mathrm{O}(g^2). \tag{10}$$

Comparing (9) and (10), we see that, in the limit $L\to\infty$, the corrections to (1) are independent of the value of g and are of the universal form

$$E_n-E_0\sim(2\pi/L)(x_n+d_n/\ln L)+\mathrm{O}(L^{-1}(\ln L)^{-2}) \tag{11}$$

where $d_n=2b_n/b$. It should be stressed that this behaviour will set in for strip widths $L\gg\exp(1/bg)$, which may be very large if g is small. An interesting special case is when $\phi_n=\phi$, corresponding to an energy gap scaling asymptotically like $4\pi/L$. In that case the logarithmic correction is as in (11) with $x_n=d_n=2$.

Next we discuss the corrections to E_0. It is convenient to consider the free energy per unit area $f=E_0/L$. The change δf in this quantity may be calculated in an expansion in g involving the correlation functions of ϕ evaluated at the fixed point:

$$\delta f=-g\langle\phi\rangle-\frac{g^2}{2!}\sum_r\langle\phi(r)\phi(0)\rangle_c-\frac{g^3}{3!}\sum_{r_1,r_2}\langle\phi(r_1)\phi(r_2)\phi(0)\rangle_c+\ldots. \tag{12}$$

In general, operators may be subtracted so that $\langle\phi\rangle=0$ in the bulk, and conformal invariance then implies that the first term vanishes in the strip also. However, operators in the conformal block of the unit operator transform anomalously (Belavin *et al* 1984) and will have a non-zero expectation value in the strip. The most important is the irrelevant operator $L_{-2}\bar{L}_{-2}\mathbf{1}$, which has $x=4$. This will give $O(L^{-4})$ corrections to f which may be expected to have a large amplitude, since they occur in first order. Indeed, such corrections are observed to be important in numerical calculations (Blöte and Nightingale 1982). They may mask the more interesting corrections we are studying here. In the higher-order terms, the sums over r may be replaced by integrals, and the correlation functions by their continuum limits, if we impose a short distance cutoff $r>1$, etc. (Here, as throughout this letter, we take the lattice spacing to give the unit of length.) The correlation functions are to be calculated in the strip geometry and are determined in terms of the infinite plane correlation functions by using the conformal mapping (Cardy 1984) $w=(L/2\pi)\ln z$. However it is simpler to transform the integrals back to the z plane. We first examine the $O(g^2)$ term. This is, for general x,

$$\delta f_2 = -\tfrac{1}{2}g^2(2\pi/L)^{2x-2}I_2(x, 2\pi/L) \tag{13}$$

where

$$I_2(x,\varepsilon) = \int d^2z\,|z^{x-2}|z-1|^{-x}\theta(|z-1|-\varepsilon). \tag{14}$$

For $0<x<1$, $I_2(x,0)$ converges and is equal to (Hentschke *et al* 1986)

$$\frac{\pi\Gamma(x/2)^2\Gamma(1-x)}{\Gamma(1-x/2)^2\Gamma(x)}. \tag{15}$$

The leading $\varepsilon\to 0$ behaviour of $I_2(x,\varepsilon)$ is easily extracted. We then find that

$$I_2(x,\varepsilon) = I_2(x,0) - \frac{\pi\varepsilon^{2-2x}}{1-x} + O(\varepsilon^{6-2x}). \tag{16}$$

Note that this is valid up to the next pole of $I_2(x,0)$, which occurs at $x=3$. The second term in (16) gives a contribution to the bulk free energy. It is non-universal because it depends on the details of the cutoff procedure. The first term contributes to the universal L^{-2} term. If g is irrelevant ($x>2$), we see that the leading correction to the free energy has the form

$$f = A(g) - \frac{\pi}{6L^2}(c + Bg^2L^{4-2x} + O(L^{6-3x})) \tag{17}$$

where B is a constant which is positive for $2<x<3$. The expression in parentheses defines an effective $c(L)$, which approaches its asymptotic value from above.

However, from (15) we see that this correction actually vanishes at $x=2$. We therefore must consider the next term. This is

$$\delta f_3 = \tfrac{1}{6}bg^3(2\pi/L)^{3x-4}I_3(x, 2\pi/L) \tag{18}$$

where

$$I_3(x,0) = \int |z_1|^{x-2}|z_2|^{x-2}|z_1-1|^{-x}|z_2-1|^{-x}|z_1-z_2|^{-x}\,d^2z_1\,d^2z_2. \tag{19}$$

This converges for $0<x<\frac{4}{3}$. The pole at $x=\frac{4}{3}$ corresponds, by the same argument as above, to a contribution to the bulk energy. The term we want is given by the analytic continuation of $I_3(x,0)$ to $x=2$. An integration by parts shows that

$$I_3(x,0) = \frac{2-x}{4-3x}\int F(z_1,z_2)(|z_1-1|^2+|z_1|^2-1)|z_1|^{-2}\,d^2z_1\,d^2z_2 \tag{20}$$

where F is the same integrand as in (19). The integral now converges for $0<x<2$ and it is relatively straightforward to determine the residue of the pole at $x=2$. The final result is $I_3(2,0)=-\pi^2$, so that

$$\delta f = -2\pi^4 b g^3/3L^2 + O(g^4). \tag{21}$$

The renormalisation group (Blöte and Nightingale 1982) now implies that g in the above expression should be replaced by $g(\ln L) \sim 1/\pi b \ln L$. This, of course, means that we should pick up logarithmic dependences on ε at higher orders in g. Although conformal invariance does not in general completely predict the four- and higher-point functions, one can see, by using the operator product expansion, from where these factors must arise. Note that no such factors arise which would correspond to replacing g by $g(\ln L)$ in the non-universal bulk term, in agreement with Blöte and Nightingale (1982). The final result for the free energy is

$$f = E_0/L = A(g) - \frac{\pi}{6L^2}\left(c + \frac{4}{b^2}(\ln L)^{-3} + O((\ln L)^{-4})\right). \tag{22}$$

Note that the effective $c(L)$ defined by (22) approaches its asymptotic value from above.

For the case of the four-state Potts model, we have obtained the operator product expansion coefficients $-b$ and $-b_\varepsilon$, where ε is the energy density, by taking the limit $q \to 4$ of the results of Dotsenko and Fateev (1985). We find $b = 4/\sqrt{3}$ and $b_\varepsilon = \sqrt{3}/2$. The ratio of these, as they appear in the renormalisation group equations (3) and (5), agrees with the ratio of coefficients found by Cardy *et al* (1980) (who used a different normalisation of the marginal scaling field g) by comparison with exact results. This result forms a non-trivial check of the work of Dotsenko and Fateev. We thus see that $d_\varepsilon = \frac{3}{4}$. In a similar way, we find that $d_\sigma = \frac{1}{16}$, where σ is the leading magnetisation operator.

Blöte and Nightingale (1982) and Nightingale and Blöte (1983) have calculated numerically the free energy per site f, and the gaps $E_\varepsilon - E_0$ and $E_\sigma - E_0$, corresponding to the leading energy and magnetisation operators, repectively, for the nearest-neighbour isotropic $q=4$ Potts model for strips up to width $L_{\max} \leq 11$. As discussed in Nightingale and Blöte (1983) and Blöte *et al* (1986), the extrapolated values of x_ε, x_σ and c agree fairly well with their expected exact values of $\frac{1}{2}$, $\frac{1}{8}$ and 1, respectively, although the agreement is not so good as for other values of q, where no logarithmic corrections are present. We now show that if these exact values are taken for granted, then the finite-size deviations found by Blöte and Nightingale are accounted for by logarithmic corrections of the type discussed above. In calculating these it is not appropriate to use the asymptotic forms in (11) and (22), but rather to replace g in (10) and (21) by $g(\ln L)$. We used the free energy data to calculate $g(\ln L)$ and then used this value to predict the deviations in the other eigenvalues. Although the correction δf is typically down by a factor of 10^{-4} on the leading behaviour, the results of Blöte and Nightingale are sufficiently precise to allow us to extract these values easily. Our results for the deviations in the gaps are shown in table 1 and compared with the same quantities derived from the data displayed in Nightingale and Blöte (1983). The agreement is rather good, despite the fact that that we have ignored corrections of higher order in $g(\ln L)$. Consideration of these terms shows that the effective expansion parameter is $\pi g(\ln L) \approx 0.1$, so that the discrepancies are of the expected order of magnitude. A plot of $g(\ln L)^{-1}$ against $\ln L$ shows some curvature, indicating that the asymptotic form of (4) has not yet been reached. The observed

Table 1. Corrections to the free energy per site and the lowest gaps of the four-state Potts model, for strips of width L. Second column shows $\delta f = f(L) - f(\infty) + \pi/6L^2$, with $f(L)$ taken from Blöte and Nightingale (1982) and $f(\infty)$ from Baxter (1973). The last four columns show predicted corrections to the gaps: $\Delta_n = (E_n(L) - E_0(L))L - 2\pi x_n$, and their exact values, derived from data which is displayed in figures 2(a) and (b) of Nightingale and Blöte (1983).

L	$\delta f \times 10^5$	$g(\ln L)$	Δ_ε	Exact	Δ_σ	Exact
6	42.17	0.0466	1.59	1.67	0.132	0.114
7	23.65	0.0426	1.46	1.60	0.121	0.105
8	14.53	0.0396	1.35	1.54	0.112	0.100
9	9.55	0.0372	1.27	1.49	0.106	—
10	6.60	0.0353	1.21	1.45	0.101	—

departures may be accounted for by adding a term $O(g^3)$, with an appropriate cofficient, to the right-hand side of (3).

We conclude that the major corrections to finite-size scaling for this model are logarithmic and that their amplitudes show quantitative agreement with the predictions of conformal invariance.

More recently Alcaraz and Barber (1987) have considered the quantum Hamiltonian four-state Potts model and another Ising model which is expected to be in the same universality class. For the latter model, they find effective values of c less than unity, which would appear to contradict our result, or to imply that it is in a different universality class. However, the situation in this case is complicated by the anisotropy, which is affected in a non-trivial way by corrections to scaling. In the estimates of the ratios c/x_σ, where the anisotropy should cancel, for both models the sign of the deviations from the exact Potts values found by Alcaraz and Barber agrees with our results, although the L dependence does not. It is possible that the asymptotic region begins at larger values of L for these quantum models. Similar poor convergence for the $q = 4$ quantum chain was found by von Gehlen *et al* (1986).

I thank M P Nightingale for providing the numerical data on which figures 2(a) and (b) of Nightingale and Blöte (1983) were based, J Shapiro for assistance in evaluating the integral in (19) and A Ludwig for useful conversations. F C Alcaraz and M N Barber kindly sent me a copy of their work before publication. This work was supported by NSF Grant No PHY83-13324.

References

Affleck I 1986 *Phys. Rev. Lett.* **56** 746
Alcaraz F C and Barber M N 1987 *J. Phys. A: Math. Gen.* **20** in press
Baxter R J 1973 *J. Phys. C: Solid State Phys.* **6** L445
Belavin A A, Polyakov A M and Zamolodchikov A B 1984 *Nucl. Phys.* B **241** 333
Blöte H W J, Cardy J L and Nightingale M P 1986 *Phys. Rev. Lett.* **56** 742
Blöte H W J and Nightingale M P 1982 *Physica* **112A** 405
Cardy J L 1984 *J. Phys. A: Math. Gen.* **17** L385
—— 1986a *Phase Transitions and Critical Phenomena* vol 11, ed C Domb and J L Lebowitz (New York: Academic) to appear
—— 1986b *Nucl. Phys.* B **270** 186

Cardy J L, Nauenberg M and Scalapino D J 1980 *Phys. Rev.* B **22** 2560
Dotsenko V S and Fateev V A 1985 *Nucl. Phys.* B **251** 691
Friedan D, Qiu Z and Shenker S 1984 *Phys. Rev. Lett.* **52** 1575
Gehlen G v, Rittenberg V and Ruegg H 1986 *J. Phys. A: Math. Gen.* **19** 107
Hentschke H J, Kleban P and Akinci G 1986 *J. Phys. A: Math. Gen.* **19** 3353
Kosterlitz J M 1974 *J. Phys. C: Solid State Phys.* **7** 1046
Ludwig A 1986 unpublished
Nightingale M P and Blöte H W J 1983 *J. Phys. A: Math. Gen.* **16** L657

J. Phys. A: Math. Gen. **20** (1987) 5039. Printed in the UK

CORRIGENDUM

Logarithmic corrections to finite-size scaling in strips
Cardy J L 1986 *J. Phys. A: Math. Gen.* **19** L1093-8

The correct value for the integral $I_3(2,0)$ is $-2\pi^2$, not $-\pi^2$ as stated. Consequently the appropriate terms in (21) and (22) should be multiplied by 2. In the numerical comparison shown in table 1, the figures in the columns headed $g(\ln L)$, Δ_ε and Δ_σ should all be multiplied by $2^{-1/3}$. The details of the calculation of the integral will be given in a forthcoming paper by A Ludwig and the author.

0305-4470/87/145039+01$02.50

J. Phys. A: Math. Gen. **18** (1985) L757–L760. Printed in Great Britain

LETTER TO THE EDITOR

Universal amplitudes in finite-size scaling: generalisation to arbitrary dimensionality

John L Cardy

Department of Physics, University of California, Santa Barbara, California 93106, USA

Received 3 July 1985

Abstract. The relationship between the correlation length and critical exponents in finite width strips in two dimensions is generalised to cylindrical geometries of arbitrary dimensionality d. For $d \neq 2$ these correspond, however, to curved spaces. The result is verified for the spherical model.

A striking result of conformal invariance at critical points has been the relationship between the amplitude A of the finite-size scaling behaviour of the correlation length ξ of an infinitely long strip of width L with periodic boundary conditions, defined by $\xi^{-1} \sim A/L$, and the scaling dimension x of the corresponding scaling operator (Cardy 1984). This relation states that

$$A = 2\pi x. \tag{1}$$

It has been verified in a large number of two-dimensional models (Luck 1982, Derrida and de Seze 1982, Nightingale and Blote 1983, Privman and Fisher 1984, Penson and Kolb 1984, Alcaraz *et al* 1985), and gives a very accurate way of determining the scaling dimensions numerically. It would therefore be very useful to generalise this result to dimensionality $d \neq 2$. In this letter we describe such a generalisation. Unfortunately the result appears to be difficult to utilise for numerical work.

For the purposes of generalisation, it is convenient first to restate the two-dimensional conformal invariance argument in a fashion independent of the use of complex variables. Consider a critical theory defined on the infinite two-dimensional plane $\mathbb{R}^2$. In polar coordinates, the metric is

$$ds^2 = dr^2 + r^2\, d\theta^2. \tag{2}$$

Under the coordinate transformation

$$(r, \theta) = (\exp(u/R), \theta) \tag{3}$$

where $-\infty < u < \infty$, the metric can be written

$$ds^2 = R^{-2} \exp(2u/R)(du^2 + R^2\, d\theta^2) \tag{4}$$

which can be recognised as a conformal factor multiplying the natural metric for the space $S^1 \times \mathbb{R}^1$, i.e. the surface of a circular cylinder of radius R. The transformation (3) thus conformally relates theories defined on $\mathbb{R}^2$ and $S^1 \times \mathbb{R}^1$. At a critical point, a two-point correlation function transforms according to

$$\langle\varphi(\exp(u_1/R), \theta_1)\varphi(\exp(u_2/R), \theta_2)\rangle_{\mathbb{R}^2}$$
$$= R^{2x} \exp[-x(u_1+u_2)/R]\langle\varphi(u_1, \theta_1)\varphi(u_2, \theta_2)\rangle_{S^1\times\mathbb{R}^1}. \tag{5}$$

Since the correlation function on the left-hand side is proportional to $|\boldsymbol{r}_1-\boldsymbol{r}_2|^{-2x}$, this determines the correlation function on the cylinder and gives a correlation length $\xi = R/x$. This agrees with (1) if we note that $L = 2\pi R$.

The generalisation to $d \neq 2$ is now straightforward. The metric on $\mathbb{R}^d$ is written

$$ds^2 = dr^2 + r^2\, d\Omega^2 \tag{6}$$

where, for example, in $d = 3$,

$$d\Omega^2 = d\theta^2 + \sin^2\theta\, d\psi^2. \tag{7}$$

Under (3) the metric is transformed conformally into

$$ds^2 = R^{-2}\exp(2u/R)(du^2 + R^2\, d\Omega^2) \tag{8}$$

which, apart from a conformal factor, is the natural metric for $S^{d-1}\times\mathbb{R}^1$. Thus critical theories on $\mathbb{R}^d$ are conformally related to those on this cylindrical geometry. As before, the correlation length along the cylinder is given by $\xi = R/x$.

This result is supposed to be valid in the continuum limit, at the critical point. For it to be useful for numerically estimating the scaling dimension x, it is necessary to approximate the continuum by a sequence of lattices. It is most convenient to choose these lattices to be regular. For $d \neq 2$, however, the space $S^{d-1}\times\mathbb{R}^1$ is curved, and only a finite number of regular lattices may be embedded in the space. For S^2, for example, these correspond to the Platonic solids, the largest of which is the dodecahedron (12 lattice points). It is not clear whether this approximates the continuum sufficiently well to give accurate values for the exponents. In addition, the lattice approximation to $S^{d-1}\times\mathbb{R}^1$ should incorporate those symmetries which correspond to translations in $\mathbb{R}^d$, and mix up the spaces S^{d-1} and $\mathbb{R}^1$. This will be true if, on distance scales much less than the inverse curvature R, the lattice is isotropic in all d directions. If this last requirement is relaxed, by introducing a lattice which approximates the symmetries of S^{d-1} only, then the amplitudes A will not be universal, but ratios of them will be. In this case it may be simplest to take the anisotropic limit, and to consider an equivalent quantum Hamiltonian defined on a lattice approximating S^{d-1}, as has already been done for $d = 2$ (Penson and Kolb 1984, Alcaraz *et al* 1985).

It would of course be useful to obtain a formula for A for geometries which are more easily approximated by a lattice, for example a 2-torus $\times\mathbb{R}^1$. Such a formula cannot be obtained by conformal transformation, because these spaces are flat, and for $d \neq 2$ the group of conformal transformations of flat space into itself is too restricted.

The only simple test of our result we have been able to find is in the spherical model, equivalent to the $n\to\infty$ limit of the n-vector model, defined by the reduced Hamiltonian

$$\mathcal{H} = \int [\tfrac{1}{2}(\nabla\varphi)^2 + \tfrac{1}{2}\mu_0^2\varphi^2 + \tfrac{1}{4}(\lambda_0 n^{-1})(\varphi^2)^2]\, d^dx. \tag{9}$$

For $d = 3$ the unrenormalised propagator on $S^2\times\mathbb{R}^1$ is found by decomposing $\varphi(u, \theta, \psi)$ into normal modes $e^{iku}Y_{lm}(\theta, \psi)$. The result is

$$G^0_{lm}(k) = [k^2 + R^{-2}l(l+1) + \mu_0^2]^{-1}. \tag{10}$$

In the $n \to \infty$ limit the only effect of the interactions (Ma 1976) is to replace μ_0^2 by μ^2, where μ^2 is determined self-consistently by

$$\mu^2 = \mu_0^2 + \frac{\lambda_0}{2} \int \frac{dk}{2\pi} \frac{1}{4\pi R^2} \sum_{lm} \frac{1}{k^2 + R^{-2} l(l+1) + \mu^2}. \tag{11}$$

The sum over m gives a factor of $2l+1$. The values of μ_0^2 corresponding to the bulk critical point is given by

$$0 = \mu_0^2 + \frac{\lambda_0}{2} \int \frac{d^3 k}{(2\pi)^3} \frac{1}{k^2 + \boldsymbol{k}_\perp^2} \tag{12}$$

so that the correlation length in the cylindrical geometry is $\xi = \mu^{-1}$ where

$$\mu^2 = \frac{\lambda_0}{2} \int \frac{dk}{2\pi} \left(\frac{1}{4\pi R^2} \sum_l \frac{2l+1}{k^2 + R^{-2} l(l+1) + \mu^2} - \int \frac{d^2 k_\perp}{(2\pi)^2} \frac{1}{k^2 + \boldsymbol{k}_\perp^2} \right). \tag{13}$$

The sums over modes are separately well defined only with an ultraviolet cut-off

$$k^2 + k_\perp^2 < \Lambda^2 \qquad k^2 + R^{-2} l(l+1) < \Lambda^2. \tag{14}$$

Equation (13) determines μ^2 as a function of R and Λ. For $R\Lambda \gg 1$ the left-hand side should be independent of Λ. However, individually the sum and the integral in the large round brackets diverge after integration over k. For this divergence to cancel the sum must be a close approximation to the integral. This occurs if we take $\mu^2 \simeq 1/4R^2$, because then the large round brackets may be written

$$\frac{1}{2\pi R^2} \sum_{l=0}^{\infty} \frac{l+\frac{1}{2}}{k^2 + R^{-2}(l+\frac{1}{2})^2} - \frac{1}{2\pi} \int_0^\infty \frac{k_\perp \, dk_\perp}{k^2 + k_\perp^2} \tag{15}$$

which actually vanishes after integration over k. A more careful analysis shows that the solution of (13) is

$$\mu^2 = \frac{1}{4R^2} - \frac{4}{\pi \lambda_0 R^3} + O(R^{-4}). \tag{16}$$

For arbitrary d between 2 and 4, the same cancellation happens, with $l(l+1)$ replaced by $l(l+d-2)$, and so

$$\mu^2 \sim (d-2)^2/4R^2 \qquad |2 < d < 4|. \tag{17}$$

This verifies our general result $\xi^{-1} = x/R$ since $x = \frac{1}{2}(d-2+\eta)$, and for the $n \to \infty$ limit $\eta = 0$.

In conclusion we have shown how the universal amplitude relation for finite width two-dimensional strips generalises to higher dimensions. Whether this will provide a useful numerical approach to critical exponents remains to be seen.

The author thanks V Privman for discussions. This work was supported by NSF Grant PHY83-13324.

References

Alcaraz F C, de Felicio J R D, Koberle R and Stilck J F 1985 *Phys. Rev.* B to be published
Cardy J L 1984 *J. Phys. A: Math. Gen.* **17** L385

Derrida B and de Seze J 1982 *J. Physique* **43** 475
Luck J M 1982 *J. Phys. A: Math. Gen.* **15** L169
Ma S K 1976 *Modern Theory of Critical Phenomena* (New York: Benjamin/Cummings)
Nightingale M P and Blote H W 1983 *J. Phys. A: Math. Gen.* **16** L657
Penson K and Kolb M 1984 *Phys. Rev.* B **29** 2854
Privman V and Fisher M E 1984 *Phys. Rev.* B **30** 322